Theory of Electrophoresis and Diffusiophoresis of Highly Charged Colloidal Particles

Interface Science and Technology

Series Editor: Arthur T. Hubbard

In this Series:

Vol. 1: Clay Surfaces: Fundamentals and Applications
Edited by F. Wypych and K.G. Satyanarayana
Vol. 2: Electrokinetics in Microfluids
By Dongqing Li
Vol. 3: Radiotracer Studies of Interfaces
Edited by G. Horányi
Vol. 4: Emulsions: Structure Stability and Interactions
Edited by D.N. Petsev
Vol. 5: Inhaled Particles
By Chiu-sen Wang
Vol. 6: Heavy Metals in the Environment: Origin, Interaction and Remediation
Edited by H.B. Bradl
Vol. 7: Activated Carbon Surfaces in Environmental Remediation
Edited by Teresa J. Bandosz
Vol. 8: Tissue Engineering: Fundamentals and Applications
By Yoshito Ikada
Vol. 9: Particles at Interfaces: Interactions, Deposition, Structure
By Zbigniew Adamczyk
Vol. 10: Interface Science in Drinking Water Treatment: Theory and Applications
Edited by G. Newcombe and D. Dixon
Vol. 11: Surface Complexation Modelling
Edited by Johannes Lützenkirchen
Vol. 12: Theory of Colloid and Interfacial Electric Phenomena
By Hiroyuki Ohshima
Vol. 13: Sorbent Deformation
By A.V. Tvardovskiy
Vol. 14: Advanced Chemistry of Monolayers at Interfaces: Trends in Methodology and Technology
Edited by Toyoko Imae
Vol. 15: Macromolecules in Solution and Brownian Relativity
By Stefano A. Mezzasalma
Vol. 16: The Properties of Water and their Role in Colloidal and Biological Systems
By Carel Jan van Oss
Vol. 17: Chemistry on Modified Oxide and Phosphate Surfaces: Fundamentals and Applications
By Robson Fernandes de Farias
Vol. 18: Interface Science and Composites
By Soo-Jin Park and Min-Kang Seo
Vol. 19: Nanoparticle Technologies: From Lab to Market
By Farid Bensebaa
Vol. 20: Particles at Interfaces
By Zbigniew Adamczyk
Vol. 21: Self-Assembly Processes at Interfaces: Multiscale Phenomena
By Vincent Ball
Vol. 22: Rheology of Emulsions: Electrohydrodynamics Principles
By Aleksandar M. Spasic
Vol. 23: Tailored Thin Coatings for Corrosion Inhibition using a Molecular Approach
By Simo Olavi Pehkonen and Shaojun Yuan
Vol. 24: Charge and Energy Storage in Electrical Double Layers
By Silvia Ahualli and Ángel V. Delgado
Vol. 25: Stimuli Responsive Polymeric Membranes
By Mihir Kumar Purkait, Manish Kumar Sinha, Piyal Mondal, and Randeep Singh

Interface Science and Technology

Volume 26

Theory of Electrophoresis and Diffusiophoresis of Highly Charged Colloidal Particles

Eric Lee

National Taiwan University, Taipei, Taiwan

ELSEVIER

ACADEMIC PRESS

An imprint of Elsevier

Academic Press is an imprint of Elsevier
125 London Wall, London, EC2Y 5AS, United Kingdom
The Boulevard, Langford Lane, Kidlington, Oxford OX5 1GB, United Kingdom
50 Hampshire Street, 5th Floor, Cambridge, MA 02139, United States
525 B Street, Suite 1650, San Diego, CA 92101, United States

ISBN: 978-0-08-100865-2
ISSN: 1573-4285

Publisher: Susan Dennis
Acquisition Editor: Anita Koch
Editorial Project Manager: Amy Clark
Production Project Manager: Nilesh Kumar Shah
Cover Designer: Alan Studholme

Typeset by SPi Global, India

Dedication

This book is dedicated to my father
Mr. Dao-Tsai Lee
A gentleman forever

Contents

List of Symbols xv
Preface xix

Section A
Electrophoresis of Rigid Particles

1. Electrophoresis of a Single Rigid Particle 3

1.1 General Introduction and Basic Concepts of the Colloid Science 3
1.1.1 Definition of a Colloid: Size, Phase, and Material 3
1.1.2 Colloidal Suspensions: Aqueous Electrolyte Solution and Others 3
1.1.3 Charged Colloidal Particles in Aqueous Electrolyte Solution 4
1.1.4 Double Layer Polarization: No Double Layer Polarization, No Particle Motion 6
1.1.5 Particle Phoretic Motions 6
1.1.6 Electrokinetic Phenomena in General 6
1.2 Introduction of Electrophoresis 7
1.2.1 Historical Review of the Classic Analytical or Semianalytical Formulas 9
1.3 Theory 11
1.3.1 Electrical Potential 12
1.3.2 Ion Concentration 12
1.3.3 Flow Field 13
1.3.4 Boundary Conditions on the Particle Surface 18
1.3.5 Boundary Conditions at Infinity 18
1.3.6 Reduction to One-Dimensional Problem 20
1.3.7 Evaluation of the Particle Mobility 22
1.3.8 Charge-Regulation Phenomenon 23
1.4 Results and Discussions 25
1.4.1 Electrostatic Features When the Particle Is at Rest and Evaluation of Particle Surface Charge Density 25
1.4.2 Electrophoretic Behavior of a Single Rigid Particle 28
1.4.3 Double Layer Polarization: A Visual Demonstration With $\delta\rho^*$ 30

1.4.4 Relative Contributions of the Electro-Migration and the Convection Mechanisms to the Double Layer Polarization Effect 38
1.4.5 Effect of Peclet Numbers Upon the Particle Motion 38
1.4.6 Charge-Regulation Phenomenon 40
References 42

2. Electrophoresis in Suspensions of Rigid Spherical Particles 47

2.1 Introduction 47
2.1.1 General Effect of Neighboring Particles in a Suspension: Double Layer Overlapping and Hydrodynamic Swarming Effects 47
2.1.2 To Model a Colloidal Suspension: The Unit Cell Model Concept 48
2.1.3 Hydrodynamic Cell Model and Electrostatic Cell Models 49
2.1.4 Electrostatic Cell Models: Levine and Neale Model (L-N Model) Versus Shilov and Zharkikh Model (S-Z Model) 49
2.2 Theory 51
2.3 Results and Discussions 52
2.3.1 General Henry's Charts for Various Suspension Concentrations 53
2.3.2 Double Layer Overlapping Effect 53
2.3.3 On the Practical Upper Bound of the Volume Fraction *H* 58
2.3.4 Comparison Between the S-Z and the L-N Boundary Conditions 58
References 61

3. Electrophoretic Motion of a Rigid Spherical Particle Normal to a Solid Plane 65

3.1 Introduction 65
3.1.1 Hydrodynamic Drag Effect: Impingement Flow 65
3.1.2 Electrostatic Attraction Effect: The Image-Charge Force Analog 66
3.1.3 Boundary Confinement Effect: Geometric Deformation of the Double Layer 66
3.2 Theory 67
3.3 Results and Discussion 72
3.3.1 General Henry's Charts at Various Separation Distance 72
3.3.2 Geometric Boundary Confinement Effect 74

3.3.3 Vortex Flow Induced 75
3.3.4 Correction Factor of Separation Distance 79
Tables of Boundary Effect Expressed as a Correction Factor Γ 79
3.3.5 Charge-Regulation Phenomenon 83
References 83

4. Electrophoretic Motion of a Rigid Spherical Particle Normal to an Air-Water Interface 87

4.1 Introduction 87
4.2 Theory 88
4.3 Results and Discussion 90
4.3.1 Repulsive Image-Charge Force Analog: Electrostatic Aspect 90
4.3.2 General Henry's Charts at Various Separation Distance 94
4.3.3 Geometric Boundary Confinement Effect 98
4.3.4 Vortex Flow Induced 98
4.3.5 Charge-Regulation Phenomenon 98
4.3.6 Correction Factor at Various Separation Distances 98
Tables of Boundary Effect Expressed as a Correction Factor Γ 101
References 105

5. Electrophoretic Motion of a Rigid Spherical Particle in a Cylindrical Pore 107

5.1 Introduction 107
5.2 Theory 109
5.3 Results and Discussion 113
5.3.1 General Henry's Charts 113
5.3.2 Boundary Confinement Effect 115
5.3.3 Electroosmosis Flow of a Charged Pore 117
5.3.4 Contour Plots of Electrokinetic Interest 118
References 121

6. Dynamic Electrophoresis of Rigid Spherical Particles 123

6.1 Introduction 123
6.2 Theory 125
6.2.1 Single Particle 126
6.2.2 Suspension System 128
6.2.3 Evaluation of the Dynamic Particle Mobility 128
6.3 Results and Discussion 130
6.3.1 Single Particle 130
6.3.2 Suspension System 137
References 141

Section B
Electrophoresis of Porous Particles

7. **Electrophoresis of Porous Particles** 145

7.1 **Introduction** 145
7.2 **Theory** 148
7.2.1 Single Porous Sphere 148
7.2.2 Suspension of Porous Particles 151
7.3 **Results and Discussion** 151
7.3.1 Single Porous Particle 151
7.3.2 Suspensions 166
References 178

8. **Electrophoretic Motion of a Porous Spherical Particle Normal to Planar Boundaries** 181

8.1 **Introduction** 181
8.2 **Theory** 182
8.2.1 Solid Plane 182
8.2.2 Air-Water Interface 184
8.3 **Results and Discussion** 184
8.3.1 Solid Plane 184
8.3.2 Air-Water Interface 192
Tables of Boundary Effect Expressed as a Correction Factor Γ 197
References 199

9. **Electrophoretic Motion of a Porous Spherical Particle in a Cylindrical Pore** 201

9.1 **Introduction** 201
9.2 **Theory** 201
9.3 **Results and Discussion** 203
9.3.1 General Henry's Charts in a Narrow Cylindrical Pore With $R_b^*=1.2$ 203
9.3.2 Effect of Particle Permeability, λa 203
9.3.3 Lateral Boundary Confinement Effect 204
9.3.4 Impact of Electroosmosis Flow 206
9.3.5 Effect of Pore Radius 207
9.3.6 Contour Plots of Electrokinetic Parameters 207
References 211

10. **Dynamic Electrophoresis of Porous Spherical Particles** 213

10.1 **Introduction** 213
10.2 **Theory** 214
10.2.1 Single Porous Particle 214
10.2.2 Suspensions 216

10.3 Results and Discussion 216
10.3.1 Single Porous Particle 216
10.3.2 Suspensions 226
References 230

11. Gel-Electrophoresis of Rigid Spherical Particles in Suspension and in a Cylindrical Pore 233

11.1 Introduction 233
11.2 Theory 235
11.2.1 Single Rigid Spherical Particle in a DBB Polymeric Fluid 235
11.2.2 Suspension of Rigid Spherical Particles in a Polymeric DBB Fluid 235
11.2.3 A Rigid Spherical Particle in a Cylindrical Pore Filled With Polymeric DBB Fluid 237
11.3 Results and Discussion 237
References 242

Section C
Electrophoresis of Liquid Droplets and Gas Bubbles

12. Electrophoresis of Liquid Droplets and Gas Bubbles 247

12.1 Introduction 247
12.1.1 Definition of a Droplet 247
12.1.2 Generation of Liquid Droplets or Gas Bubbles 248
12.1.3 Some Remarks on the Emulsion Systems 249
12.1.4 The Role of a Surfactant in General 251
12.1.5 Gas Bubbles 251
12.2 Theory 254
12.2.1 Single Liquid Droplet or Gas Bubble 254
12.2.2 Liquid Droplet or Gas Bubble Suspension 257
12.3 Results and Discussion 257
References 272

13. Electrophoretic Motion of a Liquid Droplet or a Gas Bubble Normal to Planar Boundaries 275

13.1 Introduction 275
13.2 Theory 276
13.2.1 Solid Plane 276
13.2.2 Air-Water Interface 278
13.3 Results and Discussion 278
13.3.1 Solid Plane 278
13.3.2 Air-Water Interface 284
Tables of Boundary Effect Expressed as a Correction Factor Γ 284
References 291

14. Electrophoretic Motion of a Liquid Droplet and a Gas Bubble in a Cylindrical Pore 293

14.1 Introduction 293
14.2 Theory 294
14.3 Results and Discussion 295
14.3.1 General Henry's Charts 295
14.3.2 Contour Plots for a Moving Liquid Droplet or a Gas Bubble 298
14.3.3 Effect of EOF of a Charged Wall 300
14.3.4 Effect of Viscosity Ratio 300
References 301

15. Dynamic Electrophoresis of Liquid Droplets and Gas Bubbles 303

15.1 Introduction 303
15.2 Theory 304
15.3 Results and Discussion 305
15.3.1 Single Liquid Droplet and Gas Bubble 305
15.3.2 Suspensions of Liquid Droplets and Gas Bubbles 312
References 319

Section D
Diffusiophoresis of Colloidal Particles

16. Diffusiophoresis of Rigid Particles 323

16.1 Introduction 323
16.2 Theory 327
16.2.1 Single Rigid Particle 327
16.2.2 Suspensions of Rigid Particles 329
16.3 Results and Discussion 333
16.3.1 Single Rigid Particle 333
16.3.2 Suspension 342
References 356

17. Diffusiophoresis of Liquid Droplets and Gas Bubbles 359

17.1 Introduction 359
17.2 Theory 360
17.3 Results and Discussion 360
References 383

18. Diffusiophoresis of Porous Particles 385
18.1 Introduction 385
18.2 Theory 386
18.2.1 Single Porous Particle 386
18.2.2 Suspension of Porous Particles 386
18.3 Results and Discussion 386
18.3.1 Single Porous Particle 386
18.3.2 Suspensions of Porous Particles 404
References 409

Index 411

List of Symbols

$\boldsymbol{A}$	dimensionless group in a charge-regulation model representing the number of functional groups in dimensionless form
$\boldsymbol{a}$	radius of the particle
$\boldsymbol{B}$	dimensionless group in a charge-regulation model with its reciprocal representing the extent of dissociation reaction
$\boldsymbol{b}$	radius of the outer virtual surface in suspension systems
$\boldsymbol{C}$	concentration of electrolyte ions
$\boldsymbol{C}_\mathrm{D}$	proportional constant of velocity in a subproblem
$\boldsymbol{C}_\mathrm{E}$	proportional constant of electric field in a subproblem
$\boldsymbol{c}$	focal length in bi-spherical coordinates
$\boldsymbol{D}$	diffusivity of electrolyte ions
E	magnitude of the applied electric field
$\boldsymbol{E}$	applied electric field
$\boldsymbol{e}$	elementary electric charge
F	force exerted on the particle
$\boldsymbol{F}_\mathrm{D}$	hydrodynamic drag force exerted on the particle
$\boldsymbol{F}_\mathrm{E}$	electric force exerted on the particle
$\boldsymbol{f}_j$	ion flux of ion species j
$\boldsymbol{G}_j$	one-dimensional potential function representing disturbance of ion species j
$\boldsymbol{g}_j$	potential function representing the nonconcentric disturbance of ion species j
$\boldsymbol{H}$	particle volume fraction in suspensions
$\boldsymbol{h}$	dimensionless separation distance between the particle center and a planar boundary
$\mathbf{I}$	identity tensor
$\boldsymbol{I}$	ionic strength
$\mathbf{i}$	unit vector
$\boldsymbol{K}$	equilibrium constant of the surface dissociation reaction
$\boldsymbol{k}$	Boltzmann constant
$\boldsymbol{L}$	longitudinal distance to the particle center in a cylindrical pore
$\boldsymbol{N}_\mathrm{s}$	number of functional groups on the surface of a charge-regulated particle
$\boldsymbol{n}_j$	number density of ion species j
$\boldsymbol{n}_0$	number density of all the ion species in the bulk electrolyte solution
$\boldsymbol{n}_{0j}$	number density of ion species j in the bulk electrolyte solution
$\boldsymbol{p}$	pressure of the fluid
$\boldsymbol{Pe}_j$	Peclet number of ion species j
$\boldsymbol{Q}_\mathrm{fix}$	dimensionless fixed charge density within a porous particle
$\boldsymbol{R}$	R-coordinate in the cylindrical coordinates (R, Θ, Z)
$\boldsymbol{R}_\mathrm{b}$	radius of the cylindrical pore

$\boldsymbol{Re}$	Reynolds number
$\boldsymbol{r}$	r-coordinate in the spherical coordinates (r, θ, φ)
$\boldsymbol{r_o^*}$	dimensionless conceptual distance far way from a single particle where the Debye-Hückel approximation is applicable
$\boldsymbol{T}$	absolute temperature
$\boldsymbol{T_{r\theta}}$	total shear stress
$\boldsymbol{t}$	time
U	magnitude of the particle velocity
$\mathbf{U}$	particle velocity
$\boldsymbol{U_0}$	characteristic velocity in diffusiophoretic motion defined as $\frac{\varepsilon}{\eta_f a}\left(\frac{kT}{z_1 e}\right)^2$
$\boldsymbol{U_R}$	real part of the complex particle velocity $\mathbf{U}$ in dynamic electrophoresis with $\mathbf{U}=(U_R+iU_I)\,\mathbf{i}$
$\boldsymbol{U_I}$	imaginary part of the complex particle velocity $\mathbf{U}$ in dynamic electrophoresis with $\mathbf{U}=(U_R+iU_I)\,\mathbf{i}$
$\boldsymbol{u_E}$	particle velocity predicted by the Smoluchowski's theory when an electric field ζ/a is applied, i.e., $u_E=\frac{\varepsilon\zeta}{\eta_f a}$
v	magnitude of the fluid velocity relative to the moving particle
$\mathbf{v}$	fluid velocity relative to the moving particle
$\boldsymbol{x}$	cosh(η)-cos(ξ) in bi-spherical coordinates (ξ, η, φ)
$\boldsymbol{Z}$	Z-coordinate in cylindrical coordinates (R, Θ, Z)
z_j	charge valence of ion species j
$\boldsymbol{\alpha}$	valence ratio of the cation to the anion $(\alpha=-z_1/z_2)$
$\boldsymbol{\beta}$	dimensionless diffusivity difference between the cation and the anion, i.e., $\beta=\frac{D_1-D_2}{D_1+D_2}$
$\boldsymbol{\Gamma}$	correction factor of boundary effect in planar boundaries
$\boldsymbol{\gamma}$	friction coefficient in the Brinkman model
$\boldsymbol{\delta n_j}$	disturbed number density of ion species j
$\boldsymbol{\delta\rho_j}$	disturbed space charge density of ion species j
$\boldsymbol{\delta\phi}$	disturbed electric potential
$\boldsymbol{\delta\Phi}$	one-dimensional disturbed electric potential
$\boldsymbol{\varepsilon}$	permittivity of the fluid
$\boldsymbol{\zeta}$	zeta potential of the particle
$\boldsymbol{\eta}$	η-coordinate in bi-spherical coordinates (ξ, η, φ)
$\boldsymbol{\eta_0}$	$\cosh^{-1}(h/a)$ in bi-spherical coordinates (ξ, η, φ)
$\boldsymbol{\eta_d}$	viscosity of the liquid droplet
$\boldsymbol{\eta_f}$	viscosity of the suspending fluid
$\boldsymbol{\theta}$	θ-coordinate in spherical coordinates (r, θ, φ)
$\boldsymbol{\theta_p}$	phase angle of the scaled dynamic particle mobility in degrees
$\boldsymbol{\kappa^{-1}}$	Debye screening length or Debye length
$\boldsymbol{\kappa}$	Debye-Hückel parameter
$\boldsymbol{\lambda}$	Brinkman screening length characterizing the permeability of a porous particle or material
$\boldsymbol{\mu}$	particle mobility
$\boldsymbol{\mu^*}$	dimensionless particle mobility defined as $\mu^*=\frac{\frac{\mu}{u_E}}{(\phi_0/a)}$
$\boldsymbol{\mu_I}$	imaginary part of the complex particle mobility μ in dynamic electrophoresis with $\mu=\mu_R+i\mu_I$

$\boldsymbol{\mu}_{\mathbf{m,h}}$	dimensionless particle mobility at a separation distance h from a planar boundary
$\boldsymbol{\mu}_{\mathbf{m},\infty}$	dimensionless particle mobility in an infinite medium
$\boldsymbol{\mu}_{\mathbf{m}}^{*}$	scaled electrophoretic mobility of the particle defined as $\mu_m^* = \mu^*/\phi_r$
$\boldsymbol{\mu}_{\mathbf{R}}$	real part of the complex particle mobility μ in dynamic electrophoresis with $\mu = \mu_R + i\mu_I$
$\boldsymbol{\mu}_{\mathbf{s}}^{*}$	scaled electrophoretic mobility of the porous particle divided by $Q_{fix}/(\lambda a)^2 (\mu_s^* = \mu_m^*/(Q_{fix}/\lambda a)^2)$ or scaled diffusiophoretic mobility of the porous particle divided by $Q_{fix}/\beta(\lambda a)^2 (\mu_s^* = \mu_m^*/(Q_{fix}/\beta(\lambda a))^2)$
$\boldsymbol{\xi}$	ξ-coordinate in bi-spherical coordinates (ξ, η, φ)
$\boldsymbol{\rho}_{j}$	space charge density of ion species j
$\boldsymbol{\rho}_{\mathbf{f}}$	density of the suspending fluid
$\boldsymbol{\sigma}$	surface charge density in terms of number of electric charges per unit area
$\boldsymbol{\sigma}_{\mathbf{H}}$	viscosity ratio of a liquid or a gas bubble (viscosity of the external fluid to the viscosity of the internal fluid)
$\boldsymbol{\tau}_{\mathbf{D}}$	relaxation time constant of diffusion
$\boldsymbol{\tau}_{\mathbf{E}}$	relaxation time constant of the electromigration mechanism to fully respond to the input AC electric field
$\boldsymbol{\tau}_{\mathbf{H}}$	acceleration time constant of the particle before it reaches the terminal velocity
$\boldsymbol{\varphi}$	φ-coordinate in the spherical coordinate (r, θ, φ)
$\boldsymbol{\phi}$	electric potential
$\boldsymbol{\phi}_{\mathbf{r}}$	dimensionless zeta potential of the particle $(\phi_r = \zeta/\phi_0)$
$\boldsymbol{\phi}_{\mathbf{w}}$	electric potential of the cylindrical wall
$\boldsymbol{\phi}_{\mathbf{0}}$	thermal potential of the ion species 1 or cation in a binary electrolyte solution $(\phi_0 = kT/z_1 e)$
$\boldsymbol{\Psi}$	stream function
$\boldsymbol{\psi}$	one-dimensional stream function
$\boldsymbol{\omega}$	frequency of the AC electric field $\boldsymbol{E}$ in rad/sec
$\boldsymbol{\varpi}$	distance from the axis of rotation to any point
[HA]	concentration of the function group HA on the surface of a charge-regulated particle
$[A^-]$	concentration of the A^- on the surface of a charge-regulated particle undergoing the dissociation reaction of $HA \rightleftharpoons H^+ + A^-$
$[H^+]_o$	concentration of the hydrogen ion H^+ in the bulk electrolyte solution
$[H^+]_s$	concentration of the hydrogen ion H^+ on the surface of a charge-regulated particle

Operators

$\boldsymbol{D}^2$	one-dimensional form of the E^2 operator
$\boldsymbol{E}^2$	operator defined as $E^2 = \left(\frac{\partial^2}{\partial r^2}\right) + \frac{\sin\theta}{r^2}\frac{\partial}{\partial\theta}\left(\sin\theta\frac{\partial}{\partial\theta}\right)$ in spherical coordinates (r, θ, φ)
$\boldsymbol{L}^2$	one-dimensional Laplacian operator defined as $L^{*2} = \frac{d^2}{dr^{*2}} + \frac{2}{r^*}\frac{d}{dr^*} - \frac{2}{r^{*2}}$
∇	gradient operator
∇^2	Laplacian operator
$\nabla\cdot$	divergence operator
$\nabla\times$	curl operator
$\boldsymbol{E}^4$	$E^4 = E^2 E^2$

Superscript

*	dimensionless variable

Subscripts

1	cation
2	anion
e	equilibrium state
j	*j*th ion species
r	r-direction in spherical coordinates (r, θ, φ)
z	z-direction in spherical coordinates (r, θ, φ)
$\boldsymbol{\eta}$	η-direction in bi-spherical coordinates (ξ, η, φ)
$\boldsymbol{\theta}$	θ-direction in spherical coordinates (r, θ, φ)
$\boldsymbol{\xi}$	ξ-direction in bi-spherical coordinates (ξ, η, φ)

Characteristic variables

Length	a
Ion number density	n_{10}
Ion space charge density	$n_{10}z_1e$
Electrophoretic velocity	$u_E = \varepsilon\phi_0^2/\eta_f a$
Diffusiophoretic velocity	$U_0 = \frac{\varepsilon}{\eta_f a}\left(\frac{kT}{z_1 e}\right)^2$
Stream function of electrophoresis	$u_E a^2$
Stream function of diffusiophoresis	$u_E a^2$
Potential	$\phi_0 = kT/z_1e$
Electric field	ϕ_0/a

Preface

First of all, I would like to thank Dr. Arthur Hubbard for inviting me to write this book for Elsevier. It is really my honor.

This book discusses the electrophoretic and diffusiophoretic motions of various colloidal entities, such as rigid particles, liquid droplets, gas bubbles, and porous particles, focusing on the motion-deterring double layer polarization effect pertinent to highly charged particles, with the lowly charged ones serving as the limiting cases. Boundary effects such as those from a cylindrical pore, a solid plane, or an air-water interface are analyzed as well for the electrophoretic motion of various particles considered. Moreover, dynamic electrophoresis is explored and treated as well.

The contents are suitable for researchers, graduate students, or senior college students with some basic background of colloid science and transport phenomena.

As there is no closed-form analytical formula, in general, for the situation of highly charged particles, the results are presented with extensive figures and plots as well as tables under various electrokinetic situations of interest to facilitate the possible use of interested readers. After all, a picture is worth a thousand words. Note that many color figures are also available in the electronic version of the book, which present the details more comprehensively.

I would like to thank Professor Jhih-Wei Chu and Dr. Yu-Po Tang for their involvement and encouragement in the middle of last year, which really boosted the morale of my working team. Moreover, I would like to thank my graduate students, Shan-Chi Tsai, Hsin-Yun Lee, Jen-Hung Tuan, and Meng-Yu Tsai, in particular, who were directly involved in the production of the final text of the book this year by preparing figures and plots for each chapter, deriving and typing the equations, and so on. Without their efforts, there is no way this book can be completed.

I also would like to thank some previous graduate students of mine. Penny Lai and Betty Chen contributed a lot to this writing task when they were here in my lab. Yu-Fan Lee, Fang-Yi Liao, and Xiang-Xu Wang assisted me to launch the initial campaign nearly three years ago. Hsuan Fang prepared the programs for the computations needed in the book. Their contributions are deeply appreciated. In addition, I thank Fang-Yi Wu and Wen-Chun Chang for their contributions in typing the text and proofreading the manuscript, among others. Kuan-Yu Chiu and Jie-Kai Hsu helped as well.

I am deeply grateful to Amy Clark in Elsevier, the project manager of my book. I really want to thank Amy for her patience and tolerance. Amy gave me a lot of kind help and support. I also would like to thank Anita Koch in Elsevier, who contacted me about four years ago and helped me landing a publication agreement with Elsevier. I thank Nilesh Shah, the production manager, and his team as well, who helped me a lot to produce the book in its final form.

I thank my beloved families: Amber, Elton, and Jane Lee for their support and encouragement during the long, soul-searching writing time of this book.

Finally, I would like to thank Amber C, Serena W, Emily W, Eva H, and Cindy L for their devotion and dedication in various forms. It's really my privilege and pleasure to have you around in your prime time.

Eric Lee

Section A

Electrophoresis of Rigid Particles

Chapter 1

Electrophoresis of a Single Rigid Particle

1.1 GENERAL INTRODUCTION AND BASIC CONCEPTS OF THE COLLOID SCIENCE

1.1.1 Definition of a Colloid: Size, Phase, and Material

A colloidal particle is a small particle with dimension scale ranging from nanometers to micrometers, depending on the specific colloidal behavior under consideration. For instance, the upper bound has to be smaller than about five micrometers if the stability of a colloidal suspension is of concern, as beyond this limit the Brownian motion of the particles may not be strong enough to keep them from coagulation with one another [1]. On the other hand, if the separation of living cells is of concern, then the particle (cell) size can be larger than 10 μm in practice [2–5].

For the lower bound of the particle size, the basic criterion is that it has to be larger than the molecules in its environment to form heterogeneous suspensions distinctive from a true homogeneous solution [6]. In practice, both 1 nm to 1000 nm (1 μm) [6] and 5 nm to 5000 nm (5 μm) [1] have been proposed as the appropriate size range of colloidal particles.

Moreover, a colloid can assume the form of a rigid particle, a liquid drop, a gas bubble, or a porous particle in the sense that liquid can pass through it.

1.1.2 Colloidal Suspensions: Aqueous Electrolyte Solution and Others

In practice colloids are normally present in a suspension form, with the suspending medium ranging from gas, liquid, to even solid [6]. This is because without a suspending medium, the tiny colloidal particles tend to coagulate together as a bulk cluster and all of the unique and potentially useful features pertinent to colloidal particles disappear. Indeed, suspension, dilute or concentrated, is the typical system to deal with in either scientific research or practical applications regarding colloids.

Interface Science and Technology, Vol. 26. https://doi.org/10.1016/B978-0-08-100865-2.00001-1

Among various forms of suspensions, the liquid suspension is by far the most important one, in particular the aqueous electrolyte solution, due to the obvious reason that it is much easier to handle physically than the other two possible candidates, and it happens most common in nature. Although a colloidal particle can sometimes be suspended in a salt-free agent like a liquid nonpolar hydrocarbon solvent as well [7–9], the aqueous electrolyte solution is definitely the major suspension of interest, with essentially all the fundamental theoretical analyses built upon it historically.

1.1.3 Charged Colloidal Particles in Aqueous Electrolyte Solution

1.1.3.1 Shielding or Screening Effect by Electrolyte Ions

The colloidal particle in an electrolyte suspension is electrically charged in general. This may be of chemical nature such as the dissociation of functional groups on the particle surface when put in contact with a liquid solvent, or it may be of physical nature due to the adsorption of electrolyte ions on the particle surface when suspended in an aqueous electrolyte solution. Other mechanisms are also possible [6].

Whatever the causes, when a charged particle is suspended in an electrolyte solution, it will be surrounded by the electrolyte ions originally in the bulk solution. The electrostatic Coulomb force will attract the electrolyte ions of the opposite sign, the so-called counterions, toward it and at the same time expel the ions with the same sign, the so-called coions, away from it [6]. An ion cloud predominantly by the counterions is thus established in the vicinity of the colloidal particle. The closer to the particle surface is, the more dominant the presence of counterions is, and a concentric distribution of ions is normally observed when the suspension is at rest and maintains electrostatic equilibrium [6, 10, 11].

It is important to realize that a charged particle in the suspension is perfectly "shielded" or "screened" by these surrounding ions, hence behaves electrically neutral as a whole, that is, the charged particle and its surrounding ionic atmosphere. In principle it reduces to the neutral state in the bulk solution over an infinite long distance, but in practice over several particle diameters can be regarded as electronically neutral in general [12], as shall be demonstrated later. A thorough understanding of the structure of the ion distribution surrounding the particle is essential for the understanding of all the electrostatic and electrophoretic behaviors of a colloidal particle.

1.1.3.2 Double Layer Structure With Brief Historical Review

Based on the classic works of early pioneers like Helmholtz, Grahame, Stern, Gouy, and Chapman [13–17], the ions distribution surrounding a colloidal particle in an electrolyte solution historically has been modeled as of a "double layer" structure in colloid science, with an inner compact layer right adjacent to the particle surface forming a stationary or immobile layer of counterions, and an outer diffuse layer contains predominantly the counterions that are essentially free to move in response to the possible external forces [12]. This classic model forms the basis of the entire colloid science.

A complete screening (shielding) of the surface charges by a single layer of counterions was proposed initially by Helmholtz [17]. This Helmholtz single layer idea, however, does not take into account the possible motion induced by thermal energy which tends to drive the ions to move in random direction thus away from the particle. This was later improved by Gouy [13] and Chapman [14] independently who introduced the idea of an outer diffuse layer upon the inner compact layer [12]. The surface charges are assumed only partially screened or shielded by the counterions in the inner layer, so that the rest of the surface changes, the unscreened ones, can still interact with the counterions in the outer diffuse layer based on the Coulomb's electrostatic force. The mobile ions in the outer diffuse layer, which are capable of conducting random motion induced by thermal energy, hence may move away from the charged particle against the Coulomb attraction force. The farther away from the particle is, the stronger this tendency to move away is. A diffuse layer occupied predominantly by counterions is thus resulted. With more pioneering researchers joined in, the double layer structure finally was in shape with an inner immobile layer referred to as the "Stern layer" or the "compact layer," and an outer "diffuse layer" or "diffuse double layer" [15]. The structure of the double layer surrounding a colloidal particle is shown schematically in Fig. 1.1. Although the precise structure of the double layer

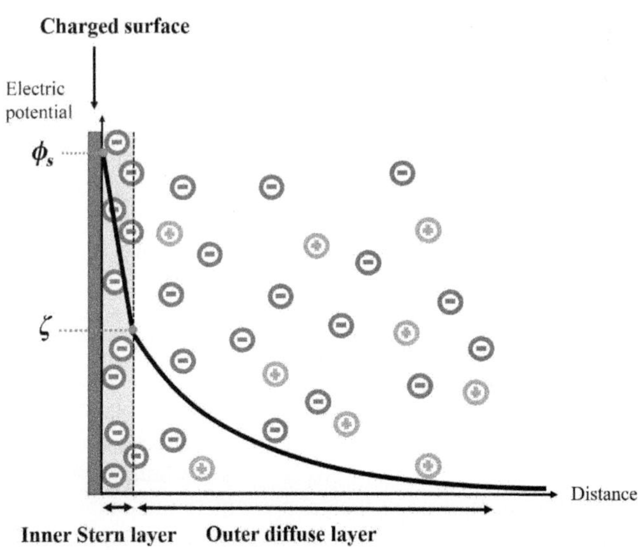

FIG. 1.1 Schematic diagram of the double layer structure surrounding a rigid particle. ϕ_r, electric potential on the particle surface; ζ, zeta potential.

has been developing as time goes by, due to the advancement of more accurate and fancy experimental apparatus coming up, the basic structure introduced here is still adequate as far as the electrokinetics is concerned, which is the major topic covered in this book. The reliability of further models is yet to be confirmed in fact.

1.1.4 Double Layer Polarization: No Double Layer Polarization, No Particle Motion

The presence of the surrounding electric double layer around the colloidal particle makes possible the various manipulations of the particle by altering the ions distribution there somehow, regardless of the varying nature of the specific particle system under consideration. As these ions are highly mobile and sensitive to field forces like electric field or concentration gradient, uneven distribution of ions can thus be generated by applying these field forces. Once this happens, the original perfect shielding or screening of the particle collapses, which means the original overall force balance upon the particle is no longer there, hence the particle starts to move. Thus we can say that the deformation of the double layer, or the double layer polarization, is the underlying driving mechanism for the particle motion. Or in short: No double layer polarization, no particle motion.

1.1.5 Particle Phoretic Motions

The motion of a colloidal particle in response to a field force is generally referred to as the phoretic motion [6, 12]. If this external force is an electric field, then the motion is referred to as the electrophoretic motion, and the phenomenon electrophoresis [18]. If it is due to a concentration gradient imposed upon a colloidal system, then it is referred to as the diffusiophoretic motion, and the phenomenon diffusiophoresis [19]. Other phoretic motions are also possible, such as the sedimentation of a particle, where the driving force is the gravity force or some kind of centrifugal force [12, 20]; or the dynamic electrophoresis, where the driving force is an alternating electric field with the particle moving back and forth, and so on [21–25].

1.1.6 Electrokinetic Phenomena in General

These phoretic motions are members of an even bigger family, the electrokinetic phenomena, with other members such as electroosmosis and streaming potential [26], to name a few. Electroosmosis is the bulk flow of the electrolyte solution due to the motion of the counterions in the double layer adjacent to a stationary charged wall [27]. Streaming potential, on the other hand, results from the fluid flow due to an applied pressure gradient upon the electrolyte solution in a capillary tube or porous plug, which generates an introduced electric potential against the motion of ions within it [26]. All of these phenomena involve the motion generated within a colloidal system, be it the colloidal particle itself or the electrolyte solution suspending it. And they all can be well explained and theoretically predicted based on the electric double layer model mentioned here.

We will focus on two major phoretic motions in this book: electrophoresis and diffusiophoresis. Electrophoresis is the most fundamental and important colloidal operation with various practical applications. Diffusiophoresis, on the other hand, is an emerging star in biomedical and biochemical fields, as we shall introduce later in this book [3, 5, 28, 29]. There are eighteen chapters altogether in this book, categorized into four sections, including Section A: Electrophoresis of Rigid Particles, Section B: Electrophoresis of Porous Particles, Section C: Electrophoresis of Liquid Drops and Gas Bubbles, and Section D: Diffusiophoresis of Colloidal Particles. Various issues of practical interest are treated in separate chapters.

1.2 INTRODUCTION OF ELECTROPHORESIS

Electrophoresis, the motion of colloidal particles in response to an externally applied electric field, is a very efficient and powerful separation technique in practice with applications ranging from DNA sequencing, protein analysis, separation of macromolecules or polyelectrolytes in biochemical and biomedical fields, to the purification of quantum dots in industrial processes, to name a few [30–34]. Moreover, new developments and novel applications have been proposed in a very fast pace. Indeed, it is a promising separation scheme with merits like fast processing time and easy to operate, and so on. In addition to the separation function, it can also be used to explore the surface electric condition of colloidal particles, hence test the validity of associated theories regarding it [35, 36]. Other technologies and advances based on electrophoresis are abundant as well in the last decade [37–43].

With such a powerful capability and great potential in practical applications, yet the underlying physical principle of electrophoresis is very simple and easy to understand: the force balance between the electric driving force and the hydrodynamic drag force. The best way to learn the principles behind is to consider the most fundamental case: the electrophoretic motion of a single rigid particle suspended in an infinite medium of electrolyte solution, which is the focus of this chapter.

As we have introduced earlier, when a charged colloidal particle is suspended in an electrolyte solution, it will be surrounded predominantly by counterions, and an electric double layer structure is established surrounding the particle. When an external electric field is applied upon the suspension, the ions in the double layer as well as the charged particle itself respond instantly and simultaneously. The particle tends to move along the direction of the electric field if it is positively charged, and against the direction of the electric field if it is negatively charged. The counterions in the diffuse double layer, on the other hand, tend to move in the opposite direction of the particle motion. A phenomenon referred to as the double layer polarization (deformation) is thus resulted. The original concentric distribution of ions is no longer there and the perfect shielding or screening of the particle charges disappears. An electric driving force is thus generated which sets the particle in motion. Note that this "double layer polarization" refers to the polarization of the outer diffuse layer only, to be precise. As within the inner Stern layer, the counterions there are so strongly bounded to the particle surface that they are essentially immobile, at least compared with those in the diffuse layer. In other words, the ions in the inner Stern layer stick to the particle surface firmly when the particle is in motion, hence the entire inner Stern layer can be regarded as part of the particle, or the effective outermost surface of the rigid particle. With perhaps some trapped solvent molecules in its vicinity, the actual or effective hydrodynamic solid-liquid interface is slightly beyond the border between the inner Stern layer and the outer diffuse layer [6]. Hence a conceptual surface referred to as the "shear plane" has been introduced, where the hydrodynamic no-slip boundary condition is supposed to be applicable. The fluid there hence assumes the same speed as the particle, in other words, the shear plane is the effective solid-liquid interface when the particle is in motion. Corresponding electric potential at the shear plane is referred to as the "electrokinetic potential" or the "zeta potential" [44]. Note that the shear plane is also regarded as the start of the outer diffuse layer in the theoretical treatment here for simplicity. After all there is no need to further distinguish these two planes so closely located in practice.

Moreover, there is no need to distinguish the actual particle surface and the shear plane here as well, since the inner Stern layer is so thin, only several molecules thick at most, and moves together with the particle. Hence in general the surface charge indicates the effective "electrokinetic charge" at the shear plane when the particle is in motion. Here the electrokinetic charge refers to the sum of the electric charges somehow on the actual particle surface and the immobile charges in the inner Stern layer due to either some surface dissociation reaction or physical adsorption, in other words, the total amount of charges behind the shear plane.

Once the particle is set to motion by the electric driving force, it is almost instantly balanced by the hydrodynamic drag force, leading to a constant steady terminal particle velocity, which can be measured experimentally. Note that both forces are functions of the particle surface charge condition as well. The particle surface charge condition hence can be evaluated should appropriate theory or treatment is established relating the particle velocity to it, which is the focus of this chapter.

1.2.1 Historical Review of the Classic Analytical or Semianalytical Formulas

Assuming the double layer is very thin, or $\kappa a \gg 1$, where κ is the Debye-Hückel parameter, and its reciprocal, κ^{-1} is referred to as the Debye screening length, or the Debye length, which is a measurement of the double layer thickness, and a is the particle radius, Smoluchowski [18] was able to derive a simple linearly proportional relationship between the particle velocity and the applied electric field:

$$\mu = \frac{U}{E} = \frac{\varepsilon\zeta}{\eta_f} \tag{1.1}$$

where U is the particle velocity, E is the strength or the magnitude of the applied electric field, μ is the electrophoretic mobility of the particle defined as U/E, the particle velocity per unit applied electric field, ε is the permittivity of the suspending medium, a material constant available in the literature, and η_f is the viscosity of the electrolyte solution. Note that this simple linear relationship is valid under a weak applied electric field only. If U is regarded as a function of E, then μ is the coefficient of the first term in the Taylor series expansion of U in terms of E. Eq. (1.1) is referred to as the Smoluchowski formula or the Smoluchowski equation, which is valid for arbitrary levels of surface potential of a rigid particle of arbitrary shape in general, as long as the double layer is very thin. Indeed it is the foundation of the entire theoretical electrokinetic analyses, with all subsequent analyses serving as its modifications or generalizations.

Hückel [45], on the other hand, derived the corresponding mobility formula when the double layer is very thick under a further limitation that the particle surface charge level has to be very low, which turns out to be two-thirds of that from the Smoluchowski formula, i.e.,

$$\mu = \frac{U}{E} = \frac{2\varepsilon\zeta}{3\eta_f} \tag{1.2}$$

Note that in Hückel's treatment, the assumption of the lowly charged condition is often referred to as the Debye-Hückel approximation, which will be mentioned later again in Section 1.3.

The above two classic formulas both neglect the impact of the double layer polarization upon the particle motion. Henry [46] was the first one to address this issue by considering the electrophoretic mobility of a lowly charged particle with arbitrary thickness of double layer under the basic assumption that the shape of the double layer remains unchanged, yet its impact on the local electric field is somehow preserved when the particle is in motion. With the polarization effect partially taken into account this way, he was able to obtain a closed-form solution, which was also presented graphically as a correction factor $f(\kappa a)$ to the Smoluchowski formula:

$$\mu = \frac{2\varepsilon\zeta}{3\eta_f} f(\kappa a) \tag{1.3}$$

Eq. (1.3) is often referred to as the Henry's formula, valid for lowly charged particle with finite double layer thickness. Note that Henry also applied the Debye-Hückel approximation in his derivation.

Later Overbeek and Booth [47–49] noticed the importance of the deformation of the double layer upon the particle motion, and formally brought about the concept of the polarization effect and the relaxation effect, with the former indicates the complete effect from the ultimate deformation of the double layer in terms of both the shape change and the corresponding modification of the local electric field. The latter, on the other hand, refers to the fact that there is a time lag between the motion of the ions within the double layer and the particle they surround. These two terminologies are essentially the same in usual phoretic motions in practice. They found that as the particle surface charge gets higher, the deviation from the Smoluchowski formula gets more significant. In other words, for a highly charged particle, a generalized Henry's formula is needed to take into account the complete effect of the double layer polarization.

Wiersema [50] was the first to use computer to solve for the particle mobility numerically, based on the idea from Overbeek and Booth [47–49]. O'Brien and White [51] later came up with a methodology based on the linear regular perturbation analysis and obtain the solution with a shooting scheme to treat the boundary conditions at infinity. Their approach turned out to be very successful and since then has become the mainstream to explore the behavior of a highly charged particle, taking full account of the double layer polarization effect.

Their computer program, however, is not readily available to the general readers. This leaves the room for other approximate analytical or semianalytical formulas valid for limiting cases to fill in the need of experimental researchers, such as those referred to in a review paper by Lyklema [44, 52–55]. In particular, Ohshima derived an approximate analytical formula valid for arbitrary levels of surface potential. When $\kappa a > 10$, the approximation error is smaller than 1% in comparison with the prediction by O'Brien and White [51]. Later he derived an even more concise analytical formula based on the approaches by Overbeek [47, 48] and Booth [49]. For $1 < \kappa a < 10$, however, the accuracy is not acceptable when the particle is highly charged [52, 56]. This range of κa for a highly charged particle turns out to be when the polarization effect of the double layer is most profound. Indeed, as Shilov and others put it: "No theory of electrophoresis can be correct without properly accounting for the polarization or relaxation of the double layer" [57].

From the above review of the classic theoretical approaches, the challenge of the mobility prediction for a highly charged particle lies mainly in the finite double layer thickness with $1 < \kappa a < 10$, where the polarization effect is most profound. Thus the major theme of this chapter is the generalization or extension of the classic Henry's formula to the highly charged particles, focusing on the effect of the double layer polarization. Visual demonstration of the exact shape of the double layer polarization, especially those of highly charged particles, certainly will shed further light on this complicated and critical mechanism underlying the electrophoretic behavior of a charged particle, as shall be presented in Section 1.4 later on.

1.3 THEORY

The system diagram is shown in Fig. 1.2, where the electrophoretic motion of a single dielectric rigid spherical particle of radius a is considered. The particle is suspended in a binary aqueous $z_1 : z_2$ electrolyte solution, z_1 and z_2 being, respectively, the valences of cations and anions, and $\alpha = -z_1/z_2$. Moreover, n_0 is the number concentration of the total electrolytes in the bulk solution, and n_{10} and n_{20} being respectively the bulk number concentrations of cations and anions.

As shown in Fig. 1.2, a constant electric field $\boldsymbol{E}$ is applied externally upon the system in the z-direction, and in response the particle moves in the z-direction with a constant velocity $\boldsymbol{U}$. The electrolyte ions in the solution respond to this external electric field as well, leading to the redistribution of ions surrounding the particle, which results in a net electric driving force and sets the particle in motion. Hence the governing equations here consist

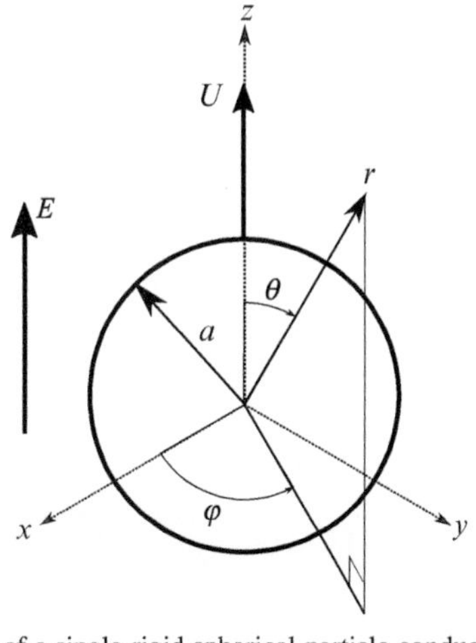

FIG. 1.2 System diagram of a single rigid spherical particle conducting electrophoretic motion.

of three basic elements: the ion flux, the electric field, and the flow field. Together they are called the general electrokinetic equations [6], which are both coupled and highly nonlinear system of differential equations in general, as will be discussed below.

1.3.1 Electrical Potential

The governing equation for the electric potential is the Poisson equation, which is based on the Gauss divergence theorem and relates the local electric potential, ϕ, and the corresponding space charge density, ρ, in the electrolyte solution as follows:

$$\nabla^2\phi = -\frac{\rho}{\varepsilon} \tag{1.4}$$

where ε is the permittivity of the solution. Eq. (1.4) is sometimes referred to as the Gauss divergence theorem or Gauss equation, as its physical essence is simply

$$\nabla \cdot \boldsymbol{E} = \frac{\rho}{\varepsilon} \tag{1.5}$$

with $\boldsymbol{E} = -\nabla\phi$, the local electric field. Note that Eq. (1.4) or (1.5) can be derived rigorously from the fundamental electrostatic Coulomb's law in vector form [6]. It indicates the physical reality that the electric charge is the source of the electric field and the electric potential. Hence if there is a redistribution of the ions in the double layer, so will be the corresponding electric field around the particle.

1.3.2 Ion Concentration

The transport of ions in the solution is governed by the well-known Nernst-Planck equation as follows:

$$\boldsymbol{f}_j = -D_j\left(\nabla n_j + \frac{z_j e n_j}{kT}\nabla\phi\right) + n_j\boldsymbol{v} \tag{1.6}$$

where $\boldsymbol{f}_j$, n_j, and D_j are respectively the ion flux, the ion number density, and the diffusivity of the ionic species j, e the elementary charge density, k the Boltzmann constant, T the absolute temperature, and v the fluid velocity. Note that the first term, $-D_j\nabla n_j$, refers to the diffusion mechanism as the Fick's law [44]. The second term, $-D_j\left(\frac{z_j e n_j}{kT}\right)\nabla\phi$, represents the electrostatic response of the ions to the local electric field, hence sometimes is referred to as the "conduction" [6] or "electro-migration" [52] mechanism of ion flux. It is due to the Coulomb electrostatic force. The third term, $n_j\boldsymbol{v}$, is due to the convection flow of the fluid. Note that while the first two mechanisms are independent of the frame of reference, the third one is. Note also that due to the nonlinear terms of $n_j\nabla\phi$ in the electro-migration term and the $n_j\boldsymbol{v}$ in the convection term, the governing equations of $\nabla\phi$, n_j, and $\boldsymbol{v}$ thus are nonlinear and coupled. Here nonlinear terms refer to products of dependent variables or their derivatives. The other nonlinear term is in the governing equation of the flow field, which will be introduced below.

Eq. (1.6) is often referred to as the ion flux equation, with obvious reason. The conservation of the ionic species yields:

$$\frac{\partial n_j}{\partial t} = -\nabla \cdot \boldsymbol{f}_j \tag{1.7}$$

Eq. (1.7) is sometimes called the ion conservation equation. Note that for a system under quasisteady state, which will be discussed later, $\frac{\partial n_j}{\partial t} = 0$ and Eq. (1.7) reduces to

$$\nabla \cdot \boldsymbol{f}_j = 0 \tag{1.8}$$

1.3.3 Flow Field

The liquid phase of the electrolyte solution is treated as an incompressible Newtonian fluid with constant physical properties. Hence the momentum transfer there is governed by the famous Navier-Stokes equation in general, which is the Newton's second law in fluid [58]. When the flow is in the creeping flow regime, the inertia term drops out and it reduces to the Stokes equation with an extra electric body force term, $-\rho\nabla\phi$ as follows:

$$\eta_f \nabla^2 \boldsymbol{v} - \nabla p - \rho \nabla \phi = \boldsymbol{0} \tag{1.9}$$

under the incompressibility constraint:

$$\nabla \cdot \boldsymbol{v} = 0 \tag{1.10}$$

where p and η_f are pressure and viscosity of the fluid, respectively. Whether a fluid flow is in the creeping flow regime is determined by the corresponding dimensionless Reynolds number:

$$Re = \frac{\rho_f v d}{\eta_f} \tag{1.11}$$

where ρ_f is the fluid density, v is the magnitude of the fluid velocity, and d is the characteristic length scale in the system under consideration, such as the particle diameter here. Note that the Reynolds number can be viewed as the ratio of the inertia force to the viscous force in the fluid:

$$Re = \frac{\text{inertia force}}{\text{viscous force}} \tag{1.12}$$

When $Re \ll 1$, or the inertia force is much smaller in comparison with the viscous force, the inertia term in the left-hand-side of the dimensionless Navier-Stokes equation, $Re\left(\frac{\partial \boldsymbol{v}^*}{\partial t^*} + \boldsymbol{v}^* \cdot \nabla^* \boldsymbol{v}^*\right)$, can be dropped leading to the simplified

Stokes equation, Eq. (1.9), and the system is referred to as in the creeping flow regime. Note that for a small particle, its velocity does not need to be very slow to stay in the creeping flow regime. For instance, for a particle of 1 μm diameter moving with a speed of 1 μm/s in water, the corresponding Reynolds number is of the order 10^{-6}. This means translates that it takes roughly only one-millionth of the viscous force on the right-hand-side of the Navier-Stokes equation to fully speed up the particle to its steady terminal speed [6]. The system is hence referred to as at quasisteady state in that the particle reaches its steady terminal velocity almost instantly in response to the external field force applied upon it, such as in electrophoresis, where the particle speed is normally dozens of particle diameters per second at most. This implies the balance of the electric driving force and the hydrodynamic drag force when the particle is in electrophoretic motion.

Note that an extra term representing the electric body force due to the presence of electrolyte ions in the solution, $-\rho\nabla\phi$, is added to the conventional Stokes equations in purely hydrodynamics. Hence Eq. (1.9) is sometimes referred to as the modified Stokes equation as well. Note also that in essence the modified Stokes equation is simply Newton's second law, $\Sigma\boldsymbol{F} = m\boldsymbol{a}$, in fluid with the inertia term $m\boldsymbol{a}$ dropped out to yield $\Sigma\boldsymbol{F} = \boldsymbol{0}$, where $\boldsymbol{a}$ is the acceleration of an object.

In practice, an alternative form of Eq. (1.9) in terms of stream function is often used instead for a two-dimensional flow field, such as the axisymmetric one consider here, so that the incompressibility constraint $\nabla\cdot\boldsymbol{v} = 0$ is satisfied automatically. Taking the curl of Eq. (1.9) to drop the pressure term by the vector identity $\nabla\times\nabla p = 0$ and introducing the stream function for an axisymmetric problem in spherical coordinates as follows:

$$v_r = -\frac{1}{r^2\sin\theta}\frac{\partial\Psi}{\partial\theta} \tag{1.13}$$

$$v_\theta = \frac{1}{r\sin\theta}\frac{\partial\Psi}{\partial r} \tag{1.14}$$

we have:

$$E^4\Psi = \frac{r\sin\theta}{\eta_f}\nabla\times(\rho\nabla\phi)\cdot\mathbf{i}_\varphi \tag{1.15}$$

where $\mathbf{i}_\varphi$ is the unit vector in the φ-direction, $E^4\Psi = E^2(E^2\Psi)$, and $E^2 = \left(\frac{\partial^2}{\partial r^2}\right) + \frac{\sin\theta}{r^2}\frac{\partial}{\partial\theta}\left(\frac{1}{\sin\theta}\frac{\partial}{\partial\theta}\right)$.

Eqs. (1.4), (1.8), (1.9), and (1.10) form a set of governing equations valid for electrokinetic motions in general, hence is referred to as the "general electrokinetic equations" [6]. Sometimes it is further referred to as the "standard electrokinetic equations" emphasizing the fact that the possible surface conductivity on the particle surface is not taken into account or simply null there [19, 44]. This set of equations, which are both highly nonlinear and coupled, is the theoretical basis of this book. The presence of the nonlinear term $-\rho\nabla\phi$ in the modified Stokes equation, Eq. (1.9), plus $n_j\boldsymbol{v}$ and $-D_j\frac{z_jen_j}{kT}\nabla\phi$ in the ion flux equation, Eq. (1.6), make the general analytical solution impossible. Numerical method is thus needed for the general solution of highly charged particles, where the effect of these nonlinear terms is most profound.

Note that these equations are in vector form hence are independent of the specific coordinates selected to describe the problem. All we need is to find out the corresponding mathematical expressions of the vector operators involved, such as ∇, $\nabla\cdot$, $\nabla\times$, ∇^2, and so forth. In spherical coordinates, which is the one used here in this chapter, these vector operators are expressed as follows in computation:

$$\nabla = \boldsymbol{i}_r\frac{\partial}{\partial r} + \boldsymbol{i}_\theta\frac{1}{r}\frac{\partial}{\partial\theta} \tag{1.16}$$

$$\nabla\cdot\boldsymbol{v} = \frac{1}{r^2}\frac{\partial}{\partial r}(r^2 v_r) + \frac{1}{r\sin\theta}\frac{\partial}{\partial\theta}(v_\theta\sin\theta) \tag{1.17}$$

$$\nabla\times\boldsymbol{v} = \boldsymbol{i}_\varphi\frac{1}{r}\left[\frac{\partial}{\partial r}(rv_\theta) - \frac{\partial v_r}{\partial\theta}\right] \tag{1.18}$$

$$\nabla^2 = \frac{1}{r^2}\frac{\partial}{\partial r}\left(r^2\frac{\partial}{\partial r}\right) + \frac{1}{r^2\sin\theta}\frac{\partial}{\partial\theta}\left(\sin\theta\frac{\partial}{\partial\theta}\right) \tag{1.19}$$

$$E^2 = \left(\frac{\partial^2}{\partial r^2}\right) + \frac{\sin\theta}{r^2}\frac{\partial}{\partial\theta}\left(\frac{1}{\sin\theta}\frac{\partial}{\partial\theta}\right) \tag{1.20}$$

Before an electric field is applied, the colloidal system is at equilibrium and the fluid is at rest, i.e., $\boldsymbol{v} = 0$. The Nernst-Planck equation, Eq. (1.6), reduces to:

$$\nabla n_{je} + \frac{z_je}{kT}n_{je}\nabla\phi_e = 0,\ \ j = 1,2 \tag{1.21}$$

Hence $n_{je} = n_{j0}\exp\left[-\left(\frac{z_je}{kT}\right)\phi_e\right]$, which is the famous Boltzmann distribution of electrolytic ions at equilibrium. Inserting it into the Poisson equation, Eq. (1.4), with

$$\rho = -\sum_{j=1}^{N} z_jen_j \tag{1.22}$$

we have:

$$\nabla^2\phi_e = -\sum_{j=1}^{N}\frac{z_je}{\varepsilon}n_{j0}\exp\left[-\frac{z_je}{kT}\phi_e\right] \tag{1.23}$$

Eq. (1.23) is the famous Poisson-Boltzmann equation which governs the relationship between the electric potential and the respective space charge distribution of the electrolyte ions in a solution at equilibrium.

Suppose a weak electric field $\boldsymbol{E}$ is applied upon the system. Under this weak electric field, each electrokinetic variable of interest here is subject to an incremental change in response, for instance, $\phi = \phi_e + \delta\phi$, $\rho = \rho_e + \delta\rho$, $n_j = n_{je} + \delta n_j$, and so forth. Here the subscript e refers to the equilibrium value and the symbol δ indicates an incremental change, which is much smaller than the corresponding equilibrium value under a weak electric field, say of the order of one percent only. Hence the standard regular linear perturbation analysis can be applied here [51, 59], assuming all the variables still obey the same set of electrokinetic equations when the particle is in motion [51]. Multiply out all the products of these variables in the nonlinear terms, for example,

$$-\rho\nabla\phi = -(\rho_e + \delta\rho)\nabla(\phi_e + \delta\phi) = -(\rho_e\nabla\phi_e + \delta\rho\nabla\phi_e + \rho_e\nabla\delta\phi + \delta\rho\nabla\delta\phi) \tag{1.24}$$

neglect the product of incremental variables based on the order of magnitude argument, and subtract off the equilibrium terms on both sides of the governing equations, we end up with a set of linear differential equations governing these incremental variables. Some subtle trick is introduced to facilitate subsequent mathematical treatments, which will be elaborated as follows.

Note that once the particle is in motion, due to the double layer polarization, the space charge distribution is no longer concentric. Hence the Boltzmann distribution, which is of purely electrostatic nature assuming the particle is at rest, fails to apply any more. However, it does serve as a good reference point for approximation. One can treat the further deviation from this reference concentric profile of n_j by a shape function g_j, as suggested by O'Brien and White [51], to represent the nonconcentric polarization of the double layer:

$$n_j = n_{j0}\exp\left[-\frac{z_j e}{kT}(\phi_e + \delta\phi + g_j)\right],\quad j = 1,2 \tag{1.25}$$

In other words, the number concentration of ions n_j is represented by g_j now, although n_j does depend on $\delta\phi$ as well. With some further mathematical maneuvers, it can be shown that the governing equations for the electrophoretic motion of a single rigid particle considered here are:

$$\nabla^2\delta\phi = \sum_{j=1}^{2}\frac{(z_j e)^2}{\varepsilon kT}n_{je}(\delta\phi + g_j) \tag{1.26}$$

$$\nabla^2 g_j - \frac{z_j e}{kT}\nabla\phi_e\cdot\nabla g_j - \frac{1}{D_j}\nabla\phi_e\cdot\boldsymbol{v} = 0,\quad j = 1,2 \tag{1.27}$$

$$E^4\Psi = \frac{\sin\theta}{\eta_f}\left[\sum_{j=1}^{2}\frac{(z_j e)^2}{kT}n_{je}\frac{d\phi_e}{dr}\frac{\partial g_j}{\partial\theta}\right] \tag{1.28}$$

Note that the above three equations stand for the Poisson equation, the ion conservation equation, and the Stokes equation, respectively. They are just expressed in terms of perturbation variables now. The incompressibility constraint is satisfied automatically with the introduction of stream function Ψ, as mentioned earlier. Note that the above equations can be easily extended to multicomponent electrolyte solution with the 2 on the summation sign replaced by N, the total number of electrolyte species.

Characteristic variables are introduced at this stage and the governing equations are converted to the dimensionless form as follows:

$$\nabla^{*2}\phi_e^* + \frac{(\kappa a)^2}{1+\alpha}[\exp(-\phi_e^*) - \exp(\alpha\phi_e^*)] = 0 \tag{1.29}$$

$$\nabla^{*2}\delta\phi^* = \frac{(\kappa a)^2}{1+\alpha}[(\delta\phi^* + g_1^*)\exp(-\phi_e^*) + \alpha(\delta\phi^* + g_2^*)\exp(\alpha\phi_e^*)] \tag{1.30}$$

$$\nabla^{*2}g_1^* - \frac{d\phi_e^*}{dr^*}\frac{\partial g_1^*}{\partial r^*} + \frac{Pe_1}{r^{*2}\sin\theta}\frac{d\phi_e^*}{dr^*}\frac{\partial \Psi^*}{\partial\theta} = 0 \tag{1.31}$$

$$\nabla^{*2}g_2^* + \alpha\frac{d\phi_e^*}{dr^*}\frac{\partial g_2^*}{\partial r^*} + \frac{Pe_2}{r^{*2}\sin\theta}\frac{d\phi_e^*}{dr^*}\frac{\partial \Psi^*}{\partial\theta} = 0 \tag{1.32}$$

$$E^{*4}\Psi^* - \frac{(\kappa a)^2}{1+\alpha}\frac{d\phi_e^*}{dr^*}\left[\exp(-\phi_e^*)\frac{\partial g_1^*}{\partial\theta} + \alpha\exp(\alpha\phi_e^*)\frac{\partial g_2^*}{\partial\theta}\right]\sin\theta = 0 \tag{1.33}$$

where $r^* = \frac{r}{a}$, $\phi_e^* = \frac{\phi_e}{\phi_0}$, $\delta\phi^* = \frac{\delta\phi}{\phi_0}$, $g_j^* = \frac{g_j}{\phi_0}$, $\Psi^* = \frac{\Psi}{u_E a^2}$, $\rho_e^* = \frac{\rho_e}{(\varepsilon\phi_0/a^2)}$, and $\delta\rho^* = \frac{\delta\rho}{(\varepsilon\phi_0/a^2)}$. Among them, a is the particle radius, ϕ_0 is the thermal potential of the ionic species 1, the cation in a binary system, with $\phi_0 = \frac{kT}{z_1 e}$. And u_E is the magnitude of the particle velocity predicted by the Smoluchowski's formula when an electric field $\frac{\phi_0}{a}$ is applied upon a particle with surface potential ϕ_0, i.e., $u_E = \frac{\varepsilon\phi_0^2}{\eta_f a}$. In addition, the magnitude of the dimensionless external electric field and the electrophoretic velocity of the particle are expressed as $E^* = \frac{E}{(\phi_0/a)}$ and $U^* = \frac{U}{u_E}$, respectively. Moreover, as briefly introduced in Section 1.2, κ is the Debye-Hückel parameter defined as $\kappa = \sqrt{\sum_{j=1}^{N} n_{j0}\frac{(z_j e)^2}{\varepsilon kT}}$ for a multicomponent system in general [6]. It's reciprocal, κ^{-1}, on the other hand, is referred to as the Debye screening length or simply the Debye length [10], which is a measurement of the double layer thickness. Hence the reciprocal of the dimensionless quantity κ can be regarded as a dimensionless double layer thickness $(\kappa a)^{-1} = \frac{\kappa^{-1}}{a}$. The larger the value of κ is, the thinner the double layer is. A related quantity I, the ionic strength in an electrolyte solution, is defined as $I = \frac{1}{2}\sum_{j=1}^{N} z_j^2 c_{j0}$, where c_{j0} is the molarity (mol/liter) of the ionic species j in the solution, which is equal to $n_{j0}/(6.02 \times 10^{26})$. $Pe_j = \frac{u_E a}{D_j}$

is the Peclet number of the ionic species j, and stands for the ratio of the convection element to the diffusion element in general. Here it is just the reciprocal of the dimensionless diffusivity, however, as u_E is just a constant reference value with no direct connection to the real particle velocity or the fluid velocity. $\phi_r = \frac{\zeta}{\phi_0}$ is the dimensionless surface potential of the particle, or the zeta potential when it is in motion. Note that n_{10} is chosen as the characteristic number density of ions, hence, $n_j^* = n_j/n_{10}$, which does not appear in the final governing equations though, as its role is replaced by g_j.

1.3.4 Boundary Conditions on the Particle Surface

The corresponding boundary conditions on the particle surface are:

$$\phi_e^* = \phi_r \tag{1.34}$$

In other words, a particle with constant surface potential is considered here. And

$$\frac{\partial \delta \phi^*}{\partial r^*} = 0 \tag{1.35}$$

which implies the particle is nonconducting, valid for most dielectric particles in practice. Moreover, as the particle is a rigid impervious one, i.e., it is not penetrable by ions, we have:

$$\frac{\partial g_j^*}{\partial r^*} = 0, \quad j = 1,2 \tag{1.36}$$

which is obtained by taking the derivative $\frac{\partial}{\partial r^*}$ of Eq. (1.25) and substituting Eq. (1.35) into it. In addition,

$$v_r = -\frac{1}{r^2 \sin\theta} \frac{\partial \Psi}{\partial \theta} = 0 \tag{1.37}$$

$$v_\theta = \frac{1}{r \sin\theta} \frac{\partial \Psi}{\partial r} = 0 \tag{1.38}$$

which is the usual hydrodynamic no-slip boundary conditions with the coordinates frame set at the center of the moving particle.

1.3.5 Boundary Conditions at Infinity

At infinity, the impact of the particle surface potential should reduce to zero, i.e.,

$$\phi_e^* = 0 \tag{1.39}$$

and

$$\frac{d\phi_e^*}{dr^*} = 0 \tag{1.40}$$

And local electric field there should equal to the imposed external electric field:

$$\frac{\partial\delta\phi^*}{\partial r^*} = -E^*\cos\theta \tag{1.41}$$

$$\delta\phi^* = -E^* r^* \cos\theta \tag{1.42}$$

Due to the Boltzmann distribution of ions and Eq. (1.39), n_j^* at infinity should equal to the bulk value n_{j0}^*. In other words, $\delta n_j = 0$, or:

$$\delta\phi^* + g_j^* = 0, \quad j = 1,2 \tag{1.43}$$

Note this condition is satisfied even when the particle is in motion, i.e., $n_j^* = n_{j0}^*$ even when there is a net electric current across the bulk electrolyte solution at infinity.

A relative fluid velocity at infinity opposite to the direction of the particle motion is observed from the moving particle:

$$\Psi^* = \frac{1}{2}U^* r^{*2} \sin^2\theta \tag{1.44}$$

Eqs. (1.39)–(1.44) are the analytical boundary conditions at infinity. However, as infinity is just a mathematical concept and cannot be treated directly with numerical scheme in general, some special treatment on this issue is needed, as elaborated below [51]. We assume that at a large enough distance, r_o^*, the electric potential ϕ_e^* becomes so small that the Debye-Hückel approximation can be used to simplify the original governing nonlinear Poisson-Boltzmann equation as follows:

$$\nabla^{*2}\phi_e^* = (\kappa a)^2 \phi_e^* \tag{1.45}$$

Moreover, substituting the above analytical boundary conditions at infinity, Eqs. (1.39)–(1.44), into the governing equations, Eqs. (1.30)–(1.33), we have:

$$\nabla^{*2}\delta\phi^* = \nabla^{*2} g_j^* = E^{*4}\Psi^* = 0 \tag{1.46}$$

Analytical solutions are thus obtained for these linear equations as follows:

$$\phi_e^* = \frac{c}{r^*}\exp[-(\kappa a) r^*] \tag{1.47}$$

$$\delta\phi^* = \left(-E^* r^* + \frac{c_0}{r^{*2}}\right)\cos\theta \tag{1.48}$$

$$g_j^* = \left(E^* r^* + \frac{c_j}{r^{*2}}\right)\cos\theta, \quad j = 1,2 \tag{1.49}$$

$$\Psi^* = \left(\frac{1}{2}U^* r^{*2} + c_3 r^* + \frac{c_4}{r^*}\right)\sin^2\theta \tag{1.50}$$

where c and c_0 to c_4 are constants to be determined. Note that they can be eliminated by taking further derivation with respect to r^* and substituting them back again to the resulted equations, for example:

$$\frac{d\phi_e^*}{dr^*} = -\left[\frac{(\kappa a)r^*+1}{r^{*2}}\right] c\exp[-(\kappa a)r^*] = -\left[\frac{(\kappa a)r^*+1}{r^*}\right]\phi_e^* \tag{1.51}$$

Similarly, we have:

$$\frac{\partial \delta\phi^*}{\partial r^*} + \frac{2}{r^*}\delta\phi^* = -3E^*\cos\theta \tag{1.52}$$

$$\frac{\partial g_j^*}{\partial r^*} + \frac{2}{r^*}g_j^* = 3E^*\cos\theta, \quad j=1,2 \tag{1.53}$$

$$\frac{\partial^2 \Psi^*}{\partial r^{*2}} + \frac{1}{r^*}\frac{\partial \Psi^*}{\partial r^*} - \frac{1}{r^{*2}}\Psi^* = \frac{3}{2}U^*\sin^2\theta \tag{1.54}$$

$$\frac{\partial^3 \Psi^*}{\partial r^{*3}} - \frac{3}{r^{*2}}\frac{\partial \Psi^*}{\partial r^*} + \frac{3}{r^{*3}}\Psi^* = -\frac{3}{2r^*}U^*\sin^2\theta \tag{1.55}$$

The details can be found elsewhere [60].

1.3.6 Reduction to One-Dimensional Problem

The two-dimensional axisymmetric problem here can be further reduced to a one-dimensional problem with the following conversion:

$$\delta\phi^*(r^*,\theta) = \delta\Phi^*(r^*)\cos\theta \tag{1.56}$$

$$g_j^*(r^*,\theta) = G_j^*(r^*)\cos\theta, \quad j=1,2 \tag{1.57}$$

$$\Psi^*(r^*,\theta) = \psi^*(r^*)\sin^2\theta \tag{1.58}$$

Eqs. (1.56)–(1.58) are inspired by the boundary conditions at infinity, i.e., Eqs. (1.48)–(1.50). For example, Eq. (1.48) indicating there is only one normal mode, $\cos\theta$, in the Fourier series expansion of the separated function of θ in the θ-direction, which is resulted from the technique of the separation of variables for linear partial differential equations, as Eqs. (1.30)–(1.33) here. On the other hand, there are two normal modes in Eq. (1.58) as $\sin^2\theta = \frac{1}{2} - \frac{1}{2}\cos 2\theta$ [61]. Eqs. (1.30)–(1.33) are thus reduced to:

$$L^{*2}\delta\Phi^* = \frac{(\kappa a)^2}{1+\alpha}[(\delta\Phi^* + g_1^*)\exp(-\phi_e^*) + \alpha(\delta\phi^* + g_2^*)\exp(\alpha\phi_e^*)] \tag{1.59}$$

$$L^{*2}G_1^* - \frac{d\phi_e^*}{dr^*}\frac{dG_1^*}{dr^*} + \frac{2Pe_1}{r^{*2}}\frac{d\phi_e^*}{dr^*}\psi^* = 0 \tag{1.60}$$

$$L^{*2}G_2^* + \alpha\frac{d\phi_e^*}{dr^*}\frac{dG_2^*}{dr^*} + \frac{2Pe_2}{r^{*2}}\frac{d\phi_e^*}{dr^*}\psi^* = 0 \tag{1.61}$$

$$D^{*4}\psi^* + \frac{(\kappa a)^2}{1+\alpha}[\exp(-\phi_e^*)G_1^* + \alpha\exp(\alpha\phi_e^*)G_2^*] = 0 \quad (1.62)$$

where

$$L^{*2} = \frac{d^2}{dr^{*2}} + \frac{2}{r^*}\frac{d}{dr^*} - \frac{2}{r^{*2}} \quad (1.63)$$

$$D^{*4} = D^{*2}D^{*2} = \left(\frac{d^2}{dr^{*2}} - \frac{2}{r^{*2}}\right)\left(\frac{d^2}{dr^{*2}} - \frac{2}{r^{*2}}\right) \quad (1.64)$$

are the corresponding one-dimensional operators of ∇^{*2} and E^{*4}, respectively. Corresponding one-dimensional boundary conditions on the particle surface, $r^* = 1$, are:

$$\frac{d\delta\Phi^*}{dr^*} = 0 \quad (1.65)$$

$$\frac{dG_j^*}{dr^*} = 0, \quad j = 1, 2 \quad (1.66)$$

$$\psi^* = 0 \quad (1.67)$$

$$\frac{d\psi^*}{dr^*} = 0 \quad (1.68)$$

And far away from the particle, $r^* = r_o^*$, the boundary conditions there are:

$$\frac{d\delta\Phi^*}{dr^*} + \frac{2}{r^*}\delta\Phi^* = -3E^* \quad (1.69)$$

$$\frac{dG_j^*}{dr^*} + \frac{2}{r^*}G_j^* = 3E^*, \quad j = 1, 2 \quad (1.70)$$

$$\frac{d^2\psi^*}{dr^{*2}} + \frac{1}{r^*}\frac{d\psi^*}{dr^*} - \frac{1}{r^{*2}}\psi^* = \frac{3}{2}U^* \quad (1.71)$$

$$\frac{d^3\psi^*}{dr^{*3}} - \frac{3}{r^{*2}}\frac{d\psi^*}{dr^*} + \frac{3}{r^{*3}}\psi^* = -\frac{3}{2r^*}U^* \quad (1.72)$$

The electric field and the flow field can thus be solved numerically by applying the pseudo-spectral method based on Chebyshev polynomials to discretize the governing equations, Eq. (1.29) and Eqs. (1.59)–(1.62). Note that these boundary conditions, Eqs. (1.69)–(1.72), are converted to algebraic equations relating dependent electrokinetic variables, ϕ_e^*, $\delta\phi^*$, g_j^*, and Ψ^* as well by the numerical discretization procedure. The details of the numerical treatment can also be found elsewhere [62–65].

1.3.7 Evaluation of the Particle Mobility

The particle mobility is obtained based on the force balance between the electric driving force $\boldsymbol{F_E}$ and the corresponding hydrodynamic drag force $\boldsymbol{F_D}$ under the condition of quasisteady state:

$$\Sigma \boldsymbol{F} = \boldsymbol{F_D} + \boldsymbol{F_E} = 0 \tag{1.73}$$

The magnitude of the electric driving force along the z-direction can be evaluated by

$$F_E = \iint_S \sigma(-\nabla\phi \cdot \boldsymbol{i}_z)dS \tag{1.74}$$

where σ is the surface charge density of the particle, which can be evaluated by the two-dimensional Gauss theorem across the particle surface under the assumption that the permittivity of a dielectric particle is much smaller than that of the aqueous solution, which is normally true. In other words, $\sigma = -\varepsilon \frac{d\phi_e}{dr}\big|_{r=a}$ for the rigid spherical particle considered here. And we have:

$$F_E = 2\pi\varepsilon \int_0^{\pi} \left[r^2 \left(\frac{d\phi_e}{dr} + \frac{\partial \delta\phi}{\partial r} \right) \cos\theta + r \frac{\partial \delta\phi}{\partial \theta}(-\sin\theta) \right]_{r=a} d\theta \tag{1.75}$$

On the other hand, there are two elements in the hydrodynamic drag force: the purely hydrodynamic one, and the electrostatic one due to the flow of electrolyte ions within the fluid:

$$F_D = \pi \int_0^{\pi} \left[\eta_f r^4 \sin^3\theta \frac{\partial}{\partial r} \left(\frac{E^2 \Psi}{r^2 \sin^2\theta} \right) - \rho r^2 \sin^2\theta \frac{\partial \delta\phi}{\partial \theta} \right]_{r=a} d\theta \tag{1.76}$$

The expressions of F_E and F_D in Eqs. (1.75), (1.76) are further converted to dimensionless one-dimensional form with the characteristic variable of force, $\varepsilon\phi_0^2$, or equivalently, $u_E \eta_f a$. Note that the θ-dependences in the force evaluations of Eqs. (1.75), (1.76) are simply integrated out. Hence we have:

$$F_E^* = \frac{4}{3}\pi \left[\frac{d\phi_e^*}{dr^*} \frac{d(r^{*2}\delta\Phi^*)}{dr^*} \right]_{r^*=1} \tag{1.77}$$

$$F_D^* = \frac{4}{3}\pi \left\{ r^{*2} \frac{(\kappa a)^2}{1+\alpha} [\exp(-\phi_e^*) - \exp(\alpha\phi_e^*)]\delta\Phi^* + r^{*4} \frac{d}{dr^*} \left(\frac{D^{*2}\psi^*}{r^{*2}} \right) \right\}_{r^*=1} \tag{1.78}$$

According to O'Brien and White [51], the problem can be divided into two subproblems with one of them corresponding to a stationary charged particle subject to an applied electric field $\boldsymbol{E}$, and the other a moving charged particle with constant velocity $\boldsymbol{U}$, both under the same boundary conditions of the original problem otherwise. The original problem is recovered by the superposition of these two subproblems mathematically. In practice this is done by setting $\boldsymbol{U}=0, \boldsymbol{E}=1$ and $\boldsymbol{U}=1, \boldsymbol{E}=0$, respectively and calculate the summation of forces, ΣF, in each subproblem.

Note that the total force upon the particle in either subproblem is not zero in general, as there is electrostatic stress field as well as hydrodynamic stress field upon the particle due to the electrolyte ions in the moving fluid. The electrolyte solution does flow in subproblem due to the applied electric field, although the particle is assumed somehow held stationary there. Note that as the frame of the coordinates is set at the center of the particle, the electrolyte solution in either subproblem flows against the particle.

Under the assumption of weak applied electric field, however, we have:

$$\Sigma F = F_D + F_E = C_D U + C_E E \tag{1.79}$$

based on the Taylor series expansion and neglect higher order nonlinear terms, where C_D and C_E are proportional constants to be determined. Thus the particle mobility, which is derived if $\Sigma F = 0$, takes the form as:

$$\mu = \frac{U}{E} = -\frac{C_E}{C_D} = -\frac{(\Sigma F)_{\mathrm{I}}}{(\Sigma F)_{\mathrm{II}}} \tag{1.80}$$

where $(\Sigma F)_{\mathrm{I}}$ and $(\Sigma F)_{\mathrm{II}}$ are the summations of forces in subproblem I and II, respectively. Or in dimensionless form:

$$\mu^* = \frac{U^*}{E^*} = -\frac{(\Sigma F)_{\mathrm{I}}^*}{(\Sigma F)_{\mathrm{II}}^*} \tag{1.81}$$

where $\Sigma F^* = F_E^* + F_D^*$.

1.3.8 Charge-Regulation Phenomenon

Sometimes specific kind of ions can penetrate into the inner immobile Stern layer and have reaction with the functional groups on the particle surface, such as in the case of a protein particle, whose surface charge condition is very sensitive to the pH value of the solution. The ions that can penetrate into the Stern layer and have some surface reaction there are called the specific ions, or the potential-determining ions [44]. It may also be called the charge-determining ions [66–68], as the total amount of surface charges is changed by the reaction. The dehydrated specific ions may also conduct physical adsorption there in the Stern layer due to the strong affinity with the particle surface [6]. There is no need in practice to distinguish these two mechanisms of charge contribution from specific ions in the Stern layer. On the other hand, ions that cannot penetrate into the Stern layer are called the indifferent ions [44]. They affect the particle motion nonetheless by contributing to the value of κa, or the double layer thickness. The change of surface charges, or the net amount of the total charges behind the border between the Stern layer and the outer diffuse layer to be precise, due to the interaction of a charged surface and its electrostatic environment is referred to as the charge-regulation phenomenon [6]. A particle with its surface capable of exhibiting charge-regulation phenomenon is called a charge-regulated particle, and its surface a charge-regulated surface.

As a representative case, we consider a particle containing functional groups of HA which is capable of undergoing the following simple dissociation reaction:

$$HA \rightleftharpoons H^+ + A^- \tag{1.82}$$

The dissociation constant of this reversible reaction is given as follows:

$$K = \frac{[H^+]_s[A^-]}{[HA]} \tag{1.83}$$

where $[HA]$ and $[A^-]$ are surface number density of HA and A^-, respectively. And $[H^+]_s$ is the concentration of the hydrogen ion, H^+, in the nearby Stern layer, or the activity of H^+ on the particle surface, to be precise [6].

As the specific ion H^+ can penetrate into the Stern layer and the thickness of the Stern layer is only several molecules at most, $[H^+]_s$ can be further assumed to equal the value of $[H^+]$ on the starting plane of the outer diffuse layer, where $[H^+]$ obeys the Boltzmann distribution [6]. Thus we have:

$$[H^+]_s = [H^+]_0 \exp\left(-\frac{e\phi_e}{kT}\right) \tag{1.84}$$

where $[H^+]_0$ is the concentration of the hydrogen ion in the bulk electrolyte solution.

Suppose N_s is the number density of the functional groups originally on the particle surface, then:

$$N_s = [HA] + [A^-] \tag{1.85}$$

Substituting Eqs. (1.84), (1.85) into Eq. (1.83), we have:

$$[A^-] = \frac{N_s}{1 + \frac{[H^+]_0}{K_a} \exp\left(-\frac{e\phi_e}{kT}\right)} \tag{1.86}$$

The surface charge density σ of the particle is given by:

$$\sigma = -e[A^-] \tag{1.87}$$

where e is the elementary charge. In other words, the particle is negatively charged here in this demonstration example. Substituting Eq. (1.87) into Eq. (1.86), we have:

$$\sigma = -\left[\frac{eN_s}{1 + \frac{[H^+]_0}{K_a} \exp\left(-\frac{e\phi_e}{kT}\right)}\right] \tag{1.88}$$

For the dielectric rigid particle considered here,

$$\sigma = -\varepsilon \left[\frac{d\phi_e}{dr} \right]_{r=a} \tag{1.89}$$

Combining Eqs. (1.88) and (1.89), we have:

$$\left[\frac{d\phi_e^*}{dr^*} \right]_{r^*=1} = \frac{A}{1 + B\exp(-\phi_e^*)} \tag{1.90}$$

where $A = eN_s a/\varepsilon\phi_0$ is the dimensionless group representing the amount of functional groups on the particle surface, and $B = [H^+]_0/K$ is the dimensionless group representing the concentration of the hydrogen ion in the bulk solution. Large value of A implies highly charged particle, and small value of B implies nearly complete dissociation reaction, or negligible charge-regulation phenomenon.

Note that when the particle is lowly charged, Eq. (1.90) can be simplified to a linear boundary condition as follows:

$$\left[\frac{d\phi_e^*}{dr^*} \right]_{r^*=1} = \frac{A}{1+B} + \frac{AB}{(1+B)^2} \phi_e^* \tag{1.91}$$

Of course more complicated dissociation reactions can happen in practice [6, 69, 70]. The methodology of the theoretical treatment demonstrated above, however, is applicable there nonetheless.

Note that the charge-regulation mechanism is an electrostatic phenomenon with the actual particle surface potential determined prior to the motion of the particle. The application of an external electric field does not further alter the surface electric condition due to the weak electric field assumption.

In general, the electrolyte solution of KCl is chosen as the benchmark solution suspending the particle in this chapter and essentially the rest of the book for demonstration purpose, as the Peclet numbers of the cation and the anion, K^+ and Cl^-, are equal, which greatly simplify the interpretation of computational results.

1.4 RESULTS AND DISCUSSIONS

1.4.1 Electrostatic Features When the Particle Is at Rest and Evaluation of Particle Surface Charge Density

In the absence of an external electric field, the system is at equilibrium and the particle is at rest or motionless. The Poisson-Boltzmann equation, Eq. (1.23), yields the equilibrium electrostatic potential distribution in the electrolyte solution suspending the particle. The ions concentration is obtained by the Boltzmann distribution as well. They both are concentric due to the geometric nature of the system, as shown in Figs. 1.3 and 1.4, where the contour plots of

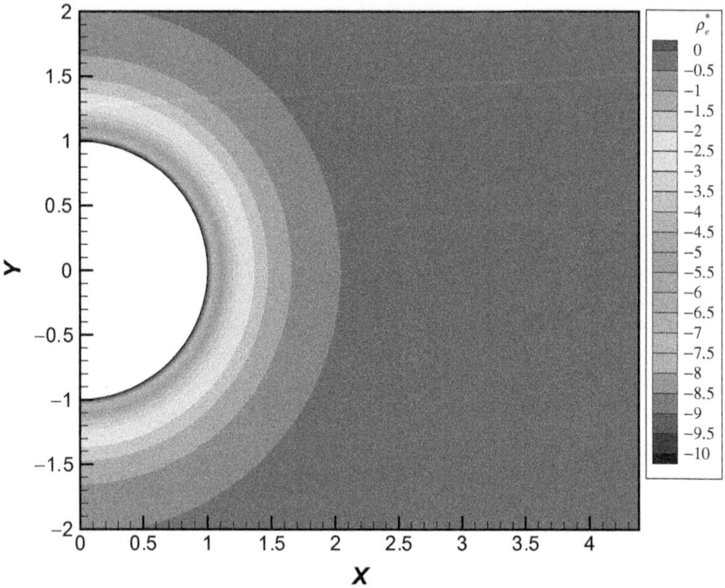

FIG. 1.3 Contour plot of the dimensionless equilibrium space charge density ρ_e^* for a rigid particle with $\phi_r = 3$ and $\kappa a = 1$.

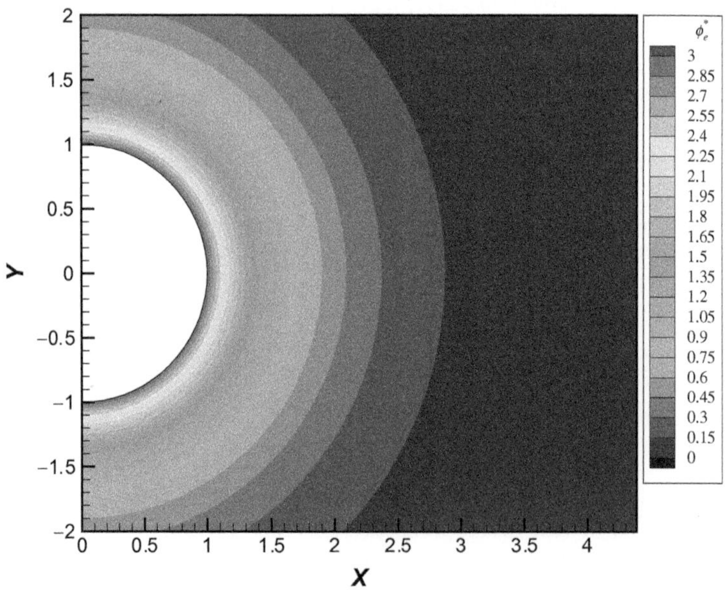

FIG. 1.4 Contour plot of the dimensionless equilibrium space charge density ρ_e^* for a rigid particle with $\phi_r = 3$ and $\kappa a = 1$.

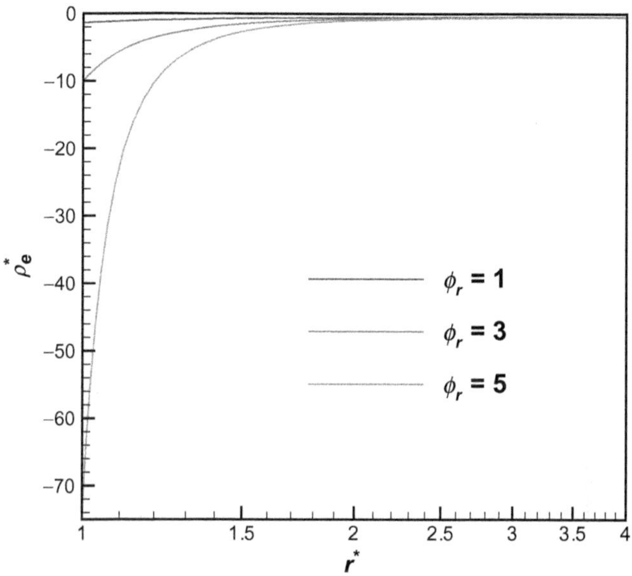

FIG. 1.5 Radial profile of the dimensionless equilibrium space charge density ρ_e^* for a rigid particle at various ϕ_r with $\kappa a = 1$.

the dimensionless space charge density and electrostatic potential, ρ_e^* and ϕ_e^*, are presented respectively. Fig. 1.3 is the famous Boltzmann distribution of electrolyte ions in dimensionless form. The situation considered in these two figures corresponds to a highly charged particle with $\phi_r = 3$ at $\kappa a = 1$, the double layer thickness where the polarization effect is regarded as most profound if the particle is in motion.

Corresponding one-dimensional profiles along the radial direction are shown in Figs. 1.5 and 1.6 for three representative surface potentials of $\phi_r = 1$, 3, and 5, respectively, which stand for lowly, highly, and very highly charged conditions, respectively. As shown in Fig. 1.5, counterions cluster in the neighborhood of the particle surface. The higher the particle surface potential is, the more profound this phenomenon is, due to the obvious reason of the stronger Coulomb attraction force. Moreover, the visible thickness of the double layer is only several particle diameters, as shown more clearly in Fig. 1.6, about three particle diameters here for $\kappa a = 1$. Note that this thickness is strongly dependent on the ionic strength, or alternatively κa, of the solution. The stronger the ionic strength is, the thinner the double layer becomes, which is sometimes referred to as the "double layer suppression effect" [44].

Steep gradients on the particle surface are observed in general for the mobility profiles of highly charged particles in Fig. 1.6. The negatives of these potential profile gradients indicate the corresponding local electric fields

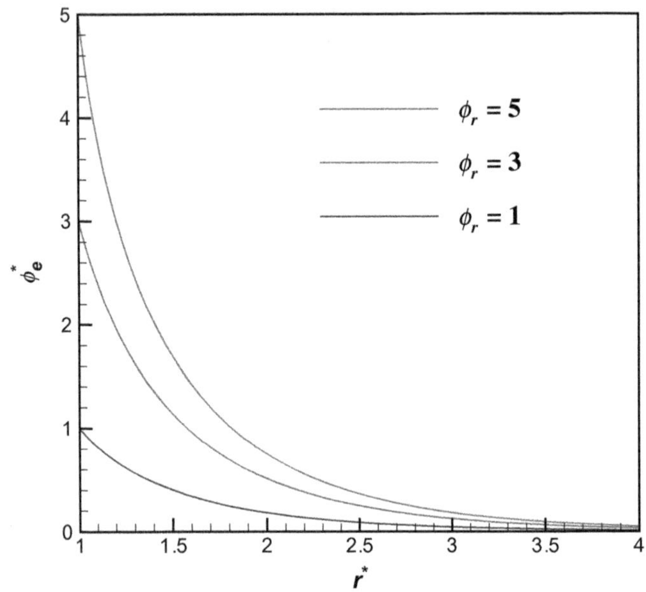

FIG. 1.6 Radial profile of the dimensionless equilibrium potential ϕ_e^* for a rigid particle at various ϕ_r with $\kappa a = 1$.

on the particle surface. If they are further multiplied by the permittivity of the solution, ε, corresponding surface charge densities, σ, are then obtained. Respective electric driving force in electrophoresis thus can be evaluated as discussed in Section 1.3. Hence the ratio of the surface charges between $\phi_r =$ 1, 3, and 5, for example, is about 2:7:15 when $\kappa a = 1$, as shown in Fig. 1.7.

Fig. 1.8 shows the percentage of the counterions within the double layer in contrast to the total amount of ions there. Nearly no appreciable coions exist in the vicinity of the particle surface for highly charged situations with $\phi_r = 3$ and 5; whereas still over 10% coions lingering there for lowly charged particle with $\phi_r = 1$. The counterions are indeed predominant in the double layer, but not exclusively, especially when the surface charge is low.

We now go on to show the electrokinetic results when the particle is in motion.

1.4.2 Electrophoretic Behavior of a Single Rigid Particle

1.4.2.1 General Henry's Chart

Fig. 1.9 shows the general Henry's chart extended to cover various highly charged situations with ϕ_r ranging from 1 to 5, respectively. The dashed line represents the classic prediction of the particle mobility by Henry [44] for a lowly charged particle using Ohshima's approximating formula [56]. Note that both the Smoluchowski's and Hückel's analytical results

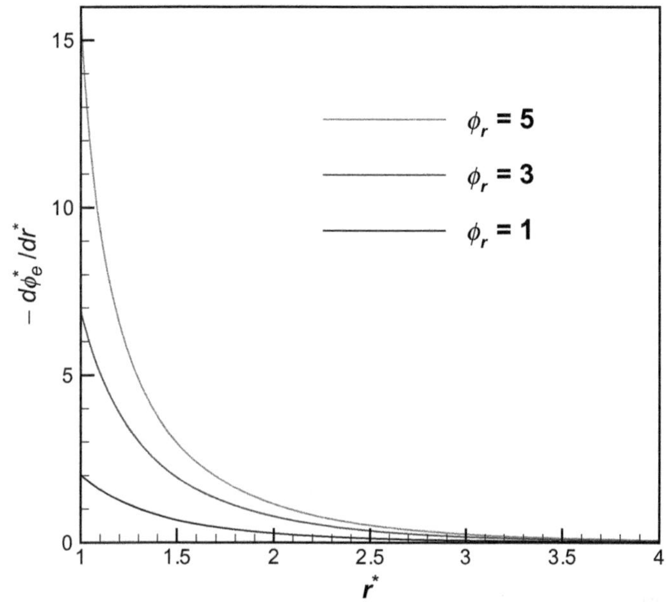

FIG. 1.7 Radial profile of the minus derivative of the dimensionless equilibrium potential, $-\frac{d\phi_e^*}{dr^*}$, for a rigid particle at various ϕ_r with $\kappa a = 1$.

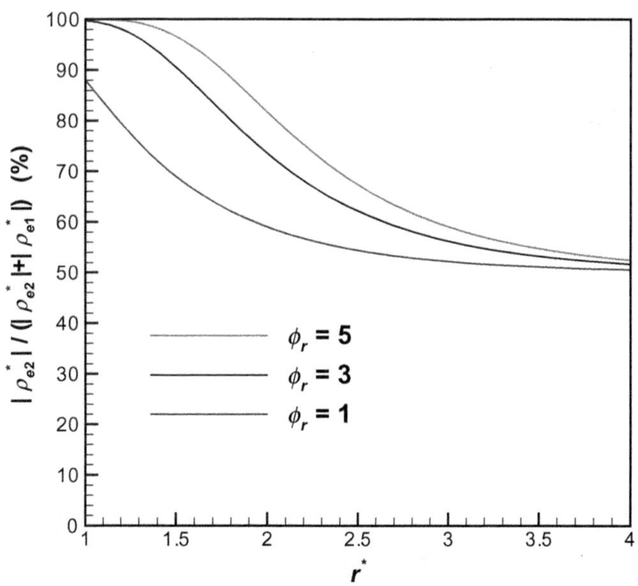

FIG. 1.8 The percentage of counterions in the double layer for a rigid particle at various ϕ_r with $\kappa a = 1$.

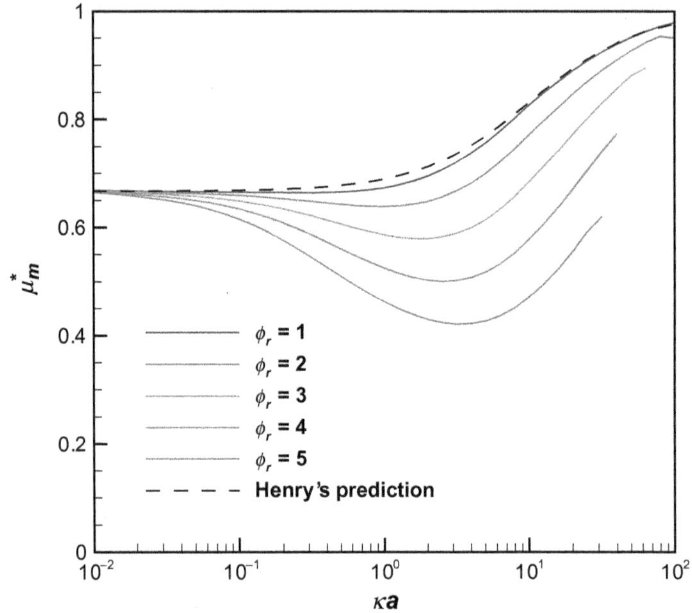

FIG. 1.9 Scaled mobility $\mu_m^*(=\mu^*/\phi_r)$ as a function of κa for a rigid particle at various ϕ_r.

are limiting cases of Henry's prediction here, which reduce to 1 and 2/3 in Fig. 1.9, respectively.

Fig. 1.9 indicates that $\phi_r = 1$ turns out to be a pretty good representation of a lowly charged particle to justify the use of Henry's formula. Appreciable deviation is observed for $\phi_r = 2$, however, and it gets more and more significant as ϕ_r increases even further, indicating there is a fundamental electrokinetic difference between a lowly charged particle and a highly charged one. This is shown below in the corresponding contour plots of the dimensionless disturbed space charge density, $\delta\rho^*$, at various levels of ϕ_r. Note that $\delta\rho^*$ here can be regarded as the dimensionless double layer polarization as well. The basic assumption of Henry's analysis is that the shape of the double layer remains unchanged when the particle is in motion, that is, no double layer polarization takes place and the double layer remains concentric in shape. Hence the contour plots of $\delta\rho^*$ here serve as the measurement of how significant this assumption deviates from reality.

1.4.3 Double Layer Polarization: A Visual Demonstration With $\delta\rho^*$

1.4.3.1 Double Layer Thickness Comparable to the Particle Radius ($\kappa a = 1$)

Fig. 1.10A–E depicts the exact shape of the double layer polarization successively with contour plots of the dimensionless disturbed space charge density, $\delta\rho^*$ for ϕ_r ranging from 1 to 5, respectively. Hardly any polarization of the double layer can be noticed when $\phi_r = 1$, and that's why Henry's prediction prevails in this situation. When $\phi_r = 2$, however, slight but appreciable polarization of the double layer starts to appear, which implies that one can still use the Henry's prediction in practice, but with the understanding that some non-negligible error in mobility prediction is bound to happen. When ϕ_r reaches 3

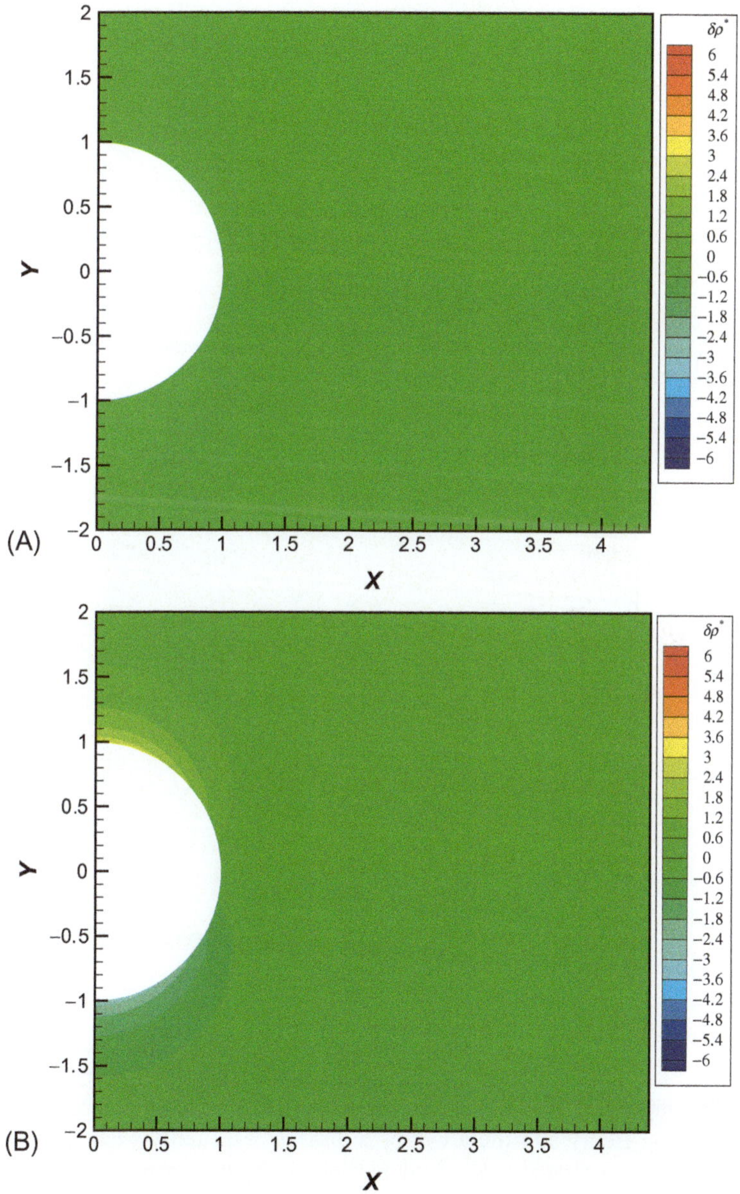

FIG. 1.10 Contour plot of the dimensionless disturbed space charge density $\delta\rho^*$ for a rigid particle with $\kappa a = 1$. (A) $\phi_r = 1$; (B) $\phi_r = 2$; (C) $\phi_r = 3$; (D) $\phi_r = 4$; (E) $\phi_r = 5$.

(Continued)

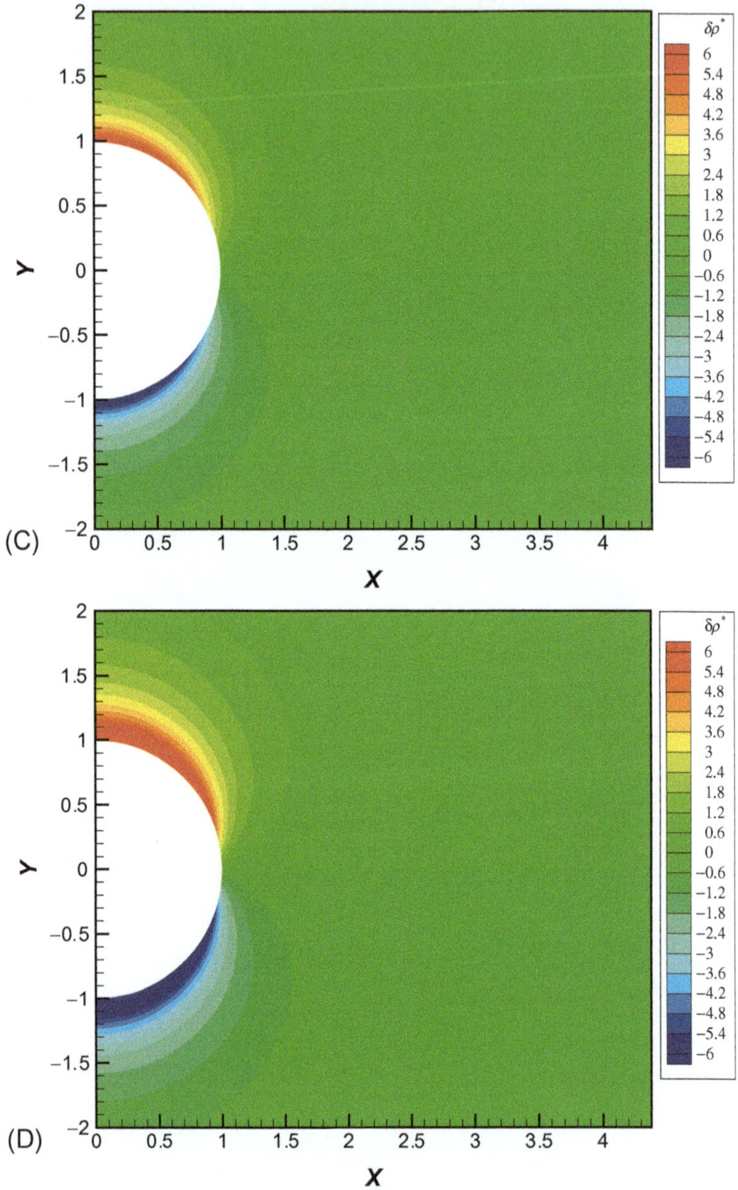

FIG. 1.10—CONT'D

or even higher, however, significant polarization becomes clearly visible. Hence using Henry's formula will be a disaster, especially when ϕ_r is around 5. Significant overestimation of the particle mobility happens by Henry's formula due to the motion-deterring double layer polarization effect in practice, as explained in detail below.

(E) 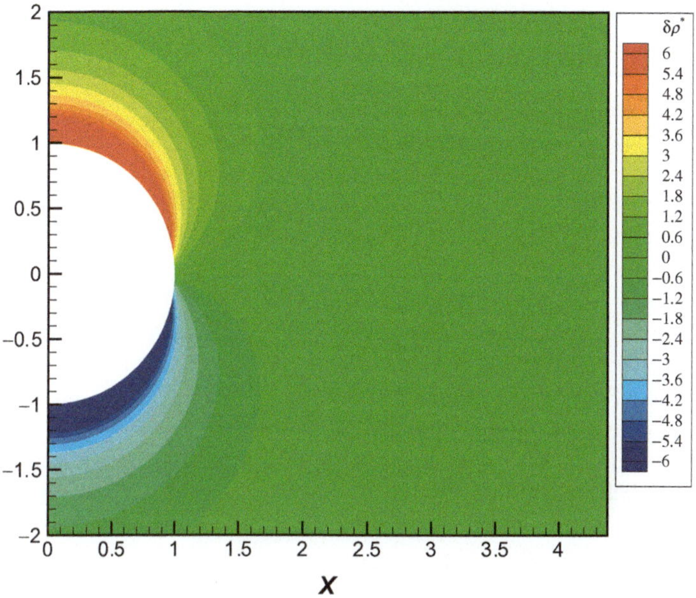

FIG. 1.10—CONT'D

When an external electric field is applied upon a particle suspension system, the electro-migration mechanism drives the counterions from the front region of the particle to its rear due to the electrostatic Coulomb force. This leads to the buildup of extra counterions ($\delta\rho^*$) in the rear hemisphere and a depletion of the same amount in the front one. An induced electric field against the particle motion is thus generated which slows down the particle motion in general. Moreover, this electro-migration movement of the counterions brings the hydrated solvent molecules with it simultaneously, and the bulk fluid is further dragged by these hydrated ions as well. This results in an osmosis flow of hydrodynamic nature which also slows down the particle motion. Moreover, a closer look at the Nernst-Planck equation reveals that, relative to the moving particle, or if the observer is going with the moving particle, the bulk fluid flows against the particle with a speed at infinity identical to the particles. This convection flow carries the ions within it, especially the counterions near the particle. Hence it slows down the particle motion further both hydrodynamically and electrostatically as well, in addition to the motion-deterring electro-migration mechanism.

The diffusion mechanism, on the other hand, tends to alleviate the buildup of counterions by inducing a backward flux against the further electro-migration to prevent it from going overboard. Hence both the electro-migration and the convection mechanisms result in the motion-deterring double layer polarization effect. The diffusion mechanism, on the other hand, is not motion-deterring in nature. Instead, it is involved in the determination of the precise extent of the double layer polarization eventually hence the corresponding particle mobility. It plays the role of a braking mechanism.

In short, when a particle is conducting the electrophoretic motion, an induced motion-deterring electric field is generated in addition to the hydrodynamic drag force. And the higher the particle charge condition is, the more significant this mobility reduction effect is, as shown in Fig. 1.9, where $\phi_r = 5$ assumes the lowest profile. Hence the actual electrophoretic mobility is much smaller than the one predicted by the Henry's formula when the particle is highly charged. An antisymmetric distribution, or cos, dependence to be precise, is generally observed for the double layer polarization, expressed as $\delta\rho^*$ here. Its motion-deterring effect is referred to as the double layer polarization effect, which is the cornerstone of essentially all the electrokinetic motions of highly charged particles.

Note that the highly charged particles here still move faster than the corresponding lowly charged ones though. It is just not as fast as predicted by the Henry's formula. Note also that as scaled dimensionless mobility μ_m^* is used in Fig. 1.9, which is defined as $\mu_m^* = \frac{\mu^*}{\phi_r}$, the actual dimensionless mobility μ^* should be multiplied by ϕ_r.

Moreover, once an external electric field is applied upon the suspension system, the rigid particle itself responds to it instantly as well. The relaxation time before it reaches the steady state terminal velocity, which is of pure hydrodynamic nature, is even shorter than the movement of the counterions in the double layer in general [57, 71]. It has to wait for the counterions to catch up to reach the quasisteady state as a whole eventually. Yes, the charged particle itself does respond to the external electric field, with or without the perfect or imperfect shielding or screening effect of the surrounding ions, as the electric field is superimposed upon the entire system simultaneously. The motion of the particle is not the sheer consequence of the redistribution of the mobile ions, not in terms of the only cause.

Note that local minima of mobility profiles are observed in general in Fig. 1.9 for highly charged particles at κa value roughly between 1 and 10, or when the double layer thickness is comparable to the particle radius. This is because when the double layer is very thin, the counterions would be confined to a region too close to the strongly charged particle surface to have any meaningful freedom of movement. When the double layer is very thick, on the other hand, the ions near the particle surface are too sparse to induce a significant polarization effect as well. As a result, $\kappa a = 1$ is chosen here as the representative case to examine the double layer polarization effect, which is supposed to be most profound in general.

The surface charge density of the particle, σ, is determined solely by the electrostatic environment surrounding the particle, as we have elaborated in Section 1.3 with $\sigma = -\varepsilon\frac{d\phi_e}{dr}$. Once an external electric field is applied upon the system, the polarization of the double layer is established almost instantly as discussed above, and the entire region of the double layer is affected by it, to some degree at least. In particular, the polarization at the concentric shear plane ($\delta\rho^*$) induces a corresponding disturbance of the electric potential ($\delta\phi^*$) via the Gauss theorem, whose gradient there then determines the overall electric driving force of the particle, as indicated by Eq. (1.77). From this point of view, the precise double layer polarization on the shear plane sets the particle in motion.

An induced electric field against the particle motion, on the other hand, is also generated simultaneously by all the counterions in the polarized double layer together. This motion-deterring effect is the one referred to as the double layer polarization effect in the theoretical electrokinetic analysis. Hence strictly speaking, the polarization of the double layer has two separate effects: one of them drives the particle in motion, the surface polarization at the concentric shear plane, which is also regarded as the starting location of the outer diffuse layer; the other retards the particle motion to some extent by generating an induced electric field against the particle motion, which is the joint effect by essentially all the ions within the polarized double layer when the particle is in motion.

Note that the electric driving forces are the same for either hemisphere of a single particle considered here, with the front one pulling and the rear one pushing the particle to move. Note also that the double layer polarization when the particle is in motion does not change the surface charge density σ, as $\delta\rho^* \ll \rho_e^*$ under the assumption that the applied electric field is very weak in comparison with the intrinsic electric field established across the double layer when the particle is at rest. Within the inner Stern layer, the intrinsic electric field there would be even more stronger. It does contribute to the generation of the electric driving force in a subtle way: Even a very small incremental change of the charge density at the shear plane, hence the electric potential there, can trigger a significant corresponding electric field, which is the gradient of the incremental change of the electric potential there.

1.4.3.2 Thin Double Layer (κa=10) and Very Thin Double Layers (κa = 20, 50, and 100)

As the role of the double layer polarization is so crucial in electrophoresis, we go on to further explore the situation when the double layer is thin, $\kappa a = 10$, or even thinner. As shown in Fig. 1.11A–D, where the double layer polarization is shown for $\kappa a = 10$, 20, 50, and 100 for a representative highly charged particle with $\phi_r = 3$. Stunning multilayer structure of ion distribution is observed for all of the cases examined. An extra outer layer of polarization with opposite charge sign appears upon the inner one in each thin double layer situation. The confinement of counterions within a thin or very thin layer due to the original double layer polarization attracts the coions nearby to form a secondary concentric layer. This secondary layer also induces an electric field which offsets the original one induced by the inner layer, leading to an apparent negligible or very weak motion-deterring double layer polarization effect.

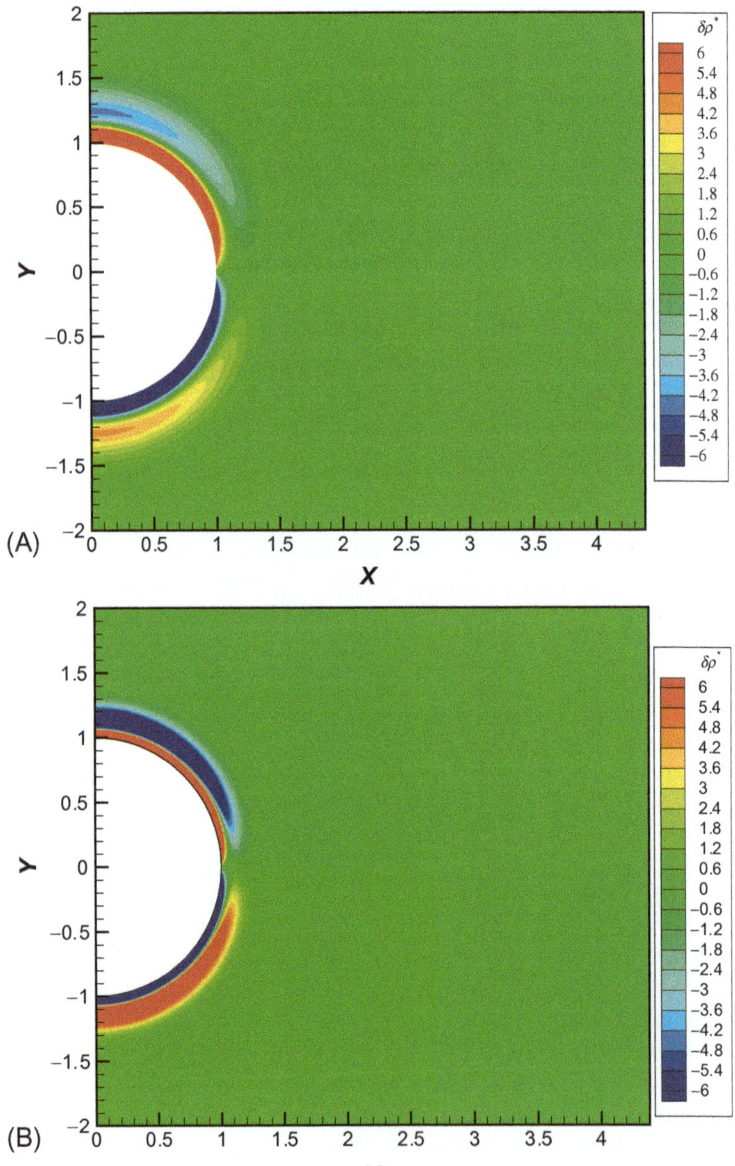

FIG. 1.11 Contour plot of the dimensionless disturbed space charge density $\delta\rho^*$ for a rigid particle with $\phi_r = 3$. (A) $\kappa a = 10$; (B) $\kappa a = 20$; (C) $\kappa a = 50$; (D) $\kappa a = 100$.

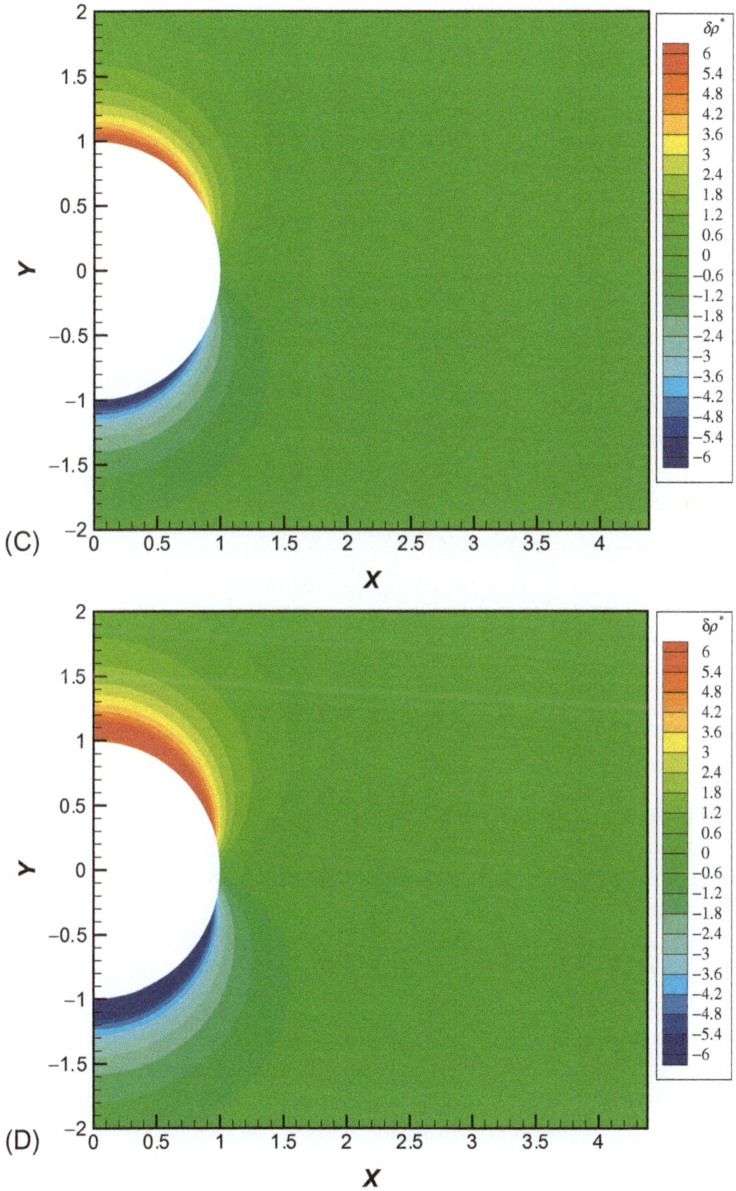

FIG. 1.11—CONT'D

Corresponding plot for $\phi_r = 5$ at $\kappa a = 100$ shows similar structure in Fig. 1.12. This observation of two-layer structure overthrows the traditional assumption that there is no polarization of a very thin double layer. There is. It is just made invisible by this unexpected two-layer structure.

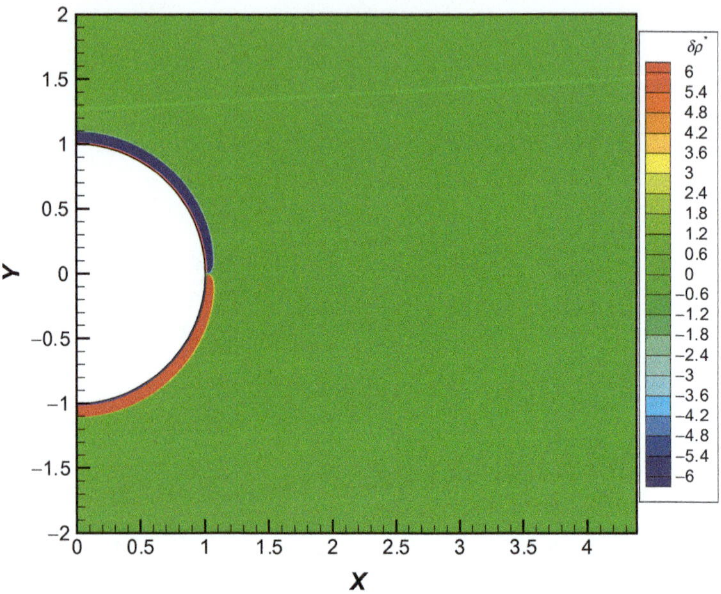

FIG. 1.12 Contour plot of the dimensionless disturbed space charge density $\delta\rho^*$ for a rigid particle with $\phi_r = 5$ and $\kappa a = 100$.

1.4.4 Relative Contributions of the Electro-Migration and the Convection Mechanisms to the Double Layer Polarization Effect

Between the electro-migration and the convection mechanisms of ion flux, which one has more profound impact upon the particle mobility? This question may be answered by setting the Peclet numbers to zero in computation, which means only the electro-migration mechanism is considered. The results are shown in Fig. 1.13 in dashed lines, which indicate that electro-migration is by far the more dominant factor here in the double layer polarization effect, especially when the particle is lowly charged. More than 85% of the particle mobility can be determined solely by this factor over the entire range of κa and ϕ_r considered, as shown in Fig. 1.14, where the percentage is evaluated by the comparison between calculations with and without taking Peclet numbers into account. In other words, the convection mechanism stands for less than 15% of the contribution to the particle mobility, or the motion-deterring double layer polarization effect for the rigid particle considered here.

1.4.5 Effect of Peclet Numbers Upon the Particle Motion

Note that the convection mechanism is electrolyte-dependent due to the ion diffusivity involved in the dimensionless Peclet number in the Nernst-Planck equation. Then how sensitive is the particle mobility with respect to different

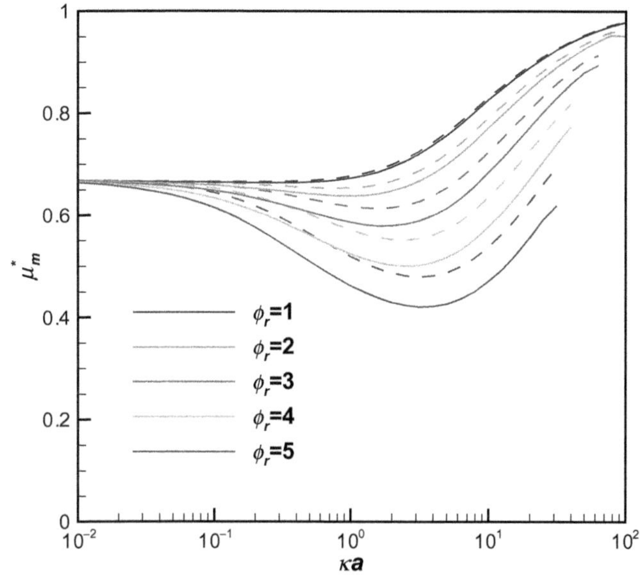

FIG. 1.13 Scaled mobility μ_m^* as a function of κa for a rigid particle at various ϕ_r in a KCl electrolyte solution ($Pe_1 = Pe_2 = 0.26$). *Dashed lines*: $Pe_1 = Pe_2 = 0$.

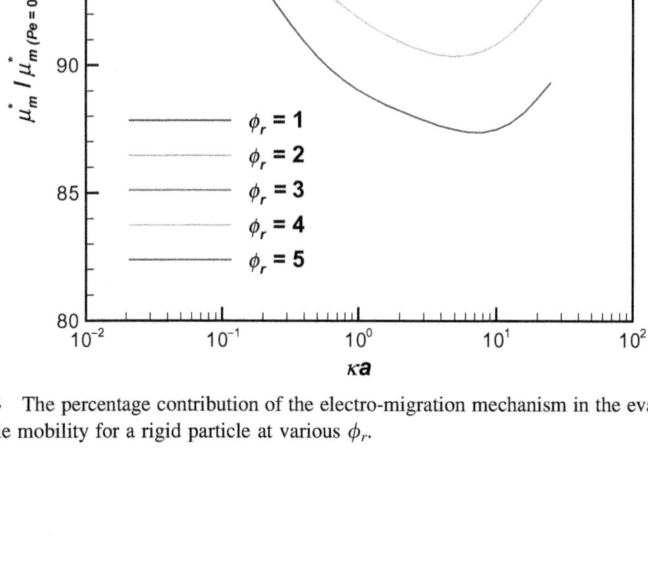

FIG. 1.14 The percentage contribution of the electro-migration mechanism in the evaluation of the particle mobility for a rigid particle at various ϕ_r.

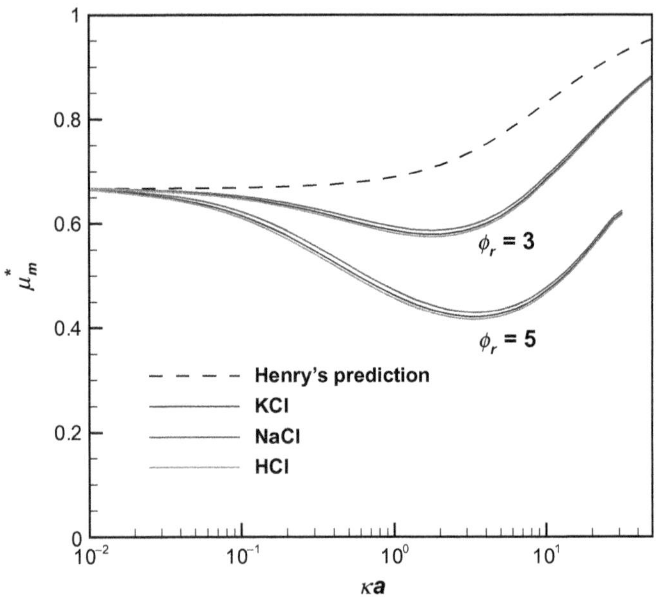

FIG. 1.15 Scaled mobility μ_m^* as a function of κa in various electrolyte solutions (KCl, NaCl, and HCl) for a rigid particle with $\phi_r = 3$ and 5.

electrolyte ions in the system? In general, these Peclet numbers are positive and less than one. For binary systems frequently encountered such as KCl, NaCl, and HCl, Fig. 1.15 shows that their mobilities are essentially the same over the entire range of κa examined for highly charged particles with $\phi_r = 3$ or 5. This indicates that the choice of KCl as the benchmark system is indeed representative, at least for a binary system.

Note also that the electro-migration mechanism is electrolyte independent, on the other hand. It depends only on the amount of electric charges carried by the ions.

1.4.6 Charge-Regulation Phenomenon

Finally, we show the results of a charge-regulated particle in Figs. 1.16 and 1.17 for the dependence of actual surface potential on the electrostatic environment in terms of κa. The parameters chosen are based on the representative case discussed in Section 1.3. Both figures indicate that the actual particle surface potential can be so sensitive to its electrostatic environment, represented by κa, that it may change drastically from lowly or moderately charged to highly or very highly charged condition with varying κa. Note that this dependence takes place when the suspension system is at equilibrium. Once the particle is in electrophoretic motion, no further change of the surface potential happens any more as the applied electric field is very

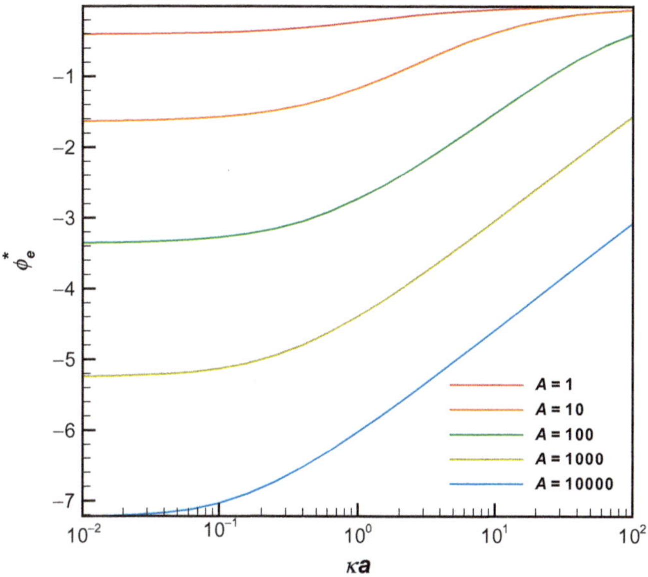

FIG. 1.16 Dimensionless equilibrium potential ϕ_e^* as a function of κa for a charge-regulated rigid particle at various A with $B = 1$.

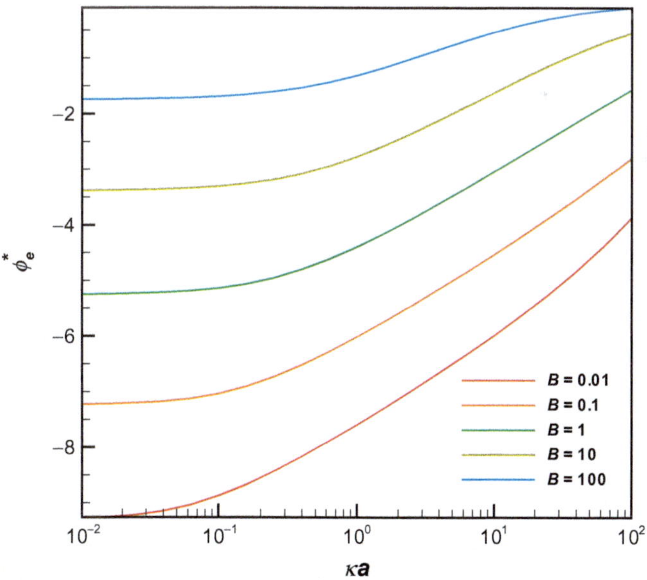

FIG. 1.17 Dimensionless equilibrium potential ϕ_e^* as a function of κa for a charge-regulated rigid particle at various B with $A = 1000$.

weak compared with the intrinsic electric field within the double layer, as we have mentioned earlier. In other words, the charge-regulation phenomenon is purely an electrostatic phenomenon. The surface charge condition remains unchanged in electrophoretic motion.

REFERENCES

[1] Van de Ven TG. Colloidal hydrodynamics. London and New York: Academic Press; 1989.
[2] Seksek O, Bolard J. Nuclear pH gradient in mammalian cells revealed by laser microspectrofluorimetry. J Cell Sci 1996;109(1):257–62.
[3] Fukuyama T, et al. Directing and boosting of cell migration by the entropic force gradient in polymer solution. Langmuir 2015;31(46):12567–72.
[4] Kim SH, Fujii T. Efficient analysis of a small number of cancer cells at the single-cell level using an electroactive double-well array. Lab Chip 2016;16(13):2440–9.
[5] Hartman SV, Božič B, Derganc J. Migration of blood cells and phospholipid vesicles induced by concentration gradients in microcavities. N Biotechnol 2018;47:60–6.
[6] Hunter RJ. Foundations of colloid science. New York: Oxford University Press; 2001.
[7] Chiang CP, et al. Electrophoresis of a spherical dispersion of polyelectrolytes in a salt-free solution. J Phys Chem B 2006;110(3):1490–8.
[8] He YY, Wu E, Lee E. Electrophoresis in suspensions of charged porous spheres in salt-free media. Chem Eng Sci 2010;65(20):5507–16.
[9] Fuoss RM. Polyelectrolytes. Science 1948;108(2812):545–50.
[10] Lyklema J. Interactions in interface and colloid science. In: Fundamentals of interface and colloid science, New York: Academic Press; 1993.
[11] Ohshima H. Electrophoretic mobility of spherical colloidal particles in concentrated suspensions. J Colloid Interface Sci 1997;188(2):481–5.
[12] Masliyah JH. Electrokinetic transport phenomena. vol. 12. Alberta, Oil Sands Technology and Research Authority; 1994.
[13] Gouy M. Sur la constitution de la charge électrique à la surface d'un électrolyte. J Phys Theor Appl 1910;9(1):457–68.
[14] Chapman DL. Li. A contribution to the theory of electrocapillarity. Lond Edinb Dublin Philos Mag J Sci 1913;25(148):475–81.
[15] Stern O. The theory of the electrolytic double-layer. Z Elektrochem 1924;30(508):1014–20.
[16] Grahame DC. The electrical double layer and the theory of electrocapillarity. Chem Rev 1947;41(3):441–501.
[17] Helmholtz HV. Ueber einige gesetze der vertheilung elektrischer ströme in körperlichen leitern, mit anwendung auf die thierisch-elektrischen versuche (schluss.). Ann Phys 1853;165(7):353–77.
[18] Smoluchowski MV. Grundrißder koagulationskinetik kolloider lösungen. Colloid Polym Sci 1917;21(3):98–104.
[19] Dukhin S, Derjaguin B. Surface and colloid science. vol. 7. New York: Wiley; 1974. 36p.
[20] Lee E, Chu JW, Hsu JP. Sedimentation potential of a concentrated spherical colloidal suspension. J Chem Phys 1999;110(23):11643–51.
[21] Minor M, et al. Dynamic aspects of electrophoresis and electroosmosis: a new fast method for measuring particle mobilities. J Colloid Interface Sci 1997;189(2):370–5.

[22] Lee E, Yen FY, Hsu JP. Dynamic electrophoretic mobility of concentrated spherical dispersions. J Phys Chem B 2001;105(30):7239–45.
[23] Hsu JP, Lee E, Yen FY. Dynamic electrophoretic mobility in electroacoustic phenomenon: concentrated dispersions at arbitrary potentials. J Phys Chem B 2002;106(18):4789–98.
[24] Lee E, Fu CH, Hsu JP. Dynamic electrophoretic mobility of a concentrated dispersion of particles with a charge-regulated surface at arbitrary potential. J Colloid Interface Sci 2002;250(2):327–36.
[25] Lee E, Min WL, Hsu JP. Dynamic electrophoresis of a droplet in a spherical cavity. Langmuir 2006;22(8):3920–8.
[26] Hunter RJ. Zeta potential in colloid science: principles and applications. vol. 2. London and New York: Academic Press; 2013.
[27] Kuhn R, Hoffstetter-Kuhn S. Capillary electrophoresis: principles and practice. Berlin: Springer Science & Business Media; 2013.
[28] Popescu MN, Uspal WE, Dietrich S. Self-diffusiophoresis of chemically active colloids. Eur Phys J Spec Top 2016;225(11-12):2189–206.
[29] Shin S, et al. Membraneless water filtration using CO_2. Nat Commun 2017;8:15181.
[30] Breadmore MC, et al. Recent advances in enhancing the sensitivity of electrophoresis and electrochromatography in capillaries and microchips (2014–2016). Electrophoresis 2017;38(1):33–59.
[31] Westermeier R. Electrophoresis in practice: a guide to methods and applications of DNA and protein separations. Weinheim: John Wiley & Sons; 2016.
[32] Zhao SS, Chen DD. Applications of capillary electrophoresis in characterizing recombinant protein therapeutics. 2014;35(1):96–108.
[33] Wilkins DJ, Ottewill RH. Flocculation of sheep leucocytes: 1. Electrophoretic studies. J Theor Biol 1962;2(2):165.
[34] Heard DH, Seaman GVF. The influence of pH and ionic strength on the electrokinetic stability of the human erythrocyte membrane. J Gen Physiol 1960;43(3):635–54.
[35] He H, et al. Surface decoration of ZnO nanorod arrays by electrophoresis in the Au colloidal solution prepared by laser ablation in water. Langmuir 2010;26(11):8925–32.
[36] Pont L, et al. Analysis of transthyretin in human serum by capillary zone electrophoresis electrospray ionization time-of-flight mass spectrometry. Application to familial amyloidotic polyneuropathy type I. Electrophoresis 2015;36(11–12):1265–73.
[37] Chiou C-S, et al. Usefulness of pulsed-field gel electrophoresis profiles for the determination of *Salmonella serovars*. Int J Food Microbiol 2015;214:1–3.
[38] Crecente RMP, et al. Ultrasound-assisted magnetic solid-phase extraction for the determination of some transition metals in Orujo spirit samples by capillary electrophoresis. Food Chem 2015;190:263–9.
[39] Deng Y, et al. A non-invasive genomic diagnostic method for bladder cancer using size-based filtration and microchip electrophoresis. Talanta 2015;144:136–44.
[40] Liang P, et al. Determination of carbohydrates in honey and milk by capillary electrophoresis in combination with graphene-cobalt microsphere hybrid paste electrodes. Food Chem 2016;190:64–70.
[41] Nuchtavorn N, et al. Recent applications of microchip electrophoresis to biomedical analysis. J Pharm Biomed Anal 2015;113:72–96.
[42] Ramón-Sierra JM, Ruiz-Ruiz JC, de la Luz Ortiz-Vázquez E. Electrophoresis characterisation of protein as a method to establish the entomological origin of stingless bee honeys. Food Chem 2015;183:43–8.

[43] Wang L, Li J, Zhang L. Determination of polyphosphates in fish and shrimp muscles by capillary electrophoresis with indirect UV detection after phosphatase inhibition using high pressure pretreatment. Food Chem 2015;185:349–54.
[44] Hunter R. Zeta potential in colloid science. New York: Academic Press; 1981. p.69.
[45] Hückel E. The electrophoresis of spherical colloid. Phys Zeit 1924;25:204–10.
[46] Henry D. The cataphoresis of suspended particles. Part I. The equation of cataphoresis. Proc R Soc Lond A: Math Phys Eng Sci 1931;133(821):106–29.
[47] Overbeek JTG. Theorie der elektrophorese. Fortschrittsberichte über Kolloide und Polymere 1943;54(7):287–364.
[48] Overbeek JTG. Quantitative interpretation of the electrophoretic velocity of colloids. Adv Colloid Sci 1950;3:797–823.
[49] Booth F. The cataphoresis of spherical, solid non-conducting particles in a symmetrical electrolyte. Proc R Soc Lond A: Math Phys Eng Sci 1950;203(1075):514–33.
[50] Wiersema P, Loeb A, Overbeek JTG. Calculation of the electrophoretic mobility of a spherical colloid particle. In: J Colloid Interface Sci 1966;22(1):78–99.
[51] O'Brien RW, White LR. Electrophoretic mobility of a spherical colloidal particle. J Chem Soc Faraday Trans 2: Mol Chem Phys 1978;74:1607–26.
[52] Delgado AV, et al. Measurement and interpretation of electrokinetic phenomena. J Colloid Interface Sci 2007;309(2):194–224.
[53] Deryagin B, Dukhin S, Korotkova A. Diffusiophoresis in electrolyte solutions and its role in mechanism of film formation from rubber latexes by method of ionic deposition. Kolloidnyi Zhurnal 1961;23(1):53.
[54] O'Brien RW, Hunter RJ. The electrophoretic mobility of large colloidal particles. Can J Chem 1981;59(13):1878–87.
[55] Ohshima H. A simple expression for Henry's function for the retardation effect in electrophoresis of spherical colloidal particles. J Colloid Interface Sci 1994;168(1):269–71.
[56] Ohshima H. Theory of colloid and interfacial electric phenomena. vol. 12. Amsterdam: Elsevier; 2006.
[57] Shilov V, et al. Polarization of the electrical double layer. Time evolution after application of an electric field. J Colloid Interface Sci 2000;232(1):141–8.
[58] Bird RB, Stewart WE, Lightfoot EN. Transport phenomena. New York: John Wiley & Sons; 1960.
[59] Lee E, Chu JW, Hsu JP. Electrophoretic mobility of a spherical particle in a spherical cavity. J Colloid Interface Sci 1997;196(2):316–20.
[60] Fang W, Lee E. Diffusiophoretic motion of an isolated charged porous sphere. J Colloid Interface Sci 2015;459:273–83.
[61] Kreyszig E. Advanced engineering mathematics. New York: John Wiley & Sons; 2010.
[62] Lee E, Chu JW, Hsu JP. Electrophoretic mobility of a sphere in a spherical cavity. J Colloid Interface Sci 1998;205(1):65–76.
[63] Gottlieb D, Orszag SA. Numerical analysis of spectral methods: theory and applications. vol. 26. Philadelphia, PA: SIAM; 1977.
[64] Canuto C, et al. Spectral methods in fluid dynamics. New York: Springer Science & Business Media; 2012.
[65] Chu JW. Electrophoretic behavior of a spherical particle in a spherical cavity [M.S. thesis]. Taipei, Taiwan: National Taiwan University; 1987.
[66] Stuart MC, et al. Adsorption of ions, polyelectrolytes and proteins. Adv Colloid Interface Sci 1991;34:477–535.

[67] Lyklema J. Points of zero charge in the presence of specific adsorption. J Colloid Interface Sci 1984;99(1):109–17.

[68] Lyklema J. Colloidal models. A bit of history. J Colloid Interface Sci 2015;446:308–16.

[69] Lee E, Chiang CP, Hsu JP. Electrophoresis of Carreau-drop dispersions with a charge-regulated surface. Langmuir 2006;22(4):1911–8.

[70] Hsu JP, Lee E, Yen FY. Electrophoresis of concentrated spherical particles with a charge-regulated surface. J Chem Phys 2000;112(14):6404–10.

[71] Landau LD, Lifshits EM. Fluid mechanics. Oxford, UK: Pergamon Press; 1959.

Chapter 2

Electrophoresis in Suspensions of Rigid Spherical Particles

2.1 INTRODUCTION

2.1.1 General Effect of Neighboring Particles in a Suspension: Double Layer Overlapping and Hydrodynamic Swarming Effects

In Chapter 1, electrophoretic motion of a single charged spherical particle immersed in an infinite medium of electrolyte solution is analyzed theoretically. In reality, however, a single particle is seldom encountered. Instead, it is the colloidal suspension, either dilute or concentrated, that needs to be considered, with the single particle serving as the limiting case of a very dilute suspension.

In a colloidal suspension, the particle motion is influenced by the neighboring ones both electrostatically and hydrodynamically in terms of double layer overlapping and extra hydrodynamic drag force due to the interaction with neighboring particles. The more concentrated the suspension is, the more severe these two effects are, which tend to reduce the particle mobility in general. It is kind of like the situation of a swimmer swimming in water crowded with people. The double layer overlapping effect reduces the polarization of double layer surrounding each individual rigid particle, hence reduces the electric driving force. The extra drag force due to the presence of the neighboring particles is obviously going to slow down the particle motion as well, referred to as the hydrodynamic swarming effect here. Hence the electrophoretic mobility in a suspension is normally lower than that of a single particle, at least for a rigid particle. In addition, the more concentrated the suspension is, the lower the particle mobility is.

How to model a suspension geometrically is a challenging task and the heart of the entire theoretical treatment. The governing electrokinetic equations and the associated boundary conditions on the individual particle surface, on the other hand, should be intuitively the same as those for a single particle treated in Chapter 1. With this in mind, we discuss this issue as follows.

Interface Science and Technology, Vol. 26. https://doi.org/10.1016/B978-0-08-100865-2.00002-3

2.1.2 To Model a Colloidal Suspension: The Unit Cell Model Concept

In order to treat a suspension properly, one has to figure out a way to represent its homogeneous nature correctly, or logically at least, both electrostatic wise and hydrodynamic wise. The fundamental requirement of an appropriate model is that it has to be able to unify the apparent homogeneity of a suspension from a local microscopic point of view, and a logical way of applying external driving force upon the system from a global macroscopic point of view. Many ways are possible, including the rigorous approach in terms of statistic mechanics [1], which, however, is too complicated and effort-consuming sometimes. And more often than not, unrealistic hence useless in practical applications.

Another way, which is very popular in many other physical disciplines as well, is the unit cell model concept, which treats the entire suspension as a characteristic unit cell composed of a representing rigid particle surrounded by a concentric shell of electrolyte solution [2]. The relative volume of the inner particle (of radius a) to the entire unit cell (of radius b), is made to equal the particle volume fraction (H) of the suspension, i.e., $H = (a/b)^3$. Hence the key parameter indicating the concentration level of a suspension is quantitatively taken into account nicely in this simple unit cell type of approach.

Note that at a first glance, it seems logical to analog the cell model as the actual environment in the vicinity of each and every individual particle in suspension. This is incorrect, however, since otherwise one will end up with a significant amount of void space unaccounted for each "spherical" unit cell. Moreover, it would be impossible to determine the correct boundary conditions on the cell surface this way, as the boundary itself is actually within the physical domain hence the conditions are unknown. The truth is, although inspired by this analog, the unit cell model is just a conceptual model, and the outer cell surface is a virtual reality surface representing the seemingly infinitely repeated uniform structure within a suspension, since from the viewpoint of a tiny colloidal particle, the entire surrounding electrolyte solution can indeed be very well treated as a concentric infinite medium. Thus the boundary conditions there need to be analyzed carefully later. The particle located in the center of the cell model, on the other hand, truly represents each and every individual particle in the solution, or their collective existence. Therefore, the boundary conditions on the particle surface can be treated exactly like those in Chapter 1.

Hence the challenging issue for the success of a cell model here is: What are the appropriate boundary conditions on the outer virtual surface, both the electrostatic and hydrodynamic ones, which can incorporate the concept of global infinity and local homogeneity as close as possible. And in the limiting case when the volume fraction approaches zero, the prediction of particle mobility can indeed reduce to that of a single particle situation, which is the ultimate self-consistency check for any proposed cell model. Note that other than this, there is no physical or mathematical basis to argue which one is correct, only that which one is better than the other in that it can interpret and explain more phenomena, at least qualitatively, and hopefully supported by some quantitative experimental data as well.

2.1.3 Hydrodynamic Cell Model and Electrostatic Cell Models

2.1.3.1 Kuwabara's Hydrodynamic Cell Model

There are already several cell models proposed in purely hydrodynamic systems [2, 3], among them the Kuwabara's unit cell model stands out as the most popular one in electrokinetics, which assumes the vorticity at the virtual surface is zero, i.e., $\nabla \times \boldsymbol{v} = \mathbf{0}$. This is a loose enforcement of boundary conditions or the Neumann type where the derivatives of the dependent variables are specified, compared with the physical reality that $\mathbf{v} = \mathbf{0}$ there at infinity as adopted in Chapter 1 [4], which is of the Dirichlet type where the dependent variables themselves are specified [4]. Hence there is some grey zone or "flexibility" of this model to unify underlying physical realities. Note that the rest condition, $\boldsymbol{v} = \mathbf{0}$, corresponds to a rigid wall surrounding each local particle under consideration, hence is unrealistic and cannot be used here. Nowadays, at least in the field of electrokinetics, $\nabla \times \boldsymbol{v} = \mathbf{0}$ is always the choice of hydrodynamic condition on the outer virtual surface in a cell model type of approach, as it implies no hydrodynamic impact between each individual local cell environment yet no ruling out of the possibility that $\boldsymbol{v}$ may actually approach zero at infinity, should the suspension is infinitely dilute. In summary, Kuwabara's hydrodynamic unit cell model is the exclusive one used in electrokinetic studies nowadays, hence is the one used here in this chapter as well.

2.1.4 Electrostatic Cell Models: Levine and Neale Model (L-N Model) Versus Shilov and Zharkikh Model (S-Z Model)

For the electrostatic boundary condition, there are two possible choices in the mainstream of current researches: Levine and Neale (L-N) [5] and Shilov and Zharkikh (S-Z) [1, 6] conditions. Both the L-N and the S-Z models treat the boundary condition of the equilibrium electric potential ϕ_e on the outer virtual surface as follows:

$$\frac{\partial \phi_e}{\partial r} = 0, \quad r = b \tag{2.1}$$

This implies that the system is electrically neutral as a whole before the application of an external electric field, a very reasonable deduction from the global point of view. It also implies that if this cell really represents the true local environment of each individual particle, then it should be electrically neutral as well from the neighboring counterparts. Indeed a very convincing electric condition good for any cell model approaches. The difference between these two models lies in the corresponding treatment when an external electric field is applied upon the system, as discussed below.

2.1.4.1 L-N Model: Merits and Shortcomings

The L-N model treats the suspension system from a macroscopic point of view with $\frac{\partial \delta\phi}{\partial r} = -E\cos\theta$ is implied with E on the outer virtual surface, where E is the magnitude of the truly applied external electric field. Obviously this condition is incorrect in the neighborhood of each individual particle, unless the suspension is infinitely dilute, as once an applied electric field is imposed upon a suspension system, it will be altered or modified locally due to the involvement of electrolyte ions in the suspension. The application of the external electric field upon the system can be precisely described this way, since the outer virtual surface is regarded as the true outer boundary of the entire suspension system in this approach.

2.1.4.2 S-Z Model: Merits and Shortcomings

The S-Z model treats the system from a microscopic (local) point of view around each individual particle and comes up with a boundary condition as $\delta\phi = -\boldsymbol{E} \cdot \mathbf{r}$ or $-Er\cos\theta$, where $\boldsymbol{E}$ is some kind of "representative" (equivalent) electric field around each individual particle (or its collective existence) which we do not know its value for sure, and E its magnitude. Note that although Shilov and his coworkers [1, 7] claimed they derived this rigorously from thermodynamics, it by no means should be regarded as the "true" or "real" electric field anywhere around a specific particle. Otherwise it encounters the unsurpassable challenge again: Then how about the void space inherently ignored in the unit "spherical" cell model? Cell model in principle is just an intuitional model inspired by geometry analog. Anything good from it is sheer luck in nature, like Prince Serendipity who always ran into good fortune when he traveled.

2.1.4.3 Evaluation of Particle Mobility in Either Models

A closer look at the S-Z boundary condition reveals that it is a loose requirement as well in the sense that $\frac{\partial \delta\phi}{\partial r} = -E\cos\theta$ is implied with E as the magnitude of some kind of electric field of choice which is neither the external electric field, nor the true local electric field actually measured around a particle if possible. It is just some kind of characteristic electric field which happens to satisfy the S-Z boundary condition. The nature of the boundary condition changes from the Neumann type (L-N) to the Dirichlet type (S-Z), but the essence remains the same except that the latter is looser in the sense that it is more general or flexible. Note that the value of E will be canceled in the evaluation of particle mobility in practice here anyway, whether it stands for the L-N or the S-Z type of electric boundary condition.

Discussions on the cell model in further details can be found in the literature [1, 6, 8, 9]. Theoretical studies of suspensions using Kuwabara's cell model coupled with either L-N or S-Z model are abundant in the literature as well [6, 8–34]. Among them, Ohshima [10] studied the lowly charged particles in particular with the L-N boundary conditions and derived an approximate analytical formula. Lee's research group [11–31, 33, 35], on the other hand, focused on the highly charged particles with mainly S-Z model, taking full account of the double layer polarization effect with numerical approach. The comparison of the results obtained here with either L-N or S-Z model will be discussed later in this chapter. Note also that the mobility derived from either model can actually be converted the other mathematically [35]. The details can be found elsewhere [30].

2.2 THEORY

The governing equations in the electrolyte solution as well as the boundary conditions on the particle surface are the same as those in Chapter 1, and so are the corresponding mathematical perturbation analysis based on the weak external electric field claim and the procedure of the numerical evaluation of the particle mobility. Hence there is no need to write them down here again one by one. They are Eqs. (1.29)–(1.38), in short. The only difference is the hydrodynamic (Kuwabara) and electrostatic (L-N or S-Z) boundary conditions on the outer virtual surface of the cell model, i.e., $r^* = b/a = \sqrt[3]{H}$ (in dimensionless form), where H is the volume fraction of the particle in the suspension system. Thus these conditions are elaborated in detail as follows.

In the cell model, we assume that the conceptual unit cell represents the entire suspension system. Before the application of an external electric field, the system is at equilibrium. Hence the electric field on the outer virtual surface is null:

$$\frac{\partial \phi_e^*}{\partial r^*} = 0, \quad r^* = b/a \tag{2.2}$$

Moreover, once an electric field is applied, as we have mentioned in Section 2.1, the corresponding electric boundary conditions on the outer virtual surface can be provided from either the L-N or the S-Z model:

$$\text{L-N}: \frac{\partial \delta\phi^*}{\partial r^*} = -E^* \cos\theta, \quad r^* = b/a \tag{2.3}$$

$$\text{S-Z}: \delta\phi^* = -\boldsymbol{E}^* \cdot \mathbf{r}^* \text{ or } -E^* r^* \cos\theta, \quad r^* = b/a \tag{2.4}$$

Moreover, physically there should be no concentration disturbance of the electrolyte ions at infinity, or $\rho = \rho_e$, as discussed in Chapter 1, which implies:

$$g_j^* + \delta\phi^* = 0 \text{ or } g_j^* = -\delta\phi^*, \quad r^* = b/a \tag{2.5}$$

The hydrodynamic conditions are obtained based on the Kuwabara's unit cell model, $\nabla \times \mathbf{v} = \mathbf{0}$. In terms of the scaled stream function Ψ^*, we have:

$$r^* \frac{\partial^2 \Psi^*}{\partial r^{*2}} + \frac{\partial^2 \Psi^*}{\partial \theta^2} - \frac{\cos\theta \partial \Psi^*}{\sin\theta \partial \theta} = 0, \quad r^* = b/a \tag{2.6}$$

With Eqs. (2.2)–(2.6) on the outer virtual surface, the flow field and electric field are solved following the same methodology in Chapter 1, and so is the subsequent evaluation of particle mobility.

The governing system, equations and associated boundary conditions, can be further reduced to one-dimensional form as treated in Chapter 1 with the replacement of the corresponding boundary conditions as follows:

$$\text{L-N}: \frac{\partial \delta \Phi^*}{\partial r^*} = -E^*, \quad r^* = b/a \tag{2.7}$$

$$\text{S-Z}: \delta\Phi^* = -E^* r^*, \quad r^* = b/a \tag{2.8}$$

$$G_j^* + \delta\Phi^* = 0 \text{ or } G_j^* = -\delta\Phi^*, \quad r^* = b/a \tag{2.9}$$

$$D^{*2}\psi^* = 0, \quad r^* = b/a \tag{2.10}$$

Once the electric and flow fields are solved, corresponding dimensionless electric driving force F_E and dimensionless hydrodynamic drag force F_D can be calculated accordingly. Following the same procedure in Chapter 1 to decompose the problem to two subproblems. The dimensionless counterpart $\mu^*(= U^*/E^*)$ can be obtained. Note that in order to highlight the double layer polarization effect, the scaled dimensionless mobility μ_m^*, defined as μ^*/ϕ_r, is actually more appreciate to demonstrate the electrophoretic behavior of particles, as it can reflect the deviation from the prediction by Smoluchowski's formula, which is free of double layer polarization. Hence all the figures here are presented with μ_m^* in Section 2.3, as we did in Chapter 1.

As for the charge-regulated particle, the only issue is to determine the surface charge level, which is of purely electrostatic nature as discussed in Chapter 1. The subsequent treatment of electrophoretic motion of the particle is exactly the same as above.

2.3 RESULTS AND DISCUSSIONS

The system diagram is shown in Fig. 2.1.

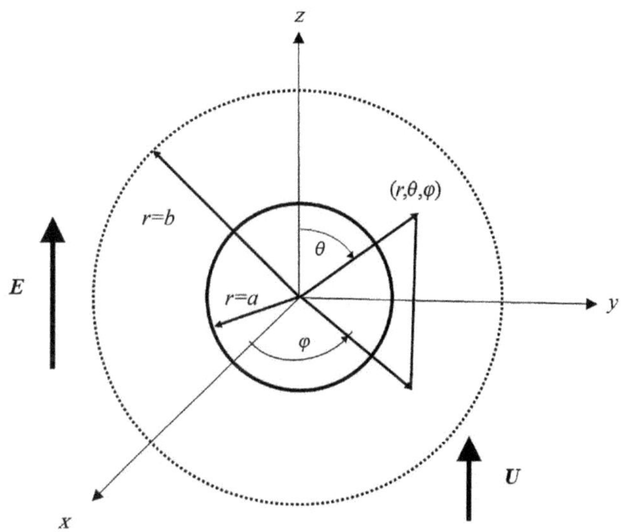

FIG. 2.1 System diagram.

2.3.1 General Henry's Charts for Various Suspension Concentrations

Fig. 2.2A–E presents the general or extended Henry's mobility charts for various volume fraction (H) of charged rigid spherical particles with ϕ_r ranging from 1 to 5, respectively. These charts provide all the information needed to access the electrophoretic behavior in suspensions of either lowly or highly charged particles. They are the highlight of this chapter and the rest is just to elaborate the various aspects of these charts.

As shown in these charts, the particle mobility decreases with increasing particle volume fraction as the suspension gets more and more concentrated. Moreover, when the double layer is very thin, the double layer overlapping effect can be negligible and the mobility reduction in this case is purely from the hydrodynamic swarming effect, such as the reduction in particle mobility at $\kappa a = 100$ in these charts. The mobility reduction is less severe in general when κa is large, or very thin double layer.

2.3.2 Double Layer Overlapping Effect

The mobility of a dilute suspension with $H = 10^{-6}$ reduces successfully to the single particle situation throughout these charts. In other words, the effect of a neighboring particle is negligible if it is located about 100 particle diameters away. On the other hand, for a dilute suspension with $H = 0.01$, or about 5 diameters between each particle, nearly complete mobility reduction to a motionless state is observed for a very thick double layer around $\kappa a = 0.01$ in these charts. This is mainly due to the significant double layer overlapping effect.

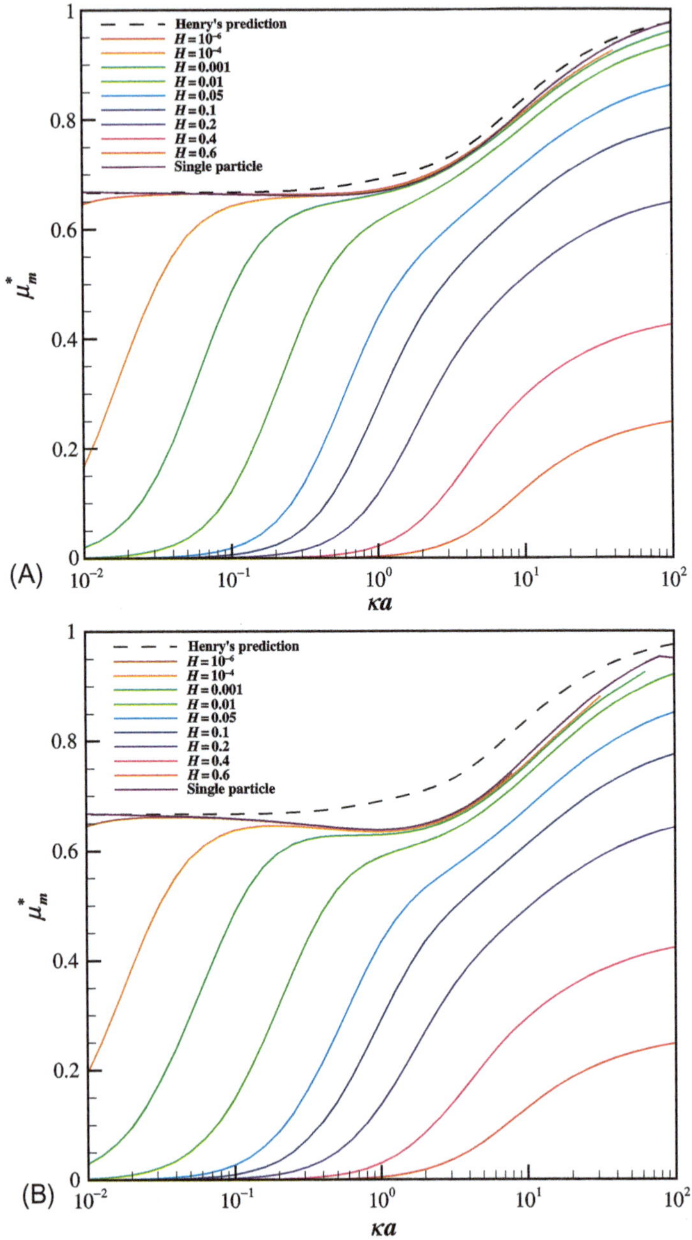

FIG. 2.2 Scaled mobility μ_m^* as a function of κa at various H with: (A) $\phi_r = 1$; (B) $\phi_r = 2$; (C) $\phi_r = 3$; (D) $\phi_r = 4$; (E) $\phi_r = 5$. Key: S-Z boundary condition.

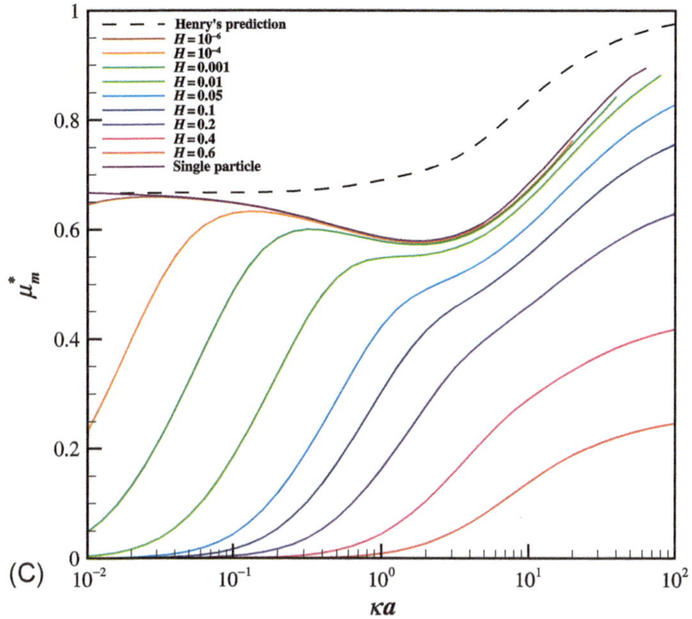

(C)

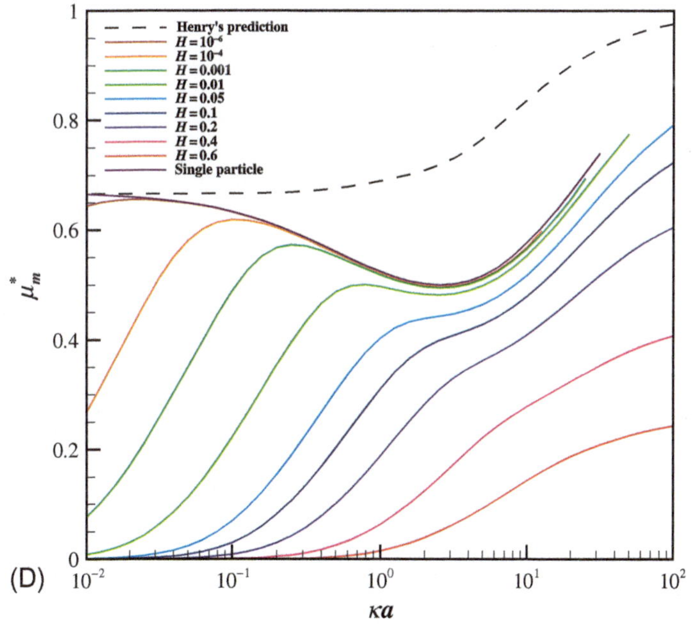

(D)

FIG. 2.2—CONT'D

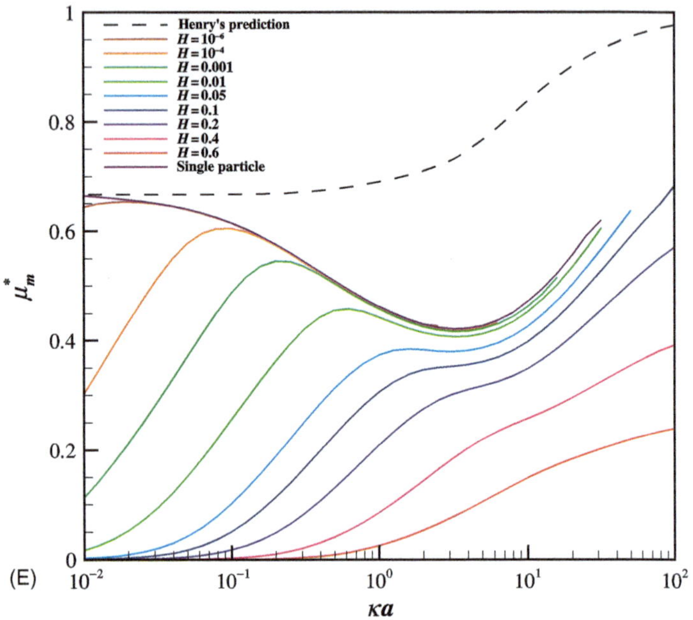

FIG. 2.2—CONT'D

As discussed in Chapter 1, the migration of counterions from the front to the rear leads to the double layer polarization, with a buildup of counterions in the rear region and a corresponding depletion of them in the front. When the double layer is thick enough to touch the ones of the neighboring particles, the counterions in these double layers start to mingle with one another. The front region of one particle becomes the rear region of another, thus buildup is offset by depletion, at least to some extent. Likewise, the surface polarization on the concentric shear plane could be drastically reduced accordingly leading to a significant reduction of the electric driving force, depending upon the extent of this double layer overlapping. Note that this double layer overlapping effect reduces the extent of bulk polarization over the entire double layer as well, hence its motion-deterring effect for highly charged particles. Fig. 2.3A shows the shape of the double layer polarization in a very dilute suspension ($H = 10^{-6}$). The shape is essentially the same as the single particle situation discussed in Chapter 1. When the particle concentration reaches $H = 0.1$, however, severe reduction of double layer polarization is observed as shown in Fig. 2.3B. And eventually complete disappearance of it is observed in Fig. 2.3C with $H = 0.4$. Note that in such a concentrated suspension, the interaction of neighboring particles is so strong and the overlapping of double layers takes place with so many layers upon one another from each and every individual neighboring particle, adjacent ones, and not adjacent ones. Indeed

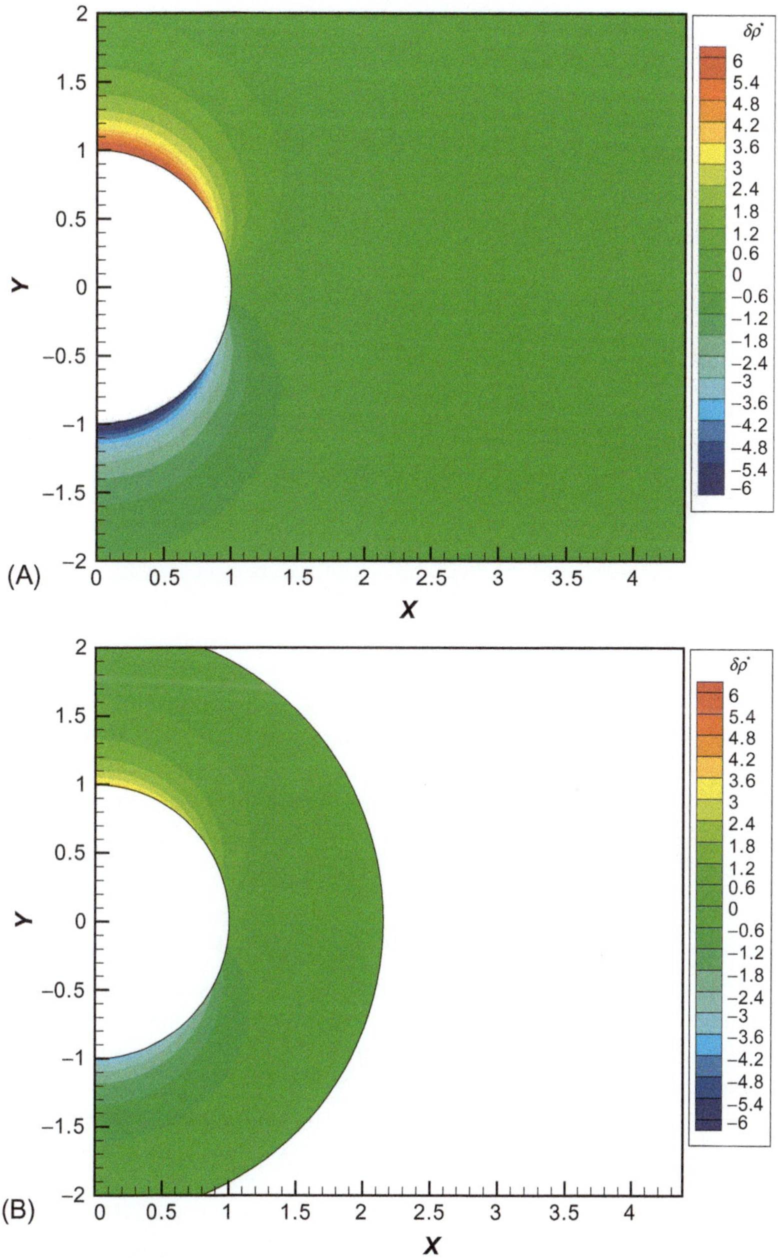

FIG. 2.3 Contour plot of the scaled disturbed charge density $\delta\rho^*$ for suspended particles with: (A) $\phi_r = 3$, $\kappa a = 1$, and $H = 10^{-6}$; (B) $\phi_r = 3$, $\kappa a = 1$, and $H = 0.1$; (C) $\phi_r = 3$, $\kappa a = 1$, and $H = 0.4$.

(Continued)

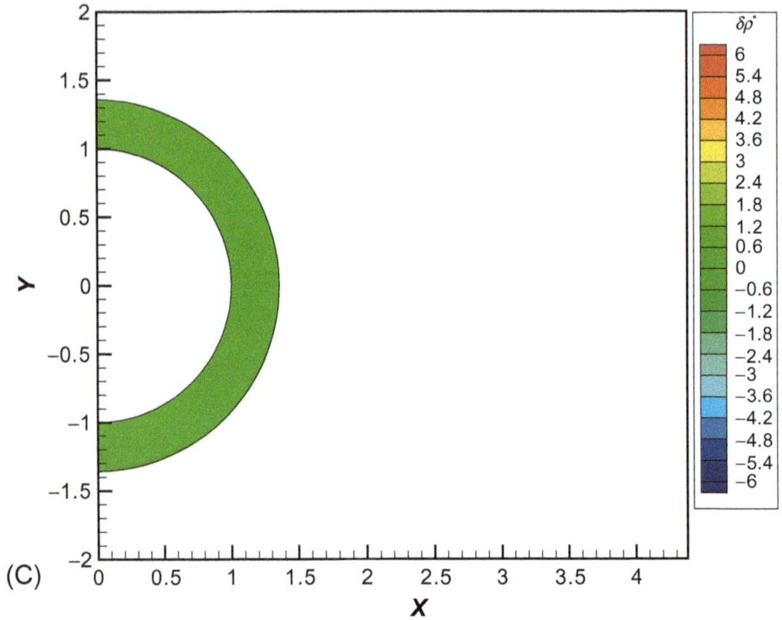

FIG. 2.3—CONT'D

the double layer overlapping effect is so nicely demonstrated by these figures. Moreover, corresponding weakening of the associated motion-deterring effect from the bulk polarization is also shown by the disappearance of local minima in mobility profiles around $\kappa a = 1$ to $\kappa a = 10$ in Fig. 2.2B–E when the suspension gets concentrated, i.e., H increases.

2.3.3 On the Practical Upper Bound of the Volume Fraction *H*

Note that due to the colloidal crystallization and other possible nonlinear interactions between particles in concentrated suspensions other than the electrostatic interactions [36–38], the upper bound of H is about 0.4 in practice to justify the theoretical treatment established here. Note that the geometric upper limit for a unit cell model is around $H = 0.74$ in terms of dense packing of spheres [36]. As far as mobility calculation is concerned, however, H can reach 1.0 here with no computational problem. We follow Ohshima's chart presentation for an L-N boundary condition which extended H level to 0.6 [10].

2.3.4 Comparison Between the S-Z and the L-N Boundary Conditions

Corresponding general Henry's charts with L-N boundary conditions are shown in Fig. 2.4A–E. The mobility behavior is quite similar to the corresponding S-Z profiles when κa is small, or thick double layer.

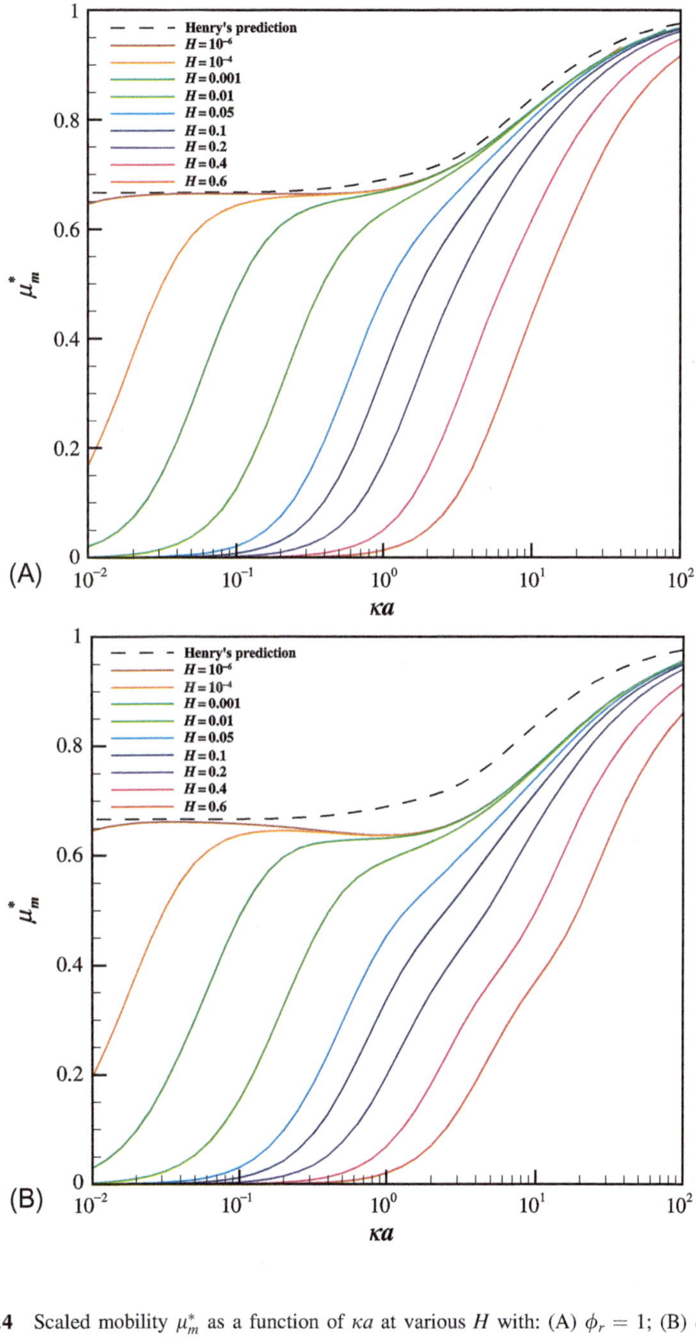

FIG. 2.4 Scaled mobility μ_m^* as a function of κa at various H with: (A) $\phi_r = 1$; (B) $\phi_r = 2$; (C) $\phi_r = 3$; (D) $\phi_r = 4$; (E) $\phi_r = 5$. Key: L-N boundary condition.

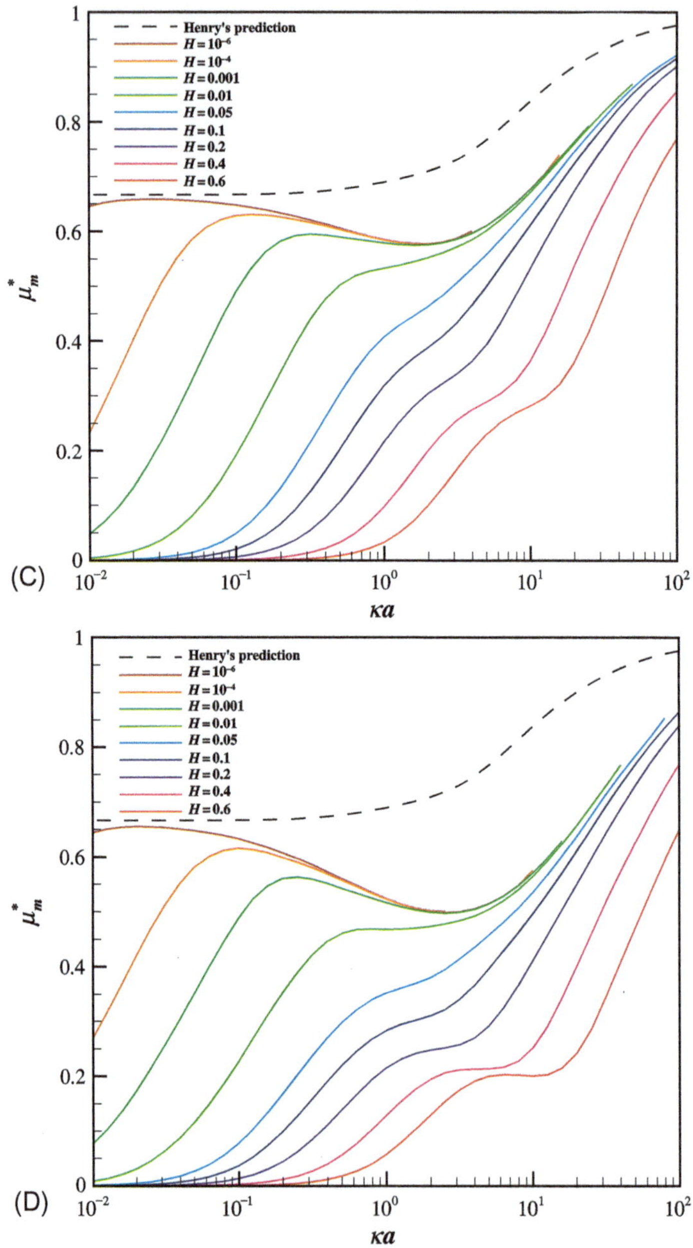

Henry's prediction
H=10⁻⁶
H=10⁻⁴
H=0.001
H=0.01
H=0.05
H=0.1
H=0.2
H=0.4
H=0.6
μ*m
κa
(C)
(D)

FIG. 2.4—CONT'D

(E)

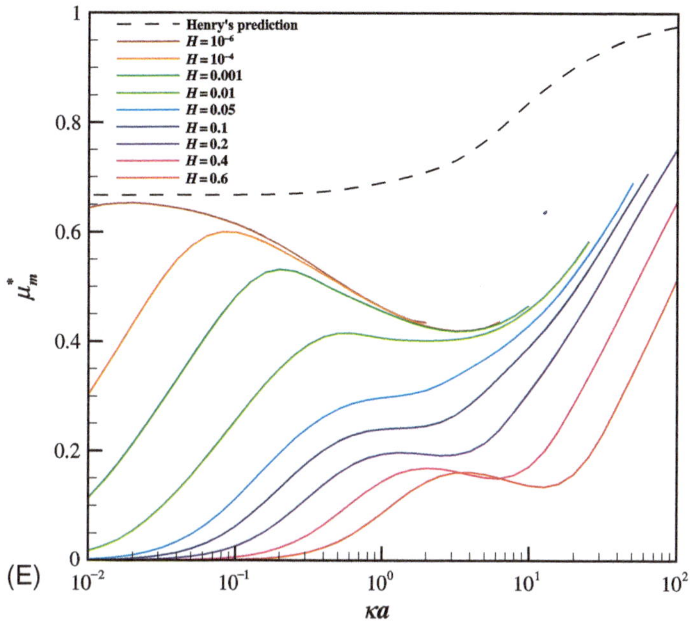

FIG. 2.4—CONT'D

When κa gets large, or thin double layer, however, all the mobility profiles there tend to approach asymptomatically to the single particle situation with $\mu_m^* = 1$. In other words, it implies that the particle moves in a suspension without feeling the presence of any other neighboring particles. This absurd behavior certainly is against both the intuition and the physical reality. At least it does not reflect the hydrodynamic swarming effect appropriately, to put it in a mild way. Hence, the S-Z model seems to be more reasonable physically in the prediction of particle mobility in a suspension, at least when the double layer is very thin.

Note that although Fig. 2.4A–E are mathematically convertible to Fig. 2.2A–E, they are presented here nonetheless for completeness as well as the convenient use for interested readers, who may favor either the S-Z or the L-N boundary conditions with their own reasons.

REFERENCES

[1] Zholkovskiy EK, et al. Hydrodynamic cell model: general formulation and comparative analysis of different approaches. Can J Chem Eng 2007;85(5):701–25.

[2] Kuwabara S. The forces experienced by randomly distributed parallel circular cylinders or spheres in a viscous flow at small Reynolds numbers. J Phys Soc Jpn 1959;14(4):527–32.

[3] Happel J. Viscosity of suspensions of uniform spheres. J Appl Phys 1957;28(11):1288–92.

[4] Kreyszig E. Advanced engineering mathematics. New York: John Wiley & Sons; 2010.

[5] Levine S, Neale GH. The prediction of electrokinetic phenomena within multiparticle systems. I. Electrophoresis and electroosmosis. J Colloid Interface Sci 1974;47(2):520–9.
[6] Zholkovskij EK, et al. Electrokinetic phenomena in concentrated disperse systems: general problem formulation and spherical cell approach. Adv Colloid Interface Sci 2007;134:279–321.
[7] Shilov V, Zharkikh N, Borkovskaya YB. The theory of nonequilibrium electrokinetic phenomena in concentrated disperse systems. 1. Application of irreversible thermodynamics to the cell model of concentrated dispersions. Colloid J (USSR) 1981;43:540.
[8] Shilov V, Borkovskaja YB, Dukhin A. Electroacoustic theory for concentrated colloids with overlapped DLs at arbitrary κa: I. Application to nanocolloids and nonaqueous colloids. J Colloid Interface Sci 2004;277(2):347–58.
[9] Dukhin A, et al. Electroacoustics for concentrated dispersions. Langmuir 1999;15(10):3445–51.
[10] Ohshima H. Electrophoretic mobility of spherical colloidal particles in concentrated suspensions. J Colloid Interface Sci 1997;188(2):481–5.
[11] Lee E, Chu JW, Hsu JP. Sedimentation potential of a concentrated spherical colloidal suspension. J Chem Phys 1999;110(23):11643–51.
[12] Lee E, Chu JW, Hsu JP. Electrophoretic mobility of a concentrated suspension of spherical particles. J Colloid Interface Sci 1999;209(1):240–6.
[13] Chih MH, et al. Sedimentation of concentrated charged spheres at low surface potentials. Langmuir 2000;16(4):1650–4.
[14] Hsu JP, Lee E, Yen FY. Electrophoresis of concentrated spherical particles with a charge-regulated surface. J Chem Phys 2000;112(14):6404–10.
[15] Lee E, Yen FY, Hsu JP. Electrophoretic mobility of concentrated spheres with a charge-regulated surface. Electrophoresis 2000;21(3):475–80.
[16] Lee E, Chih MH, Hsu JP. Conductivity of a concentrated cylindrical dispersion. Langmuir 2001;17(6):1821–5.
[17] Lee E, Chih MH, Hsu JP. Conductivity of a concentrated spherical colloidal dispersion. J Phys Chem B 2001;105(4):747–53.
[18] Lee E, Yen FY, Hsu JP. Dynamic electrophoretic mobility of concentrated spherical dispersions. J Phys Chem B 2001;105(30):7239–45.
[19] Hsu JP, Lee E, Yen FY. Dynamic electrophoretic mobility in electroacoustic phenomenon: concentrated dispersions at arbitrary potentials. J Phys Chem B 2002;106(18):4789–98.
[20] Lee E, Fu CH, Hsu JP. Dynamic electrophoretic mobility of a concentrated dispersion of particles with a charge-regulated surface at arbitrary potential. J Colloid Interface Sci 2002;250(2):327–36.
[21] Lee E, et al. Sedimentation of concentrated spherical particles with a charge-regulated surface. J Colloid Interface Sci 2002;251(1):109–19.
[22] Lin WH, Lee E, Hsu JP. Electrophoresis of a concentrated spherical dispersion at arbitrary electrical potentials. J Colloid Interface Sci 2002;248(2):398–403.
[23] Hsu JP, Lee E, Huang YF. Electrophoresis of a concentrated dispersion of spherical particles in a non-Newtonian fluid. Langmuir 2004;20(6):2149–56.
[24] Lee E, Chou KT, Hsu JP. Electrophoresis of a concentrated dispersion of spherical particles covered by an ion-penetrable membrane layer. J Colloid Interface Sci 2004;280(2):518–26.
[25] Lee E, Chang CJ, Hsu JP. Electrophoresis of a concentrated aqueous dispersion of non-Newtonian drops. J Colloid Interface Sci 2005;282(2):486–92.
[26] Lee E, Chou KT, Hsu JP. Sedimentation of a concentrated dispersion of composite colloidal particles. J Colloid Interface Sci 2006;295(1):279–90.

[27] Hsu JP, et al. Diffusiophoresis of concentrated suspensions of spherical particles with distinct ionic diffusion velocities. J Phys Chem B 2007;111(10):2533–9.
[28] Lou J, Lee E. Dynamic electrophoresis of concentrated droplet dispersions at arbitrary surface potentials. Chem Eng Sci 2007;62(22):6234–45.
[29] Chiang CP, He YY, Lee E. Sedimentation velocity and potential in a concentrated suspension of charged liquid drops. Langmuir 2008;24(20):11361–9.
[30] He YY, Lee E. Electrophoresis in concentrated dispersions of charged porous spheres. Chem Eng Sci 2008;63(23):5719–27.
[31] Lou J, Lee E. Diffusiophoresis of concentrated suspensions of liquid drops. J Phys Chem C 2008;112(32):12455–62.
[32] He YY, Wu E, Lee E. Electrophoresis in suspensions of charged porous spheres in salt-free media. Chem Eng Sci 2010;65(20):5507–16.
[33] Lou J, Shih CY, Lee E. Diffusiophoresis of concentrated suspensions of spherical particles with charge-regulated surface: polarization effect with nonlinear Poisson-Boltzmann equation. Langmuir 2010;26(1):47–55.
[34] Tsai P, Lee E. Gel electrophoresis in suspensions of charged spherical particles. Soft Matter 2011;7(12):5789–98.
[35] Carrique F, et al. Influence of cell-model boundary conditions on the conductivity and electrophoretic mobility of concentrated suspensions. Adv Colloid Interface Sci 2005;118(1–3):43–50.
[36] Mewis J, Wagner NJ. Colloidal suspension rheology. Cambridge: Cambridge University Press; 2012.
[37] Mooney M. The viscosity of a concentrated suspension of spherical particles. J Colloid Sci 1951;6(2):162–70.
[38] Ramakrishnan S, et al. Concentration fluctuations in a model colloid-polymer suspension: experimental tests of depletion theories. Langmuir 2002;18(4):1082–90.

Chapter 3

Electrophoretic Motion of a Rigid Spherical Particle Normal to a Solid Plane

3.1 INTRODUCTION

In practice the electrophoretic motion is often carried out within a confined domain with a nearby boundary to facilitate the operation as the suspension is in liquid form and a container is necessary to control the direction of the particle motion. For instance, in electrophoretic deposition, a solid plane is encountered with the particles moving toward it and depositing on its surface eventually [1–6]. It can model many situations encountered in biomedical applications as well [7–14]. Note that in comparison with the tiny size of typical colloidal particles, which is often of micro- to nanometer scale, any visible nearby finite object can be treated as a flat solid plane.

3.1.1 Hydrodynamic Drag Effect: Impingement Flow

The presence of a nearby boundary has a profound impact upon the particle motion both hydrodynamically and electrostatically. The drag force induced due to the planar boundary ahead slows down the particle motion in general. Abundant studies have been documented in the literature regarding the corresponding purely hydrodynamic problem with no involvement of electric force from the electrolyte ions, or the impingement flow [15, 16]. Analytical formula in the form of infinite series were provided for both the solid planar boundary and the free interface with a rigid spherical particle moving toward it [16]. The hydrodynamic impact is expressed as a correction factor of the Stokes law there to account for the boundary effect. In general, the closer the particle to the planar boundary is, the more significant this hydrodynamic motion-deterring drag effect is.

Interface Science and Technology, Vol. 26. https://doi.org/10.1016/B978-0-08-100865-2.00003-5

3.1.2 Electrostatic Attraction Effect: The Image-Charge Force Analog

In electrostatics, the deformation of the electric field around a charged particle of any shape due to the presence of a chargeless solid plane is equivalent to the presence of an identical charged particle located in the mirror image position (image-charge) but with opposite sign of charges. As a result, an attractive electric (Coulomb) force is generated. For a free surface or an interface between a liquid and a gas, the corresponding image-charge force is of repulsion nature, and is equivalent to that from an identical charged particle bearing the same sign. This is referred to as the "image-charge force analog" in electrostatics [17–19]. It is a direct mathematical consequence based on the consideration of the planar boundary conditions for electrostatic potential there, which will be discussed later. Here it should be noted that it is still true for a particle suspended in an electrolyte solution with counterions surrounding it when the particle is motionless and the system is at an electrostatic equilibrium.

This translates into the conclusion that the presence of a solid plane provides electric attraction force. Moreover, the closer the particle to the plane is, the stronger this attraction force is. This similarity of increasing force with decreasing separation distance between the electrostatic attraction and the hydrodynamic retardation makes the mobility reduction much weaker than what might be expected from a purely hydrodynamic point of view for an impingement flow. The combined effect still leads to a reduction of overall mobility as the particle moves toward it, as we shall discuss later.

3.1.3 Boundary Confinement Effect: Geometric Deformation of the Double Layer

In addition to the above two boundary effects of electrostatic and hydrodynamic nature respectively, the presence of the boundary also has a geometric confining effect upon the ion distribution around the particle if the double layer is thick enough to reach the boundary. This affects the electric driving force once an external electric field is applied upon the system. This perhaps is the most important issue in the study of boundary effect in electrokinetic motion, as the double layer polarization/deformation determines the ultimate particle motion in general. And obviously the thicker the double layer is, the more significant this factor is, as the deformation of the double layer can be much more profound. When the double layer is very thin, this particular effect is negligible.

Note that generally speaking, when a charged particle approaches a dielectric solid plane, polarized electric charges of opposite sign will be induced on the plane, which tend to cluster in the area ahead of the particle. Hence the electric potential of the plane will not stay as a constant, strictly speaking. And certainly for a nongrounded conducting metal plane, the surface potential there can be maintained at some nonzero constant value thus generates an electroosmosis flow which may either enhance or retard the particle motion, depending on the sign of charges on the solid plane [20–22]. The extension to cover such systems is straightforward by changing the corresponding boundary conditions at the solid plane. The details can be found elsewhere [23]. For simplicity, it is not presented here.

Many studies have been reported in the literature with different approaches regarding various limiting cases, such as very thin double layer or lowly charged particle [24–26]. The motion-deterring polarization effect is negligible in these situations. Lee's group, on the other hand, has treated the problem with bispherical coordinates coupled with pseudo-spectral method for various colloidal entities, taking full account of the double layer polarization effect, and valid for arbitrary charge levels and double layer thickness [15, 27–29]. Here in Chapter 3 we focus on the electrophoretic motion of a rigid spherical particle moving normal to a solid plane. Note that the mobility is the same in magnitude whether the particle moves toward or away from it due to the quasisteady state assumption.

3.2 THEORY

The system under consideration is shown in Fig. 3.1, where a rigid spherical particle is conducting electrophoretic motion normal to an infinitely large solid flat plane. The plane is assumed to be a conducting one, like a grounded metal plane with a constant surface potential held at zero, which is taken to be the reference value of earth. An electric field is applied in the upward

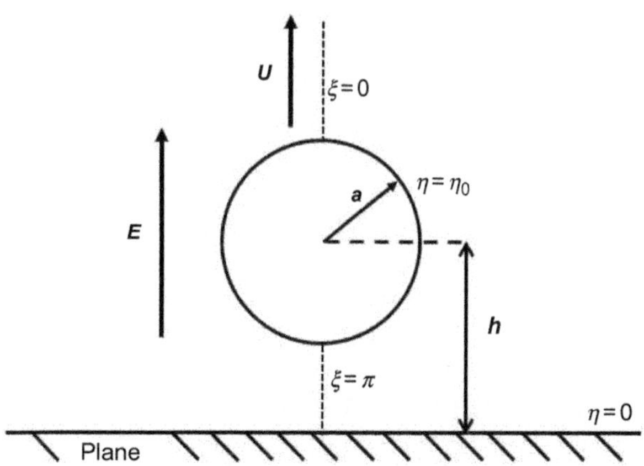

FIG. 3.1 System diagram.

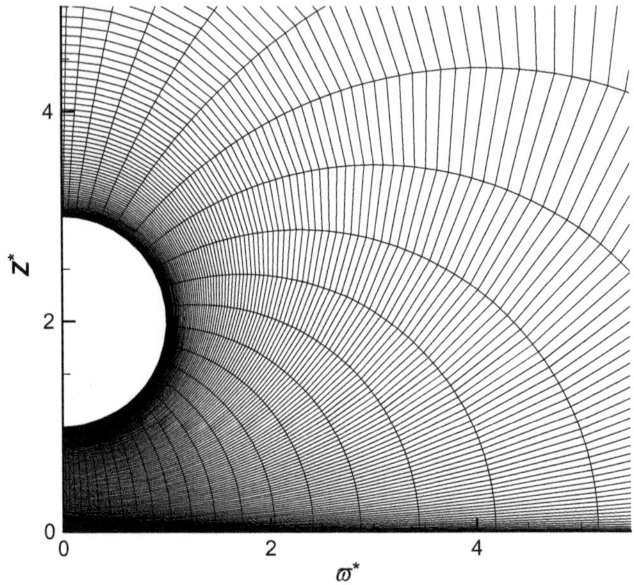

FIG. 3.2 Mesh diagram in computation, (30 × 130) nodes in (ξ, η) domain.

direction normal to the plane. In practice it can also be toward the plane. Corresponding mesh diagram in computation is shown in Fig. 3.2. Altogether 30 nodes in ξ direction and 130 nodes in η direction are used.

Note that this is a 2-D axisymmetric problem geometrically speaking, with the axis passing through the center of the spherical particle and normal to the plane. Significant simplification of mathematical deductions can thus be achieved. To treat this problem with a rigorous mathematical approach, the issue of infinity has to be handled carefully. Fortunately the well-known axisymmetric version of the orthogonal bispherical coordinates (ξ, η) is cut out for this purpose with the infinite plane treated as a large circle with an infinite radius of curvature. A circle in (x, y) coordinates is described as a contour line with η as a constant, i.e., η = const. The physical domain between the spherical particle and the solid plane can thus be mapped to a rectangular mathematical domain in terms of (ξ, η). The details can be found elsewhere [16]. Note that η used here is not to be confused with the fluid viscosity of the suspending medium. Clarification will be mentioned whenever needed.

The bispherical coordinates (ξ, η) are adopted with $\eta = 0$ and $\eta = \eta_0$ representing respectively the planar solid surface and the surface of the particle. The bispherical coordinates used are related to the familiar cylindrical coordinates (ϖ,Z) by:

$$Z = c\frac{\sinh \eta}{\cosh \eta - \cos \xi} \tag{3.1}$$

$$\varpi = c\frac{\sin \xi}{\cosh \eta - \cos \xi} \tag{3.2}$$

where c is the focal length defined as $c = a \cdot \sinh(\eta_0)$. And $0 \leq \eta \leq \eta_0$, $0 \leq \xi \leq \pi$ with $\eta_0 = \cosh^{-1}(h/a)$, where h is the separation distance between the particle center and the plane and a is the particle radius. Note that separation distance h is the crucial factor determining the boundary effect both hydrodynamically and electrostatically, including the geometric confinement effect.

The governing equations here are based on the standard electrokinetic model, which is independent of the specific coordinates used. That is, written in terms of vector symbols like ∇, ∇^2, E^2, E^4, etc., the governing equations would be exactly the same regardless of the specific coordinates used, be it spherical, cylindrical, or like the bispherical one used here. Hence the explicit form of the governing equations are exactly the same as those in Chapter 1, that is, Eqs. (1.29)–(1.33). The only difference among various coordinates is the precise definition of the various vector operators. The governing equations in Chapter 1, written in bispherical coordinates and in dimensionless form, are as follows:

$$\nabla^{*2}\phi_e^* = -\frac{(\kappa a)^2}{(1+\alpha)}[\exp(-\phi_e^*) - \exp(\alpha\phi_e^*)] \tag{3.3}$$

$$\nabla^{*2}\delta\phi^* = \frac{(\kappa a)^2}{(1+\alpha)}[\exp(-\phi_e^*)(\delta\phi^* + g_1^*) + \alpha\exp(\alpha\phi_e^*)(\delta\phi^* + g_2^*)] \tag{3.4}$$

$$\begin{aligned}&\nabla^{*2}g_1^* - \left(\frac{x}{c^*}\right)^2\left(\frac{\partial\phi_e^*}{\partial\xi}\frac{\partial g_1^*}{\partial\xi} + \frac{\partial\phi_e^*}{\partial\eta}\frac{\partial g_1^*}{\partial\eta}\right)\\ &= \frac{Pe_1}{\sin\xi}\left(\frac{x}{c^*}\right)^3\left(\frac{\partial\phi_e^*}{\partial\xi}\frac{\partial\Psi^*}{\partial\eta} - \frac{\partial\phi_e^*}{\partial\eta}\frac{\partial\Psi^*}{\partial\xi}\right) - Pe_1U^*\left(\frac{x}{c^*}\right)^2\left(\frac{\partial\phi_e^*}{\partial\xi}\frac{\partial Z^*}{\partial\xi} + \frac{\partial\phi_e^*}{\partial\eta}\frac{\partial Z^*}{\partial\eta}\right)\end{aligned} \tag{3.5}$$

$$\begin{aligned}&\nabla^{*2}g_2^* + \alpha\left(\frac{x}{c^*}\right)^2\left(\frac{\partial\phi_e^*}{\partial\xi}\frac{\partial g_2^*}{\partial\xi} + \frac{\partial\phi_e^*}{\partial\eta}\frac{\partial g_2^*}{\partial\eta}\right)\\ &= \frac{Pe_2}{\sin\xi}\left(\frac{x}{c^*}\right)^3\left(\frac{\partial\phi_e^*}{\partial\xi}\frac{\partial\Psi^*}{\partial\eta} - \frac{\partial\phi_e^*}{\partial\eta}\frac{\partial\Psi^*}{\partial\xi}\right) - Pe_2U^*\left(\frac{x}{c^*}\right)^2\left(\frac{\partial\phi_e^*}{\partial\xi}\frac{\partial Z^*}{\partial\xi} + \frac{\partial\phi_e^*}{\partial\eta}\frac{\partial Z^*}{\partial\eta}\right)\end{aligned} \tag{3.6}$$

$$\begin{aligned}E^{*4}\Psi^* = &-\frac{(\kappa a)^2}{(1+\alpha)}\frac{x\sin\xi}{c^*}\left[\left(\frac{\partial g_1^*}{\partial\xi}\exp(-\phi_e^*) + \alpha\frac{\partial g_2^*}{\partial\xi}\exp(\alpha\phi_e^*)\right)\frac{\partial\phi_e^*}{\partial\eta}\right.\\ &\left.-\left(\frac{\partial g_1^*}{\partial\eta}\exp(-\phi_e^*) + \alpha\frac{\partial g_2^*}{\partial\eta}\exp(\alpha\phi_e^*)\right)\frac{\partial\phi_e^*}{\partial\xi}\right]\end{aligned} \tag{3.7}$$

where $x = \cosh\eta - \cos\xi$, $c^* = c/a$, $Z^* = c^*\sin\eta/x$ and $Pe_j = \frac{u_E a}{D_j} = \frac{\varepsilon}{\eta_f D_j}\left(\frac{kT}{z_1 e}\right)^2$, with η_f the suspending fluid viscosity (not to be confused with the η in bispherical coordinates) and u_E is the particle velocity predicted by Smoluchowski's formula when an electric field $E = \frac{kT}{z_1 e}/a$ is applied upon the particle, i.e., $u_E = \frac{\varepsilon}{\eta_f a}\left(\frac{kT}{z_1 e}\right)^2$. Moreover, ϖ^*, the corresponding dimensionless quantity for ϖ, is defined as $\varpi^* = \varpi/a$. Note that in the above conversions, dimensionless operators defined as follows have been used:

$$\nabla^{*2} = \frac{x^2}{c^{*2}}\left[\frac{\partial^2}{\partial\eta^2} + \frac{\partial^2}{\partial\xi^2} - \frac{\sinh\eta}{x}\frac{\partial}{\partial\eta} + \frac{\cos\xi\cosh\eta - 1}{x\sin\xi}\frac{\partial}{\partial\xi}\right] \tag{3.8}$$

$$E^{*2} = \frac{x^2}{c^{*2}}\left[\frac{\partial^2}{\partial\eta^2} + \frac{\partial^2}{\partial\xi^2} + \frac{\sinh\eta}{x}\frac{\partial}{\partial\eta} + \frac{1 - \cos\xi\cosh\eta}{x\sin\xi}\frac{\partial}{\partial\xi}\right] \tag{3.9}$$

and $E^{*4} = E^{*2}E^{*2}$.

The subsequent perturbation analysis and its resulted equations are the same as those encountered in Chapters 1 and 2, thus will not be repeated here.

The corresponding physical boundary conditions of a rigid dielectric particle with constant potential at the particle surface are identical as those in Chapter 1, i.e., Eqs. (1.34)–(1.38), but have to be treated in bispherical coordinates here as follows, i.e., at $\eta = \eta_0$:

$$\phi_e^* = \phi_r \tag{3.10}$$

$$\frac{\partial\delta\phi^*}{\partial\eta} = 0 \tag{3.11}$$

$$\frac{\partial g_j^*}{\partial\eta} = 0 \tag{3.12}$$

$$\Psi^* = \frac{-1}{2}\left(\frac{c^*\sin\xi}{x}\right)^2 U^* \tag{3.13}$$

$$\frac{\partial\Psi^*}{\partial\eta} = \frac{c^{*2}}{x^3}\sin\xi\sinh\eta\ U^* \tag{3.14}$$

The precise definition of each variable can be found as well in the List of Symbols at the end of the book.

For a charge-regulated particle, on the other hand, surface potential is not a constant and determined by the ionic environment in its neighborhood. Thus Eq. (3.10) will be replaced by the following equation,

$$\frac{\partial\phi_e^*}{\partial\eta} = \frac{c^*}{x}\left(\frac{A}{1 + B\exp(-\phi_e^*)}\right) \tag{3.15}$$

where $A=\frac{aN_se^2}{\varepsilon kT}$, $B=\frac{[H^+]_0}{K_a}$ and N_s is the number of sites on the particle surface with $[H^+]_0$ the bulk concentration of cation, following the corresponding deduction in Chapter 1.

The symmetric nature of the problem leads to the following boundary conditions along the axis, i.e., at $\xi=0$ or π:

$$\frac{\partial\phi_e^*}{\partial\xi}=0 \tag{3.16}$$

$$\frac{\partial\delta\phi^*}{\partial\xi}=0 \tag{3.17}$$

$$\frac{\partial g_j^*}{\partial\xi}=0 \tag{3.18}$$

$$\Psi^*=0 \tag{3.19}$$

$$\frac{\partial\Psi^*}{\partial\xi}=0 \tag{3.20}$$

On the surface of the solid plane, the electric potential is assumed to remain as a constant assuming that the plane is a conducting one, like a metal plane for instance, as we have mentioned earlier. For simplicity, it is further assumed to be grounded to earth to reach the environmental reference value, say $\phi_e^*=0$. The hydrodynamic boundary conditions at the solid plane are the usual no-slip condition. We thus have, at $\eta=0$:

$$\phi_e^*=0 \tag{3.21}$$

$$\delta\phi^*=-\boldsymbol{E}^*\cdot\mathbf{r}^* \tag{3.22}$$

$$g_j^*=-\delta\phi^* \tag{3.23}$$

$$\Psi^*=0 \tag{3.24}$$

$$\frac{\partial\Psi^*}{\partial\eta}=0 \tag{3.25}$$

$$\frac{\Psi^*}{r^{*2}}\rightarrow 0,\quad \xi=0 \tag{3.26}$$

In these expressions, $\boldsymbol{E}^*$ is the scaled applied electric field in the Z-direction and $\boldsymbol{r}^*=c^*\left(\frac{\cosh\eta+\cos\xi}{x}\right)^{1/2}$.

Note Eq. (3.22) is the Shilov-type (S-Z) boundary condition of a cell model, which tries to unify the actual local environment of a particle with the virtual surface where an external electric field is supposed to apply, just like the concept discussed in Chapter 2 for a particle suspension. Here the entire plane ($\eta = 0$) is treated like the outer virtual surface enclosing the particle here. Eq. (3.23) reflects the fact that the electric potential remains unchanged on the planar surface when the particle is in motion. Eq. (3.26) implies that the fluid at infinity is stationary. The details can be found elsewhere [15].

Evaluation of the forces acting on the particle, i.e., the electric driving force F_E and the hydrodynamic retarding force F_D, are the same as those adopted in Chapter 1, here expressed in bispherical coordinates and in dimensionless form as follows:

$$F_E^* = -2\pi \int_0^\pi \frac{\sin\xi}{x}\frac{\partial\phi^*}{\partial\eta}\left[(1-\cosh\eta\cos\xi)\frac{\partial\phi^*}{\partial\eta} - \sinh\eta\sin\xi\frac{\partial\phi^*}{\partial\xi}\right]d\xi$$
$$+\pi\int_0^\pi \frac{\sin^2\xi}{x}\left[\frac{\partial\phi^*}{\partial\eta}\left(\sin\xi\frac{\partial\phi^*}{\partial\eta} + x\frac{\partial^2\phi^*}{\partial\eta\partial\xi}\right) + \frac{\partial\phi^*}{\partial\xi}\left(\sin\xi\frac{\partial\phi^*}{\partial\xi} + x\frac{\partial^2\phi^*}{\partial\xi^2}\right)\right]d\xi \tag{3.27}$$

$$F_D^* = \pi\int_0^\pi -\varpi^{*3}\frac{\partial}{\partial\eta}\frac{E^{*2}\Psi^*}{\varpi^{*2}}d\xi$$
$$-\frac{(\kappa a)^2}{(1+\alpha)}\pi\int_0^\pi \varpi^{*2}\left[\exp\left(-\phi_e^*\right)\left(1-\left(\delta\phi^* + g_1^*\right)\right)\right. \tag{3.28}$$
$$\left. -\exp\left(\alpha\phi_e^*\right)\left(1+\alpha\left(\delta\phi^* + g_2^*\right)\right)\right]\frac{\partial\phi^*}{\partial\xi}d\xi$$

where $\varpi^* = \varpi/a$.

The particle mobility is evaluated by the balance of these two forces for the two subproblems introduced in Chapter 1 for a single particle. The procedure is exactly the same as there.

3.3 RESULTS AND DISCUSSION

3.3.1 General Henry's Charts at Various Separation Distance

Fig. 3.3A–C shows particle mobility as a function of κa at various separation distances to the plane for $\phi_r = 1, 3$, and 5, respectively. Corresponding profile for a single particle is shown as well for comparison purpose. The presence of the solid plane reduces particle mobility in general. The closer the particle to the plane is, the slower the particle motion is.

Moreover, the thicker the double layer is, the more significant the mobility reduction as once the double layer touches the plane it will be forced to

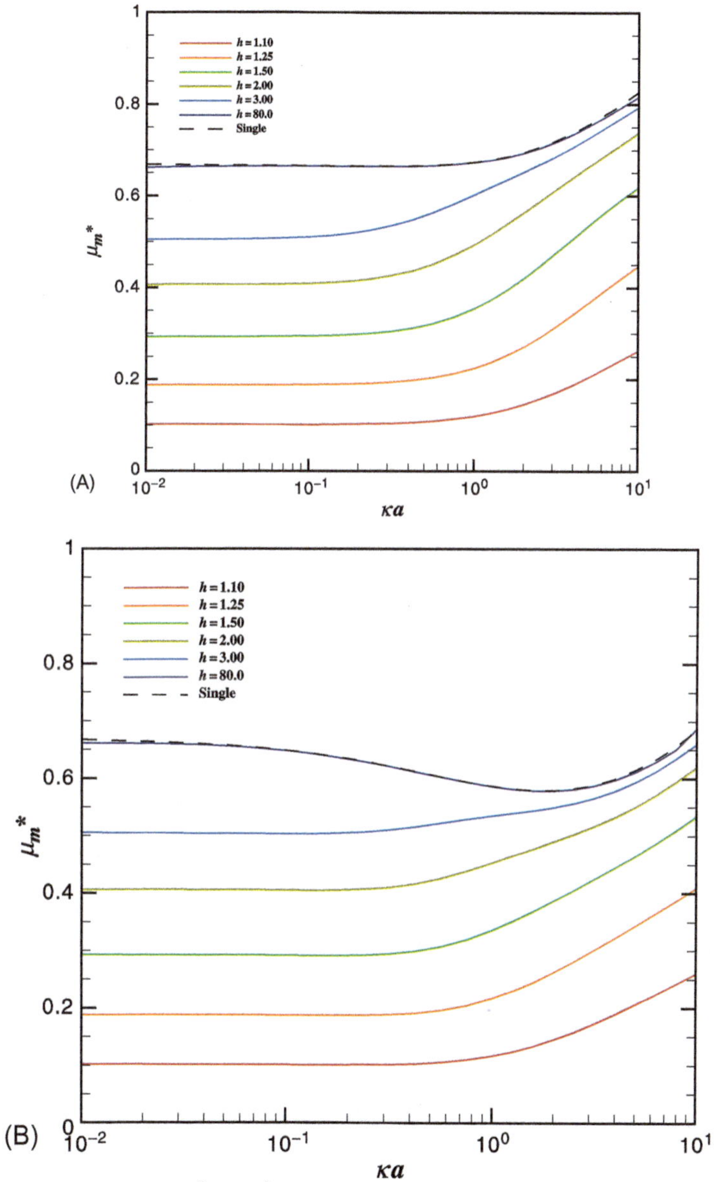

FIG. 3.3 Scaled mobility $\left(\mu_m^* = \frac{\mu^*}{\phi_r}\right)$ as a function of κa at different h with: (A) $\phi_r = 1$; (B) $\phi_r = 3$; (C) $\phi_r = 5$.

(Continued)

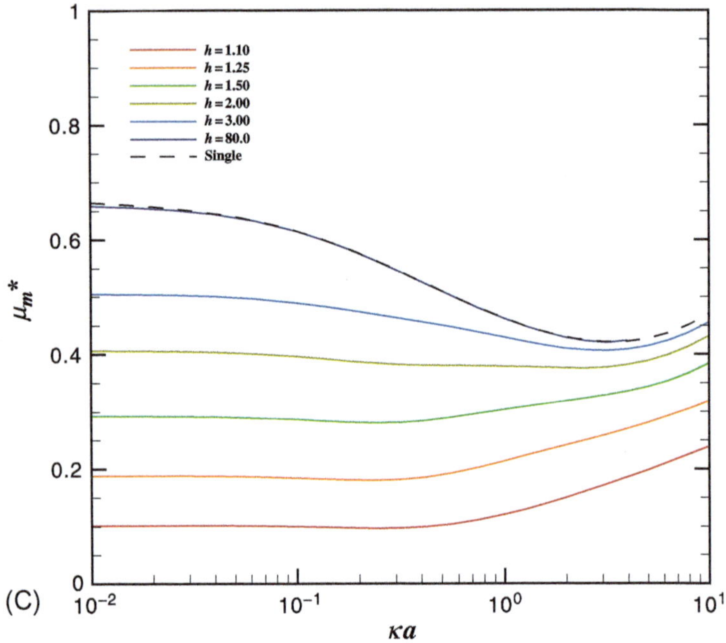

FIG. 3.3—CONT'D

reverse its original distribution somewhat in the gap region, which leads to the weakening of the original motion-deterring local electric field, or the double layer polarization effect. This deformation effect due to the boundary confinement can get so severe that the local minima of the mobility profiles of highly charged particles, $\phi_r = 3$ and 5 in Fig. 3.3B and C disappear completely as the particle gets closer to the plane. This suggests that the shape of the double layer hence its motion-deterring effect is drastically changed by the presence of the plane, which is further depicted in Fig. 3.4.

3.3.2 Geometric Boundary Confinement Effect

The boundary confinement effect is also shown with various contour plots in Fig. 3.5. As shown in Fig. 3.5A, the entire equilibrium double layer in the gap is wiped out completely as there is simply no room for the ions to exist there, where the separation distance is only one-tenth of the particle radius, or $h = 1.1$. This certainly has a strong impact upon the electrostatic environment surrounding the particle, as shown in Fig. 3.5B, where the corresponding

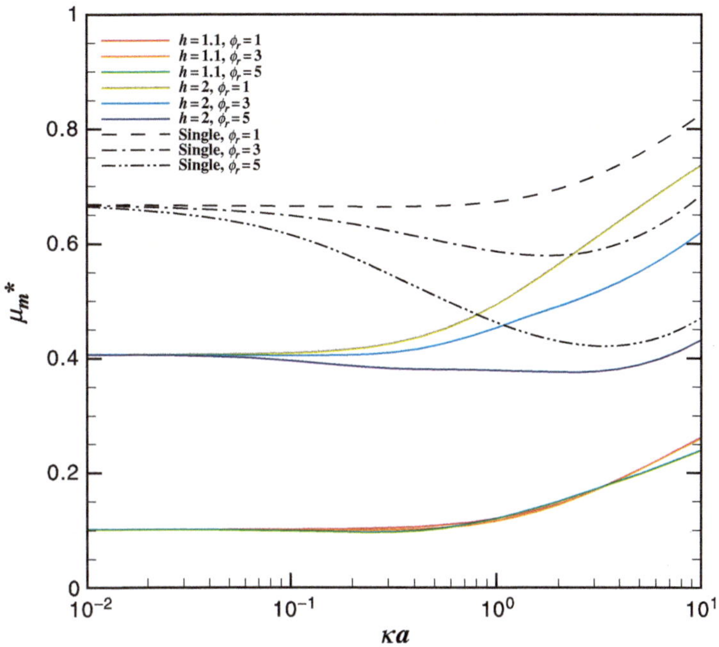

FIG. 3.4 Scaled mobility (μ_m^*) as a function of κa at different ϕ_r for $h = 1.1$, 2 in comparison with the single sphere situation.

electric potential distribution indicating a strong attractive image-charge analog force between the particle and the solid plane, as a large potential difference is established in the narrow gap there. The electric driving force, as a result, tends to increase as the particle approaches the solid plane, synchronized with the increasing hydrodynamic drag force. It turns out that the latter dominates all the way though.

The exact shape of the double layer polarization near the plane ($h = 1.1$) is shown for a highly charged particle ($\phi_r = 3$) as well in Fig. 3.5C. The polarization of the double layer in the gap disappears completely, hence so is its motion-deterring effect there. This leads to the disappearance of local minima in the mobility profiles in Figs. 3.3 and 3.4. Even more drastic shape change is shown in Fig. 3.5D for $\phi_r = 5$, where the counterions in the gap is visibly squeezed out horizontally.

3.3.3 Vortex Flow Induced

Flow fields corresponding to $h = 2$ and $h = 1.1$ are shown in Fig. 3.6A–C. The upward moving particle generates a clockwise axisymmetric vortex flow

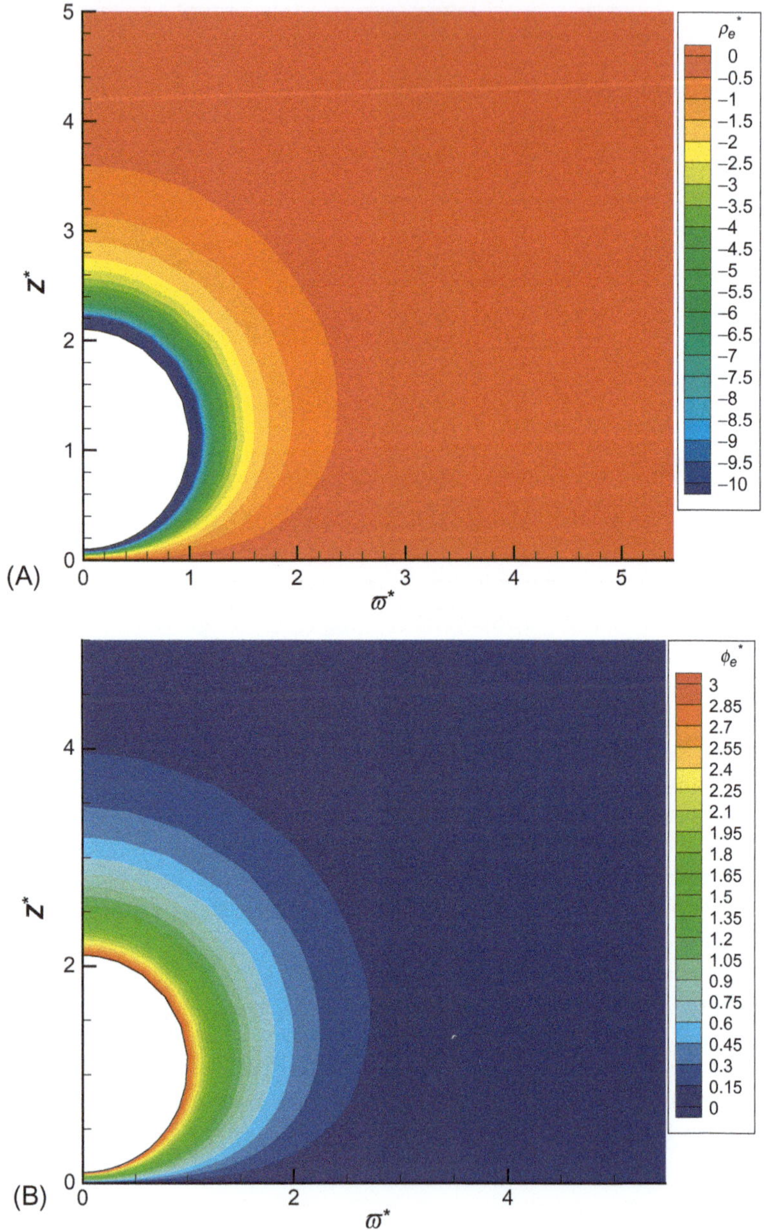

FIG. 3.5 Contour plots of different scaled variables for $\kappa a = 1$ and $h = 1.1$. (A) Scaled equilibrium charge density (ρ_e^*) at $\phi_r = 3$; (B) Scaled equilibrium electric potential (ϕ_e^*) at $\phi_r = 3$; (C) Scaled disturbed charge density ($\delta\rho^*$) at $\phi_r = 3$; (D) Scaled disturbed charge density ($\delta\rho^*$) at $\phi_r = 5$.

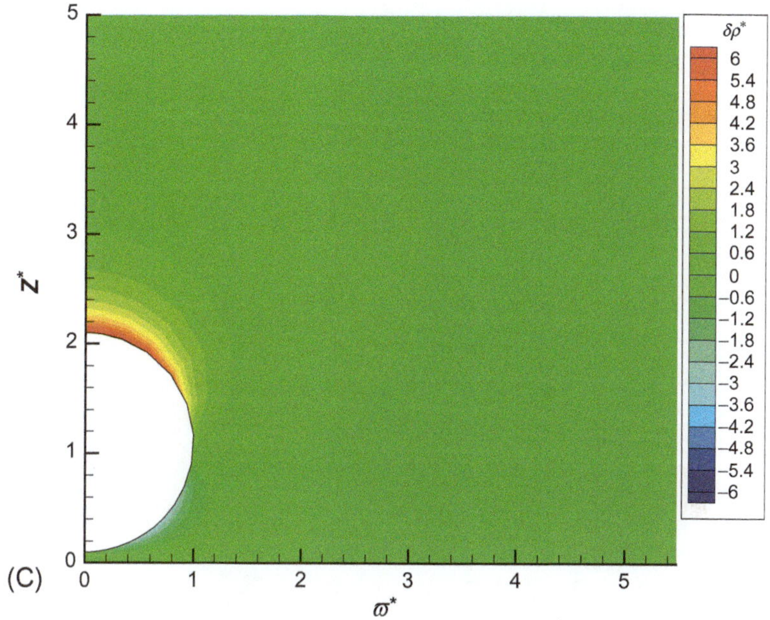

(C)
z^*
ϖ^*
$\delta\rho^*$
6
5.4
4.8
4.2
3.6
3
2.4
1.8
1.2
0.6
0
−0.6
−1.2
−1.8
−2.4
−3
−3.6
−4.2
−4.8
−5.4
−6

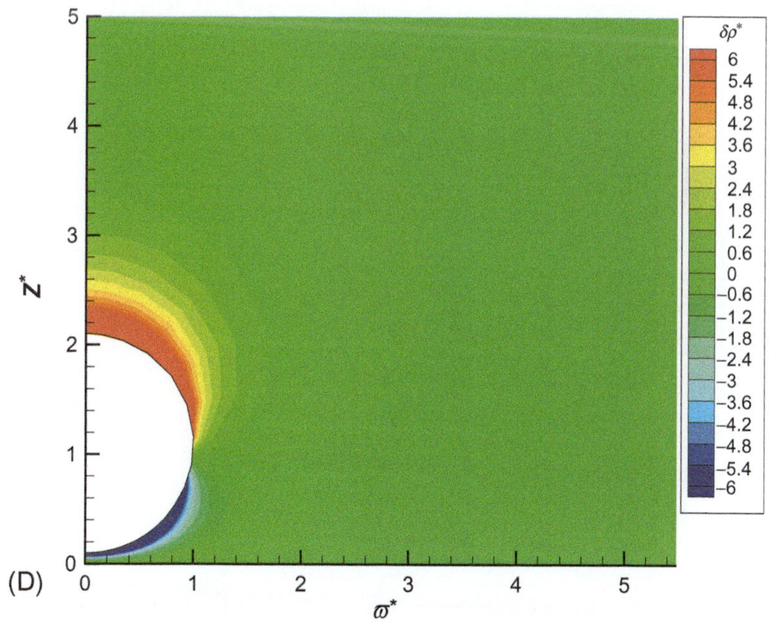

(D)
z^*
ϖ^*
$\delta\rho^*$
6
5.4
4.8
4.2
3.6
3
2.4
1.8
1.2
0.6
0
−0.6
−1.2
−1.8
−2.4
−3
−3.6
−4.2
−4.8
−5.4
−6

FIG. 3.5—CONT'D

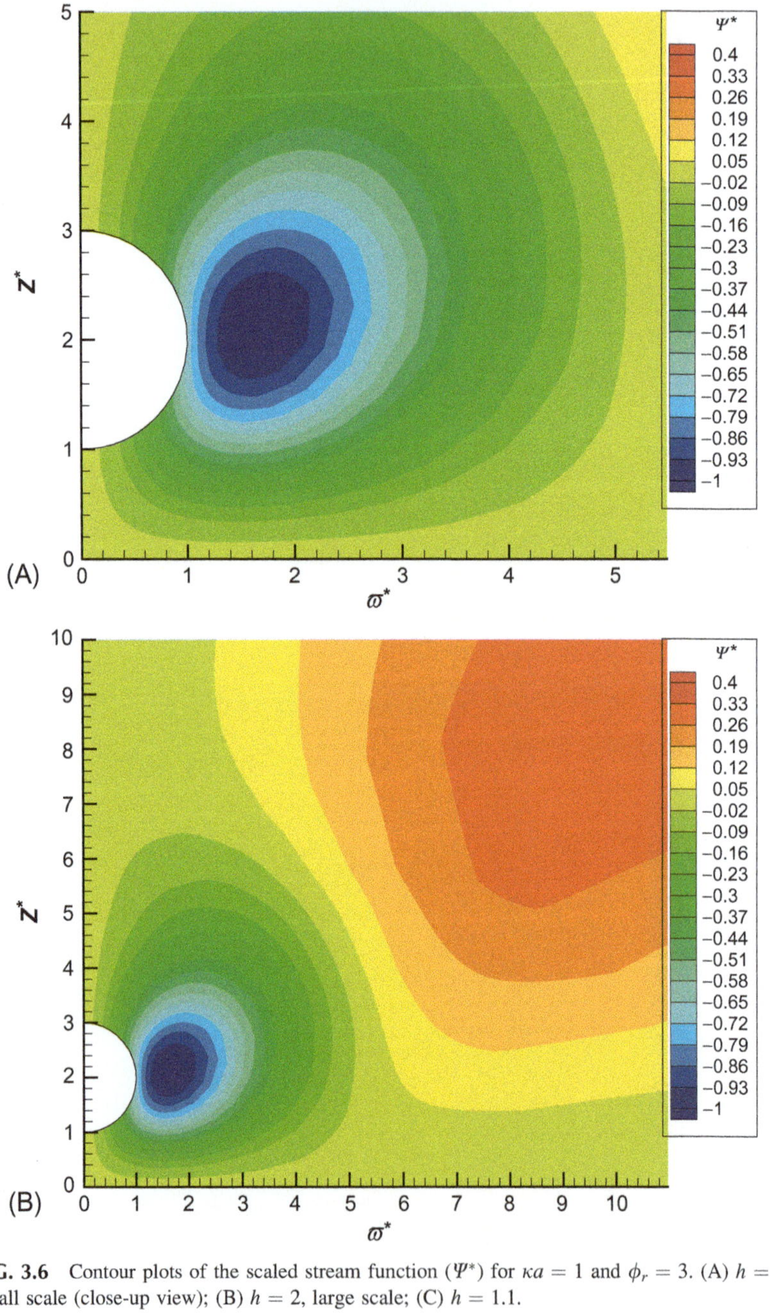

FIG. 3.6 Contour plots of the scaled stream function (Ψ^*) for $\kappa a = 1$ and $\phi_r = 3$. (A) $h = 2$, small scale (close-up view); (B) $h = 2$, large scale; (C) $h = 1.1$.

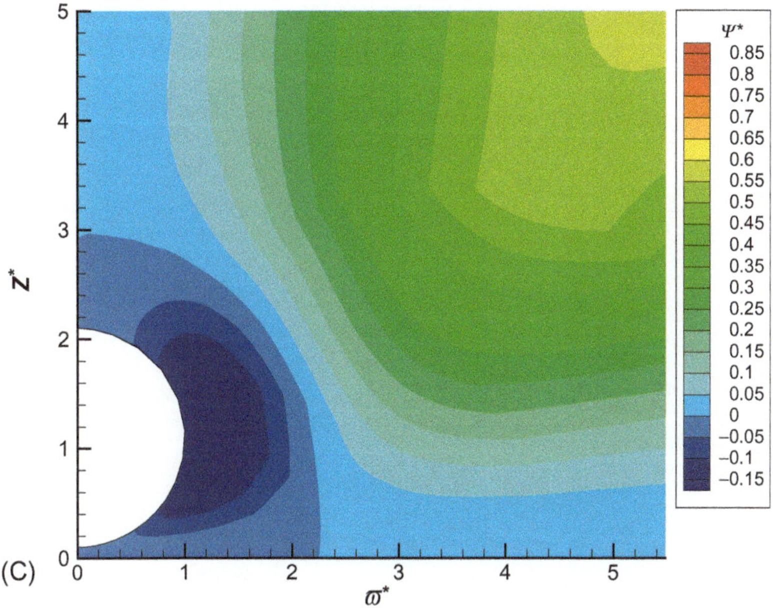

(C)

FIG. 3.6—CONT'D

in its neighborhood to draw fluid into its wake area, as indicated in Fig. 3.6A or B. This primary vortex flow is so strong that it generates a secondary vortex flow, as shown in Fig. 3.6B in a larger scale of the plot. This secondary vortex flow takes the opposite orientation of the primary one, as it should from the hydrodynamic viewpoint. When the gap is so narrow, however, there is no room for the fluid to reach there and the primary vortex flow simply breaks down, as shown in Fig. 3.6C for $h = 1.1$.

3.3.4 Correction Factor of Separation Distance

The boundary effect is also expressed as a correction factor to the single particle mobility at various separation distance in Tables 3.1–3.3 for $\phi_r = 1$, 3, and 5, respectively, at various κa. These tables are provided to facilitate the possible use of interested readers for accurate comparison with experimental data or verification of their specific theories.

Tables of Boundary Effect Expressed as a Correction Factor Γ

$\Gamma = \mu^*_h/\mu^*_\infty$, where μ^*_h refers to the particle mobility at a separation distance h, whereas refers to the mobility in an infinite medium.

TABLE 3.1 Correction Factor Γ for $\phi_r = 1$

	κa							
h	*0.01*	*0.1*	*0.5*	*1*	*2*	*5*	*7*	*10*
1.1	0.153	0.155	0.164	0.180	0.211	0.273	0.296	0.319
1.2	0.239	0.241	0.257	0.283	0.331	0.417	0.446	0.472
1.4	0.385	0.389	0.417	0.461	0.534	0.639	0.667	0.687
1.6	0.484	0.489	0.527	0.582	0.665	0.761	0.782	0.795
1.8	0.555	0.561	0.608	0.670	0.752	0.833	0.848	0.856
2.0	0.609	0.616	0.669	0.734	0.812	0.878	0.888	0.894
4.0	0.823	0.837	0.908	0.950	0.973	0.981	0.980	0.977
6.0	0.885	0.903	0.964	0.984	0.991	0.992	0.989	0.986
8.0	0.915	0.935	0.983	0.993	0.995	0.994	0.991	0.988
10.0	0.932	0.953	0.991	0.996	0.997	0.995	0.992	0.988
20.0	0.966	0.987	0.998	0.999	0.998	0.996	0.993	0.989
40.0	0.983	0.997	0.999	0.999	0.999	0.996	0.993	0.989
60.0	0.988	0.999	1.000	0.999	0.999	0.996	0.993	0.989
80.0	0.991	0.999	1.000	0.999	0.998	0.996	0.993	0.988
∞	1.000	1.000	1.000	1.000	1.000	1.000	1.000	1.000

TABLE 3.2 Correction Factor Γ for $\phi_r = 3$

	κa							
h	*0.01*	*0.1*	*0.5*	*1*	*2*	*5*	*7*	*10*
1.1	0.153	0.157	0.173	0.200	0.251	0.337	0.362	0.383
1.2	0.239	0.245	0.270	0.313	0.388	0.491	0.516	0.532
1.4	0.386	0.396	0.438	0.505	0.600	0.698	0.716	0.727
1.6	0.485	0.497	0.552	0.628	0.722	0.800	0.814	0.821
1.8	0.556	0.570	0.634	0.713	0.797	0.860	0.869	0.874
2.0	0.610	0.625	0.695	0.773	0.846	0.897	0.904	0.906
4.0	0.824	0.845	0.921	0.958	0.977	0.983	0.982	0.979
6.0	0.886	0.910	0.970	0.986	0.992	0.992	0.990	0.986
8.0	0.916	0.941	0.986	0.994	0.996	0.994	0.992	0.988
10.0	0.933	0.958	0.992	0.996	0.997	0.995	0.992	0.988
20.0	0.967	0.989	0.999	0.999	0.998	0.996	0.993	0.989
40.0	0.983	0.998	1.000	0.999	0.999	0.996	0.993	0.989
60.0	0.988	0.999	1.000	0.999	0.999	0.996	0.993	0.992
80.0	0.991	1.000	1.000	0.999	0.998	0.995	0.992	–
∞	1.000	1.000	1.000	1.000	1.000	1.000	1.000	1.000

Notes: "–" refers to the fact that Γ approaches unity.

TABLE 3.3 Correction Factor Γ for $\phi_r = 5$

	κa							
h	*0.01*	*0.1*	*0.5*	*1*	*2*	*5*	*7*	*10*
1.1	0.154	0.163	0.201	0.261	0.350	0.464	0.492	0.509
1.2	0.240	0.254	0.315	0.398	0.498	0.599	0.618	0.624
1.4	0.387	0.410	0.501	0.595	0.687	0.762	0.770	0.774
1.6	0.486	0.514	0.615	0.703	0.782	0.836	0.843	0.848
1.8	0.558	0.588	0.692	0.772	0.838	0.881	0.888	0.891
2.0	0.612	0.645	0.747	0.820	0.875	0.911	0.916	0.918
4.0	0.827	0.863	0.937	0.965	0.979	0.985	0.983	0.981
6.0	0.889	0.923	0.977	0.988	0.993	0.993	0.990	0.991
8.0	0.918	0.951	0.989	0.995	0.996	0.995	0.992	–
10.0	0.935	0.967	0.994	0.997	0.997	0.995	0.993	–
20.0	0.968	0.993	0.999	0.999	0.998	0.996	–	–
40.0	0.984	0.999	1.000	0.999	0.999	–	–	–
60.0	0.989	1.000	1.000	0.999	0.999	–	–	–
80.0	0.991	1.000	1.000	0.999	0.999	–	–	–
∞	1.000	1.000	1.000	1.000	1.000	1.000	1.000	1.000

Notes: "–" refers to the fact that Γ approaches unity.

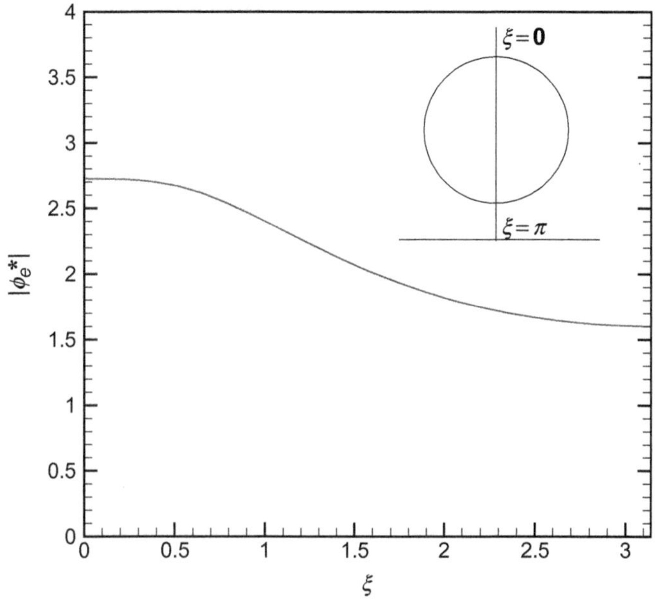

FIG. 3.7 Profile of the scaled equilibrium surface electric potential in magnitude $|\phi_e^*|$ at $\kappa a = 1$ and $h = 1.1$ with $A = 100$, $B = 1$ in the charge-regulation model.

3.3.5 Charge-Regulation Phenomenon

Finally, for a charge-regulated particle, a variation of surface charge is generated when the particle is very close to the planar boundary, as the ionic environment is no longer concentrically symmetric due to the geometric boundary confinement effect. The actual surface potential would be different in general between the hemisphere facing the gap and the one away from it, as shown in Fig. 3.7 for a representative sample case of choice. This kind of information is particularly useful in the investigation of surface adhesion by microorganisms, for instance [30–32].

REFERENCES

[1] Besra L, Liu M. A review on fundamentals and applications of electrophoretic deposition (EPD). Prog Mater Sci 2007;52:1–61.

[2] Hamaker H. Formation of a deposit by electrophoresis. Trans Faraday Soc 1940;35:279–87.

[3] Sarkar P, Nicholson PS. Electrophoretic deposition (EPD): mechanisms, kinetics, and application to ceramics. J Am Ceram Soc 1996;79:1987–2002.

[4] Komino T, Kuwabara H, Ikeda M, Yahiro M, Takimiya K, Adachi C. Droplet manipulation by an external electric field for crystalline film growth. Langmuir 2013;29:9592–7.

[5] Bouyer F, Foissy A. Electrophoretic deposition of silicon carbide. J Am Ceram Soc 1999;82:2001–10.

[6] Yun J-H, Wong RJ, Ng YH, Du A, Amal R. Combined electrophoretic deposition-anodization method to fabricate reduced graphene oxide-TiO_2 nanotube films. RSC Adv 2012;2:8164–71.

[7] Chiarot PR, Sullivan P, Mrad RB. An overview of electrospray applications in MEMS and microfluidic systems. J Microelectromech Syst 2011;20:1241–9.
[8] Hernàndez-Navarro S, Tierno P, Ignés-Mullol J, Sagués F. Ac electrophoresis of microdroplets in anisotropic liquids: transport, assembling and reaction. Soft Matter 2013;9: 7999–8004.
[9] Göpel W, Heiduschka P. Interface analysis in biosensor design. Biosens Bioelectron 1995;10:853–83.
[10] Das M, Dhand C, Sumana G, Srivastava A, Nagarajan R, Nain L, et al. Electrophoretic fabrication of chitosan-zirconium-oxide nanobiocomposite platform for nucleic acid detection. Biomacromolecules 2011;12:540–7.
[11] Dhand C, Singh S, Arya SK, Datta M, Malhotra B. Cholesterol biosensor based on electrophoretically deposited conducting polymer film derived from nano-structured polyaniline colloidal suspension. Anal Chim Acta 2007;602:244–51.
[12] Dhand C, Solanki PR, Datta M, Malhotra B. Polyaniline/single-walled carbon nanotubes composite based triglyceride biosensor. Electroanalysis 2010;22:2683–93.
[13] Dhand C, Solanki PR, Pandey MK, Datta M, Malhotra BD. Electrophoretically deposited polyaniline nanotubes based film for cholesterol detection. Electrophoresis 2010; 31:3754–62.
[14] Li Y, Wu K, Zhitomirsky I. Electrodeposition of composite zinc oxide-chitosan films. Colloids Surf A: Physicochem Eng Asp 2010;356:63–70.
[15] Chih MH, Lee E, Hsu J-P. Electrophoresis of a sphere normal to a plane at arbitrary electrical potential and double layer thickness. J Colloid Interface Sci 2002;248:383–8.
[16] Happel J, Brenner H. Low Reynolds number hydrodynamics: with special applications to particulate media. Springer Science & Business Media; 2012.
[17] Mbamala E, Von Grünberg H. Effective interaction of a charged colloidal particle with an air-water interface. J Phys: Condens Matter 2002;14:4881.
[18] Mbamala E, Von Grünberg H. Charged colloids and proteins at an air-water interface: the effect of dielectric substrates on interaction and phase behavior. Phys Rev E 2003; 67:031608.
[19] Russ C, Von Grünberg H, Dijkstra M, van Roij R. Three-body forces between charged colloidal particles. Phys Rev E 2002;66:011402.
[20] Shugai AA, Carnie SL. Electrophoretic motion of a spherical particle with a thick double layer in bounded flows. J Colloid Interface Sci 1999;213:298–315.
[21] Hsu J-P, Ku M-H, Kuo C-C. Electrophoresis of a charge-regulated sphere normal to a large disk. Langmuir 2005;21:7588–97.
[22] Zeng-Ju L. Electrostatic interaction of a spherical particle in the vicinity of a circular orifice. Chin Phys B 2010;19:058202.
[23] He Y-Y. Electrokinetic phenomena of charged porous colloidal spheres. Taiwan: Department of Chemical Engineering, College of Engineering, National Taiwan University; 2010.
[24] Morrison F, Stukel J. Electrophoresis of an insulating sphere normal to a conducting plane. J Colloid Interface Sci 1970;33:88–93.
[25] Feng JJ, Wu W-Y. Electrophoretic motion of an arbitrary prolate body of revolution toward an infinite conducting wall. J Fluid Mech 1994;264:41–58.
[26] Ennis J, Anderson J. Boundary effects on electrophoretic motion of spherical particles for thick double layers and low zeta potential. J Colloid Interface Sci 1997;185:497–514.
[27] Cheng W-L, He Y-Y, Lee E. Electrophoresis of a soft particle normal to a plane. J Colloid Interface Sci 2009;335:130–9.

[28] Lee E, Chen C-T, Hsu J-P. Electrophoresis of a rigid sphere in a Carreau fluid normal to a planar surface. J Colloid Interface Sci 2005;285:857–64.
[29] Tsai P, Huang C-H, Lee E. Electrophoresis of a charged colloidal particle in porous media: boundary effect of a solid plane. Langmuir 2011;27:13481–8.
[30] Nelson DL, Lehninger AL, Cox MM. Lehninger principles of biochemistry. Macmillan; 2008.
[31] Nakamura M, Ohshima H, Kondo T. Electrophoretic behavior of antigen-and antibody-carrying latex particles. J Colloid Interface Sci 1992;149:241–6.
[32] Ohshima H, Kondo T. Electrophoretic mobility and Donnan potential of a large colloidal particle with a surface charge layer. J Colloid Interface Sci 1987;116:305–11.

Chapter 4

Electrophoretic Motion of a Rigid Spherical Particle Normal to an Air-Water Interface

4.1 INTRODUCTION

In practice, the air-water interface has been used as a two-dimensional platform for many important scientific and industrial operations, for instance, the two-dimensional crystallization of proteins and other colloids [1–3]. It is also proposed as a stage for self-assembly operations [4]. Indeed new novel applications have been proposed in various fields due to its intrinsic nature of the two-dimensional confinement of particles getting there, and the inertness toward physical and chemical changes of its properties.

In electrokinetics, air-water interface is much more than just yet another planar boundary with perhaps lower hydrodynamic drag force as the fluid there can move horizontally. When a colloidal particle approaches the air-water interface, an upward capillary force is generated due to the deformation of the interface shape, which is much stronger than the particle weight hence the particle had a strong tendency to be trapped there [5, 6]. In reality, however, particles in a suspension never pop up to the interface automatically as a whole. Some kind of energy barrier apparently exists for this thermodynamically heavily favored process. Further studies about the detailed relationship of the interface deformation and the resulted capillary force were conducted by various research groups [7–11]. In particular, the electrostatic interactions between a charged particle and the air-water interface are discussed which indicates that the repulsive Coulomb force from image-charge force analog may provide the explanation of this energy barrier [12–14].

Obviously, before a colloidal particle can be trapped at the interface, which is a thermodynamically heavily favored situation, the particle has to

Interface Science and Technology, Vol. 26. https://doi.org/10.1016/B978-0-08-100865-2.00004-7

be driven there somehow. As colloidal particles are normally charged, electrophoresis is the most natural selection with various advantages such as easy to apply and control externally, and so forth [15, 16]. Understanding of electrostatic interaction between a charged particle and the air-water interface hence becomes the key to the success of applying electrophoresis in this particular system, as the energy barrier against the approaching particle motion is precisely of electrostatic nature. Corresponding studies in this regard are abundant [13, 17–20], which are highlighted by the works of Mbamala and von Grünberg [13], who used the Poisson-Boltzmann theory to calculate the ions distribution and concluded that the involvement of ions surrounding the particle does not change the basic observation in electrostatics that the presence of an air-water interface is equivalent to put an identical charge in the mirror image position. It only decreases the magnitude of this repulsive force as the particle is partially shielded by the ion cloud. This is briefly referred to as the "image-charge force analog," which is introduced in the previous chapter as well.

Electrophoretic motion of a charged particle normal to an air-water interface was explored experimentally by various groups based on different motivations [15, 16, 21–24]. Theoretical investigation, on the other hand, has been very scarce, perhaps due to the challenging task involved in its mathematical treatment and subsequent numerical solution. Lee's group in particular has studied the electrophoresis phenomena under this geometric configuration for a spherical rigid particle of either constant zeta potential or charge-regulation nature [25, 26]. Corresponding motion for a liquid droplet or a bubble as well as a porous particle was treated as well [27, 28], which will be addressed in chapters to come.

In conclusion, many colloidal operations use air-water interface as a platform for either two-dimensional self-assembly or crystallization, which often involves the usage of electrophoretic motion to drive the charged particle toward the vicinity of the interface first, where the overwhelming capillary attraction force will trap the particle there for intended further surface operations. As shown by the experimental study of surface adhesion [20], for instance, very often the rate-limiting step is the electrophoretic motion of the particle. Hence it will be the focus of this chapter.

4.2 THEORY

The system diagram is shown in Fig. 4.1. Corresponding mesh diagram in computation is the same as Fig. 3.2. The geometric configuration is the same as the rigid sphere normal to a solid plane treated in Chapter 3. Hence bispherical coordinates are used here as well. The governing equations and the boundary conditions at the particle surface are exactly the same

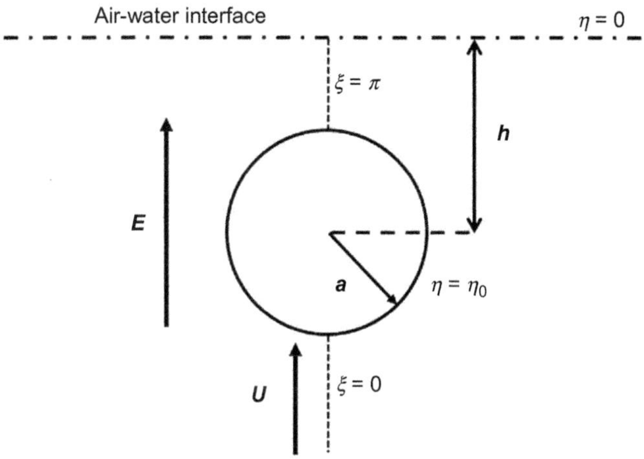

FIG. 4.1 System diagram.

as those in Chapter 3 too, both in vector form and in explicit bispherical coordinates, i.e., Eqs. (3.3)–(3.7) and Eqs. (3.10)–3.20. Hence no need to repeat them explicitly here.

The boundary conditions at infinity are the same as those in Chapter 3 as well, i.e., Eqs. (3.23), (3.26).

The only different situation is the boundary condition at the interphase ($\eta = 0$), where the fluid is allowed to move horizontally, but the normal velocity still remains zero, as obviously the water cannot penetrate into the air. Hence the corresponding hydrodynamically boundary conditions here are:

$$v_\eta = 0 \tag{4.1}$$

and

$$T_{r\theta} = 0 \tag{4.2}$$

where $T_{r\theta}$ is the total shear stress with contributions from both the electric field, the so-called Maxwell stress tensor, and the hydrodynamic flow field [29].

The null total shear stress, $T_{r\theta}$, is based on the fact that the air viscosity is much smaller than the water viscosity ($\eta_{f,\text{air}} : \eta_{f,\text{water}} = 1 : 48$ at 25°C [30]), implying a null total shear stress on the air side, and continuity of shear stress across the interface is further assumed in the water side. Thus, we have the boundary conditions in bispherical coordinates at $\eta = 0$ as follows:

$$\frac{\partial \Psi^*}{\partial \xi} = 0 \tag{4.3}$$

and

$$\left\{\varepsilon\left(\boldsymbol{EE}-\frac{1}{2}|\boldsymbol{E}|^2\mathbf{I}\right)+\eta_f\left[\nabla\boldsymbol{U}+(\nabla\boldsymbol{U})^T\right]\right\}\cdot\mathbf{i}_\eta\cdot\mathbf{i}_\xi=0 \tag{4.4}$$

where $\boldsymbol{E} = -\nabla\phi$ and $\mathbf{i}_\eta$, $\mathbf{i}_\xi$ are the unit vectors along the η and ξ direction, respectively. Note that the first term in Eq. (4.4) stands for the Maxwell stress tensor, where $|\boldsymbol{E}|$ is the magnitude of the electric field and $\mathbf{I}$ is the identity tensor.

Or equivalently:

$$\frac{\partial\delta\phi^*}{\partial\eta}\frac{\partial\phi_e^*}{\partial\xi}-\frac{x}{c^*\sin\xi}\frac{\partial^2\Psi^*}{\partial\eta^2}=0 \tag{4.5}$$

The air phase is treated here as a nonconducting medium as the dielectric constant of air is much smaller than that of water ($\varepsilon_{air} : \varepsilon_{water} = 1 : 78$ 25°C [31]), hence $\varepsilon_{air} \approx 0$. Moreover, the interface is considered as chargeless due to the repulsive "image-charge force analog" in electrostatics considered in Chapter 3, which is also confirmed by Ohshima's analysis on the surface tension of electrolyte solutions [32]. As a result, corresponding electric physical boundary conditions at $\eta = 0$ are:

$$\frac{\partial\phi_e^*}{\partial\eta}=0 \tag{4.6}$$

and

$$\frac{\partial\delta\phi^*}{\partial\eta}=-E^* \tag{4.7}$$

where E is the magnitude of the normal electric field applied across the interface.

Moreover,

$$g_j^*+\delta\phi^*=0 \tag{4.8}$$

or

$$g_j^*=-\delta\phi^* \tag{4.9}$$

at the interface ($\eta = 0$) as well, assuming ϕ_e^* remains unchanged there in electrophoresis.

4.3 RESULTS AND DISCUSSION

4.3.1 Repulsive Image-Charge Force Analog: Electrostatic Aspect

We first show the electrostatic results when the particle is motionless or at rest. Fig. 4.2A–C depicts the ions distribution in the double layer for highly charged situation at various separation distances from the interface.

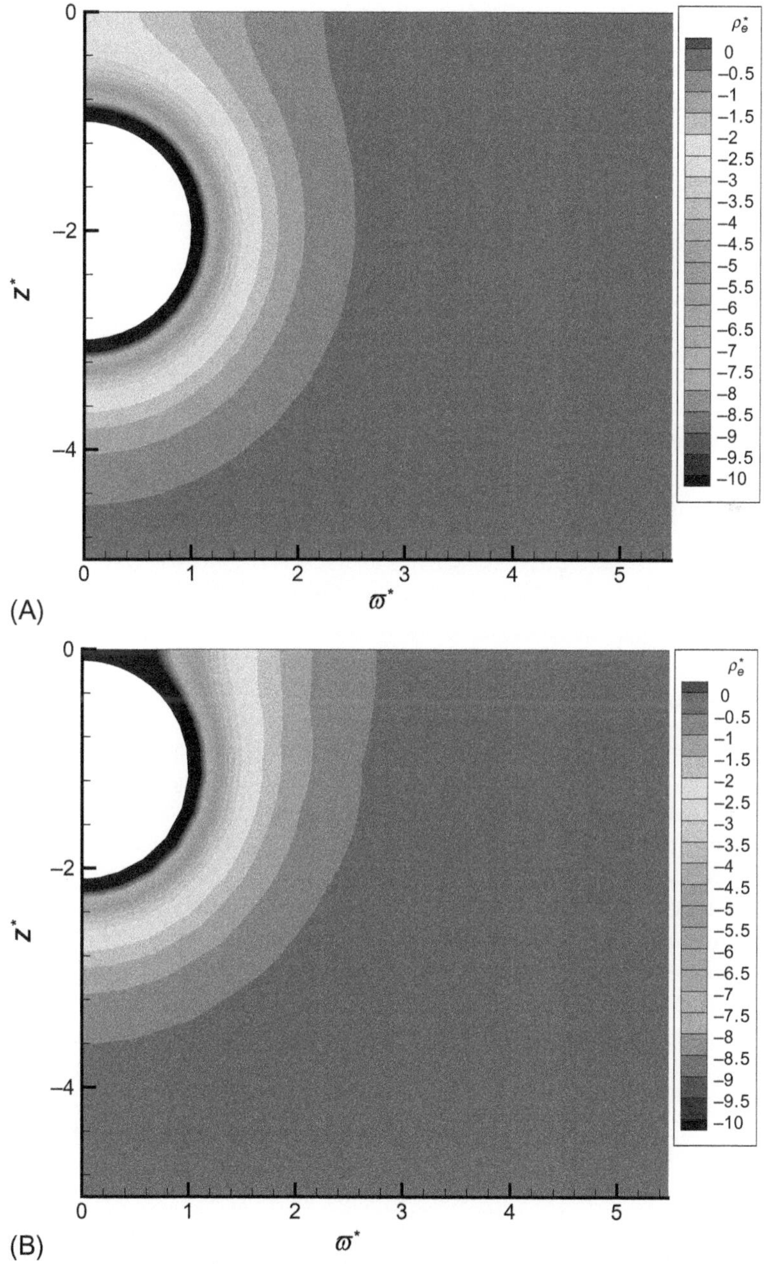

FIG. 4.2 Contours plots of the scaled equilibrium charge density (ρ_e^*) for $\kappa a = 1$. (A) $\phi_r = 3$, $h = 2$; (B) $\phi_r = 3$, $h = 1.1$; (C) $\phi_r = 5$, $h = 1.1$.

Continued

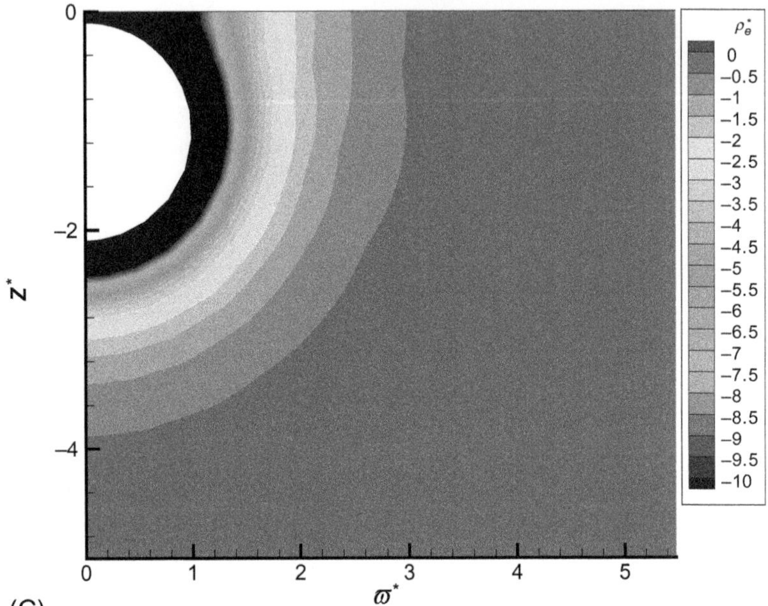

(C)

FIG. 4.2—CONT'D

Unlike the corresponding solid plane, the double layers assume a bell-like shape now once it touches the interface, as shown in Fig. 4.2A ($h = 2$) with pretty wide opening along the interface. This bell-like shape is similar except with an even wider opening when the particle gets even closer to the interface, as shown in Fig. 4.2B ($h = 1.1$). Particle even more highly charged (ϕ_r=5) assumes even wider opening, as shown in Fig. 4.2C. Note that this wide opening or bell shape has nothing to do with the capability of horizontal flow at the interface, since the particle is still motionless. Hence the underlying mechanism is of electrostatic nature.

Corresponding equilibrium electric potential distributions are shown in Fig. 4.3A–C with similar observation. The strong local electric field induced by the image-charge force analog in the gap region when the particle gets close to the interface attracts counterions in the double layer toward it, leading to a redistribution in favor of the front region. The abundant counterions there repulse one another hence spill over to the horizontal neighborhood, leading to a bell-like shape double layer. Hence the corresponding electric potential assumes similar shape due to the governing Poisson equation.

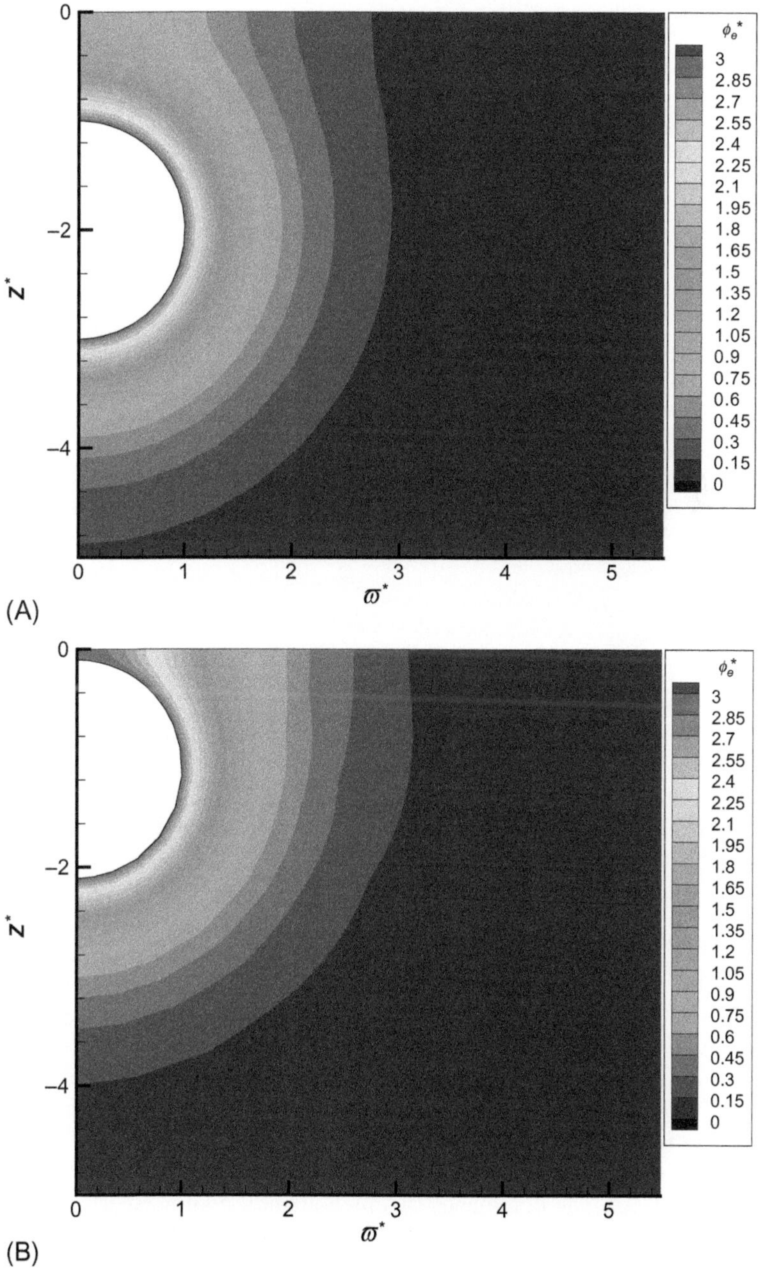

FIG. 4.3 Contours plot of the scaled equilibrium electric potential (ϕ_e^*) for $\kappa a = 1$. (A) $\phi_r = 3$, $h = 2$; (B) $\phi_r = 3$, $h = 1.1$; (C) $\phi_r = 5$, $h = 1.1$.

Continued

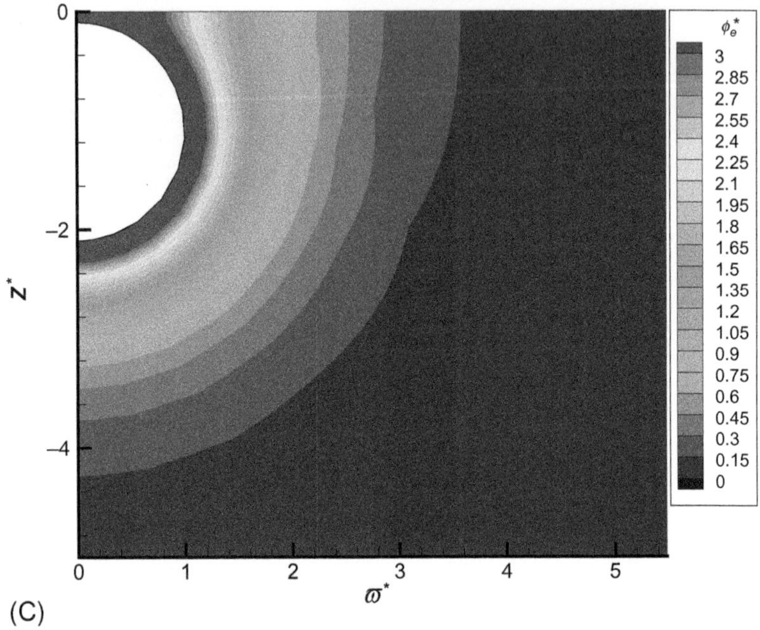

FIG. 4.3—CONT'D

Corresponding one-dimensional profiles of the equilibrium electric potential along the air-water interface are shown in Fig. 4.4A and B. The closer the particle approaches the interface, the stronger this repulsive electrostatic image-charge force ahead is, leading to a higher repulsive electric potential right ahead of the particle as shown.

4.3.2 General Henry's Charts at Various Separation Distance

The general Henry's charts of particle mobility are shown in Fig. 4.5A–C for $\phi_r = 1, 3$, and 5 respectively at various separation distance. The general behavior is quite similar to the corresponding solid plane situation, qualitatively speaking. The mobility reduction is most severe when the double layer is thick as the boundary confinement effect would be most drastic. Moreover, the local minima disappear completely when the particle is very close to the interface ($h = 1.1$), reveals that the double layer must be severely deformed by this confinement effect. The dominance of this boundary confinement effect is also shown in Fig. 4.6 for clarity.

Moreover, as the electrolyte concentration gets higher, or larger κa, the approaching speed of the particle toward the interface increases accordingly.

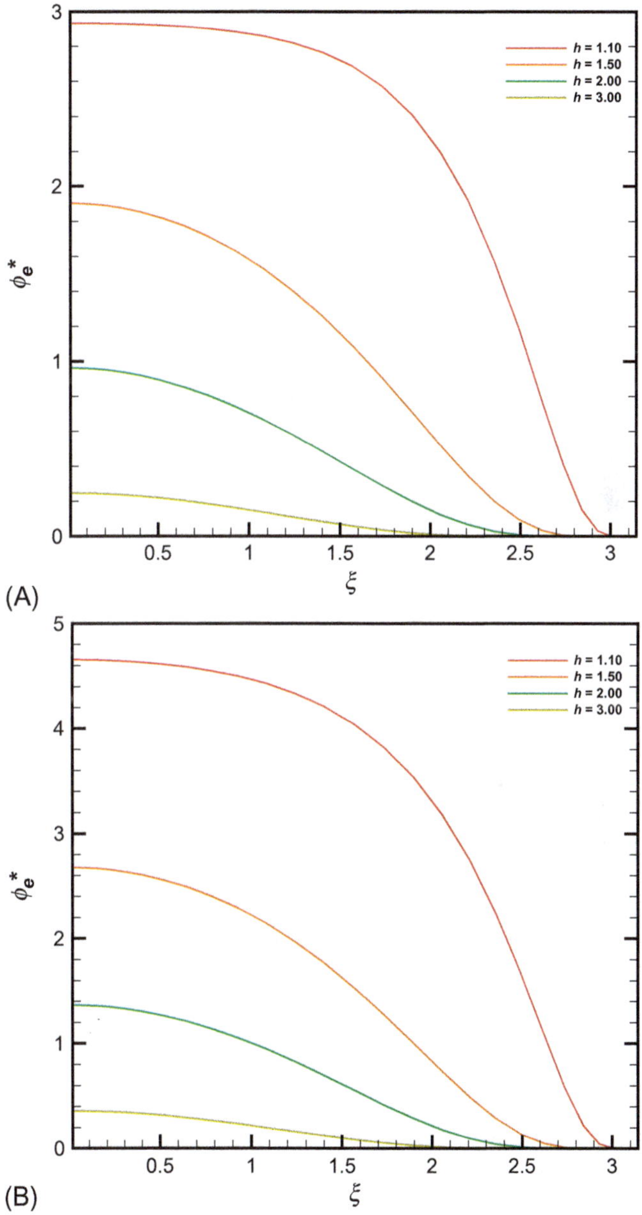

FIG. 4.4 Distribution of the scaled equilibrium electric potential (ϕ_e^*) along the air-water interface at different h for $\kappa a = 1$. (A) $\phi_r = 3$; (B) $\phi_r = 5$.

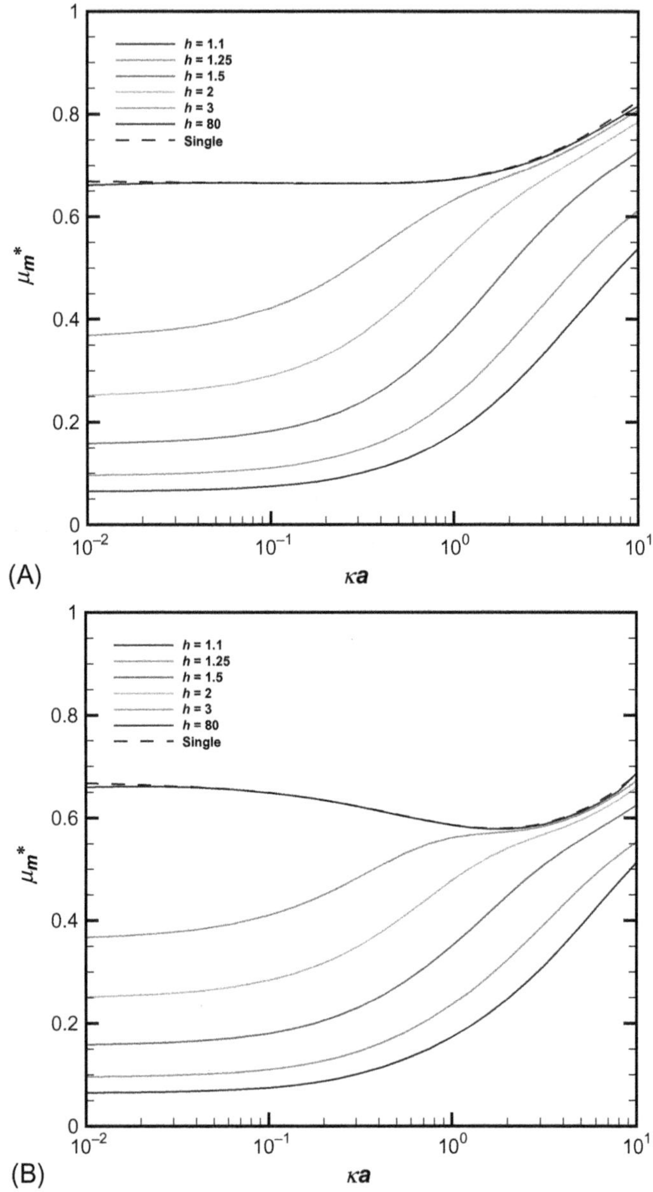

FIG. 4.5 Scaled mobility $\left(\mu_m^* = \frac{\mu^*}{\phi_r}\right)$ as a function of κa at different h. (A) $\phi_r = 1$; (B) $\phi_r = 3$; (C) $\phi_r = 5$.

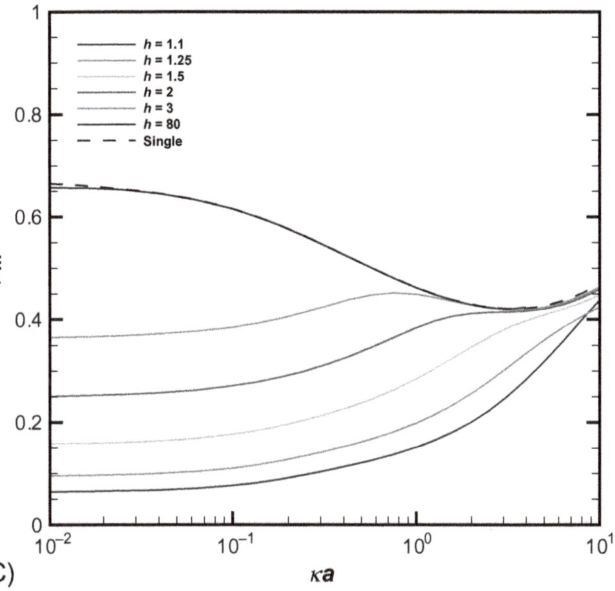

(C)

FIG. 4.5—CONT'D

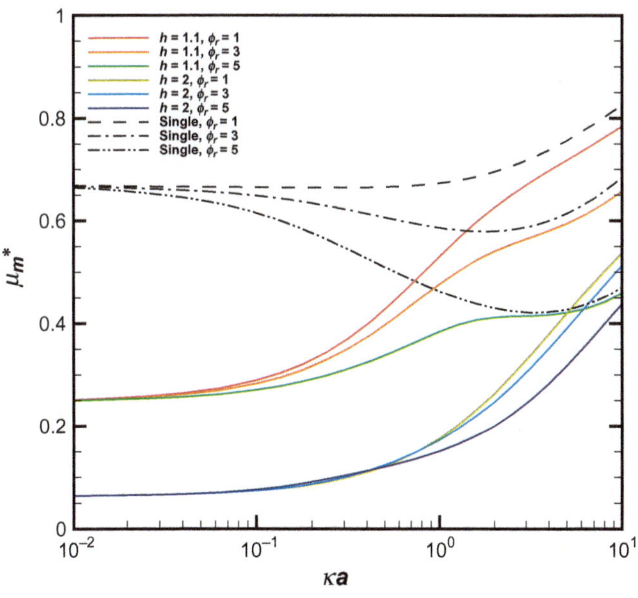

FIG. 4.6 Scaled mobility (μ_m^*) as a function of κa at different ϕ_r for $h = 2$, 1.1 and in comparison with the single sphere situation.

This implies that a higher electrolyte concentration by adding salt, say, may improve the intended operation at the interface, at least in terms of production rate, as the electrophoretic particle motion is usually the rate-limiting step of the entire process [33].

4.3.3 Geometric Boundary Confinement Effect

Representative plots of double layer polarization are shown in Fig. 4.7A and B. Boundary confinement in the gap region nearly wipeout the double layer polarization there completely when the particle gets very close to the interface, as shown in Fig. 4.7B with $h = 1.1$, or one-tenth of the particle radius.

4.3.4 Vortex Flow Induced

Corresponding plots of flow field are shown in Fig. 4.8A–C. In each case, a separate primary vortex flow near the particle with clockwise orientation is observed, which induces an outer secondary counterclockwise vortex flow, similar to the solid plane situation observed in Chapter 3.

4.3.5 Charge-Regulation Phenomenon

Variation of the surface potential for a charge-regulated particle is shown in Fig. 4.9. The variation is due to the drastic deformation of the electrolyte environment around the particle by the boundary confinement effect of the interface. Again the behavior is qualitatively similar to the solid plane situation.

4.3.6 Correction Factor at Various Separation Distances

To facilitate the possible use of interested readers, Tables 4.1–4.3 provide the correction factors in terms of mobility reduction at various separation distance for $\phi_r = 1$, 3, and 5, respectively, at various electrolyte strength.

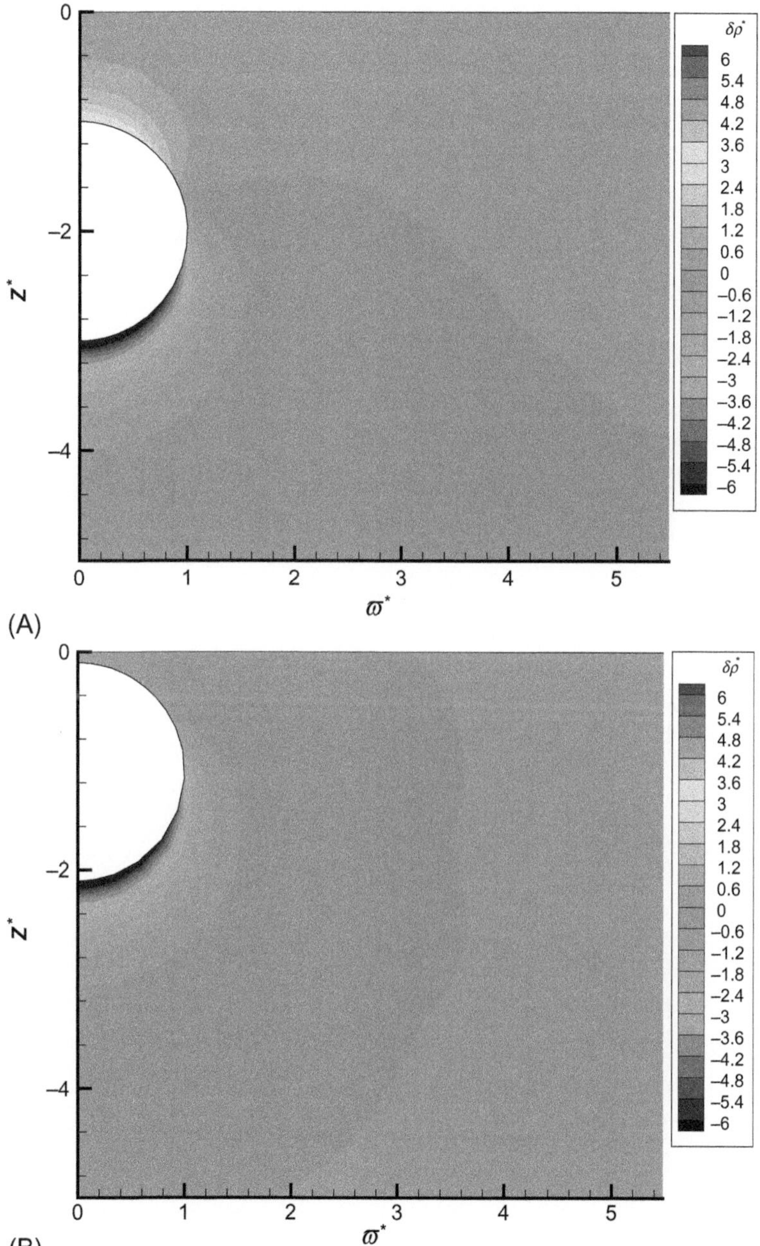

FIG. 4.7 Contour plots of the scaled disturbed charge density ($\delta\rho^*$) for $\kappa a = 1$ and $\phi_r = 3$. (A) $h = 2$; (B) $h = 1.1$.

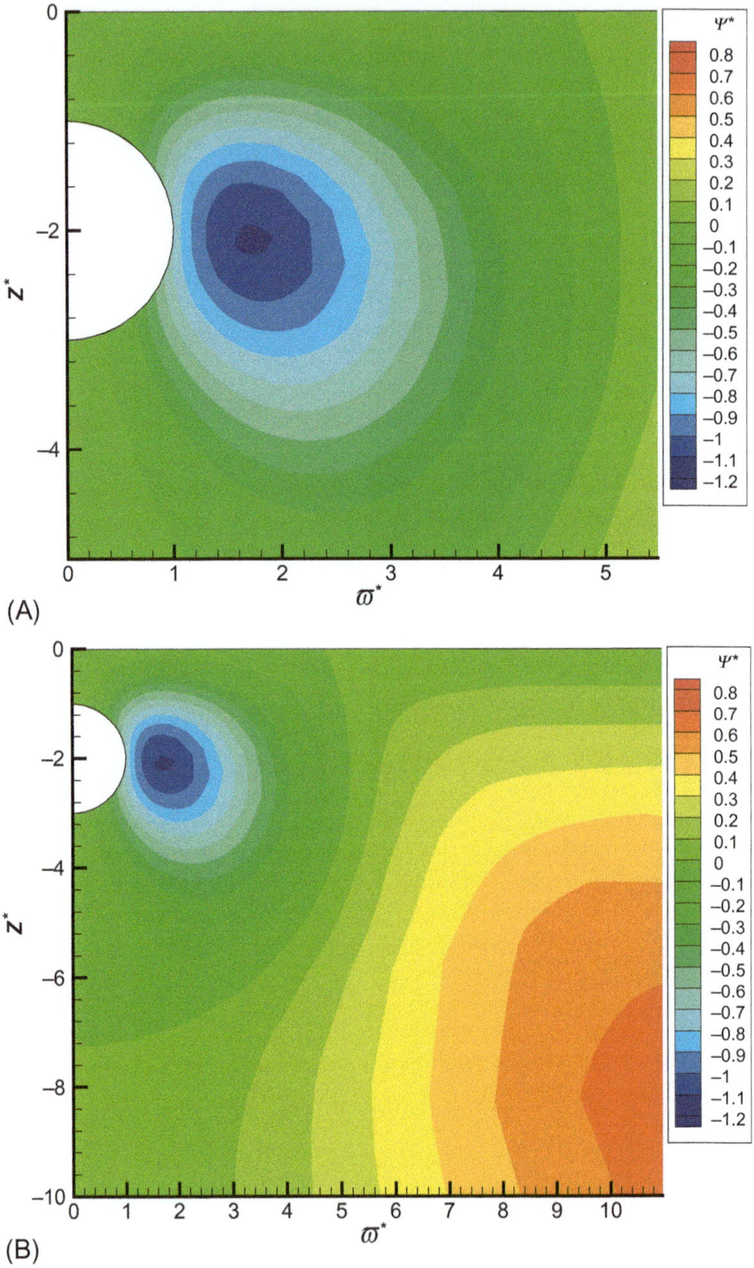

FIG. 4.8 Contour plots of the scaled stream function (Ψ^*) for $\kappa a = 1$ and $\phi_r = 3$. (A) $h = 2$, small scale (close-up view); (B) $h = 2$, large scale; (C) $h = 1.1$.

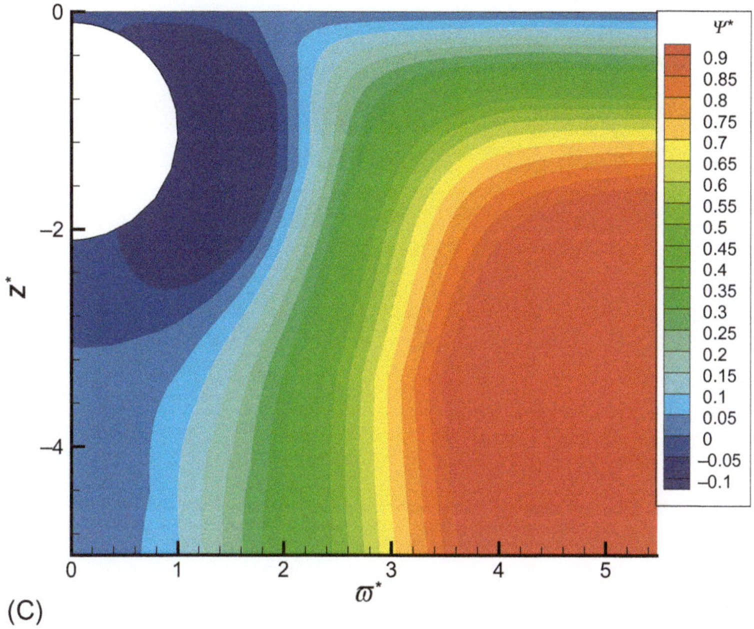

(C)

FIG. 4.8—CONT'D

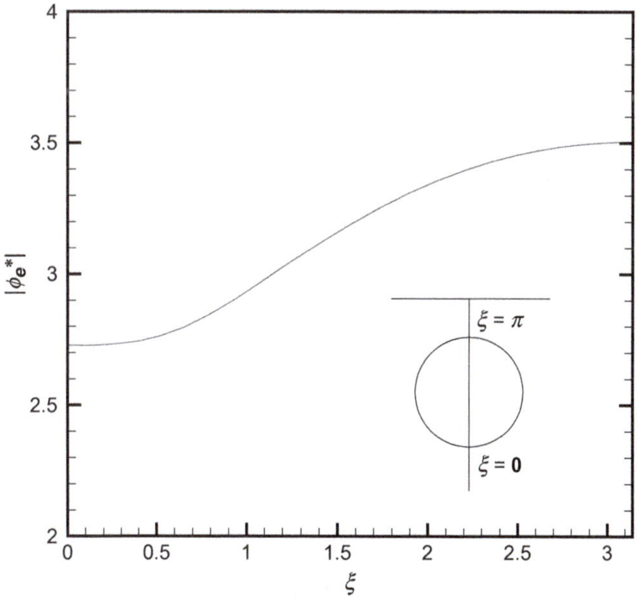

FIG. 4.9 Profile of the scaled equilibrium surface electric potential in magnitude $|\phi_e^*|$ for $\kappa a = 1$ and $h = 1.1$ with $A = 100$, $B = 1$ in the charge-regulation model.

TABLES OF BOUNDARY EFFECT EXPRESSED AS A CORRECTION FACTOR Γ

$\Gamma = \mu^*_h / \mu^*_\infty$, where μ^*_h refers to the particle mobility at a separation distance h, whereas refers to the mobility in an infinite medium.

TABLE 4.1 Correction Factor Γ for $\phi_r = 1$

	κa							
h	*0.01*	*0.1*	*0.5*	*1*	*2*	*5*	*7*	*10*
1.1	0.097	0.112	0.184	0.263	0.380	0.556	0.609	0.652
1.2	0.121	0.139	0.225	0.316	0.443	0.609	0.650	0.677
1.4	0.202	0.234	0.368	0.498	0.657	0.809	0.833	0.844
1.6	0.270	0.311	0.479	0.627	0.781	0.891	0.901	0.905
1.8	0.327	0.378	0.570	0.721	0.856	0.930	0.934	0.934
2.0	0.378	0.436	0.644	0.790	0.903	0.951	0.952	0.951
4.0	0.652	0.746	0.932	0.979	0.991	0.991	0.989	0.985
6.0	0.762	0.862	0.981	0.995	0.996	0.994	0.992	0.988
8.0	0.822	0.917	0.993	0.998	0.998	0.995	0.993	0.989
10.0	0.858	0.947	0.997	0.998	0.998	0.996	0.993	0.989
20.0	0.934	0.991	0.999	0.999	0.999	0.996	0.993	0.989
40.0	0.972	0.999	1.000	0.999	0.999	0.996	0.993	0.989
60.0	0.984	1.000	1.000	0.999	0.999	0.996	0.993	0.989
80.0	0.989	1.000	1.000	0.999	0.998	0.996	0.992	0.988
∞	1.000	1.000	1.000	1.000	1.000	1.000	1.000	1.000

TABLE 4.2 Correction Factor Γ for $\phi_r = 3$

	κa							
h	*0.01*	*0.1*	*0.5*	*1*	*2*	*5*	*7*	*10*
1.1	0.097	0.115	0.205	0.295	0.427	0.636	0.701	0.750
1.2	0.121	0.142	0.246	0.348	0.489	0.684	0.729	0.752
1.4	0.203	0.237	0.389	0.530	0.705	0.872	0.888	0.888
1.6	0.270	0.314	0.498	0.656	0.826	0.933	0.935	0.931
1.8	0.327	0.380	0.585	0.748	0.895	0.958	0.956	0.952
2.0	0.377	0.437	0.656	0.815	0.935	0.970	0.968	0.963
4.0	0.651	0.743	0.943	0.988	0.995	0.993	0.990	0.986
6.0	0.761	0.860	0.988	0.998	0.998	0.995	0.992	0.988
8.0	0.820	0.917	0.997	0.999	0.998	0.996	0.993	0.989
10.0	0.857	0.948	0.999	0.999	0.998	0.996	0.993	0.989
20.0	0.933	0.993	1.000	0.999	0.999	0.996	0.993	0.989
40.0	0.971	0.999	1.000	0.999	0.999	0.996	0.993	0.989
60.0	0.983	1.000	1.000	0.999	0.999	0.996	0.993	0.992
80.0	0.989	1.000	1.000	0.999	0.999	0.995	0.992	–
∞	1.000	1.000	1.000	1.000	1.000	1.000	1.000	1.000

Notes: "–" refers to the fact that Γ approaches unity.

TABLE 4.3 Correction Factor Γ for $\phi_r = 5$

	κa							
h	*0.01*	*0.1*	*0.5*	*1*	2	*5*	*7*	*10*
1.1	0.097	0.124	0.237	0.327	0.467	0.745	0.849	0.931
1.2	0.121	0.152	0.275	0.375	0.527	0.776	0.838	0.863
1.4	0.203	0.248	0.413	0.551	0.740	0.935	0.948	0.941
1.6	0.270	0.323	0.513	0.672	0.858	0.970	0.968	0.960
1.8	0.327	0.386	0.594	0.762	0.923	0.981	0.976	0.970
2.0	0.377	0.440	0.662	0.830	0.958	0.986	0.981	0.975
4.0	0.649	0.734	0.954	0.996	0.998	0.995	0.992	0.988
6.0	0.758	0.854	0.995	1.000	0.999	0.996	0.993	0.993
8.0	0.817	0.915	1.001	1.000	0.999	0.996	0.993	–
10.0	0.853	0.949	1.001	1.000	0.999	0.996	0.994	–
20.0	0.929	0.996	1.000	0.999	0.999	0.996	–	–
40.0	0.969	1.000	1.000	0.999	0.999	–	–	–
60.0	0.982	1.000	1.000	0.999	0.999	–	–	–
80.0	0.988	1.000	1.000	0.999	0.999	–	–	–
∞	1.000	1.000	1.000	1.000	1.000	1.000	1.000	1.000

Notes: "–" refers to the fact that Γ approaches unity.

REFERENCES

[1] Painter O, Lee R, Scherer A, Yariv A, O'brien J, Dapkus P, et al. Two-dimensional photonic band-gap defect mode laser. Science 1999;284:1819–21.

[2] Jiang P, Bertone J, Hwang K, Colvin V. Single-crystal colloidal multilayers of controlled thickness. Chem Mater 1999;11:2132–40.

[3] Vlasov YA, Bo XZ, Sturm JC, Norris DJ. On-chip natural assembly of silicon photonic bandgap crystals. Nature 2001;414:289–93.

[4] Zalineeva A, Baranton SV, Coutanceau C, Jerkiewicz G. Electrochemical behavior of unsupported shaped palladium nanoparticles. Langmuir 2014;31:1605–9.

[5] Kralchevsky PA, Denkov ND. Capillary forces and structuring in layers of colloid particles. Curr Opin Colloid Interface Sci 2001;6:383–401.

[6] Bresme F, Oettel M. Nanoparticles at fluid interfaces. J Phys Condens Matter 2007;19:413101.

[7] Nicolson M. The interaction between floating particles. In: Mathematical Proceedings of the Cambridge Philosophical Society, Cambridge: Cambridge University Press; 1949. p. 288–95.

[8] Hinsch K. Holographic study of liquid surface deformations produced by floating particles. J Colloid Interface Sci 1983;92:243–55.

[9] Stamou D, Duschl C, Johannsmann D. Long-range attraction between colloidal spheres at the air-water interface: the consequence of an irregular meniscus. Phys Rev E 2000;62:5263.

[10] Kralchevsky PA, Nagayama K. Capillary interactions between particles bound to interfaces, liquid films and biomembranes. Adv Colloid Interface Sci 2000;85:145–92.

[11] Oettel M, Dietrich S. Colloidal interactions at fluid interfaces. Langmuir 2008;24:1425–41.

[12] Mbamala E, von Grünberg H. Charged colloids and proteins at an air-water interface: the effect of dielectric substrates on interaction and phase behavior. Phys Rev E 2003; 67:031608.

[13] Mbamala E, Von Grünberg H. Effective interaction of a charged colloidal particle with an air-water interface. J Phys: Condens Matter 2002;14:4881.

[14] Findenegg G. JN Israelachvili: Intermolecular and surface forces (with applications to colloidal and biological systems). Ber Bunsenges Phys Chem 1986;90:1241–2.

[15] Sugiyama Y, Inoue T, Ikematsu M, Iseki M, Sekiguchi T. Controlling the orientation of purple membrane fragments on an air/water interface by a new method of direct electric field application during purple membrane spreading. Jpn J Appl Phys 1997;36:5674.

[16] Arisawa S, Yamamoto R. Control of molecular orientations by a pulsating electric field applied to a monolayer at the air-water interface. Langmuir 1993;9:1028–30.

[17] Pieranski P. Two-dimensional interfacial colloidal crystals. Phys Rev Lett 1980;45:569.

[18] Stillinger Jr FH. Interfacial solutions of the Poisson-Boltzmann equation. J Chem Phys 1961;35:1584–9.

[19] Hurd AJ. The electrostatic interaction between interfacial colloidal particles. J Phys A: Math Gen 1985;18:L1055.

[20] Carl A, Bannuscher A, von Klitzing R. Particle stabilized aqueous foams at different length scales: synergy between silica particles and alkylamines. Langmuir 2015;31:1615–22.

[21] Gao Y, Li D. Translational motion of a spherical particle near a planar liquid-fluid interface. J Colloid Interface Sci 2008;319:344–52.

[22] Usui S, Healy T. Zeta potential of insoluble monolayer of long-chain alcohol at the air-aqueous solution interface. J Colloid Interface Sci 2001;240:127–32.

[23] Huddleston R, Smith L, Akers R. Foams. London: Academic Press; 1976.

[24] Hayes C. Water-air interface in the presence of an applied electric field. J Phys Chem 1975;79:1689–93.
[25] Tsai P, Fang H, Lee E. Electrophoresis of a charge-regulated sphere normal to an air-water interface. J Phys Chem B 2011;115:6484–94.
[26] Tsai P, Lou J, He YY, Lee E. Electrophoresis of a spherical particle normal to an air-water interface. Electrophoresis 2010;31:3363–71.
[27] Hsu H-P, Lee E. Electrophoresis of a charged porous sphere normal to an air-water interface. Phys Chem Chem Phys 2012;14:15729–38.
[28] Liao C, Lee Y-F, Tsai S-C, Fang W, Lee E. Electrophoretic motion of a liquid droplet and a bubble normal to an air-water interface. Colloids Surf A: Physicochem Eng Asp 2016;507:124–33.
[29] Bird RB. Transport phenomena. Appl Mech Rev 2002;55:R1–4.
[30] Green DW, Perry RH. Perry's chemical engineers' handbook; 1973.
[31] Speight JG. Lange's handbook of chemistry. New York: McGraw-Hill; 2005.
[32] Ohshima H, Matsubara H. Surface tension of electrolyte solutions. Colloid Polym Sci 2004;282:1044–8.
[33] MacRitchie F, Alexander A. Kinetics of adsorption of proteins at interfaces. Part III. The role of electrical barriers in adsorption. J Colloid Sci 1963;18:464–9.

Chapter 5

Electrophoretic Motion of a Rigid Spherical Particle in a Cylindrical Pore

5.1 INTRODUCTION

Traditionally electrophoresis has often been conducted in a slit between two planar plates, such as in the separation of proteins in a slab packed either with or without gel-like porous polymeric materials [1]. Such arrangements, however, have severe problems regarding the control of heat generation from the Joule heating effect [2]. The passing of electric current through the suspending medium during electrophoretic operations results in a heat dissipation proportional to the square of the electric current [2]. Without an efficient way to transfer the heat to the surrounding environment, the building-up of heat inside the medium will generate a buoyancy-driven vortex flow perpendicular to the particle moving direction, which causes hydrodynamic mixing in micro-scale hence blurs the resolution of the experimental outcome. The moving front of the colloidal suspension in early days is severely damaged in this way, for instance. Moreover, the separation performance along the direction of the electric field is destroyed as well, if the heat transfer is really bad. As a result, the heat transfer to the surrounding must be both effective and easy to apply in all the practical applications of electrophoresis.

Based mainly on this major concern, the capillary electrophoresis (CE), where electrophoresis is conducted within a very narrow capillary, be it a cylindrical pore or a rectangular duct in geometry, has become the dominant, if not exclusive, device designed for essentially all the practical applications of electrophoresis since 1980s when it was first introduced for either the mobility measurement or the separation of charged particles [3–7]. Indeed CE has now become essentially the synonym of electrophoresis, at least in lab analysis operations. The narrow cross-section in a capillary makes the dissipation of heat very easy to handle [8]. For instance, simply putting it in cooling water, moving or not, will do [9]. The sample can thus be maintained in

Interface Science and Technology, Vol. 26. https://doi.org/10.1016/B978-0-08-100865-2.00005-9

constant temperature and the thermally driven vortex flow from Joule heating effect can be suppressed completely, a merit unmatchable by the traditional slab arrangement.

Moreover, coupled with the fast advancement of microfluidic and nanofluidic technologies and their applications, CE has found enormous potential applications in various fields of interest [10–14]. In environmental protection, for instance, CE has the merits such as very small consumption of sample quantities, very fast operational time, suitability for in situ operations, and so forth [8]. Indeed CE has become the bridge connecting what can be done in lab and what has to be handled and reckoned with in the real world. In fact, CE has been integrated to sample preparing section prior to it and quantitative analysis of compositions after it to form a so-called μTAS, micron-total-analysis-system device in the past two decades [15]. This has later developed to an even more complicated system called lab-on-a-chip (LOC) by integrating with it other electrokinetically driven subdevices, such as micro-pumps and pressure-injection actuators, for instance, to enhance the overall speed and efficiency in practical applications [10,11]. With the quantum leap in miniaturization technology, the CE-based LOC has become a portable and powerful analyzing device which finds wide applications in biomedical and biochemical fields in particular, such as the in situ analysis of proteins and DNAs in medical clinics [12], or other applications that require fast analysis of the composition of a small amount of target sample, such as a contaminated or polluted substance [13]. Therefore, the understanding of the electrophoretic motion of a particle in a channel becomes crucial in the design and engineering of modern LOCs involving CE [14].

Here we focus on the cylindrical pore for simplicity, as it can be described with only one parameter, the diameter of the pore. The machinery of a rectangular duct may be cheaper in cost, but the impact of the wall is difficult to determine as the aspect ratio is a varying parameter as well and vortex flow in the corner region may happen leading to an overall flow field too complicated for a meaningful theoretical investigation [16].

Capillary diameter used in CE was very large in the past, ranging from 25 μm to 200 μm [17]. The technology in the fabrication of capillary has, however, developed very fast in the last decade or so, both in precision and durability. The commercial availability of capillary diameter has been downsized to 2 μm or even 0.2 μm [18,19], which means the boundary effect would be very profound for a particle of 1 μm of diameter in a pore of 2 μm, and the particle would simply be blocked from entering the smaller one, 0.2 μm, whereas a small or particle of radius 100 nm, or 0.1 μm, still yields a ratio of radii of 0.5. The boundary effect is no longer negligible as might be expected from the traditional duct in CE with a width about 150 μm [20]. As a matter of fact, in the research field of inorganic membrane, pore sizes around 20 nm can be obtained easily nowadays [18], and machinery fabrication of duct size of several nanometers has been reported since nearly a decade ago [21]. As a result, the study of boundary effect within a confined device becomes essential in the future advancement of LOC involving CE. Here we focus on the electrophoretic motion of a rigid colloidal particle along the centerline of a cylindrical pore with comparable radial dimension to lay the foundation of general theoretical analysis.

The confinement effect of a cylindrical pore is more significant than a planar boundary in general, at least from a geometric point of view. The presence of the pore wall affects the particle motion both hydrodynamically and electrostatically, similar to the situations encountered previously in Chapters 3 and 4, except that the particle motion is now along the centerline of the pore instead of normal to it. For the corresponding purely hydrodynamic problem, where the creeping flow around a spherical particle moving along the centerline of a cylindrical pore, exact analytical results are available and the retardation effect upon the particle motion can be expressed as a correction factor to the Stokes law at various separation distance to the pore wall [22–24].

A charged pore wall, meanwhile, generates an electroosmosis flow (EOF) which may either enhance or retard the particle motion, depending on the sign of charges it carries. This drastically elevates the electrostatic impact to a new level.

Moreover, the electric double layer surrounding the particle is squeezed once it touches the pore wall. This boundary confinement effect, which mainly takes place in the lateral direction, adds yet another factor of the ultimate particle motion, in addition to the hydrodynamic retardation effect and electrostatic effect of a channel wall.

Many theoretical studies have been reported regarding the electrophoretic motion of a rigid spherical particle in a cylindrical pore as considered here, under various levels of simplifications of either the particle surface potentials or its double layer thickness [25–28]. Lee's group further investigated other related problems as well, such as the motion of a liquid droplet [25], a porous or a soft sphere [26,27], and a rigid sphere in a polymeric Debye-Bueche-Brinkman (DBB) fluid [28], taking full account of the double layer polarization effect with no limitation of the particle charge level or double layer thickness. Some of them will be discussed in the chapters to come. The problem formulation and subsequent mathematical and numerical treatment is presented in Section 5.2.

5.2 THEORY

The system under consideration is shown in Fig. 5.1, where a spherical colloidal particle of radius a moves along the axis of a cylindrical pore of radius R_b with velocity $\mathbf{U}$ subject to an applied electric field $\mathbf{E}$ in the Z-direction. The cylindrical pore is filled with the KCl binary electrolyte solution as treated in the previous chapters with the same definition of symbols which are not shown here for brevity.

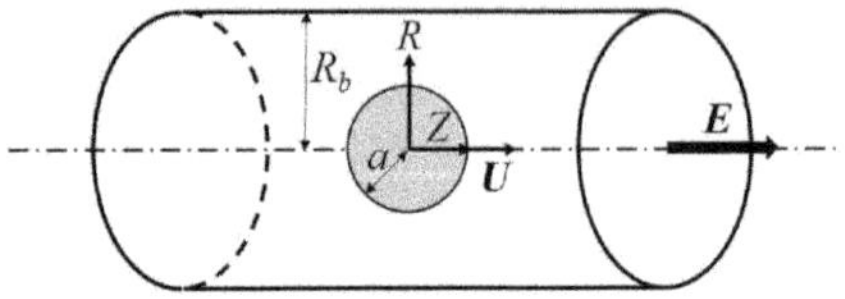

FIG. 5.1 System diagram.

Note that the governing equations and the corresponding boundary conditions at the particle surface are exactly the same as those encountered in Chapter 1, both in vector form or in spherical coordinates, which is the one chosen here. The boundary conditions at the pore surface, plus the upstream and the downstream far away from the particle, however, needs to be further examined. Hence Eqs. (1.29)–(1.33) and Eqs. (1.34)–(1.38) in Chapter 1 are directly adopted here.

For a dielectric solid cylindrical pore with constant surface potential, we have, in dimensionless form,

$$\phi_e^* = \phi_w^*, \quad R^* = \frac{R_b}{a} \tag{5.1}$$

$$\frac{\partial \delta\phi^*}{\partial R^*} = 0, \quad R^* = \frac{R_b}{a} \tag{5.2}$$

$$\frac{\partial g_j^*}{\partial R^*} = 0, \quad R^* = \frac{R_b}{a} \tag{5.3}$$

where the last two conditions imply the nonconducting and nonpenetrable nature of the pore wall.

The usual no-slip hydrodynamic boundary condition on a solid surface yields:

$$\Psi^* = 0, \quad R^* = \frac{R_b}{a} \tag{5.4}$$

$$\frac{\partial \Psi^*}{\partial R^*} = 0, \quad R^* = \frac{R_b}{a} \tag{5.5}$$

Far away from the particle, either downstream or upstream, the flow field and the corresponding electric field approach the situation in the absence of the particle. One-dimensional fully developed EOF and corresponding electric potential profile, $\phi_{e,\infty}^*(R^*)$, can thus be obtained numerically for a charged pore wall assuming the ion distribution follows the Poisson-Boltzmann distribution [28]. In other words,

$$\Psi^* = \frac{1}{2}U^* R^{*2} + \Psi_\infty^*(R^*), \quad Z^* = L^* \tag{5.6}$$

$$\phi_e^* = \phi_{e,\infty}^*(R^*), \quad Z^* = L^* \tag{5.7}$$

where $L^*\left(=\frac{L}{a}\right)$ stands for the scaled distance large enough so that the particle mobility does not change with increasing L^* any more. In calculation practice, $L^* = 10$ satisfies this convergence requirement and is used here. Moreover,

$$\frac{\partial \delta \phi^*}{\partial Z^*} = -E^*, \quad Z^* = L^* \tag{5.8}$$

where E^* stands for the magnitude of the scaled applied external electric field along the axis of the cylindrical pore. And

$$\frac{\partial \Psi^*}{\partial Z^*} = 0, \quad Z^* = L^* \tag{5.9}$$

reflects the fully developed nature of the flow field there. Furthermore,

$$g_j^* = -\delta\phi^* \quad \text{or} \quad g_j^* + \delta\phi^* = 0, \quad Z^* = L^* \tag{5.10}$$

implies that the electric field is not disturbed far away from the particle. Note that for a charge less wall, ϕ_e^* is zero and the flow field corresponds to a parabolic profile with approaching velocity $\boldsymbol{U}$ at the centerline.

The major challenge here, however, is that we have two sets of natural coordinates to fully describe the entire physical system considered here: The spherical coordinates (r, θ) for the particle as well as the electrolyte solution it is immersed; and the cylindrical coordinates (R, Z) suited for the pore surface and the upstream and the downstream boundaries as shown above.

In practice, the spherical coordinates are chosen to be the working horse here in computation, with some further geometric mapping as discussed later. The expression in cylindrical coordinates of boundary conditions shown above, on the other hand, are converted to the corresponding ones in spherical coordinates via the following conversion formula (in dimensionless form) between the two coordinates:

$$\frac{\partial}{\partial R^*} = \sin\theta \frac{\partial}{\partial r^*} + \frac{\cos\theta}{r^*}\frac{\partial}{\partial \theta} \tag{5.11}$$

$$\frac{\partial}{\partial Z^*} = \cos\theta \frac{\partial}{\partial r^*} - \frac{\sin\theta}{r^*}\frac{\partial}{\partial \theta} \tag{5.12}$$

Another challenge is the irregular physical domain under consideration, which is neither spherical nor cylindrical. The axisymmetric nature of the problem reduces it to a two-dimensional half domain, as shown in Fig. 5.2:

The pseudo-spectral method cannot be applied directly to this region though as the method requires an orthogonal geometric domain to lay out collocation points in each direction for further mathematical treatment and simplification [29]. Moreover, it is difficult to express in explicit form the precise locations of a cylindrical pore and the upstream or downstream planes in corresponding spherical coordinates.

FIG. 5.2 Two-dimensional physical domain.

To reconcile with this irregular domain problem as well as the expression of the boundary location, the two-dimensional physical domain in Fig. 5.2 is further decomposed into three separate subregions as shown in Fig. 5.3:

The mesh is further transformed to the corresponding rectangular region (x, y) as shown in Fig. 5.4 via a conformal mapping. The detail can be found elsewhere [30].

Fig. 5.4 is the ultimate mathematical domain for computation with pseudo-spectral method adopted in each subdomain respectively, and continuity of each variable and its derivative are enforced at each interface of the three subregions shown. This type of treatment is referred to as the "patched pseudo-spectral method" [29], which is an extension of the original classic pseudo-spectral method. Details of this mapping and can be found elsewhere [26].

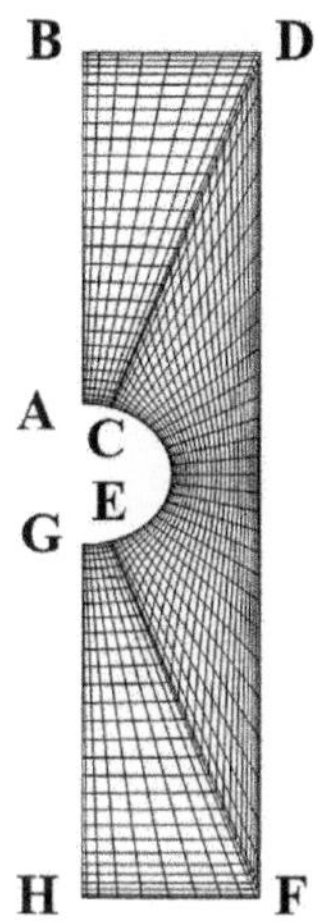

FIG. 5.3 Mesh diagram of subregions.

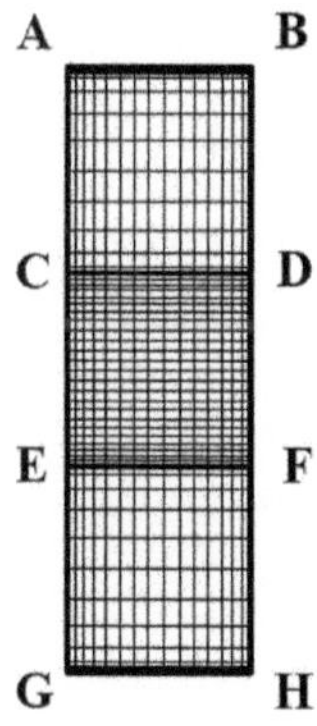

FIG. 5.4 Orthogonal mesh diagram of subregions after conformal mapping.

The governing equations and corresponding boundary conditions are all converted accordingly as well with this mapping (or variable change) between (r, θ) and (x, y). Note that boundary conditions in either cylindrical or spherical coordinates originally are all in rectangular Cartesian coordinates now, hence it can be applied straightforwardly on the outer boundaries in Fig. 5.4. The particle mobility can be calculated following the same procedure in Chapter 1, that is, Eqs. (1.77)–(1.81), once the electrokinetic field is obtained.

5.3 RESULTS AND DISCUSSION

5.3.1 General Henry's Charts

We first show in Fig. 5.5A–C the general Henry's charts for a rigid particle in a cylindrical pore at various separation distance to the wall for $\phi_r = 1, 3$, and 5,

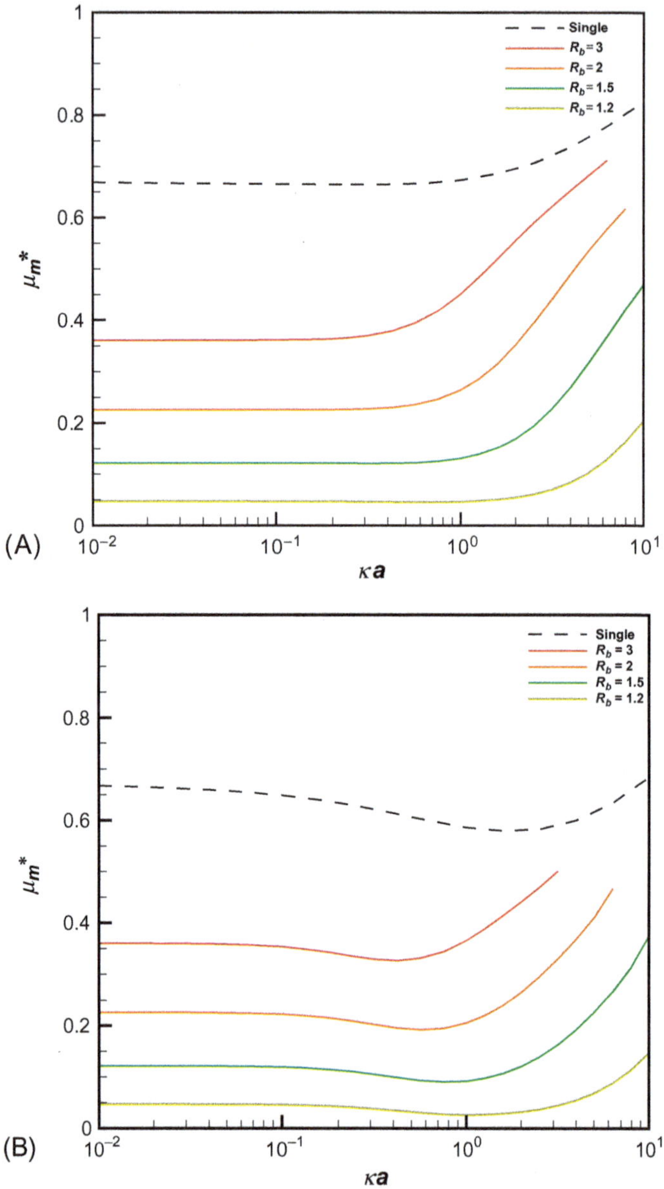

FIG. 5.5 Scaled mobility μ_m^* versus κa for different R_b with the dash line representing a single sphere situation. (A) $\phi_r = 1$; (B) $\phi_r = 3$; (C) $\phi_r = 5$.

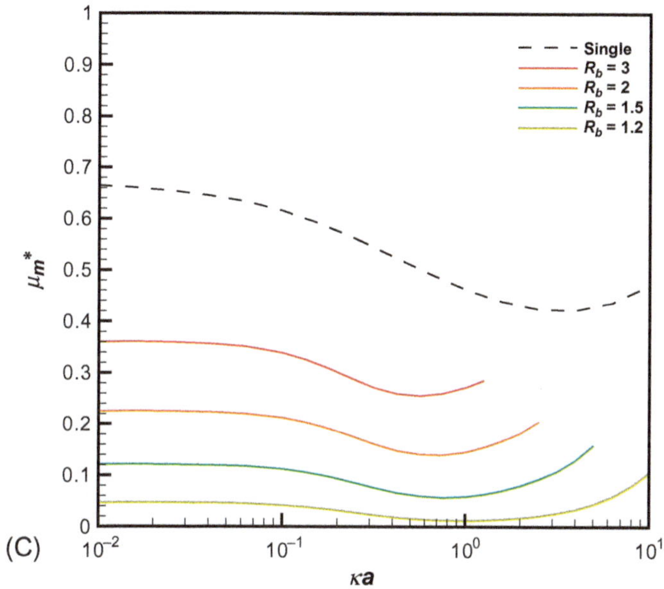

FIG. 5.5—CONT'D

respectively. The reduction of particle mobility is much more severe compared with the previous systems in Chapters 3 and 4 with planar boundaries at the same separation distance. This is because the geometric boundary confinement effect is much more significant here, which completely encloses the particle from every lateral direction.

It is interesting to note, however, that the local minima of mobility profiles are present no matter how close the particle is to the pore wall, indicating the double layer polarization effect is basically intact in a pore. This is because the double layer polarization due to the external electric field takes place mainly along the electric field direction of the longitudinal particle motion, hence hardly any present in the lateral neighborhood to be confined by the nearby pore wall, as shown in Fig. 5.6A. See also Fig. 1.9C in Chapter 1 of a single particle for comparison.

This is also clearly envisioned in Fig. 5.7, where the local minima in profiles of highly charged particles remain intact at such a narrow separation distance ($R_b = 1.2$), indicating the structure of the double layer polarization remains essentially intact.

5.3.2 Boundary Confinement Effect

The nearby wall of the cylindrical pore does affect the electrostatic aspect of the particle motion in a subtle way though via its impact upon the equilibrium

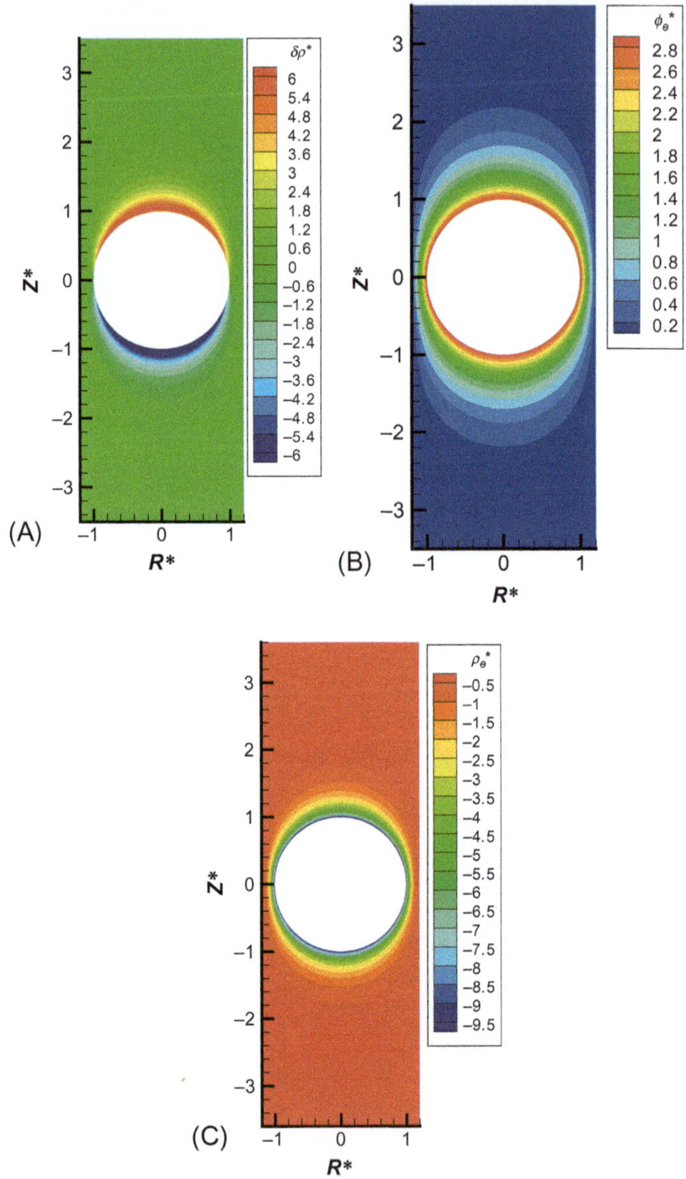

FIG. 5.6 Contour plots of different scaled variables at $\kappa a = 1$, $R_b = 1.2$, and $\phi_r = 3$. (A) Scaled disturbed charge density ($\delta\rho^*$); (B) Scaled equilibrium electric potential (ϕ_e^*); (C) Scaled equilibrium charge density (ρ_e^*).

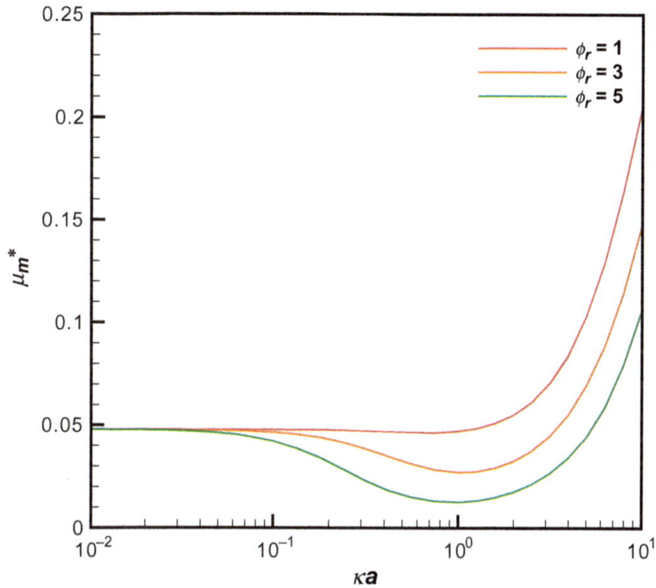

FIG. 5.7 Scaled mobility μ_m^* versus κa for different ϕ_r at $R_b = 1.2$.

double layer shape hence the corresponding equilibrium electrostatic electric potential profile surrounding the particle, as shown in Fig. 5.6B and C. The original concentric shapes are destroyed now by the lateral confinement effect, leading to a varying distribution of counterions along the θ direction hence the electric field at the particle surface. The electric driving force of the particle is thus changed accordingly leading to the severe mobility reduction observed in Fig. 5.5A–C when the double layer is very thick (small κa), where the boundary confinement effect is most significant.

5.3.3 Electroosmosis Flow of a Charged Pore

All the above analyses are based on a chargeless wall to focus on the boundary confinement effect of geometric nature. In reality, however, a charged wall induces EOF in response to the applied external electric field as well. This flow may enhance or retard the particle motion depending on whether the wall assumes the same charge sign as that of the particle. For example, a positively charged wall has negative counterions near it. These counterions migrate against the electric field direction and generate an EOF which will retard the motion of a positively charged particle as considered here. A negatively charged wall, on the other hand, enhances the particle motion

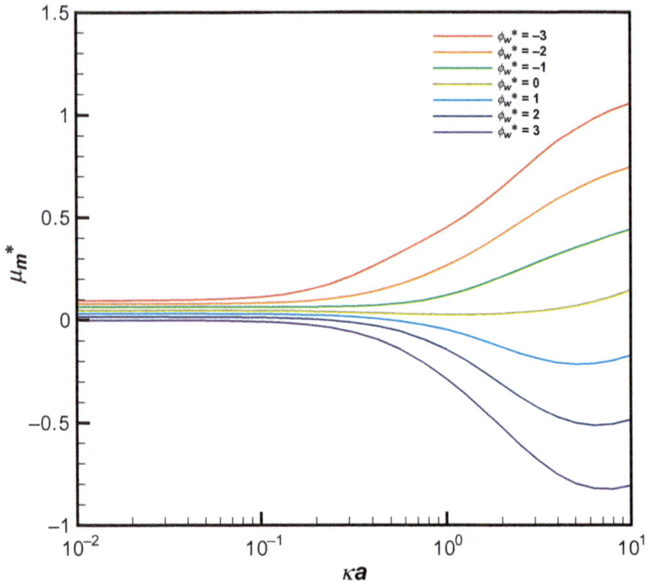

FIG. 5.8 Scaled mobility μ_m^* versus κa for different ϕ_w at $\phi_r = 3$ and $R_b = 1.2$.

leading to an increase of particle mobility. Fig. 5.8 shows the impact of a charged wall upon a representative highly charged particle ($\phi_r = 3$) at a very narrow separation distance ($R_b = 1.2$). The enhancement or reduction of particle mobility clearly follows the reasoning above.

5.3.4 Contour Plots of Electrokinetic Interest

Relevant contour plots are shown in Fig. 5.9A–C when the wall and the particle are equally charged at $\phi_r = 3$ and in Fig. 5.10A–C the pore wall charge is $\phi_r = -3$. In these plots, the ions in the double layer of the particle mingle with those from the charged pore wall and yield an axisymmetric distribution at equilibrium in the absence of external electric field, and so do their corresponding electric potentials. Corresponding polarization plots are axisymmetric as well.

Corresponding flow fields for a charged wall with either $\phi_r = 3$ or -3 are presented in Fig. 5.11A and B, respectively. Indeed the electrophoretic behavior of a highly charged particle in a charged cylindrical pore could be very complicated. These plots are helpful to shed some light into this important system.

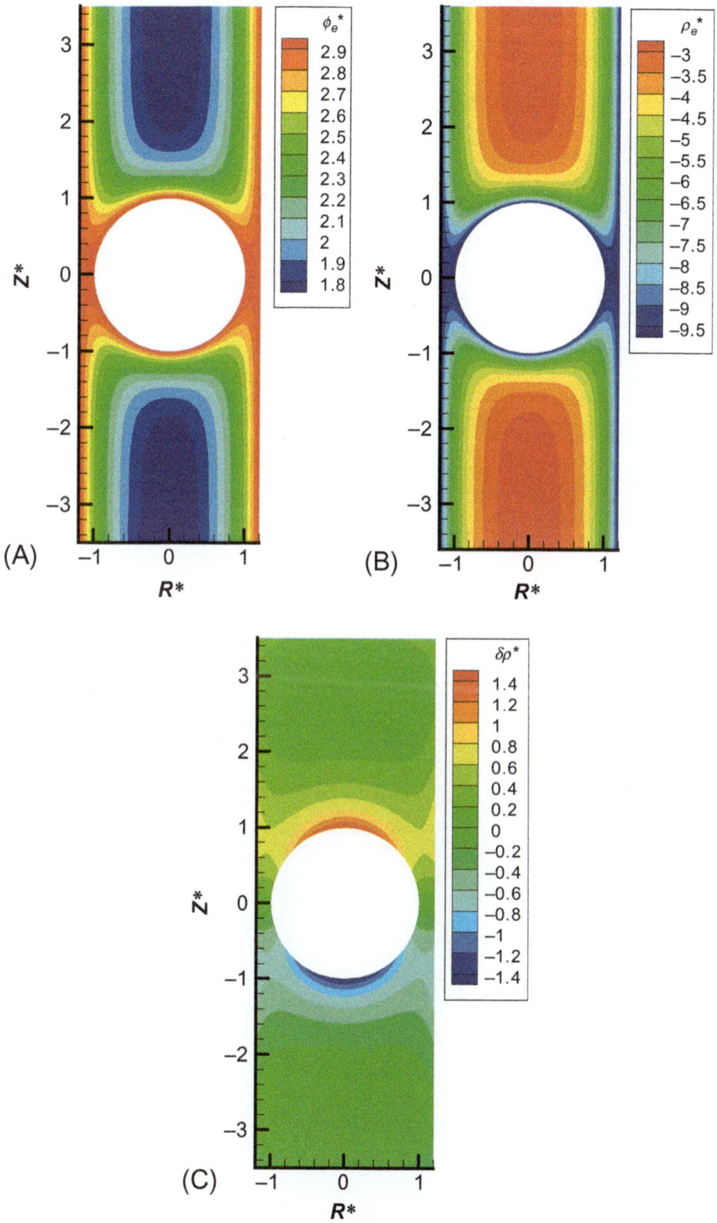

FIG. 5.9 Contour plots of different scaled variables at $\phi_r = 3$, $\kappa a = 1$, $R_b = 1.2$, and $\phi_w = 3$. (A) Scaled disturbed charge density (ϕ_e^*); (B) Scaled equilibrium charge density (ρ_e^*); (C) Scaled equilibrium electric potential ($\delta\rho^*$).

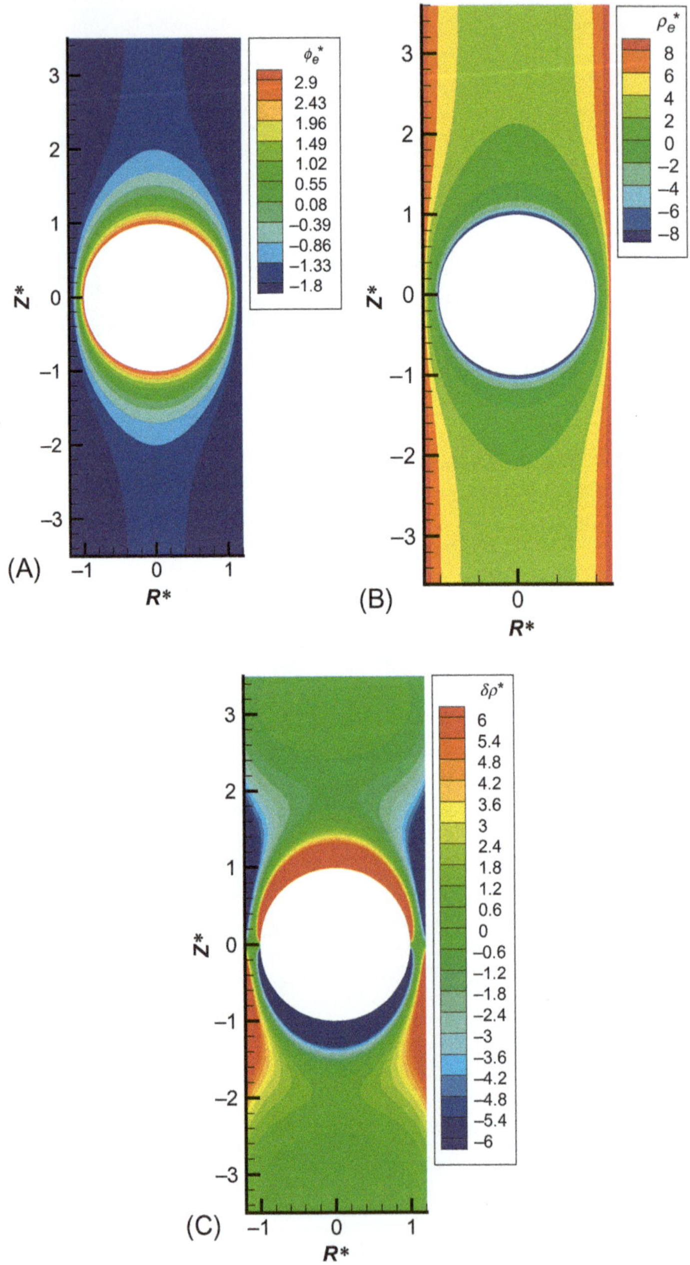

FIG. 5.10 Contour plots of different scaled variables at $\phi_r = 3$, $\kappa a = 1$, $R_b = 1.2$, and $\phi_w = -3$. (A) Scaled disturbed charge density (ϕ_e^*); (B) Scaled equilibrium charge density (ρ_e^*); (C) Scaled equilibrium electric potential ($\delta\rho^*$).

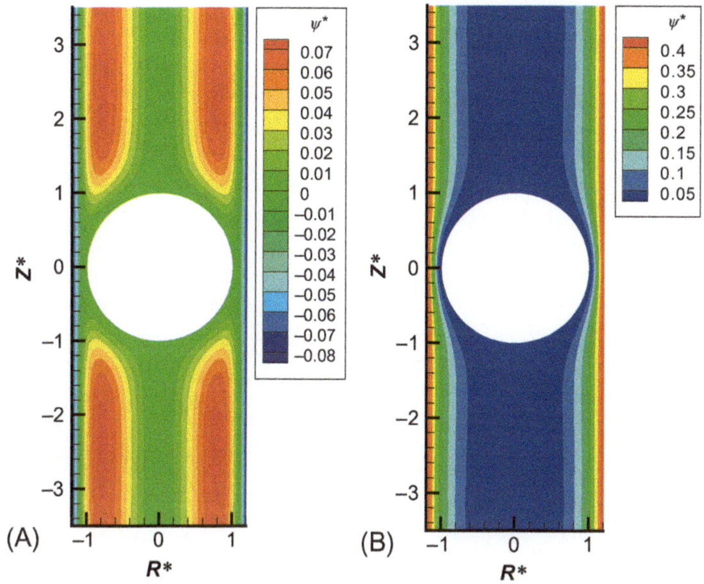

FIG. 5.11 Contour plots of the scaled stream function Ψ^* at $\phi_r = 3$, $\kappa a = 1$, and $R_b = 1.2$. (A) The scaled stream function Ψ^* with $\phi_w = 3$; (B) The scaled stream function Ψ^* with $\phi_w = -3$.

REFERENCES

[1] Kostal V, Arriaga EA. Recent advances in the analysis of biological particles by capillary electrophoresis. Electrophoresis 2008;29(12):2578–86.

[2] Xuan X. Joule heating in electrokinetic flow. Electrophoresis 2008;29(1):33–43.

[3] Ren J, Fang N, Wu D. Inverse-flow derivatization for capillary electrophoresis of DNA fragments with laser-induced fluorescence detection. Anal Chim Acta 2002;470(2):129–35.

[4] Heller C. Capillary electrophoresis of proteins and nucleic acids in gels and entangled polymer solutions. J Chromatogr A 1995;698(1–2):19–31.

[5] Kao YY, Liu KT, Huang MF, Chiu TC, Chang HT. Analysis of amino acids and biogenic amines in breast cancer cells by capillary electrophoresis using polymer solutions containing sodium dodecyl sulfate. J Chromatogr A 2010;1217(4):582–7.

[6] Li Z, Yang R, Wang Q, Zhang D, Zhuang S, Yamaguchi Y. Electrophoresis of periodontal pathogens in poly (ethyleneoxide) solutions with uncoated capillary. Anal Biochem 2015;471:70–2.

[7] Song X, Li L, Qian H, Fang N, Ren J. Highly efficient size separation of CdTe quantum dots by capillary gel electrophoresis using polymer solution as sieving medium. Electrophoresis 2006;27(7):1341–6.

[8] Jorgenson JW, Lukacs KD. Zone electrophoresis in open-tubular glass capillaries. Anal Chem 1981;53(8):1298–302.

[9] Weinberger R. Practical capillary electrophoresis. Elsevier; 2000. ISBN 0080539343.

[10] Kovarik ML, Jacobson SC. Nanofluidics in lab-on-a-chip devices. Anal Chem 2009;81(17):7133–40.

[11] Abgrall P, Gue AM. Lab-on-chip technologies: making a microfluidic network and coupling it into a complete microsystem a review. J Micromech Microeng 2007;17(5):R15.

[12] Srinivasan V, Pamula VK, Fair RB. An integrated digital microfluidic lab-on-a-chip for clinical diagnostics on human physiological fluids. Lab Chip 2004;4(4):310–5.

[13] Kang Y, Li D. Electrokinetic motion of particles and cells in microchannels. Microfluid Nanofluid 2009;6(4):431–60.

[14] Lagerqvist J, Zwolak M, Di Ventra M. Fast DNA sequencing via transverse electronic transport. Nano Lett 2006;6(4):779–82.

[15] Manz A, Graber N, Widmer HM. Miniaturized total chemical analysis systems: a novel concept for chemical sensing. Sens Actuators B: Chem 1990;1(1-6):244–8.

[16] Ennis J, Anderson JL. Boundary effects on electrophoretic motion of spherical particles for thick double layers and low zeta potential. J Colloid Interface Sci 1997;185(2):497–514.

[17] Grossman PD, Colburn JC. Capillary electrophoresis: theory and practice. Academic Press; 2012. ISBN 0323138195.

[18] Woudenberg Jr RC. Anhydrous proton conducting materials for use in high temperature polymer electrolyte membrane fuel cells. University of Massachusetts Amherst; 2007. ISBN 0549330569.

[19] molex. Product overview polymicro technologies; 2016.

[20] Keh HJ, Chiou JY. Electrophoresis of a colloidal sphere in a circular cylindrical pore. AIChE J 1996;42(5):1397–406.

[21] Ali I, Aboul-Enein HY, Gupta VK. Microchip-based nano chromatography and nano capillary electrophoresis in genomics and proteomics. Chromatographia 2009;69(1):13–22.

[22] Happel J, Brenner H. Low Reynolds number hydrodynamics (Martinus Nijhoff, The Hague, 1983). Google Scholar; 1983;322–31.

[23] Bohlin T. Terminal velocities of solid spheres in cylindrical enclosures. Transactions of the Royal Institute of Technology; 1960 Tech. Rep.

[24] Haberman WL, Sayre RM. Motion of rigid and fluid spheres in stationary and moving liquids inside cylindrical tubes. Washington, DC: David Taylor Model Basin; 1958 Report.

[25] Huang CH, Lee E. Electrophoretic motion of a liquid droplet in a cylindrical pore. J Phys Chem C 2012;116(28):15058–67.

[26] Huang CH, Hsu HP, Lee E. Electrophoretic motion of a charged porous sphere within micro-and nanochannels. Phys Chem Chem Phys 2012;14(2):657–67.

[27] Huang CH, Cheng WL, He YY, Lee E. Electrophoresis of a soft particle within a cylindrical pore: polarization effect with the nonlinear Poisson-Boltzmann equation. J Phys Chem B 2010;114(31):10114–25.

[28] Lee YF, Huang YF, Tsai SC, Lai HY, Lee E. Electrophoretic and electroosmotic motion of a charged spherical particle within a cylindrical pore filled with Debye-Bueche-Brinkman polymeric solution. Langmuir 2016;32(49):13106–15.

[29] Orszag SA. Spectral methods for problems in complex geometrics. In: Numerical methods for partial differential equations, Elsevier; 1979. p. 273–305.

[30] Huan CT. Electrophoretic motion of colloidal particle in a micro- or nano-channel. Ph.D. thesis, National Taiwan University; 2012.

Chapter 6

Dynamic Electrophoresis of Rigid Spherical Particles

6.1 INTRODUCTION

Electroacoustics is a phenomenon associated with the reversible interaction between the alternative current (AC) and the sonic wave in a colloidal suspension [1]. It was first used by Debye to measure the mass of electrolyte ions in its solution and later extended by others to explore the surface properties of a charged particle and the charge level as well as the particle size [2–10].

It consists of two mechanisms: colloidal vibrational potential (CVP) and electrokinetic sonic amplitude (ESA). When a (super) sonic wave is introduced into a colloidal suspension, the charged colloidal particles vibrate accordingly with this pressure wave hence generate an AC electric current and associated AC electric potential which can be measured between any two positions in the solution, should the two positions be short-circuited externally [11]. The potential measured is called the CVP which is usually very weak, hence is not often used in practical applications [12].

On the other hand, when an AC electric current is introduced into a colloidal suspension, the particles oscillate accordingly as well due to the charges carried and generate a measurable sonic wave due to the density difference between the particle and the solution. This is called the ESA, where amplitude refers to the strength of the sonic wave detected and recorded by a suitable device capable of detecting the pressure wave signals in the suspension. Note that in practice the frequency of the input AC, hence its resulted sonic wave, is often around 10^7 Hz, an ultra-sonic region way beyond the human hearing capability [13].

CVP and ESA are reverse phenomena to each other. However, the signal of sonic wave measured in ESA is much stronger than the corresponding measurable electric potential in CVP, as the energy carried by an input AC current is much stronger than that by an input sonic wave. Moreover, the detection of a pressure signal is much easier and more accurate than short-circuiting two positions in a suspension with electric probes. Hence ESA has been by far the favored one in practical applications of the electroacoustic phenomena.

Interface Science and Technology, Vol. 26. https://doi.org/10.1016/B978-0-08-100865-2.00006-0

Note that the oscillatory motion of the particle in response to an AC electric current applied externally in ESA is the counterpart of electrophoresis in a dynamic way, hence it is often referred to as the "dynamic electrophoresis" or "AC electrophoresis" [2]. Unlike its corresponding conventional static electrophoresis (or DC electrophoresis), dynamic electrophoresis is not at quasisteady state in general. Instead, its magnitude is of dynamic nature in the sense that the magnitude of the particle motion varies with time in a periodic way of the same pattern as the input AC current.

Note that not only the rigid particle itself moves in response to the input AC electric field, so do the electrolyte ions in the double layer surrounding it and they are not synchronized with each other in general. The motion-deterring double layer polarization effect hence becomes time-dependent as well.

With this drag of surrounding double layer, the particle may move with a phase lag or an apparent phase lead ahead of the input AC electric current. Hence both the magnitude and the phase of the dynamic mobility are the key factors to examine in corresponding theoretical analyses [14–21].

There are many superior merits of the AC electrophoresis to the conventional DC electrophoresis. Unlike the experimental measurement of the particle mobility in DC electrophoresis, for instance, which requires sufficient dilution in prior to keep track the movement of an isolated single particle, even with a modern micro-electrophoresis apparatus, the mainstream experimental setup nowadays [22], no dilution is needed at all with AC electrophoresis. This is a major merit as the zeta potential may be drastically different between the dilute and concentrated suspensions, as we have discussed in Chapters 1 and 2. Moreover, many macromolecules like proteins and DNAs, or even living cells, are very sensitive and vulnerable to the dilution procedure normally conducted in practice, which very often involves the usage of chemical agents that might be detrimental to the samples. AC electrophoresis resolves this dilemma completely and makes possible the measurement of zeta potential in concentrated suspensions as they are, if that is precisely the system of interest.

The other major merit is the suppression of the unwanted Joule heating effect which normally takes place in DC electrophoresis, as background electrolyte ions (or run buffer) are often added to the system to maintain sufficient current across the suspension to enhance the separation performance [23]. The Joule heating effect is proportioned to the square of the passing current, and an increase of four degrees in Celsius will kill human cells, among other living microorganisms [24]. Moreover, the heat generated often leads to buoyancy-driven vortex flows perpendicular to the particle moving direction thus blurs the separation resolution [12]. With AC electrophoresis, no extra run buffer is needed as no net particle motion takes place. And as long as the signal detector is sensitive enough, the original particle vibration is normally good enough. With all these merits, however, the success of its practical application relies heavily upon the accurate relationship established between the surface charge condition and the measurable data from the dynamic electrophoresis experiment, such as the magnitude and phase angle of the particle motion with varying frequency of the input AC electric current, which is the focus of this chapter.

There are a series of theoretical analyses available in the literature on dynamic electrophoresis of rigid particles [2, 25, 26]. For a very thin double layer, O'Brien was able to derive an analytical formula for a single rigid particle which takes the form as a corresponding Smoluchowski formula in AC electrophoresis, with a correction factor depending on the input AC frequency if there is no involvement of surface conductivity of the particle [2]. The results are good for dilute suspension up to 4 volume percent under the constraint that $\kappa a > 20$. Ohshima [27], on the other hand, derived an approximate analytical formula valid for arbitrary double layer thickness when the particle is lowly charged, say $\phi_r = 1$ in practice, which can be taken as the corresponding Henry's formula in dynamic electrophoresis. Indeed these two classic contributions provide the basic knowledge of the particle behavior in dynamic electrophoresis.

For concentrated particle suspensions, Ohshima obtained a formula valid for lowly charged particle based on Kuwabara's unit cell model adopting the Levine and Neale (L-N) electric boundary conditions [28, 29]. Lee's group, on the other hand, treated the concentrated suspensions of particles with arbitrary surface potentials and double layer thickness using the same Kuwabara hydrodynamic cell model but with S-Z electric boundary condition instead which has a better performance in general [14, 19, 21, 30]. Double layer polarization effect was fully taken into account in these analyses and the complete set of electrokinetic equations was solved directly with no further simplifications or assumptions at all. Hence these results are so far the most reliable and general information on dynamic electrophoresis of highly charged particles in suspensions.

With this framework in mind, we treat the dynamic electrophoresis of a single rigid particle with arbitrary zeta potentials and double layer thickness first. Figures of dynamic mobility are presented to investigate the particle behavior in general. Discussions are given to shed light upon this powerful experimental technique, especially for $\kappa a < 20$ when no approximate analytical formula is available. Extension to the suspension system, either dilute or concentrated, is also given to complete the treatment of the topic.

6.2 THEORY

The system diagrams are shown in Fig. 6.1A and B or a single particle and a suspension system, respectively.

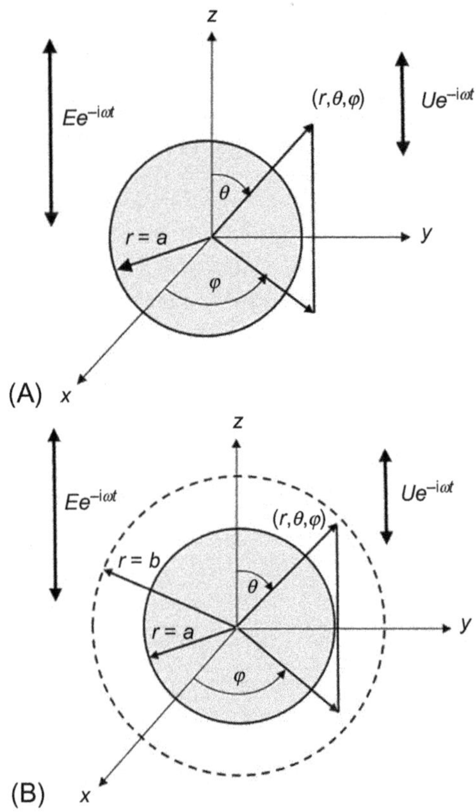

FIG. 6.1 System diagram. (A) System diagram for a single rigid particle; (B) System diagram for the suspension of rigid particles.

6.2.1 Single Particle

We consider the dynamic electrophoretic motion of a single spherical rigid particle suspended in an electrolyte solution with parameters set exactly as those in Chapter 1 except that an AC electric field of the form $\boldsymbol{E} = Ee^{-i\omega t}\mathbf{i_z}$ is applied instead, where $\mathbf{i_z}$ is the unit vector in the z-direction, t is time, E and ω are the magnitude and frequency of $\boldsymbol{E}$, respectively, and $\mathrm{i} = \sqrt{-1}$.

Let $\boldsymbol{U} = Ue^{-i\omega t}\mathbf{i_z} = (U_R + iU_I)\mathbf{i_z}$ be the particle dynamic electrophoretic velocity in complex form, U_R and U_I being respectively the real and the imaginary part of $\boldsymbol{U}$, which in polar form expression further yields $\boldsymbol{U} = |\boldsymbol{U}|e^{-i\omega t}$ where $|\boldsymbol{U}|$ is the magnitude of the dynamic particle velocity, i.e., $|\boldsymbol{U}| = \sqrt{U_R{}^2 + U_I{}^2}$. Note that the complex factor $e^{-i\omega t}$, which by Euler's formula yields $e^{-i\omega t} = \cos\omega t - i\sin\omega t$, is used to express the sinusoidal behavior of the AC electric field as well as the physical quantities affected by it in general. The real parts stand for the actual physical variables under consideration, the imaginary part, on the other hand, is just a convenient mathematical auxiliary counterpart to facilitate subsequent mathematical manipulation. It is just the polar form of a complex number indicating the phase angle of a periodic behavior. There is no physical meaning directly associated with it. The magnitude and the phase angle are the two physical quantities to investigate.

Compared with the DC electrophoresis, the electrokinetic variables are now time-dependent in general, and assume the same periodic pattern as the input AC electric field as follows:

$$\phi(r,\theta,t) = \phi_e(r) + \delta\phi(r,\theta)e^{-i\omega t} \tag{6.1}$$

$$n_j(r,\theta,t) = n_{je}(r,\theta) + \delta n_j(r,\theta)e^{-i\omega t} \tag{6.2}$$

$$\mathbf{v}(r,\theta,t) = 0 + \delta\mathbf{v}(r,\theta)e^{-i\omega t} \tag{6.3}$$

$$\rho(r,\theta,t) = \rho_e(r,\theta) + \delta\rho(r,\theta)e^{-i\omega t} \tag{6.4}$$

Corresponding variables at equilibrium state when the particle is motionless before the application of the external electric field, however, are the same as those encountered in Chapter 1. Hence we have

$$\nabla^{*2}\phi_e^* = -\frac{(\kappa a)^2}{1+\alpha}\left[\exp(-\phi_e^*) - \exp(\alpha\phi_e^*)\right] \tag{6.5}$$

Note that dimensionless terms are used above to simplify the system. The definitions of all the dimensionless terms below are same as those in Chapter 1 as well. Based on the time-dependent periodic perturbation terms, corresponding governing equations of them can be derived following the same methodology as adopted in Chapter 1. They are, for a binary electrolyte solution:

$$\begin{aligned} &\nabla^{*2}(\delta\phi^*) - \frac{(\kappa a)^2}{1+\alpha}\left[\exp(-\phi_e^*) + \alpha\exp(\alpha\phi_e^*)\right]\delta\phi^* \\ &= \frac{(\kappa a)^2}{1+\alpha}\left[\exp(-\phi_e^*)g_1^* + \alpha\exp(\alpha\phi_e^*)g_2^*\right] \end{aligned} \tag{6.6}$$

$$\nabla^{*2}g_1^* - \nabla^*\phi_e^*\cdot\nabla^*g_1^* - Pe_1\delta v^*\cdot\nabla^*\phi_e^* + i\frac{\omega a^2}{D_1}(\delta\phi^* + g_1^*) = 0 \tag{6.7}$$

$$\nabla^{*2}g_2^* + \alpha\nabla^*\phi_e^*\cdot\nabla^*g_2^* - Pe_2\delta v^*\cdot\nabla^*\phi_e^* + i\frac{\omega a^2}{D_2}(\delta\phi^* + g_2^*) = 0 \tag{6.8}$$

Note that Eqs. (6.7) and (6.8) are the dynamic governing equations for g_j, $j = 1, 2$ for a binary system. Corresponding dynamic counterpart of the governing equations of flow field is

$$E^{*4}\Psi^* + i\frac{\rho_f \omega a^2}{\eta_f} E^{*2}\Psi^* = \frac{(\kappa a)^2}{1+\alpha}\left[\frac{\partial g_1^*}{\partial \theta}\exp(-\phi_e^*) + \alpha\frac{\partial g_2^*}{\partial \theta}\exp(\alpha\phi_e^*)\right]\frac{d\phi_e^*}{dr^*}\sin\theta \tag{6.9}$$

where

$$E^{*4} = E^{*2}E^{*2}, \quad E^{*2} \equiv \frac{\partial^2}{\partial r^{*2}} + \frac{\sin\theta}{r^{*2}}\frac{\partial}{\partial \theta}\left(\frac{1}{\sin\theta}\frac{\partial}{\partial \theta}\right) \tag{6.10}$$

Note that the stream function conversion has been introduced in the above derivation as in Chapter 1. Eqs. (6.5)–(6.9) are the governing equations for the AC electrophoresis considered here.

Associated boundary conditions on the particle surface are exactly the same as those in Chapter 1, based on the same underlying physical essence, except that all the variables now are in complex form with real and imaginary parts.

The treatment of boundary conditions at infinity are the same as those in Chapter 1 as well, i.e., replacement with Debye-Hückel approximation at a sufficiently large distance r_o^*. They are Eqs. (1.26)–(1.33). Note that the variables are in complex form now.

6.2.2 Suspension System

Kuwabara's unit cell model coupled with S-Z boundary conditions is used to describe a suspension system. The governing equations and associated boundary conditions on the particle surface are exactly the same as treated above for a single particle situation. The boundary conditions on the virtual surface, on the other hand, is replaced by the counterparts in Chapter 2. They are Eqs. (2.2)–(2.6).

The resulted system is further reduced to one-dimensional version as treated in Chapter 1. In practice, it is this one-dimensional form of the governing equations that are solved numerically with the pseudo-spectral method. The details can be found elsewhere [14, 15].

6.2.3 Evaluation of the Dynamic Particle Mobility

Once the electric field and flow field are solved, corresponding electric driving force upon the particle as well as the hydrodynamic drag force can be obtained via the following equations:

$$F_E^* = \frac{8}{3}\pi\left[r^*\frac{d\phi_e^*}{dr^*}(\Phi_R + i\Phi_I)e^{-i\omega t}\right]_{r^*=1} \tag{6.11}$$

$$F_D^* = \frac{4}{3}\pi \left[r^{*4} \frac{\partial}{\partial r^*} \frac{D^2(\Psi_R + i\Psi_I)}{r^{*2}} e^{-i\omega t} \right]_{r^*=1} \\ + \frac{4}{3}\pi \left[r^{*2} \left(\exp\left(-\phi_e^*\right) - \exp\left(\alpha\phi_e^*\right) \right) (\Phi_R + i\Phi_I) e^{-i\omega t} \right]_{r^*=1} \tag{6.12}$$

Note that the divergence theorem has been applied in the derivation of these two equations to convert volume integral to surface integral which greatly simplifies the calculations. And Φ_R, Φ_I, ψ_R, ψ_I are the real parts and imaginary parts of the corresponding one-dimensional form of $\delta\phi^*$ and ψ^*, respectively. Note also the second term in Eq. (6.12) comes from the extra drag force due to the flow of the electrolyte solution, hence is in the nature of electric force. The details can be found elsewhere [14, 15].

The dynamic particle mobility is defined as follows:

$$\mu = \mu_R + i\mu_I = \frac{U}{E} \tag{6.13}$$

Note that both U and E are in complex form, hence a complex the dynamic mobility is resulted, which has real and imaginary parts, μ_R and μ_I, respectively. The complex particle velocity, $U = U_R + iU_I$, is obtained via the Newton's second law as follows:

$$F_D + F_E = \frac{4}{3}\pi a^3 \left(\rho_p - \rho_f\right) \frac{d\left(Ue^{-i\omega t}\right)}{dt} \tag{6.14}$$

or in dimensionless terms,

$$F_D^* + F_E^* = -i\frac{4}{3}\pi\omega^* \left(\frac{\rho_p - \rho_f}{\rho_f} \right) \left(U_R^* + iU_I^*\right) e^{-i\omega^* t^*} \tag{6.15}$$

where $\omega^* = \frac{\rho_f \omega a^2}{\eta_f}$ is the dimensionless frequency.

Note that the fluid buoyancy force is already considered here. The dimensionless dynamic mobility $\mu^* = \mu_R^* + i\mu_I^*$ is thus obtained by Eqs. (6.13), (6.14) once F_E^* and F_D^* are available. Note again as treated in Chapter 1, the scaled mobility $\mu_m^* = \frac{\mu^*}{\phi_r}$ is used in all the subsequent presentation of results.

In practice, μ_m^* is often expressed in polar form instead, i.e., $\mu_m^* = |\mu_m^*| e^{i\theta_p}$, where $|\mu_m^*| = \sqrt{\mu_R^{*2} + i\mu_I^{*2}}$ is the magnitude of the dynamic mobility, and $\theta_\mathrm{p} = \tan^{-1}\frac{\mu_I^*}{\mu_R^*}$ is the corresponding phase angle. $\theta_p > 0$ implies a phase lag behind the input AC electric field, whereas $\theta_p < 0$ a phase lead.

The dynamic electrophoresis of charge-regulated particle, on the other hand, is exactly the same as above once the particle surface charge level is determined, which is a purely electrostatic issue as discussed in Chapter 1.

6.3 RESULTS AND DISCUSSION

6.3.1 Single Particle

In the conventional DC or static electrophoresis, only one data of mobility measurement can be obtained for a specific colloidal system under consideration. In dynamic or AC electrophoresis, however, the dynamic mobility is a function of the input AC frequency as well, i.e., $\mu^* = \mu^*(\omega)$, hence a set of experimental measurement of particle mobility can be obtained at various frequency. A complete range of mobility profile as a function of AC frequency can thus be established accordingly, in terms of both its magnitude and its phase angle. This certainly greatly increases the reliability of the experimental data, as possible human or instrumental errors in operations are minimized in principle. Moreover the particle behavior is more clearly demonstrated this way with more information revealed.

Once an external AC electric field is applied upon the system, the colloidal suspension responds immediately to it. Note that not only the charged particle, the electrolyte ions within the double layer respond simultaneously as well. In fact even the ions in the bulk solution moves back and forth accordingly, but they have no impact upon the particle motion, which is the major source of the output sonic wave, hence is out of the picture of the theoretical treatment here. Therefore, the mechanism of interest here is the acceleration of the rigid particle itself in response to the input AC electric field, which leads to the back and forth oscillation around the original rest location, subject to the modification by the surrounding counterions in the double layer. On the other hand, the surface charges of the particle move with it with no delay whatsoever as they are strongly bounded to the particle in the inner compact Stern layer. They move together with the rigid particle as a whole.

The acceleration time of the particle before it reaches the terminal velocity, as balanced by the retarding hydrodynamic drag force, is of the order: $\tau_H = \frac{\rho_f a^2}{2\eta_f}$ [31], where ρ_f and η_f are the density and viscosity of the suspending medium and the particle radius. Hence for a particle of 1 μm in radius suspended in an aqueous electrolyte solution at room temperature, τ_H is around 5×10^{-7} seconds. And for a particle of radius 100 nm, it is 5×10^{-9} seconds. Thus the particle essentially reaches the terminal velocity instantly. Note that this time constant is obtained solely from the particle inertial force and the retarding viscous force hence is of purely hydrodynamic nature. It is the major factor determining the velocity field around the moving particle. It has nothing to do with the presence of the ions in the suspension, whether it is within the double layer, or in the bulk phase. They simply do not appear in the expression of τ_H at all. Note that τ_H is the same whether the suspension is dilute or concentrated. And the smaller the particle size is, the faster it reaches the terminal velocity, as the required inertia from acceleration for it is less.

Moreover, it tells us that the charged particle will move, with or without the motion of the surrounding ions, as the external electric field is superimposed upon the entire suspension system directly, not via the relay of the surrounding ions. Hence even if the surrounding ions appear motionless somehow due to the failure to keep pace with the input AC electric field, an output sonic wave will still be generated by the charged particle perfectly, as long as τ_H is the shortest characteristic time in the suspension system.

The other factor is the movement of the surrounding ions within the double layer, which consists of mainly the electro-migration, the convection and the diffusion mechanisms as considered in the ion-flux equation, or the Nernst-Planck equation in Chapter 1. The electro-migration mechanism is a spontaneous movement of ions under the Coulomb electric force due to the charges they carry. Once they move, the surrounding hydration molecules move with them and drag the fluid nearby as well to generate convection flow, with the apparent backflow from the moving particle superimposed upon it. In return this convective fluid flow contributes to the ion flux as well. The diffusion mechanism, on the other hand, serves as a balancing backward flux once the buildup of ions is overboard.

In accordance with the particle motion, the ions in the double layer sweep back and forth around the particle. Thus the thicker the double layer (small κa), the longer it takes in general for the ions to regain the equilibrium state surrounding the particle. If the movement of the ions in the double layer, and the particle itself, can keep pace with the input AC electric field, in other words, both the particle and the double layer regain their original state before the applied electric field changes direction again, then the dynamic mobility observed would be exactly the same as those obtained in corresponding DC electrophoresis, as the mobility amplitude stands for the largest particle velocity achieved, which is the terminal velocity appearing before each cyclic direction change of the AC field. In this case, the double layer polarization is fully recovered each time the particle changes moving direction, hence the magnitude would be exactly the same as those predicted by the general Henry's charts in DC-electrophoresis. In other words, when the AC frequency is low, a horizontal profile of $|\mu_m^*|$ is expected, as shown in Figs. 6.2–6.5. These figures are sometimes referred to as the electroacoustic spectrometry and contain all the information an AC electrophoresis measurement could possibly provide [13]. The double layer thickness considered ranges from very thick ($\kappa a = 0.1$), comparable to the particle size ($\kappa a = 1$), thin ($\kappa a = 10$) and very thin ($\kappa a = 20$), respectively, in Figs. 6.2–6.5. Note that Ohshima's analytical prediction valid for lowly charged condition is shown as well for comparison purpose. Moreover, the phase angle γ in radians has been replaced by θ_p in degrees for convenience. The thinner the double layer (large κa) is, the longer this horizontal range extends with increasing frequency as might be expected. Moreover, no phase lag is anticipated as all the three mechanisms of ion flux can keep pace with the AC electric field.

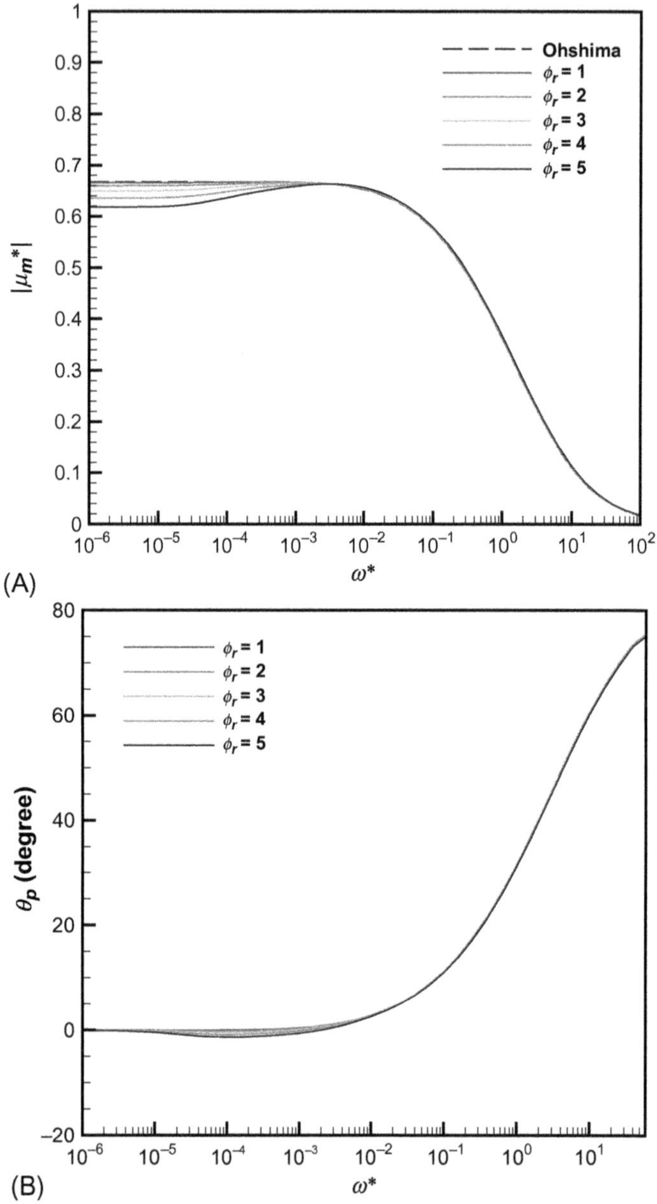

FIG. 6.2 (A) Magnitude of the scaled mobility $|\mu_m^*|$ versus the scaled frequency ω^* at various ϕ_r with $\kappa a = 0.1$; (B) Phase angle of the scaled mobility θ_p versus the scaled frequency ω^* at various ϕ_r with $\kappa a = 0.1$.

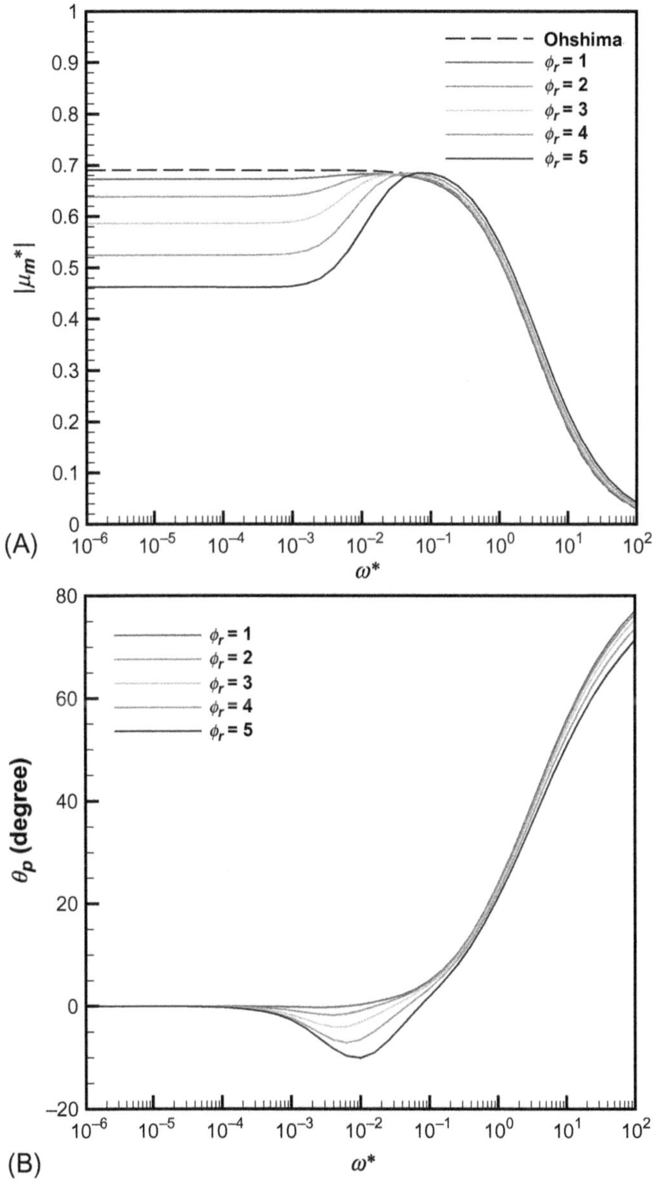

FIG. 6.3 (A) Magnitude of the scaled mobility $|\mu_m^*|$ versus the scaled frequency ω^* at various ϕ_r with $\kappa a = 1$; (B) Phase angle of the scaled mobility θ_p versus the scaled frequency ω^* at various ϕ_r with $\kappa a = 1$.

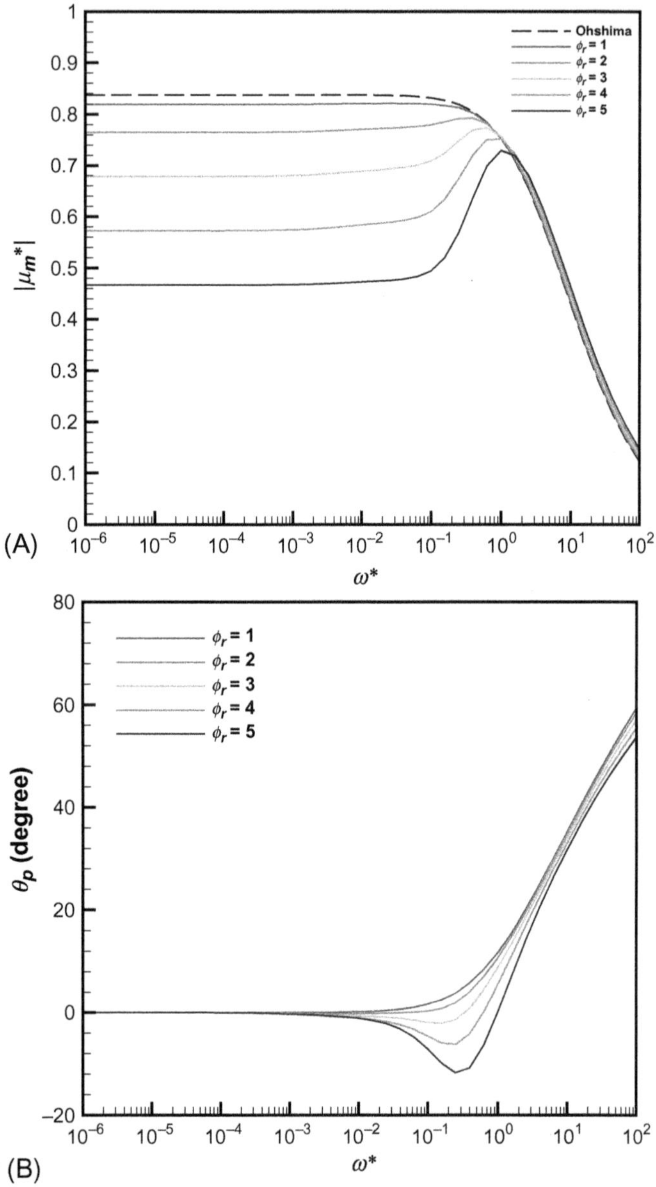

FIG. 6.4 (A) Magnitude of the scaled mobility $|\mu_m^*|$ versus the scaled frequency ω^* at various ϕ_r with $\kappa a = 10$; (B) Phase angle of the scaled mobility θ_p versus the scaled frequency ω^* at various ϕ_r with $\kappa a = 10$.

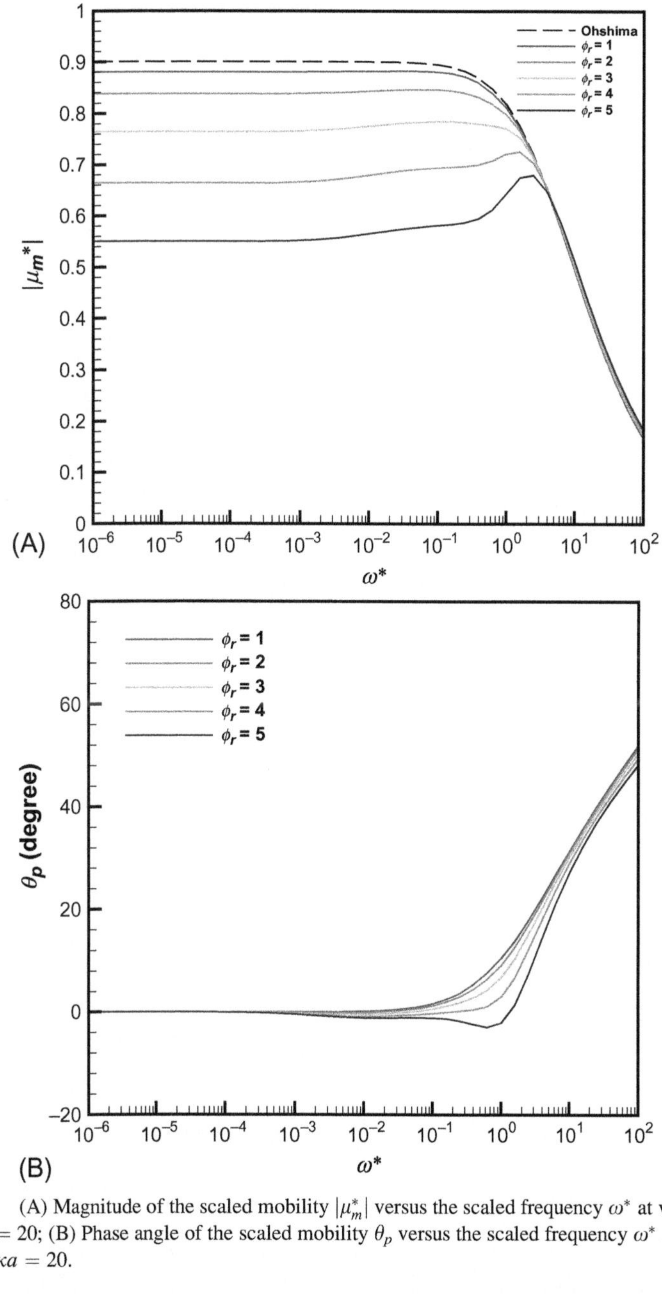

FIG. 6.5 (A) Magnitude of the scaled mobility $|\mu_m^*|$ versus the scaled frequency ω^* at various ϕ_r with $\kappa a = 20$; (B) Phase angle of the scaled mobility θ_p versus the scaled frequency ω^* at various ϕ_r with $\kappa a = 20$.

The diffusion mechanism is normally much slower than the other two mechanisms, as the former does not take place here until sufficient local buildup of ions is established by the other two mechanisms first. Hence in general the diffusion mechanism will bow out first from the stage as the AC frequency increases past some threshold value, which is too fast for it to keep pace with any more. As the diffusion mechanism functions as a brake in general for the redistribution of counterions in the double layer once an external electric field is applied upon the suspension system, without its involvement, the ions tend to overshoot from one end of the original double layer to the other, leading to an effective increase of double layer thickness, which reduces the motion-deterring double layer polarization effect significantly, as a large amount of counterions are driven away from the vicinity of the particle. This leads to the observed gradual enhancement of mobility magnitude as frequency increases, as shown in Figs. 6.2A–6.5A. The behavior is surface potential dependent, hence we know the underlying physical mechanism is the double layer polarization.

Drastic apparent phase lead, which is phase lag in disguise, is observed in the corresponding Figs. 6.2B–6.5B. Note these apparent phase lead are particle surface potential dependent, indicating double layer polarization effect is still there, but solely by the electro-migration mechanism only now. The same observation and conclusion apply to the magnitude of dynamic mobility in Figs. 6.2A–6.5A as discussed above. The increase of mobility magnitude can also be explained by the possible time lag between the motion of the particle itself and the surrounding ions conducting electro-migration only, which happens to yield a motion-enhancing polarization effect with counterions cluster in front of the moving particle. Without further conclusive evidence, however, we will just leave it there as an interesting characteristic behavior of AC electrophoresis.

Note that after reaching some maximal values, all the mobility profiles coincide with one another to form a basically unique profile corresponding to the Smoluchowski type behavior. As it is independent of the particle surface potential, we know that the double layer polarization effect is completely gone. The ions cannot keep pace with the AC electric field anymore, and the particle behaves as if it is in a salt-free system, with no effect from the surrounding ions whatsoever, since they are driven far away from the particle already before reaching this point. Certainly the particle itself can still keep on moving back and forth and sending out detectable sonic wave, as the electric field is superimposed upon it. This continues until the frequency goes so high even for the particle itself to keep pace with it. Eventually the particle will become motionless and no more sonic wave is generated and detected. Gradual diminishing of the mobility magnitude is observed in Figs. 6.2A–6.5A, as the effective acceleration time is shortened continuously with increasing frequency now. Corresponding behavior of increasing phase lag is also observed throughout Figs. 6.2B–6.5B.

Note that in practice this region of Smoluchowski type behavior is actually preferred for the measurement of dynamic mobility, as the particle surface potential can be easily obtained subsequently. No more complicated nonlinear double layer polarization effect is involved to muddy the water. That's why about 1–100 MHz of AC electric current are normally used in particle.

For the very thin double layer situation, O'Brien [2] was able to obtain a characteristic time constant, or the relaxation time constant, for the electro-migration mechanism to fully respond to the input AC field, which is of the order $\tau_E = \frac{1}{\kappa^2 D}$. As for the diffusion relaxation time constant, it is of the order $\tau_D = \frac{a^2}{D}$. Thus for a 10^{-3} M KCl solution, τ_E is of the order 10^{-7} seconds and τ_D is 10^{-3} seconds. Indeed diffusion is much slower hence takes longer time to fully respond to the field change. Note that $\frac{\tau_E}{\tau_D} = \frac{1}{(\kappa a)^2}$, hence the larger κa is (thin double layer), the faster the electro-migration mechanism in comparison with the diffusion mechanism, as less distance needs to be traveled across the double layer, and diffusion is independent of the double layer thickness.

Note also that in the definition of the dimensionless scaled frequency $\omega^* = \frac{\rho_f \omega a^2}{\eta_f} = 2\tau_H \omega$ here. In other words, $\frac{\omega^*}{\omega} = 2\tau_H$, which is dependent of particle size. Or $\omega^* = \frac{\tau_H}{\frac{1}{2\omega}}$ indicates the ratio of the full particle acceleration time to the time span of half the period of input AC electric field. Hence when ω^* is smaller than 1, full acceleration of particle response is expected each time the AC electric field changes its direction. When ω^* is larger than 1, however, the particle is getting behind and fails to keep pace with the input AC electric field.

6.3.2 Suspension System

Corresponding figures for $\kappa a = 1$ at various levels of particle concentration are shown in Figs. 6.6–6.8 for $\phi_r = 1$, 3, and 5, respectively. The profiles of $H = 10^{-6}$ reduce to the single particle situation in all these charts, indicating the success to model a suspension with the cell model here.

The particle mobility diminishes as the suspension gets more and more concentrated regardless of the frequency level. Similar to the corresponding static DC electrophoresis, this is due to the double layer overlapping effect which severely reduce the electric driving force, and the hydrodynamic swarming effect which increases the drag force. Moreover, the local maxima observed in the magnitude profiles for highly charged particles are, before the reaching of the summit, due to the enhancement of particle speed from the overshooting of the electro-migration mechanism as the diffusion mechanism bows out. This weakens the motion-deterring double layer polarization effect. Once it reaches the summit, or the local maximum, the further increase of the

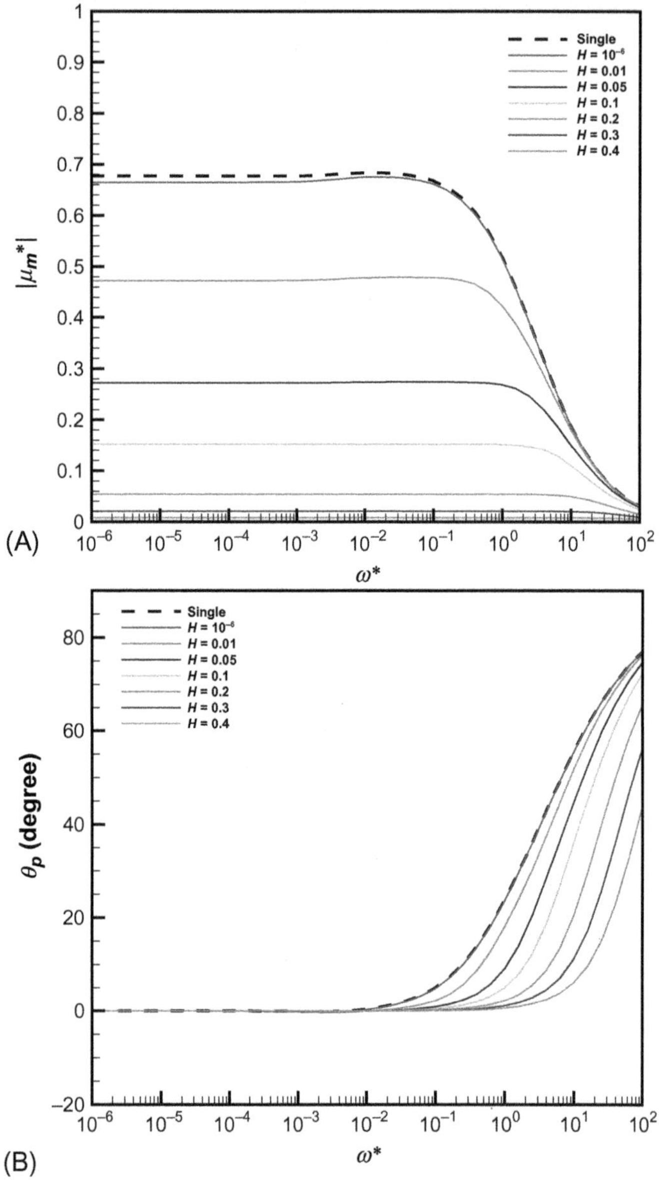

FIG. 6.6 (A) Magnitude of the scaled mobility $|\mu_m^*|$ versus the scaled frequency ω^* at various volume fraction H with $\kappa a = 1$ and $\phi_r = 1$; (B) Phase angle of the scaled mobility θ_p versus the scaled frequency ω^* at various volume fraction H with $\kappa a = 1$ and $\phi_r = 1$.

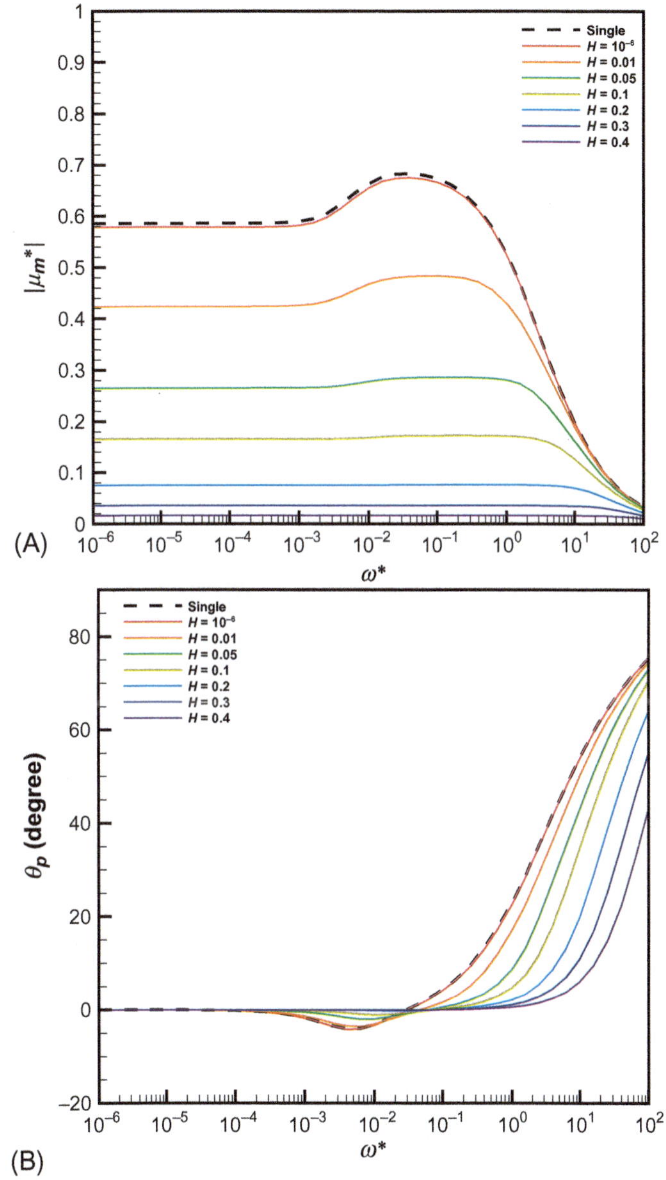

FIG. 6.7 (A) Magnitude of the scaled mobility $|\mu_m^*|$ versus the scaled frequency ω^* at various volume fraction H with $\kappa a = 1$ and $\phi_r = 3$; (B) Phase angle of the scaled mobility θ_p versus the scaled frequency ω^* at various volume fraction H with $\kappa a = 1$ and $\phi_r = 3$.

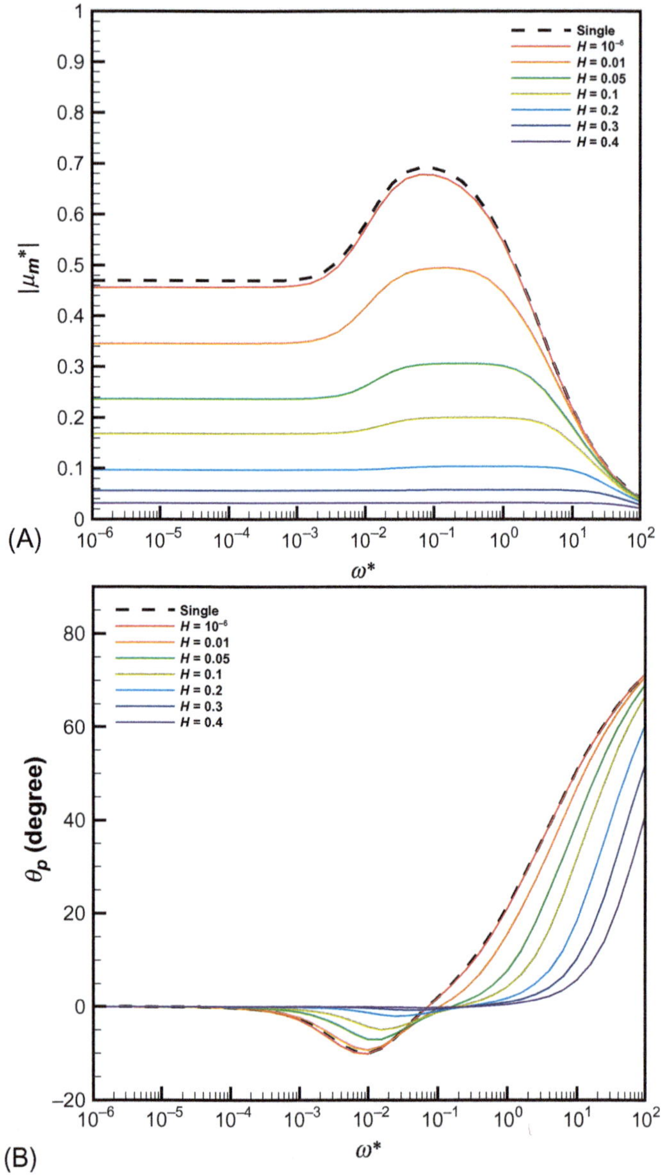

FIG. 6.8 (A) Magnitude of the scaled mobility $|\mu_m^*|$ versus the scaled frequency ω^* at various volume fraction H with $\kappa a = 1$ and $\phi_r = 5$; (B) Phase angle of the scaled mobility θ_p versus the scaled frequency ω^* at various volume fraction H with $\kappa a = 1$ and $\phi_r = 5$.

AC frequency becomes simply too fast even for the particle itself to respond, as discussed earlier in the single particle situation.

Note there is a strong dependence of particle phase angle on concentration levels, *H*. The profiles are distinctive from each other at high frequency in particular. The more concentrated the suspension is, the less the amount of phase lag, or apparent lead. This is due to the hydrodynamic swarming effect of neighboring particles, which hinders the particle movement hence reduces the relaxation time in response.

The charts or diagrams shown here in this chapter are about all the information one might want to know in order to either understand the fundamental electrokinetic behavior of the AC electrophoresis, or to use it for determining the actual particle surface potential from experimental measurement. Indeed AC electrophoresis behaves as a generalization of the DC electrophoresis with the latter serving as the zero-frequency limit.

REFERENCES

[1] Hunter RJ. Recent developments in the electroacoustic characterisation of colloidal suspensions and emulsions. Colloids Surf A: Physicochem Eng Asp 1998;141(1):37–66.

[2] O'Brien R. Electro-acoustic effects in a dilute suspension of spherical particles. J Fluid Mech 1988;190:71–86.

[3] Gibb SE, Hunter RJ. Dynamic mobility of colloidal particles with thick double layers. J Colloid Interface Sci 2000;224(1):99–111.

[4] Dukhin AS, et al. Acoustic and electroacoustic spectroscopy. Colloids Surf A: Physicochem Eng Asp 2000;173(1–3):127–58.

[5] Guffond MC, et al. The dynamic mobility of colloidal semiconducting antimony-doped tin(IV) oxide particles. J Colloid Interface Sci 2001;235(2):371–9.

[6] Mizuno D, Kimura Y, Hayakawa R. Dynamic electrophoretic mobility of colloidal particles measured by the newly developed method of quasi-elastic light scattering in a sinusoidal electric field. Langmuir 2000;16(24):9547–54.

[7] Debye P. A method for the determination of the mass of electrolytic ions. J Chem Phys 1933;1(1):13–6.

[8] Booth F, Enderby JA. On electrical effects due to sound waves in colloidal suspensions. Proc Phys Soc Sect A 1952;65(5):321.

[9] Hermans J. XXXV. Charged colloid particles in an ultrasonic field. Lond Edinb Dublin Philos Mag J Sci 1938;25(168):426–38.

[10] Booth F, Enderby JA. On electrical effects due to sound waves in colloidal suspensions. Proc Phys Soc Sect A 1951;65(5):321.

[11] Dukhin A, et al. Electroacoustic phenomena in concentrated dispersions: new theory and CVI experiment. Langmuir 1999;15(20):6692–706.

[12] Jubery TZ, Srivastava SK, Dutta P. Dielectrophoretic separation of bioparticles in microdevices: a review. Electrophoresis 2014;35(5):691–713.

[13] Delgado AV, et al. Measurement and interpretation of electrokinetic phenomena. J Colloid Interface Sci 2007;309(2):194–224.

[14] Lee E, Yen FY, Hsu JP. Dynamic electrophoretic mobility of concentrated spherical dispersions. J Phys Chem B 2001;105(30):7239–45.

[15] Hsu JP, Lee E, Yen FY. Dynamic electrophoretic mobility in electroacoustic phenomenon: concentrated dispersions at arbitrary potentials. J Phys Chem B 2002;106(18):4789–98.

[16] Lee E, Fu CH, Hsu JP. Dynamic electrophoretic mobility of a concentrated dispersion of particles with a charge-regulated surface at arbitrary potential. J Colloid Interface Sci 2002;250(2):327–36.

[17] Tung CP, Lee E, Hsu JP. Dynamic electrophoretic mobility of a sphere in a spherical cavity. J Colloid Interface Sci 2003;260(1):118–25.

[18] Lee E, Min WL, Hsu JP. Dynamic electrophoresis of a droplet in a spherical cavity. Langmuir 2006;22(8):3920–8.

[19] Min WL, Lee E, Hsu JP. Dynamic electrophoresis of a spherical dispersion of soft particles subject to a stress-jump condition. J Colloid Interface Sci 2006;299(1):464–71.

[20] Hsu JP, Min WL, Lee E. Dynamic electrophoresis of droplet dispersions at low surface potentials. J Colloid Interface Sci 2007;306(2):421–7.

[21] Lou J, Lee E. Dynamic electrophoresis of concentrated droplet dispersions at arbitrary surface potentials. Chem Eng Sci 2007;62(22):6234–45.

[22] Hunter RJ, O'Brien RW. Electroacoustic characterization of colloids with unusual particle properties. Colloids Surf A: Physicochem Eng Asp 1997;126(2–3):123–8.

[23] Hawkins BG, Kirby BJ. Electrothermal flow effects in insulating (electrodeless) dielectrophoresis systems. Electrophoresis 2010;31(22):3622–33.

[24] Voldman J. Electrical forces for microscale cell manipulation. Annu Rev Biomed Eng 2006;8:425–54.

[25] Mangelsdorf CS, White LR. Electrophoretic mobility of a spherical colloidal particle in an oscillating electric field. J Chem Soc Faraday Trans 1992;88(24):3567–81.

[26] O'Brien RW, White LR. Electrophoretic mobility of a spherical colloidal particle. J Chem Soc Faraday Trans 2: Mol Chem Phys 1978;74:1607–26.

[27] Ohshima H. Dynamic electrophoretic mobility of a spherical colloidal particle. J Colloid Interface Sci 1996;179(2):431–8.

[28] Levine S, Neale GH. The prediction of electrokinetic phenomena within multiparticle systems. I. Electrophoresis and electroosmosis. J Colloid Interface Sci 1974;47(2):520–9.

[29] Ohshima H. Dynamic electrophoretic mobility of spherical colloidal particles in concentrated suspensions: approximation of nonoverlapping double layers. Colloids Surf A: Physicochem Eng Asp 1999;159(2):293–7.

[30] Shilov V, Zharkikh N, Borkovskaya YB. Theory of non-equilibrium electrosurface phenomena in concentrated disperse systems. 1. Application of non-equilibrium thermodynamics to cell model of concentrated dispersions. Colloid J USSR 1981;43(3):434–8.

[31] Shilov V, et al. Polarization of the electrical double layer. Time evolution after application of an electric field. J Colloid Interface Sci 2000;232(1):141–8.

Section B

Electrophoresis of Porous Particles

Chapter 7

Electrophoresis of Porous Particles

7.1 INTRODUCTION

In the last section, Chapters 1–6, we discussed the electrophoretic motion of rigid particles. In reality, however, many colloidal particles are not rigid but of porous nature in the sense that they are permeable to liquid medium, such as polyelectrolytes [1–4]. Polyelectrolytes are macromolecules having many dissociable functional groups [1], such as proteins and DNAs in biochemistry and biomedical fields [1, 4–6]. It is very important to understand their electrophoretic behavior as electrophoresis technique is widely used to separate them and explore their charge conditions, among other things [5, 7–11].

Polyelectrolytes have many unique features in comparison with the previously discussed rigid particles. Electrostatically speaking, the Coulomb attractive force is so strong between the single backbone of the charged macromolecule and the many dissociated small ion groups, or the counterions, that the actual charge level of the polyelectrolyte backbone is much less than the nominal one. This is due to the "shielding" or "screening" effect of the many counterions upon the originally highly charged backbone, a phenomenon called "counterion condensation" in electrostatics of polyelectrolytes [3, 12]. Moreover, the total amount of charges of a polyelectrolyte is normally much larger than that of a rigid particle, as the fixed charges of the former distribute over the entire interior of the porous particle, whereas the surface charges of the latter distribute on the particle surface only [1]. The size or molecular weight of a polyelectrolyte is generally much larger than a normal rigid particle as well.

Hydrodynamically speaking, the structure of a polyelectrolyte is of porous nature; hence, fluid can flow through it, to some degree at least [13]. As a result, the overall hydrodynamic drag force is much smaller than the corresponding rigid particle in general. Hence, in order to model a polyelectrolyte, or a porous particle in general, both the electrostatic nature of the fixed charges distributed within the particle and the hydrodynamic permeable

Interface Science and Technology, Vol. 26. https://doi.org/10.1016/B978-0-08-100865-2.00007-2

nature which allows the suspending medium to flow through it when the particle is in motion have to be reckoned with appropriately. A natural selection is to model a polyelectrolyte particle as a charged porous sphere with fixed charges distributed uniformly within it, for simplicity or at least as a primary approach [14, 15].

The idea of a charged porous sphere comes from two origins. The first one is from geometric analogy. There are many experimental observations of some DNA or protein molecules, indicating that they actually assume globular shape under the microscope due to the curvy and twisted tendency of the linear polymer backbone [5]. More importantly, from a conceptual point of view, every possible orientation would be swept over by a polyelectrolyte particle in motion after a certain period of time. Hence, averagely speaking, the overall morphology over a period would be similar to a conceptual spherical porous particle with the actual polyelectrolyte rotating dynamically within this frame from one orientation to another. This corresponds perfectly to a charged porous sphere over time [1].

Indeed, charged porous sphere has been used to model a polyelectrolyte successfully for a long time, as pointed out by Overbeek [4]. Since then, many more publications have been available in the literature in the theoretical front as well [16–18]. Hermans and Fujita [19, 20], in particular, considered the electrophoresis of a charged porous sphere under the Debye-Hückel approximation, which assumes the particle is lowly charged and hence the shape of the double layer remains unchanged when the particle is in motion; in other words, the effect of the double layer polarization is neglected. Analytical formula was obtained there. Lee's group, on the other hand, took full account of the effect of the double layer polarization and explored the behavior of charged porous particles, either lowly or highly charged, for a variety of systems of interest [21–27]. The polarization effect was found to be drastic in comparison with the corresponding rigid particle situation as the porous particle moves much faster due to higher electric driving force coupled with lower hydrodynamic drag force. Corresponding convection-induced double layer polarization, hence, is much stronger than the rigid particle situation in general.

To model the fluid flow within a porous medium, Brinkman equation is often adopted [13]. Based on the Darcy's law [28], an extra term accounting for the extra hydrodynamic drag force the fluid experiences flowing through a porous structure is added to the Stokes equation. This extra drag force is assumed to be proportional to the local fluid velocity within the porous medium and uniformly distributed within the entire porous region. The proportional constant, γ, is called the friction coefficient or the permeability of the medium and is a material constant. Debye and Bueche, on the other hand, obtained the same form of the equation for the fluid flowing through a dilute solution of linear polymers [29]. Together, they are often referred to as the Debye-Bueche-Brinkman equation or DBB equation [30]. Adopting the conceptual spherical space analog of a polyelectrolyte, Debye and Bueche's approach is a theoretically exact rigorous deduction based on the Einstein's formula [31], whereas Brinkman's approach is of semiempirical nature only, as it was originally proposed for a fluid flow through a heterogeneous packed bed, although it fits well with the experimental data.

For the purely hydrodynamic problem of a spherical porous sphere in creeping flow motion, Neale et al. obtained an analytical formula relating fluid velocity and drag force based on the Brinkman equation, which is a modified Stokes law for a porous sphere [32]. Indeed, the application of the Brinkman equation or simply DBB equation has become the standard approach treating the hydrodynamic response in a porous structure. In particular, it has also been found to be very suitable in the theoretical treatment of electrokinetic problems involving porous material [33]. Moreover, a category of colloids named "soft particle" is closely related to the porous particle considered here in this chapter as well. The soft particle is a rigid particle covered by a concentric outer layer of porous material [34]. The most famous member of this family is the spherical polyelectrolyte brush (SPB), which has found many potential applications in various fields such as the biomedical and the biochemical in recent years [35–40]. The outer porous layer dominates the electrokinetic behavior of a soft particle with the inner rigid core serving as an extra source of hydrodynamic drag force. Thus, it is similar to the porous system considered here, with many corresponding theoretical analyses reported as well [41–44].

Note that, strictly speaking, the precise structure of the double layer does not exist in porous particle systems, as there is no inner immobile compact Stern layer at all. Hence, there is no "double" layer. There is only one "single" layer here. The concept of electrolyte strength is still applicable, however, and so is its suppression effect upon the ion cloud of counterions. The concept of counterion condensation, on the other hand, bears some subtle similarity of the inner immobile compact layer. Note that as far as the electrophoretic motion is concerned, only the outer diffuse layer is of interest. Only it can interact with the applied electric field and generate electric driving force upon the particle. There is no fundamental difference in this regard between the porous and rigid particles, except that the diffuse layer here is not confined to the "outer" region exterior to the porous particle anymore. The interior of the porous particle or actually the entire domain of the system belongs to the "single" diffuse layer now. As a result, the double or single layer polarization takes place within the porous particle as well, which turns out to be the major source of the electric driving force, as we will discuss later. For simplicity and consistency, the terminology of the "double layer" as well as the "double layer polarization (effect)" is still used here, with the understanding that it is the "single" diffuse layer that we are referring to.

Here in this chapter, both a single porous particle and the suspension systems are considered regarding their electrophoretic behavior. Related contents can also be found elsewhere [26].

7.2 THEORY

7.2.1 Single Porous Sphere

We consider here a single charged spherical porous particle permeable to the suspending electrolyte solution and with fixed immobile electric charges uniformly distributed within the particle. The system diagram is shown in Fig. 7.1A.

Compared with the rigid particle situation, an extra domain within the porous particle needs to be considered here, which consists of the governing equations there as well as the electric and hydrodynamic boundary conditions on the porous particle surface and at the particle center. The governing equations of the electrolyte solution outside the porous particle and associated boundary conditions at infinity, on the other hand, are exactly the same as those for a rigid particle situation treated in Chapter 1, hence will not be repeated here again.

The governing equations inside the porous particle are:

$$\nabla^2 \phi = -\frac{\rho + \rho_{\mathrm{fix}}}{\varepsilon} \tag{7.1}$$

$$\eta_{\mathrm{f}} \nabla^2 \boldsymbol{v} - \nabla p - \rho \nabla \phi - \gamma \boldsymbol{v} = 0 \tag{7.2}$$

$$\nabla \cdot \boldsymbol{v} = 0 \tag{7.3}$$

where ρ_{fix} is the fixed charge density carried by the porous particle, and γ is the friction coefficient inside the porous particle representing the extra drag force proportional to the local fluid velocity in a Brinkman medium. Note that γ is also a measurement of particle permeability. The smaller γ is, the higher the particle permeability. In practice, γ is often replaced by the Brinkman screening length λ, also called as the Debye-Bueche screening length, which is defined as $\lambda^{-1} = \sqrt{\frac{\eta_{\mathrm{f}}}{\gamma}}$. λ^{-1} represents the penetration length of a fluid into the porous structure, and $\lambda a = \frac{a}{\lambda^{-1}}$ is its dimensionless form. Thus, λ or λa can be regarded as a measurement of particle permeability as well. The smaller λa is, the higher the particle permeability is.

Eq. (7.1) is the Poisson equation with an extra contribution of the fixed charges carried by the porous particle, and Eq. (7.2) is the modified Brinkman equation with an extra local electric body force term $-\rho \nabla \phi$.

At equilibrium state, the particle is motionless, and Eq. (7.1) reduces to

$$\nabla^{*2} \phi_{\mathrm{e}}^{*} = -\frac{(\kappa a)^2}{1+\alpha} \left[\exp\left(-\phi_{\mathrm{e}}^{*}\right) - \exp\left(\alpha \phi_{\mathrm{e}}^{*}\right) \right] - Q_{\mathrm{fix}}, \quad 0 \le r^{*} \le 1 \tag{7.4}$$

where Q_{fix} is the dimensionless form of ρ_{fix} defined as $Q_{\mathrm{fix}} = \frac{\rho_{\mathrm{fix}} a^2}{\varepsilon \phi_0}$.

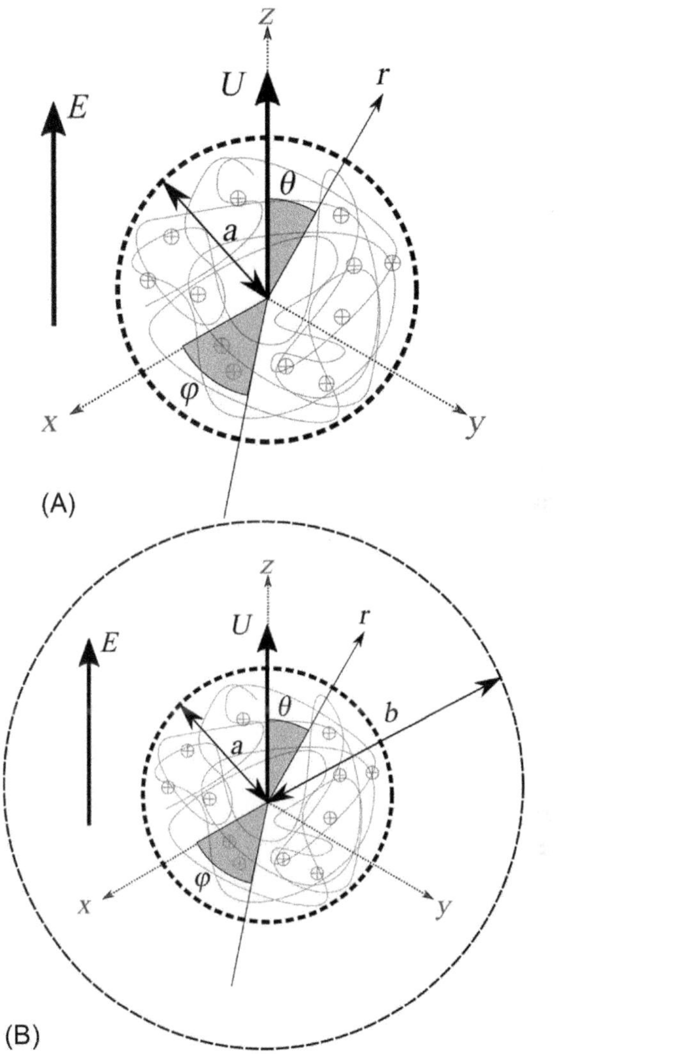

FIG. 7.1 System diagram. (A) A single porous particle; (B) suspension of porous particles.

The associated boundary conditions at $r^* = 0$ and $r^* = 1$ are respectively:

$$\frac{d\phi_e^*}{dr^*} = 0, \quad r^* = 0 \tag{7.5}$$

$$\phi_e^*\big|_{r^*=1^-} = \phi_e^*\big|_{r^*=1^+}, \quad r^* = 1 \tag{7.6}$$

$$\left.\frac{d\phi_e^*}{dr^*}\right|_{r^*=1^-} = \left.\frac{d\phi_e^*}{dr^*}\right|_{r^*=1^+}, \quad r^* = 1 \tag{7.7}$$

Note that the continuity condition across the porous surface is assumed in Eqs. (7.6) and (7.7).

The boundary condition at infinity is the same as those used in Chapter 1, where Debye-Hückel approximation is used there.

Once an external electric field is applied, we have:

$$\nabla^{*2}\delta\phi^* - \frac{(\kappa a)^2}{1+\alpha}\left[\exp(-\phi_e^*) + \alpha\exp(\alpha\phi_e^*)\right]\delta\phi^* = \frac{(\kappa a)^2}{1+\alpha}\left[\exp(-\phi_e^*)g_1^* + \alpha\exp(\alpha\phi_e^*)g_2^*\right] \tag{7.8}$$

$$\nabla^{*2}g_1^* - \nabla^*\phi_e^* \cdot \nabla^* g_1^* - Pe_1 \mathbf{v}^* \cdot \nabla^*\phi_e^* = 0 \tag{7.9}$$

$$\nabla^{*2}g_2^* + \alpha\nabla^*\phi_e^* \cdot \nabla^* g_2^* - Pe_2 \mathbf{v}^* \cdot \nabla^*\phi_e^* = 0 \tag{7.10}$$

$$E^{*4}\Psi^* - (\lambda a)^2 E^{*2}\Psi^* + \sin\theta\left(\frac{d\rho_e^*}{dr^*}\frac{\partial\delta\phi^*}{\partial\theta} - \frac{d\phi_e^*}{dr^*}\frac{\partial\delta\rho^*}{\partial\theta}\right) = 0, \quad 0 \le r^* \le 1 \tag{7.11}$$

where λa is the dimensionless permeability. Note that the extra hydrodynamic force in the Brinkman equation is proportional to the square of λa according to Eq. (7.11).

Corresponding boundary conditions at the particle center ($r^* = 0$) are:

$$\delta\phi^* = g_j^* = 0, \; j = 1,2, \quad r^* = 0 \tag{7.12}$$

$$\frac{\partial\Psi^*}{\partial\theta} = \frac{\partial\Psi^*}{\partial r^*} = 0, \; r^* = 0 \tag{7.13}$$

where $\delta\phi^*$ and g_j^* are set to zero in Eq. (7.12) as the reference potentials in Eqs. (7.8)–(7.10), and Eq. (7.13) is from the finiteness of velocity at the particle center by the definition of stream function.

On the surface of the porous particle, continuity of electrokinetic variables is assumed, they are:

$$\delta\phi^*|_{r^*=1^-} = \delta\phi^*|_{r^*=1^+} \tag{7.14}$$

$$\left.\frac{\partial\delta\phi^*}{\partial r^*}\right|_{r^*=1^-} = \left.\frac{\partial\delta\phi^*}{\partial r^*}\right|_{r^*=1^+} \tag{7.15}$$

$$\left.g_j^*\right|_{r^*=1^-} = \left.g_j^*\right|_{r^*=1^+}, \; j = 1,2 \tag{7.16}$$

$$\left.\frac{\partial g_j^*}{\partial r^*}\right|_{r^*=1^-} = \left.\frac{\partial g_j^*}{\partial r^*}\right|_{r^*=1^+}, \; j = 1,2 \tag{7.17}$$

$$\Psi^*|_{r^*=1^-} = \Psi^*|_{r^*=1^+} \tag{7.18}$$

$$\left.\frac{\partial\Psi^*}{\partial r^*}\right|_{r^*=1^-} = \left.\frac{\partial\Psi^*}{\partial r^*}\right|_{r^*=1^+} \tag{7.19}$$

$$\left.\frac{\partial^2 \Psi^*}{\partial r^{*2}}\right|_{r^*=1^-} = \left.\frac{\partial^2 \Psi^*}{\partial r^{*2}}\right|_{r^*=1^+} \tag{7.20}$$

$$\left.\left[\frac{\partial^3 \Psi^*}{\partial r^{*3}} - (\lambda a)^2 \frac{\partial \Psi^*}{\partial r^*}\right]\right|_{r^*=1^-} = \left.\frac{\partial^3 \Psi^*}{\partial r^{*3}}\right|_{r^*=1^+} \tag{7.21}$$

Note that Eq. (7.20) stands for the continuity of shear stress. Eq. (7.21), on the other hand, stands for the continuity of normal stress. The rigorous mathematical derivation can be found elsewhere [26, 32].

The governing system in the extra domain discussed above is solved simultaneously with the outside governing system of equations plus boundary conditions at infinity, after the separation of variables and conversion to the one-dimensional form as treated in Chapter 1. Once the flow field and the electric field are solved, corresponding hydrodynamic drag force and electric driving force, in dimensionless form, can be evaluated as follows:

$$F_{\mathrm{D}}^* = \iiint_{r^* \le 1} (\lambda a)^2 \mathbf{v}^* \cdot \mathbf{i}_z dV = -\frac{8}{3}\pi (\lambda a)^2 [\psi^*]_{r^*=1} \tag{7.22}$$

$$F_{\mathrm{E}}^* = \iiint_{r^* \le 1} Q_{\mathrm{fix}} (-\nabla^* \delta\phi^*) \cdot \mathbf{i}_z dV = -\frac{4}{3}\pi Q_{\mathrm{fix}} [\delta\Phi^*]_{r^*=1} \tag{7.23}$$

The mobility is then obtained via the same methodology of decomposition into two subproblems as treated in Chapter 1.

7.2.2 Suspension of Porous Particles

The system diagram is shown in Fig. 7.1B, where Kuwabara's hydrodynamic unit cell model is used coupled with the Shilov-Zharkikh (S-Z) electric boundary conditions on the outer virtual surface, as adopted in Chapter 2.

The governing equations and associated boundary conditions are otherwise the same as those discussed above for a single porous particle situation, and the particle mobility can, thus, be obtained following the same treatment as above.

7.3 RESULTS AND DISCUSSION

7.3.1 Single Porous Particle

We first discuss the electrostatic feature of a porous particle, that is, when the particle is at rest. Due to the permeability of the particle, the counterions may reside within the porous particle, and hence, reduce the nominal fixed charge density there, which is referred to as the "counterion condensation effect," as discussed in the Introduction section earlier. This is shown in Fig. 7.2A–C for a highly permeable porous particle ($\lambda a = 1$) at various charge levels

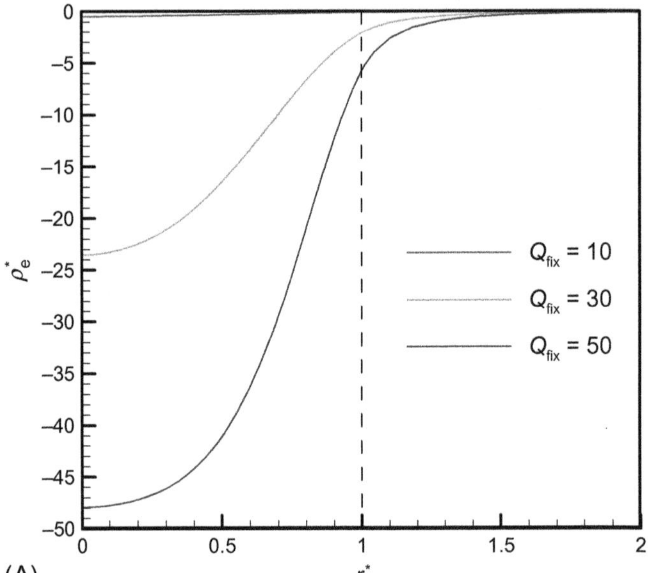

(A)

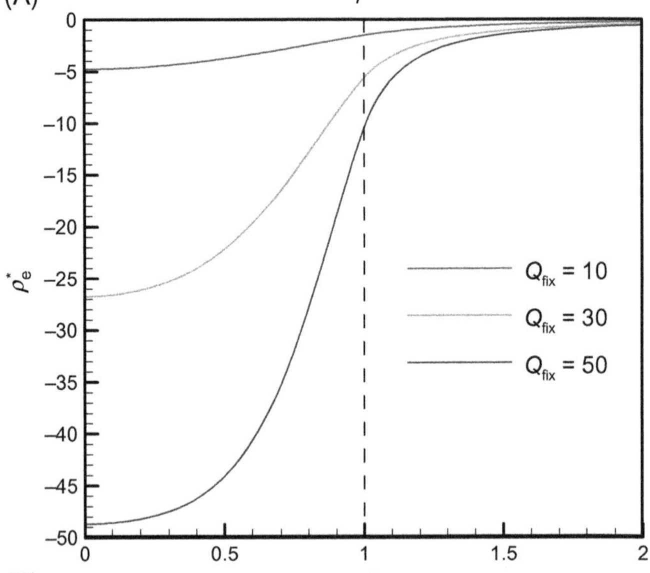

(B)

FIG. 7.2 Scaled equilibrium charged density (ρ_e^*) as a function of dimensionless radius (r^*) for a highly permeable porous particle ($\lambda a=1$) at various fixed charge density (Q_{fix}) and κa. (A) $\kappa a=0.1$; (B) $\kappa a=1$; (C) $\kappa a=10$.

($Q_{\text{fix}}=10$, 30, and 50, respectively) and double layer thickness ($\kappa a=0.1$, 1, and 10, respectively). Nearly complete screening or shielding of the particle fixed charges is observed due to the powerful suppression effect of the strong electrolyte strength, or equivalently very thin double layer ($\kappa a=10$) in

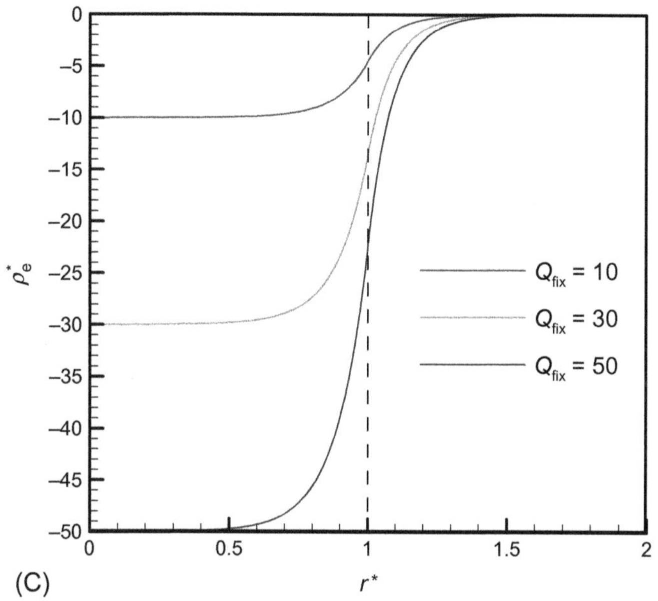

(C)

FIG. 7.2—CONT'D

Fig. 7.2C. The inner core of the interior region can actually be regarded as chargeless due to this overwhelming shielding effect. Only a small portion near the boundary bears some remaining effective particle charges: the summation of the fixed charges carried by the porous particle and the space charges contained in the electrolyte solution penetrating into the particle interior. This is much smaller than the nominal fixed charges, especially for a highly charged particle at large κa.

Note that this shielding effect is only an electrostatic feature when the particle is at rest. Once an external electric field is applied, redistribution/polarization of the counterions takes place all over the particle interior, whether it is in the shielded inner core or the relatively non-shielded outer region. This is due to the fact that the electrolytes in the suspending medium after all are mobile in nature, thus will respond to the superimposed electric field to a certain extent; hence, a local polarization/deformation of the original equilibrium ions distribution is resulted nonetheless. This in turn generates an induced electric field, which sets the particle in motion. Hence, the electrostatic counterion condensation effect does not nullify the electrophoretic motion of a porous particle. It only affects the particle motion indirectly via its impact upon the equilibrium distribution of counterions within the porous particle; hence, the induced local electric field, as indicated by Eq. (7.23). Indeed, the situation is complicated and full of subtlety.

Corresponding profiles of the scaled electrostatic potential are shown in Fig. 7.3A–C. Note that the counterion condensation effect or shielding effect

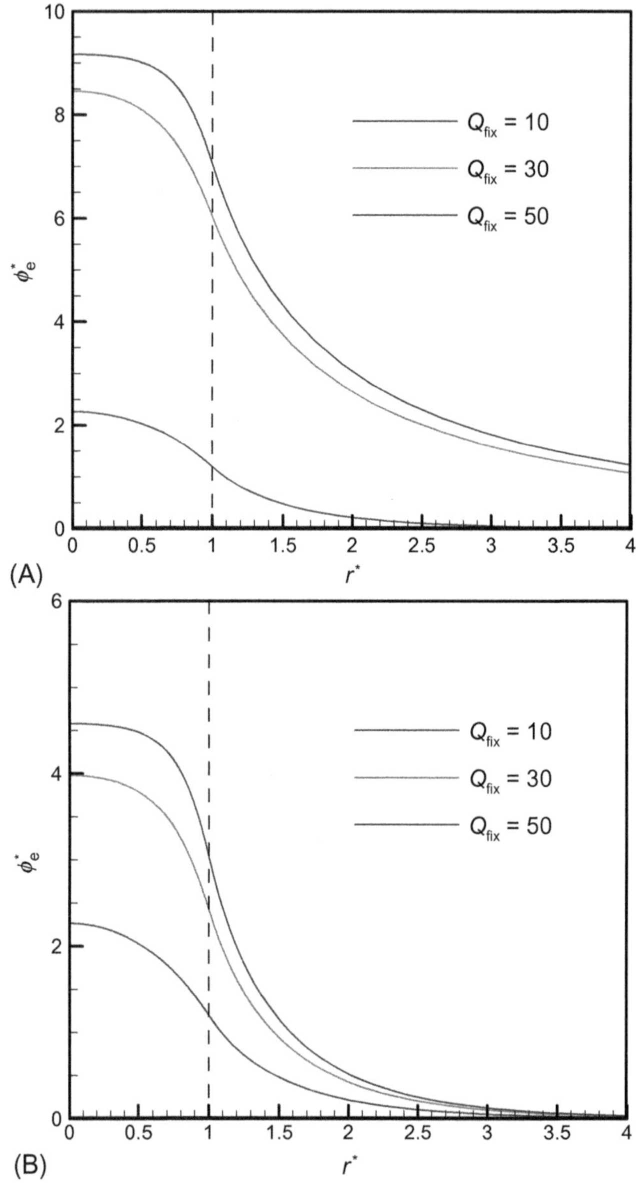

FIG. 7.3 Scaled equilibrium potential (ϕ_e^*) as a function of dimensionless radius (r^*) for a highly permeable porous particle ($\lambda a = 1$) at various fixed charge density (Q_{fix}) and κa. (A) $\kappa a = 0.1$; (B) $\kappa a = 1$; (C) $\kappa a = 10$.

suppresses the fixed charges so completely at large κa in Fig. 7.3C that the electric potentials are almost null there.

As for the electrophoretic motion of a porous particle, some comments need to be addressed first. First of all, there is no such a concept as surface

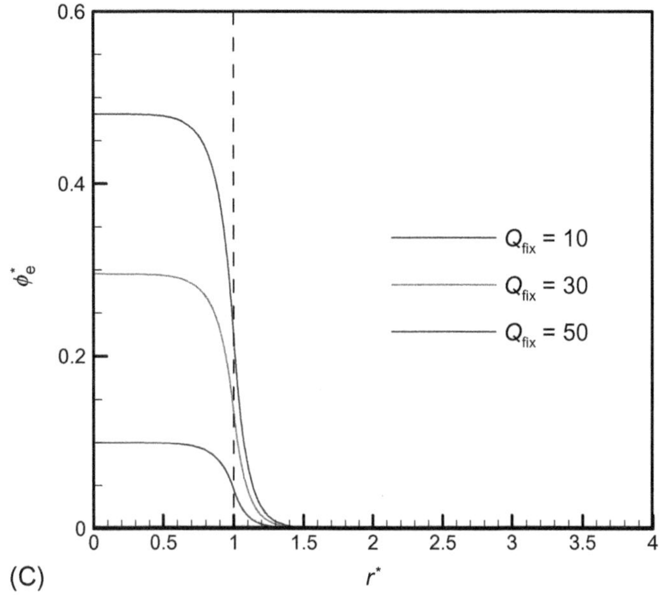

(C)

FIG. 7.3—CONT'D

potential or zeta potential in porous particles. The electric charges are distributed instead within the entire interior region of the particle characterized by the fixed charge density, ρ_{fix}, which in dimensionless form is Q_{fix}. Thus, Q_{fix} replaces the role of ϕ_r as the measurement of the charge level of a porous particle. Secondly, there is no such a concept as "shear plane" in porous particles either. The fluid can simply flow through a porous particle freely in principle; no shear plane and no need for it. Thirdly, there is no immobile inner compact Stern layer in porous particles. Thus, there is no "double layer" structure as well in porous particles. There is only one single diffuse layer of predominantly counterions, strictly speaking, which in practice is spreading over the entire system, both inside and outside the porous particle.

Finally, the role or physical meaning of κa in rigid particles fails in porous particles. Due to the particle permeability, both the electric driving force and hydrodynamic drag force are generated within the porous particle, the center stage of electrophoretic motion. Larger κa, or stronger electrolyte strength, implies more counterions will be driven into the particle interior to contribute to the particle motion. The counterions left outside are essentially useless except making some trivial motion-deterring polarization effect. The major polarization effect that retards the particle motion takes place mainly within the particle where the antisymmetric local polarization of counterions drives the particle in motion and, at the same time, generates a collective induced bulk electric field against the particle motion to some extent. Overall, however, the more counterions are driven into the interior of the porous particle, the faster the particle moves in general.

Note also that the dimensionless term $Q_{fix}/(\lambda a)^2$ stands out as a natural characteristic particle mobility in general, judging from the expressions of the dimensionless electric driving force, F_E^*, and the hydrodynamic drag force, F_D^*, in Eqs. (7.22) and (7.23). When the two forces are balanced and the particle is conducting electrophoretic motion, we have:

$$\frac{Q_{fix}}{(\lambda a)^2} = \frac{\iiint_{r^* \leq 1} \boldsymbol{v}^* \cdot \mathbf{i_z} dV}{\iiint_{r^* \leq 1} (-\nabla^* \delta\phi^*) \cdot \mathbf{i_z} dV} \tag{7.24}$$

whose exact physical meaning will be discussed later. Moreover, as we mentioned in the Introduction section, Hermans and Fujita [19] derived an analytical formula for a porous particle under Debye-Hückel approximation assuming the shape of double layer remains unchanged during motion, i.e., $\delta\rho^* = 0$. This is valid for a lowly charged particle where the motion-deterring double layer polarization effect is negligible. The exact form of the formula goes like the following:

$$\mu_{HF}^* = \frac{Q_{fix}}{(\lambda a)^2}\left(1 - \left\{\frac{1}{\kappa a} - \frac{\kappa a + 1}{(\kappa a)^2 - v^2}\left[\frac{1}{\Gamma(\kappa a)} - \frac{v^2}{(\kappa a)^2}\frac{1}{\Gamma(v)}\right]\right\} v^2 \Gamma(\kappa a) e^{-\kappa a} \cosh(\kappa a)\right) \tag{7.25}$$

where $v = \lambda a\sqrt{\frac{2}{3}}$ and $\Gamma(x) = 1 - \frac{\tanh(x)}{x}$. Again, the dimensionless term $Q_{fix}/(\lambda a)^2$ appears explicitly in this analytical formula.

Besides, as the dimensionless particle mobility, $\mu^* = \frac{U^*}{E^*}$, is a strong function of particle permeability λa, in the way that the smaller λa is, the faster (much faster!) the particle moves, it would be difficult to present mobility profiles within one frame for different particles with permeability varying a lot. Hence, we decide to define a scaled particle mobility as a ratio to $Q_{fix}/(\lambda a)^2$ for convenience:

$$\mu_s^* = \frac{\mu^*}{Q_{fix}/(\lambda a)^2} \tag{7.26}$$

Moreover, the ratio of μ^* to μ_{HF}^* is also adopted to express particle mobility whenever necessary, which is a correction factor highlighting the nonlinear behavior of highly charged porous particles. Those being said, we go on to investigate the electrophoretic motion of a charged porous particle.

Fig. 7.4A–D presents the general Henry's charts for a porous particle with various permeability ($\lambda a = 0.3$, 1, 3, and 10, respectively). Note that $(\lambda a)^2$ goes over four orders of magnitude in terms of extra drag force generated. Ratio to the corresponding analytical prediction by Hermans and Fujita [19], μ^*/μ_{HF}^*, is picked to reveal the underlying nonlinear physical essence. The ratio, or the correction factor in nature, is always less than unity as the neglected polarization effect by Hermans and Fujita is motion-deterring in general. Moreover, they used the Helmholtz linear equation under the

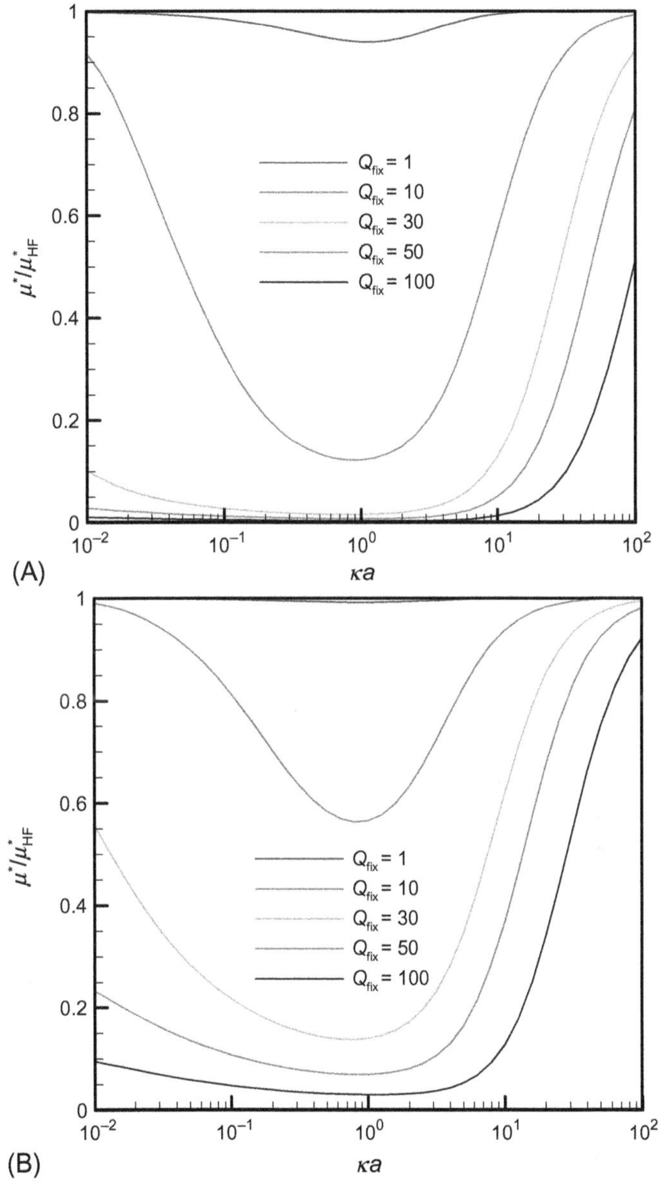

FIG. 7.4 Ratio of the scaled mobility to the analytical prediction by Hermans and Fujita (μ^*/μ^*_{HF}) as a function of κa for a porous particle with various permeability (λa). (A) $\lambda a = 0.3$; (B) $\lambda a = 1$; (C) $\lambda a = 3$; (D) $\lambda a = 10$.

Debye-Hückel approximation, which tends to underestimate the Coulomb attraction force between the charged particle and the counterions around it, compared with the rigorous Poisson-Boltzmann equation. Hence, the counterion condensation effect, or the shielding effect of the nominal charge

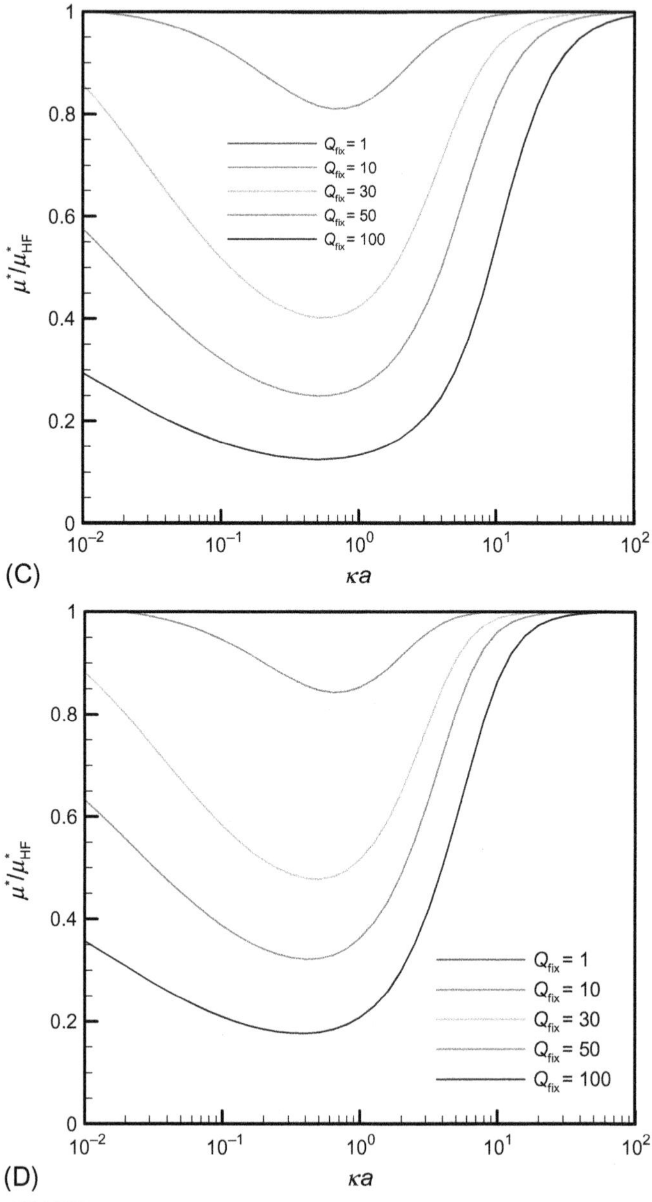

FIG. 7.4—CONT'D

level by the driven-in counterions, cannot be taken into count properly, leading to an overestimated effective charge level of the porous particle, hence its mobility. As a result, the actual mobility of a highly charged porous particle is drastically smaller than the one predicted by the analytical formula, especially when the particle is also highly permeable, as shown in Fig. 7.4A with $\lambda a = 0.3$.

Note that it is difficult to separate these two factors precisely though, as electrostatic behavior is indeed the foundation of the corresponding electrophoretic behavior. An in-depth analysis has been provided by Lee et al. [26] and is recommended for further reading by interested readers. In practice, however, there is no need to further separate these two closely related nonlinear factors. Hence, the deviation from unity can be regarded as the extent of the double layer polarization effect plus the counterion condensation or the shielding effect, or in general the nonlinear effects. Severe nonlinear effect is observed in general for particles both highly charged and permeable. Moreover, based on these figures, we can conclude that Q_{fix} greater than or around 30 can be regarded as highly charged, whereas λa lesser than or around unity can be taken as highly permeable. Corresponding reduction of particle mobility is closely related to the fast particle motion for sure. As mentioned earlier in the Introduction section, the significant increase of total charges due to the distribution of fixed charges within the entire interior of the porous particle results in a strong electric driving force in general for a porous particle. On the other hand, drastic reduction of hydrodynamic drag force is resulted for a highly permeable particle as well. Combination of these two factors leads to a much faster particle motion for a porous particle both highly charged and highly permeable, which leads to a much profound motion-deterring double layer polarization effect in general compared with the rigid particle situation. This is further demonstrated in Table 7.1, where ratios of velocities of a porous particle to a reference rigid particle with $\phi_r = 1$ are shown under various electrokinetic conditions. More than 20 times faster speed is observed for a highly permeable particle ($\lambda a = 1$) when Q_{fix} is 30 and κa is 10, for instance.

Moreover, all the mobility profiles in Fig. 7.4 approach asymptotically to unity as κa increases to around 100. This implies they all approach $Q_{\text{fix}}/(\lambda a)^2$, the so-called free-drainage situation [20, 22] when the electrophoretic mobility of a polymer polyelectrolyte reduces to the mobility of an individual polymer segment or monomer. In other words, if a polyelectrolyte is modeled as a charged porous sphere, then $Q_{\text{fix}}/(\lambda a)^2$ stands for the mobility of a monomer composing it. Under this limiting condition, the applied electric field as well as the flow field is not distorted by the presence of the porous particle at all [19]. The nearly complete shielding effect by the nearly perfect counterion condensation when κa is large explains the underlying physical mechanism [19, 20]. Note that when there is no extra electric driving force from Q_{fix}, there will be no extra hydrodynamics drag force from γ (or $(\lambda a)^2$) in response and the free-drainage situation is established there.

Corresponding contour plots of the scaled double layer polarization ($\delta\rho^*$) are shown in Fig. 7.5A–C for a highly permeable particle with $\lambda a = 1$ at $\kappa a = 1$ and various Q_{fix} levels. Hardly any visible polarization is observed for $Q_{\text{fix}} = 1$ and thus is not shown here, whereas the clearly noticeable extent

TABLE 7.1 Ratio of the Porous Particle Velocity to a Constant Reference Velocity of a Rigid Particle With $\phi_r = 1$

	Q_{fix}					
κa	*1*	*10*	*20*	*30*	*50*	*100*
$\lambda a = 1$						
0.1	1.9	15.2	12.2	10.1	8.9	1.9
1	1.7	9.5	7.0	5.8	5.0	1.7
10	1.3	11.8	23.4	23.2	16.2	1.3
$\lambda a = 3$						
0.1	0.5	4.9	8.2	8.4	8.4	0.5
1	0.3	2.8	4.4	4.6	4.6	0.3
10	0.1	1.5	4.1	6.0	8.0	0.1
$\lambda a = 10$						
0.1	0.3	3.3	6.0	6.7	7.2	0.3
1	0.2	1.5	2.6	3.1	3.6	0.2
10	0.0	0.2	0.5	0.9	1.6	0.0

for $Q_{fix} = 10$ is shown in Fig. 7.5A. Severe antisymmetric polarization is observed in Fig. 7.5B for $Q_{fix} = 30$ as well as in Fig. 7.5C for $Q_{fix} = 50$. Note that the double layer polarization for a porous particle mainly exists within the particle interior, unlike the corresponding rigid particle situation where the double layer polarization, and actually the entire double layer itself, exists exclusively in the exterior to the particle.

Fig. 7.6A–C presents the corresponding scaled disturbed electric potential ($\delta\phi^*$) when an external electric field is applied. Note that the gradient of these contour plots stands for the induced local electric field ($\nabla^*\delta\phi^*$), and collectively, the overall electric driving force $F_E^* = \iiint_{r^* \leq 1} Q_{fix}(-\nabla^*\delta\phi^*) \cdot \mathbf{i}_z dV$ in Eq. (7.23) is thus determined. Note also that the nominal particle charge level Q_{fix} is involved directly, instead of the effective charge level considering the electrostatic counterion condensation or the shielding effect. The shielding effect affects directly the equilibrium distribution of ions ρ_e^* thus $\delta\rho^*$, and indirectly the local electric field $\nabla^*\delta\phi^*$ by determining the local $\delta\phi^*$ according to the Poisson equation or Gauss law. For a relatively lowly charged particle ($Q_{fix} = 10$), the gradients are essentially along the electric field, whereas for relatively highly charged particles ($Q_{fix} = 30$ and $Q_{fix} = 50$), local maxima and local minima of $\delta\phi^*$ are observed in either hemisphere, respectively. As a result, some extent of the reduction of electric driving force happens

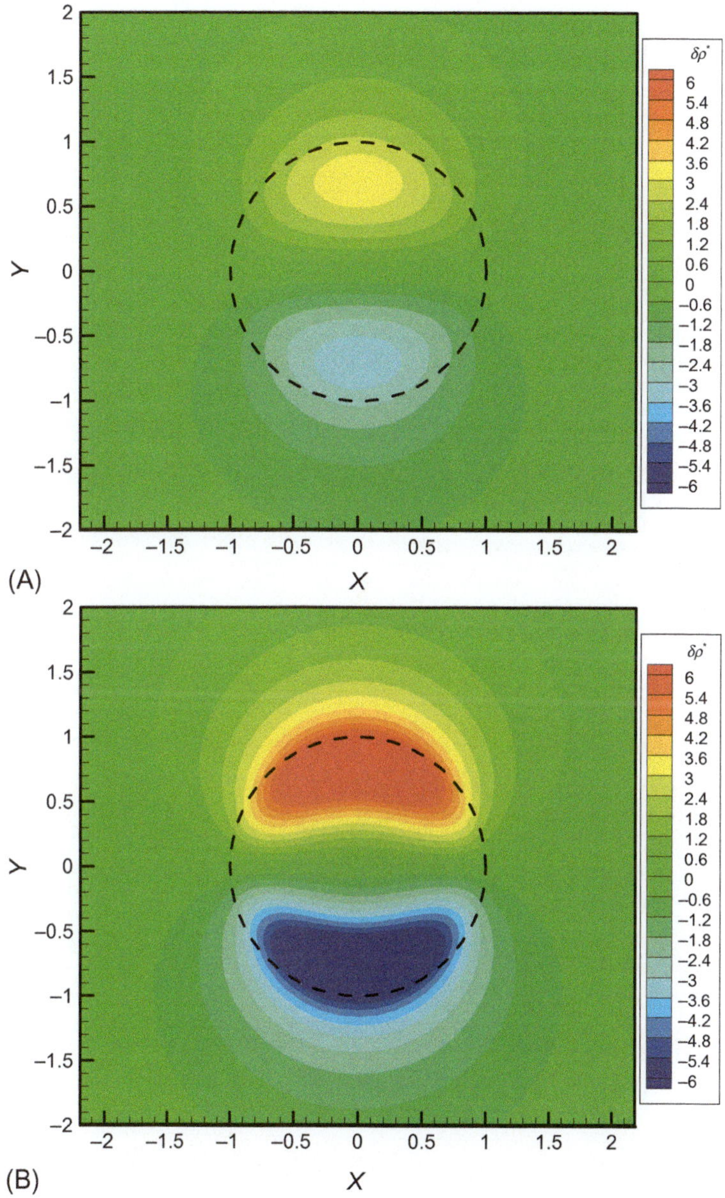

FIG. 7.5 Contour plots of the scaled double layer polarization ($\delta\rho^*$) for a highly permeable porous particle ($\lambda a = 1$) at $\kappa a = 1$ and various Q_{fix}. (A) $Q_{\text{fix}} = 10$; (B) $Q_{\text{fix}} = 30$; (C) $Q_{\text{fix}} = 50$.

due to the appearance of some gradients pointing against the direction of the applied electric field. Hence, a less charged porous particle may actually move faster than a highly charged one in principle, depending on how severe this reduction is. This will be discussed later again.

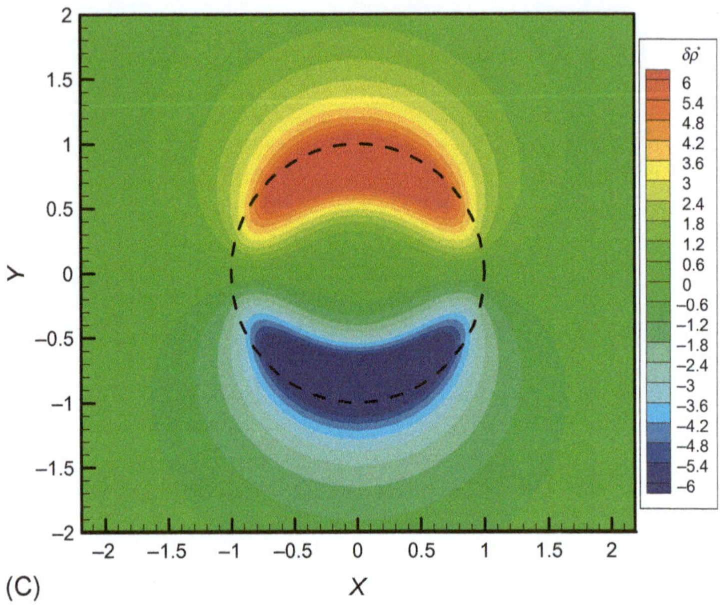

FIG. 7.5—CONT'D

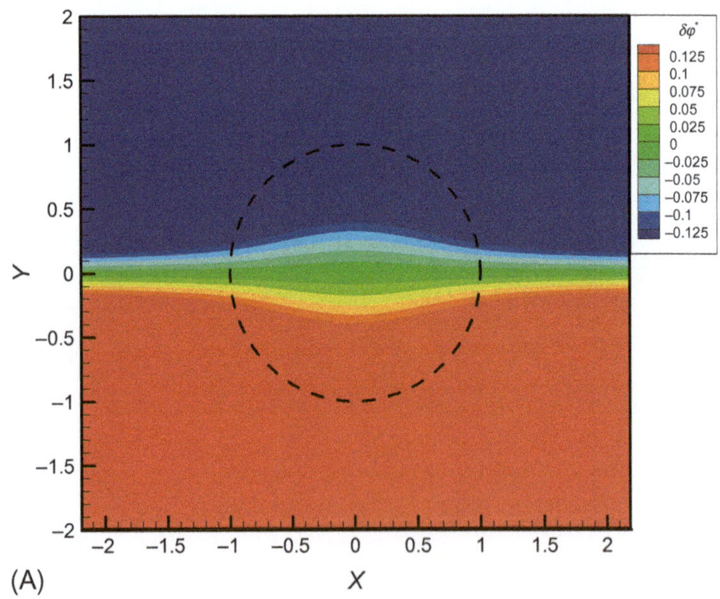

FIG. 7.6 Contour plots of the scaled disturbed potential ($\delta\phi^*$) for a highly permeable porous particle ($\lambda a = 1$) at $\kappa a = 1$ and various $Q_{\rm fix}$. (A) $Q_{\rm fix} = 10$; (B) $Q_{\rm fix} = 30$; (C) $Q_{\rm fix} = 50$.

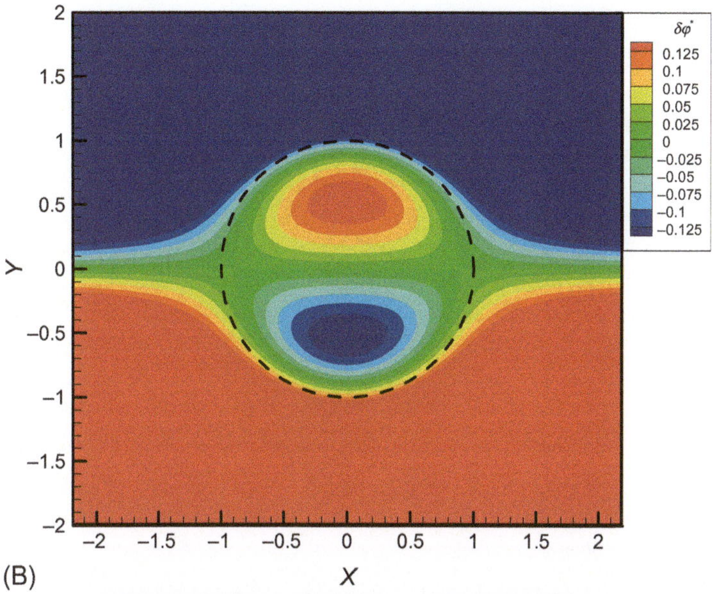

(B)

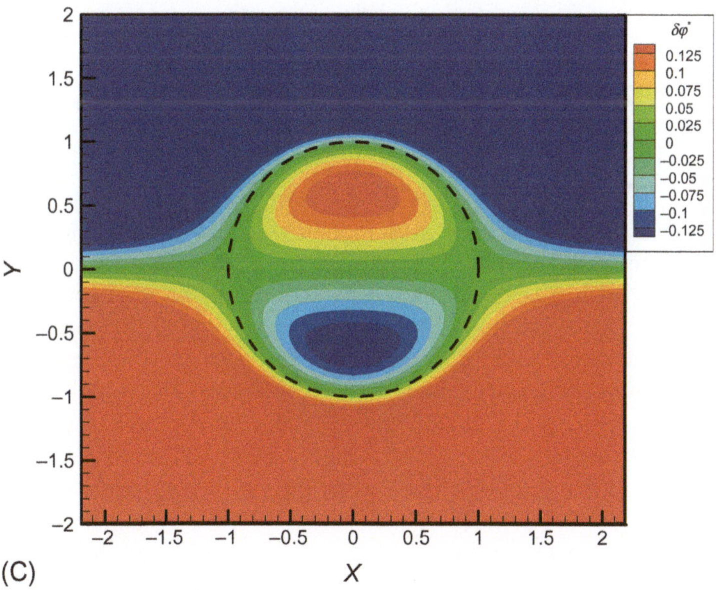

(C)

FIG. 7.6—CONT'D

Fig. 7.7 demonstrates the impact of the particle permeability ranging from nearly rigid ($\lambda a = 10$) to extremely permeable ($\lambda a = 0.3$) for a highly charged particle with $Q_{\text{fix}} = 30$. The particle mobility as well as its polarization effect increases with increasing permeability as we have discussed earlier in

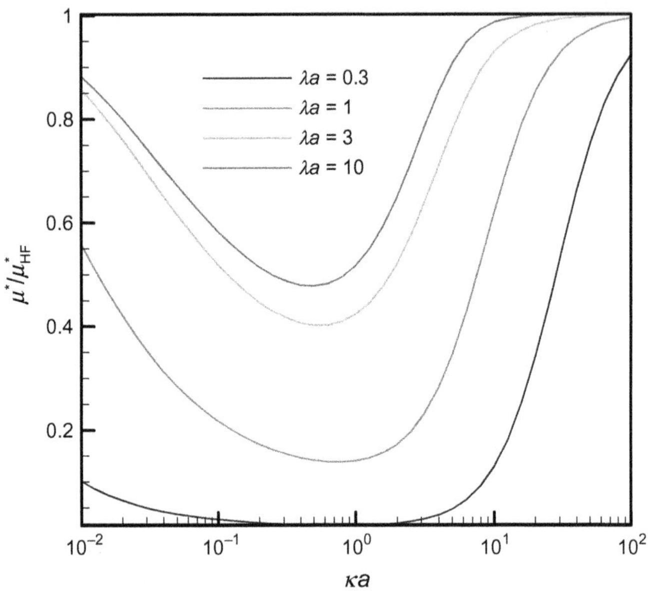

FIG. 7.7 Ratio of the scaled particle mobility to the analytical prediction by Hermans and Fujita (μ^*/μ^*_{HF}) as a function of κa for a highly charged particle ($Q_{\text{fix}}=30$) with various permeability (λa).

Fig. 7.4. Unlike the other nonlinear effect of counterion condensation, which is purely an electrostatic effect taking place when particle is at rest and remains that way even when it is in motion, the impact of particle permeability here takes place only when the particle is in motion. Hence, its impact on particle motion is mainly via the double layer polarization effect. The higher the permeability is, the faster the particle moves due to the decrease of the extra hydrodynamic drag force. Note that $(\lambda a)^2$ goes over four orders of magnitude and it would be awkward to present with true particle mobility. This is further demonstrated in Fig. 7.8A as well where the profiles of the dimensionless particle mobility ($\mu^* = \frac{U^*}{E^*}$) of a highly permeable particle ($\lambda a=1$) are compared with the imaginary ones when the Peclet numbers are set to zero. In other words, the convection mechanism is not considered at all in the Nernst-Planck ion flux equation, hence the convection-induced polarization effect is eliminated. The counterion condensation or shielding effect is retained there as the nonlinear Poisson-Boltzmann equation is used nonetheless when the particle is at rest. For a highly charged particle with $Q_{\text{fix}}=100$ at $\kappa a=10$, for instance, roughly ten times fastness is observed in particle velocity without this motion-deterring effect from convection. In other words, the convection mechanism is dominant in the overall polarization effect for porous particles. Complete reverse of the role is observed here in comparison

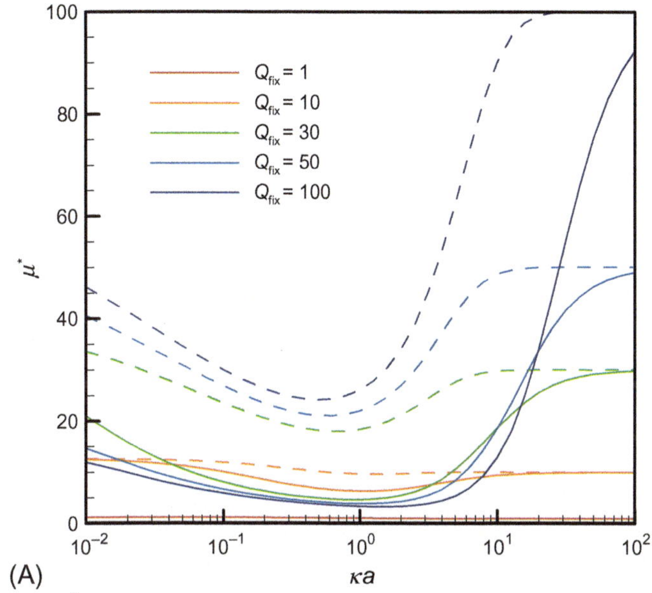

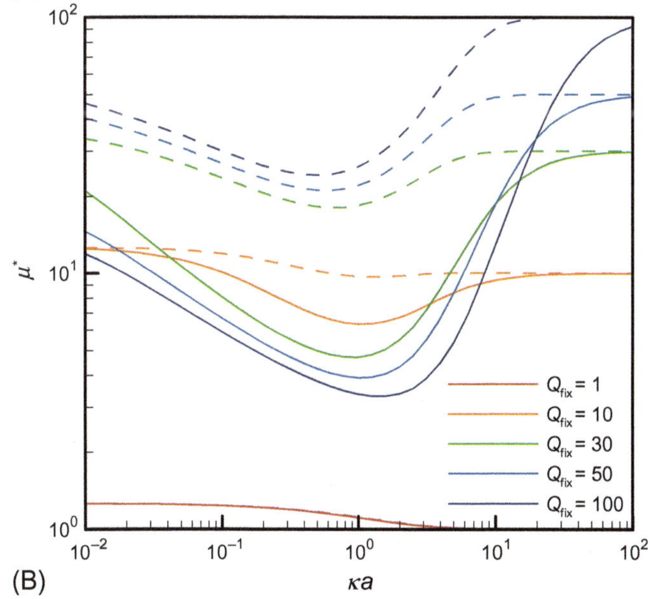

FIG. 7.8 Dimensionless particle mobility μ^* as a function of κa for a highly permeable porous particle ($\lambda a = 1$) with various fixed charge density (Q_{fix}). *Dashed lines*: $Pe_1 = Pe_2 = 0$. (A) Normal scale ordinate; (B) log scale ordinate; (C) ratio to μ^*_{HF}.

with the rigid particle where the electro-migration mechanism dominates. For clarity and completeness purpose, the figure is presented in log scale of the particle mobility again in Fig. 7.8B, and as the ratio to the corresponding Hermans and Fujita's prediction in Fig. 7.8C.

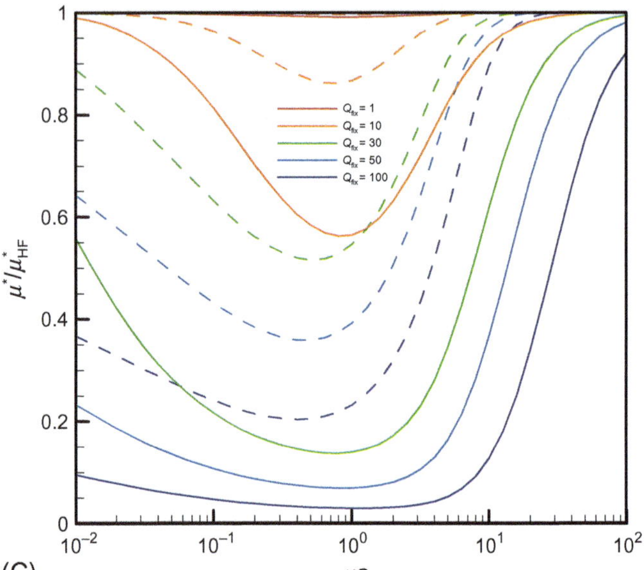

(C)

FIG. 7.8—CONT'D

Note that it is clearly shown in Fig. 7.8B around $\kappa a = 1$ that a less charged porous particle may move faster than a highly charged one at κa around unity. The reason has been attributed to the nonlinear effect of highly charged particles and graphically elaborated in Figs. 7.5 and 7.6. Here we check the behavior of particle motion at two other representative double layer thicknesses as well ($\kappa a = 0.1$ and 10) for particles with different permeability ($\lambda a = 1$, 3, and 10, respectively). The results are shown in Fig. 7.9A–C where particle mobilities are expressed as functions of Q_{fix} for $\kappa a = 0.1$, 1, and 10, respectively. The intriguing phenomenon that a less charged particle may move faster than a highly charged one is also observed at $\kappa a = 0.1$ and 10 for a highly permeable particle ($\lambda a = 1$), as shown in Fig. 7.9A. In fact, the higher the particle charge level is, the slower the particle moves, with a local maximum or minimum in each profile, depending on the sign of Q_{fix}. No such phenomenon for particles less permeable is shown in Fig. 7.9B ($\lambda a = 3$) and Fig. 7.9C ($\lambda a = 10$).

7.3.2 Suspensions

As shown in Fig. 7.10A–C, where the general Henry's charts are presented for a highly permeable ($\lambda a = 1$) porous particle with various charge levels ($Q_{\text{fix}} = 10$, 30, and 50) and particle volume fractions ($H = 10^{-6}$ to 0.4), the behavior is very complicated. First of all, the dilute suspension with $H = 10^{-6}$ is one of a kind with a wide local minimum over a large portion of κa examined. Moreover, it reduces to the single particle situation, which

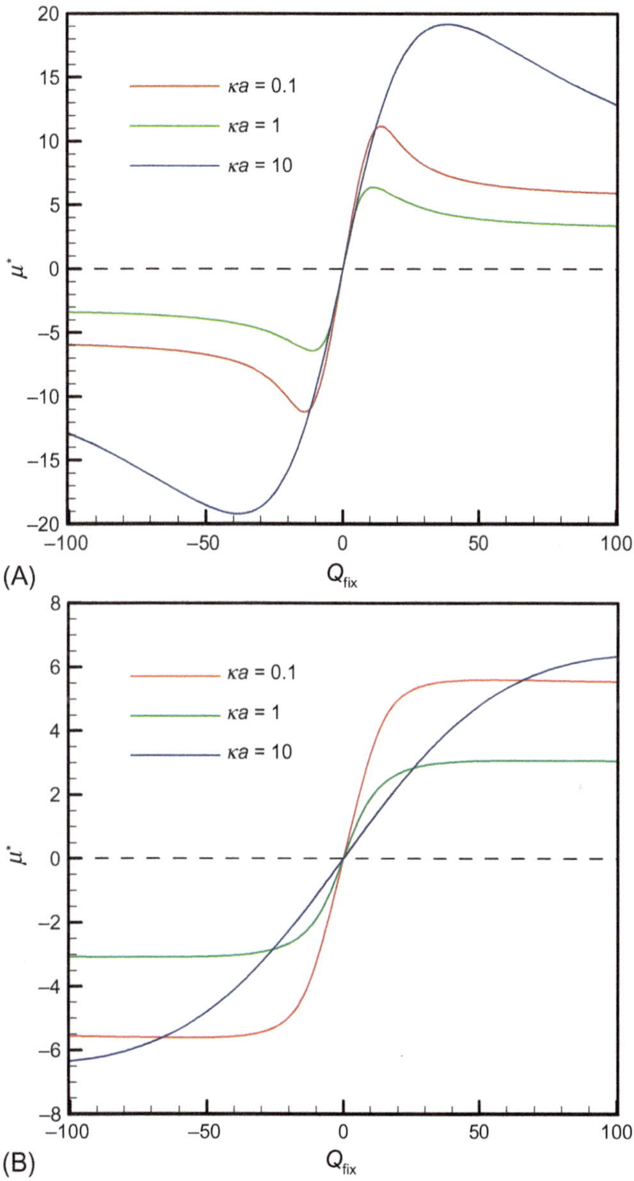

FIG. 7.9 Dimensionless particle mobility μ^* as a function of Q_{fix} at different κa for a porous particle with various permeability (λa). (A) $\lambda a = 1$; (B) $\lambda a = 3$; (C) $\lambda a = 10$.

indicates the accuracy of the cell model used here. However, when the suspension gets more concentrated, the range of this concave local minimum gets narrower as well and horizontal plateaus are generally observed when κa is small. Note that scaled particle mobility relative to $Q_{\text{fix}}/(\lambda a)^2$ is used here.

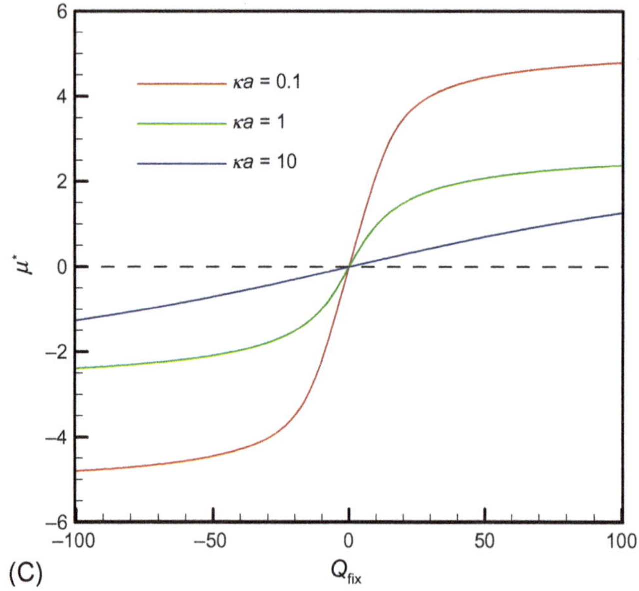

FIG. 7.9—CONT'D

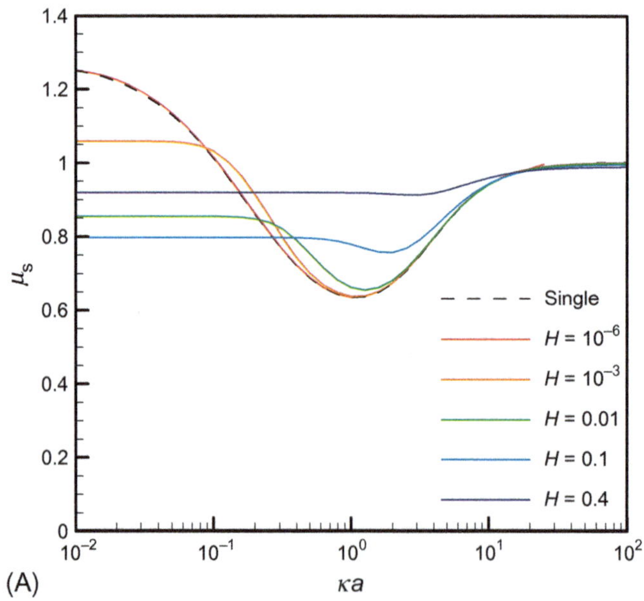

FIG. 7.10 Scaled mobility profiles relative to $Q_{fix}/(\lambda a)^2$ as a function of κa in various suspensions of highly permeable porous particles ($\lambda a = 1$) with various fixed charge density (Q_{fix}). *Dashed line*: single porous particle. (A) $Q_{fix} = 10$; (B) $Q_{fix} = 30$; (C) $Q_{fix} = 50$.

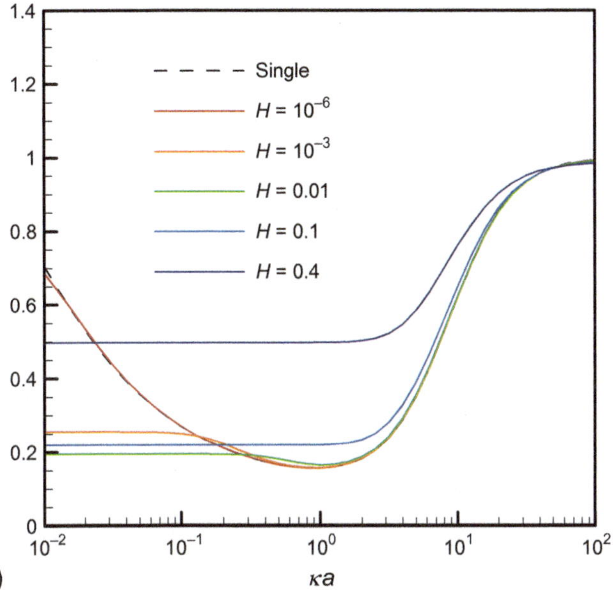

(B)

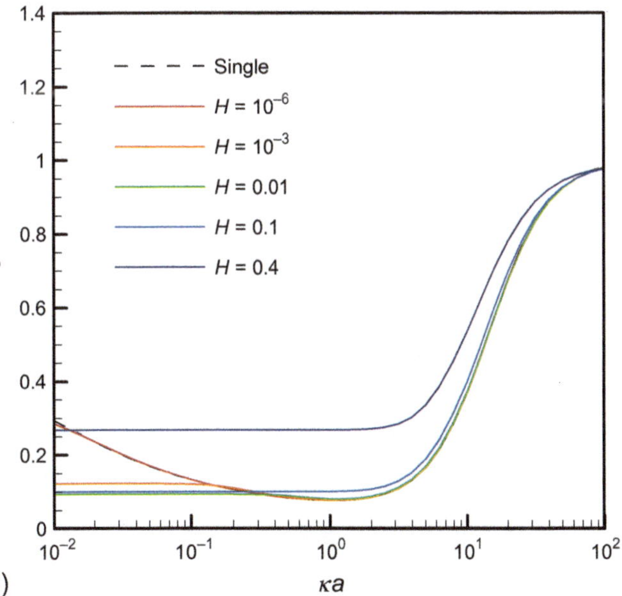

(C)

FIG. 7.10—CONT'D

Probably, the most eye-catching phenomenon is that the more concentrated the suspension is, the higher the particle mobility becomes, at least for κa around unity; indeed, a bazaar behavior against the intuition at first glance. Further check on the corresponding contour plots of the scaled double layer polarization ($\delta\rho^*$), as shown in Fig. 7.11A–C, indicates that the

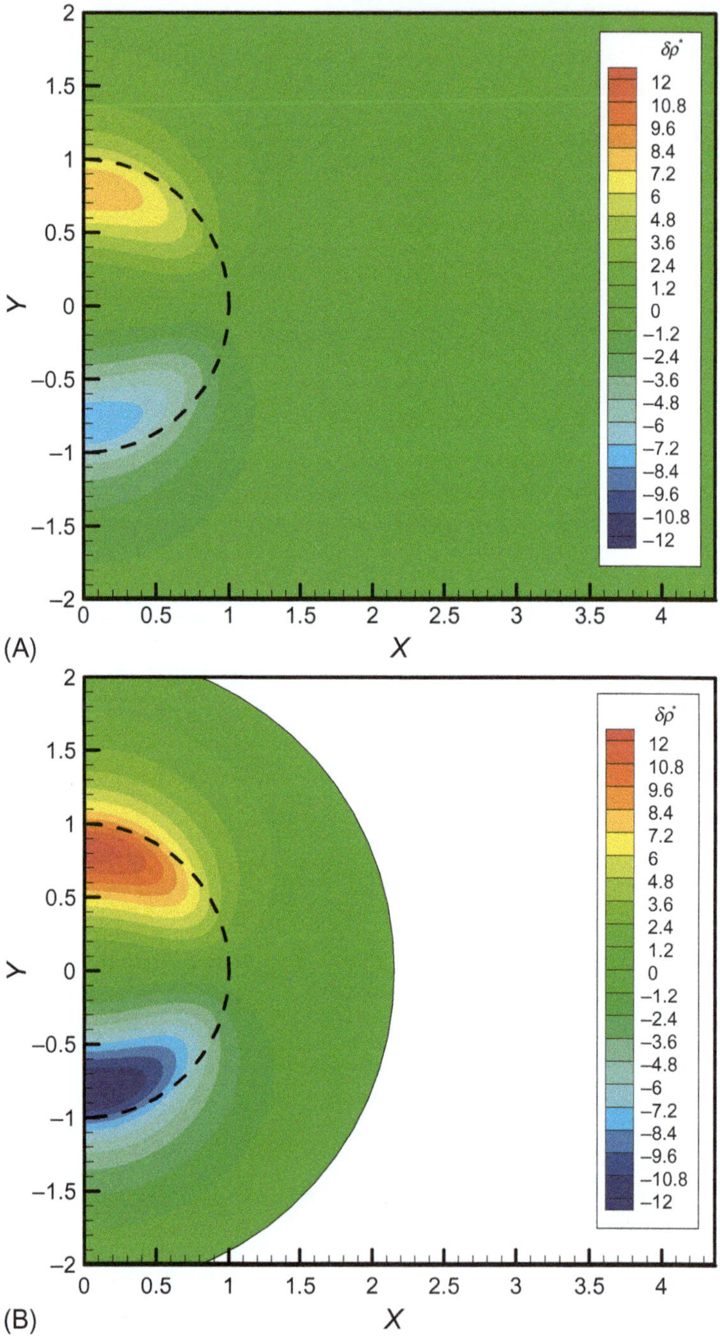

FIG. 7.11 Contour plots of the scaled double layer polarization ($\delta\rho^*$) in suspensions of highly permeable ($\lambda a=1$) and highly charged ($Q_{\rm fix}=30$) porous particles at $\kappa a=1$. (A) $H=10^{-6}$; (B) $H=0.1$; (C) $H=0.4$.

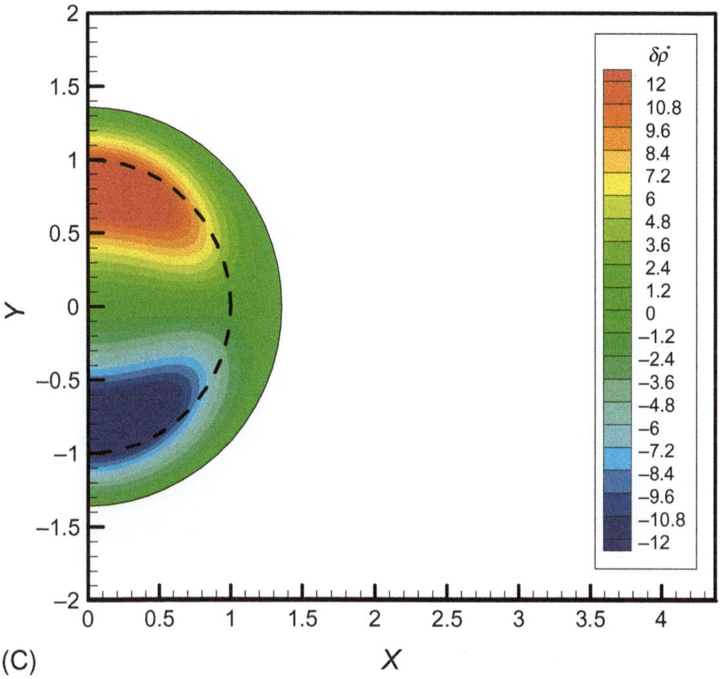

(C)

FIG. 7.11—CONT'D

counterions outside the porous particles diminish with increasing volume fraction. They are driven into the interior of the neighboring porous particles by the strong Coulomb attraction force to enhance the overall antisymmetric polarization there; thus, the electric driving force. Overall, the mobility of the particles in the suspension increases as a result. The more concentrated the particle volume fraction is, the closer these strongly attractive neighboring particles are to the counterions lingering around. This explains the intriguing phenomenon mentioned above.

Corresponding contour plots of the scaled disturbed electric potential ($\delta\phi^*$) are shown in Fig. 7.12A–C as well for completeness. The boundary confinement effect from the neighboring particles upon the lingering around counterions is so severe that apparently a significant portion of the region within the particle is occupied by the penetrating bulk electrolyte solution in Fig. 7.12C.

Moreover, all the mobility profiles in Fig. 7.10 approach asymptotically to unity as κa gets very large to around $\kappa a = 100$. This is similar to the behavior observed earlier for a single particle, but happens here at smaller κa. This is because concentrated suspensions increase the double layer suppression effect.

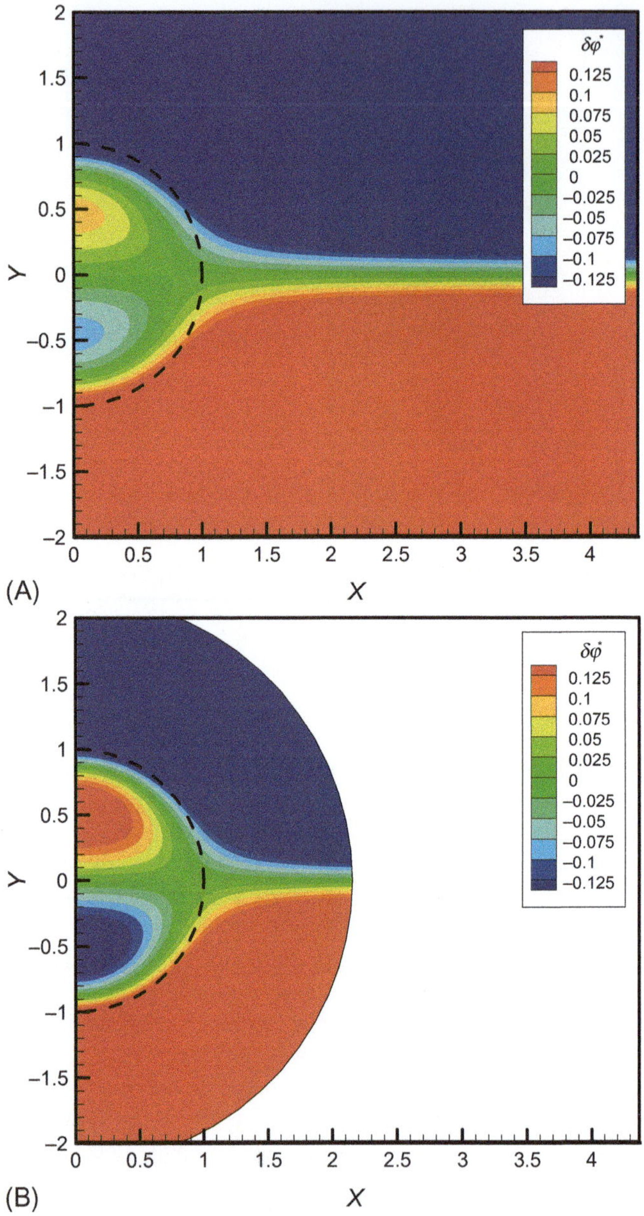

FIG. 7.12 Contour plots of the scaled disturbed potential ($\delta\phi^*$) in suspensions of highly permeable ($\lambda a = 1$) and highly charged ($Q_{\text{fix}} = 30$) porous particles at $\kappa a = 1$. (A) $H = 10^{-6}$; (B) $H = 0.1$; (C) $H = 0.4$.

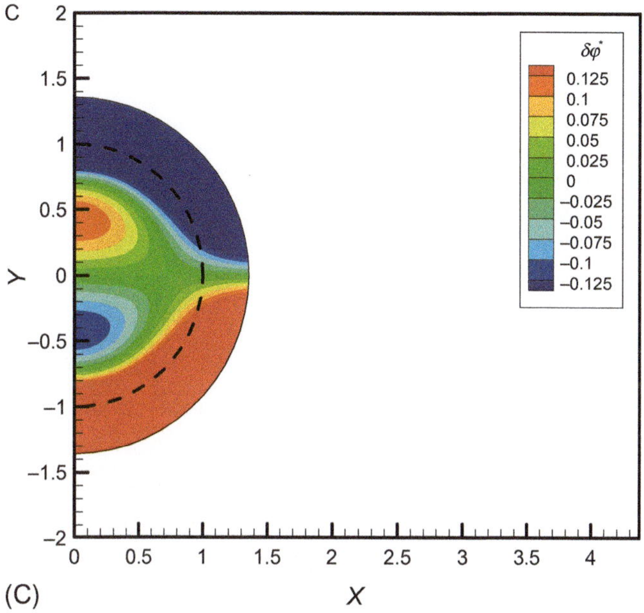

(C)

FIG. 7.12—CONT'D

Fig. 7.13 presents the same mobility profiles in Fig. 7.10, but expressed as a ratio to μ^*_{HF} to highlight the nonlinear effect pertinent to highly charged particles. The overall behavior is consistent with what observed in Fig. 7.10.

Fig. 7.14A–C again present the mobility profiles in Fig. 7.10, but as a function of the particle volume fraction H for $\kappa a = 0.1$, 1, and 10, respectively. The electrophoretic behavior of porous particles is wonderfully presented this way, so insightful and so clear! Really, a picture is worth a thousand words! The behavior when the double layer is thick ($\kappa a = 0.1$) is complicated because more counterions linger around outside the porous particles, as we have elaborated earlier in Fig. 7.10. Note that they all converge to the free-drainage situation when H approaches unity, as all the counterions would be confined completely into the interior by the geometric confinement effect of neighboring particles. Moreover, certainly there is no extra hydrodynamic drag force now as the fluid flow is actually absent when H approaches unity.

Fig. 7.15A–C shows the mobility profiles as a function of particle permeability, λa, in a suspension with $H = 0.1$. Ratio to the Hermans and Fujita's analytical prediction is used as the ordinate with the dashed lines representing the corresponding profiles when Peclet numbers of the electrolytes are set to zero, i.e., no involvement of convection mechanism in the Nernst-Planck ion flux equation. Again, the mobility behavior is wonderfully represented this way. Indeed, the correct way to analyze the suspensions of porous particles

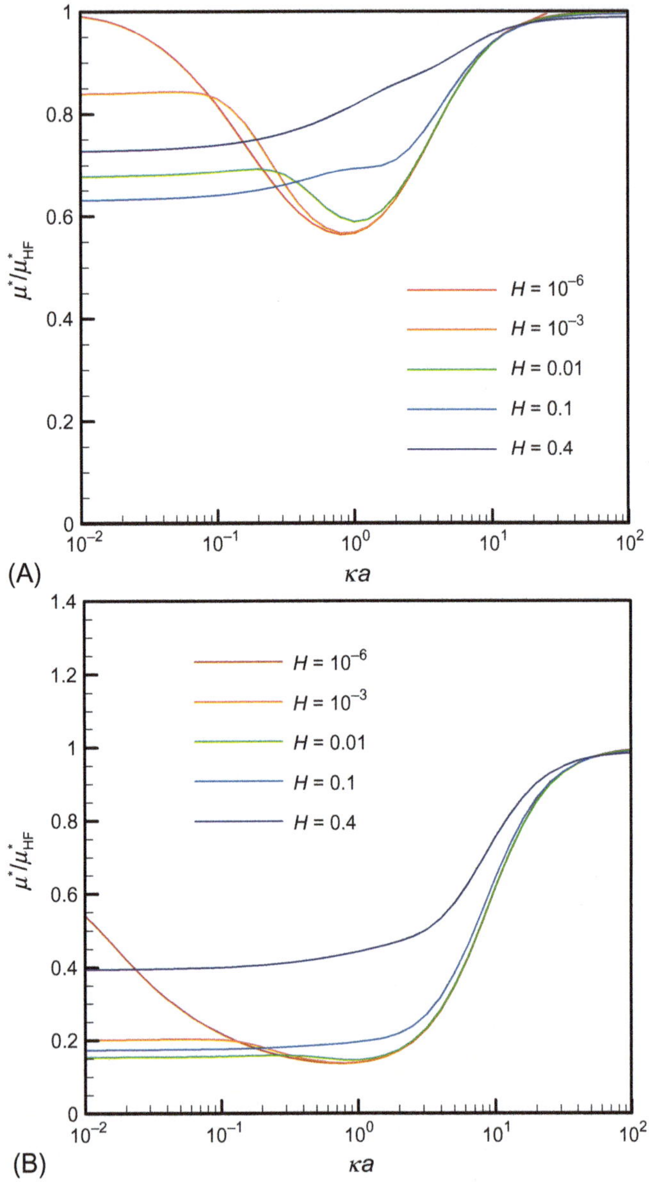

FIG. 7.13 Ratio of μ^* to μ^*_{HF} as a function of κa in various suspensions of highly permeable porous particles ($\lambda a = 1$) with various fixed charge density (Q_{fix}). (A) $Q_{fix} = 10$; (B) $Q_{fix} = 30$; (C) $Q_{fix} = 50$.

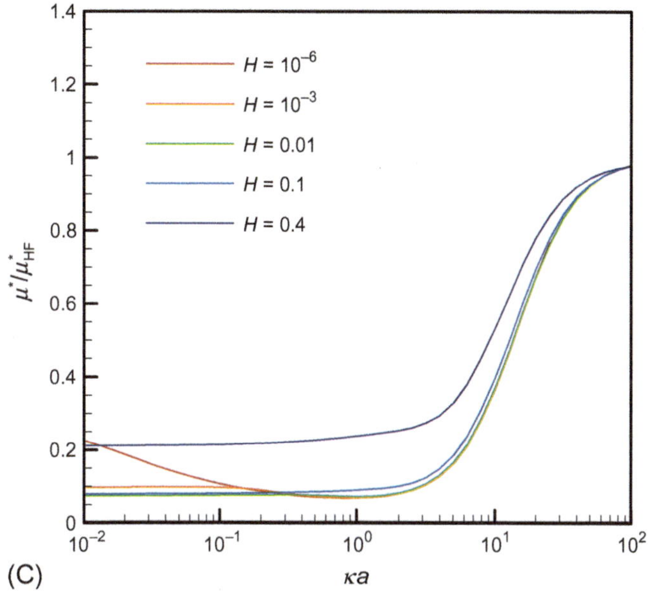

FIG. 7.13—CONT'D

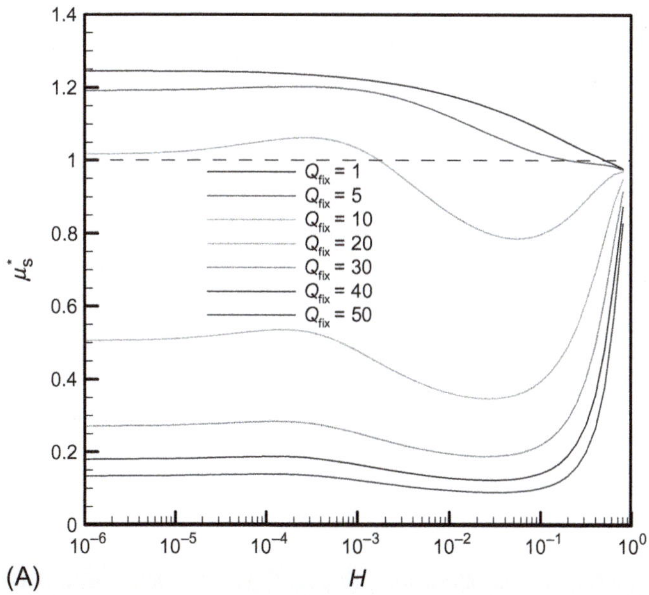

FIG. 7.14 Scaled mobility profiles relative to $Q_{\text{fix}}/(\lambda a)^2$ as a function of H in suspensions of highly permeable porous particle ($\lambda a = 1$) with various fixed charged density (Q_{fix}) at different κa. *Dashed line*: horizontal unity line. (A) $\kappa a = 0.1$; (B) $\kappa a = 1$. (C) $\kappa a = 10$.

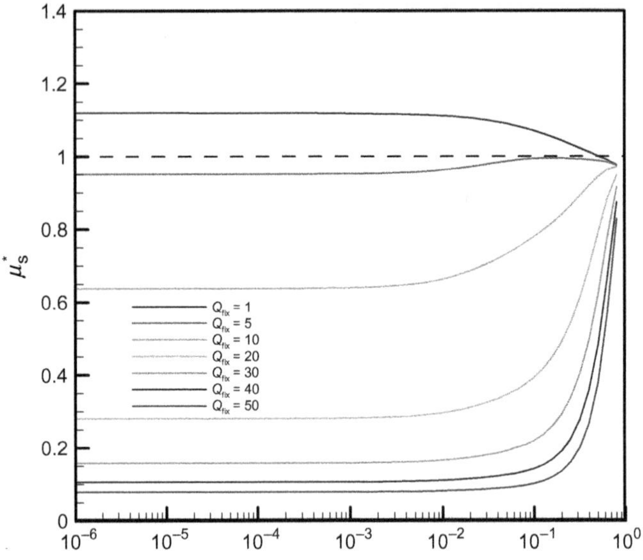

(B)

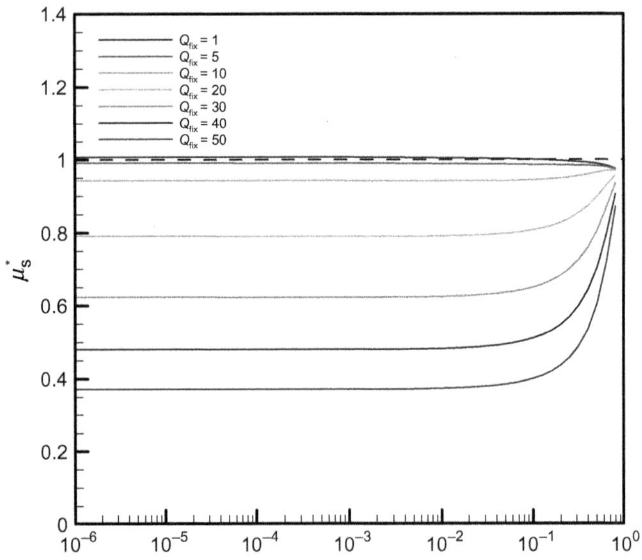

(C)

FIG. 7.14—CONT'D

is to set the κa value as a constant first, and then the dependence of the rest of the electrokinetic parameters would be clearly demonstrated. This is because the redistribution of the counterions lingering around exterior to porous particles in a suspension is the major factor in determining the particle electrophoretic behavior there. The motion-deterring effect of the convection-induced

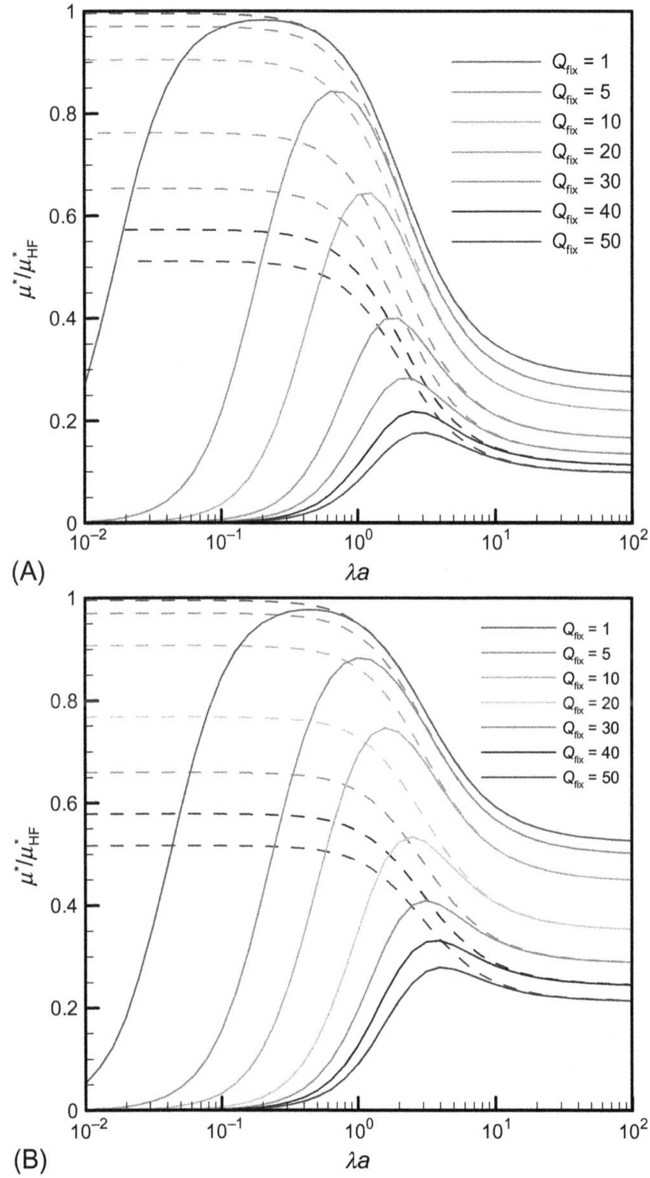

FIG. 7.15 Ratio of μ^* to μ^*_{HF} as a function of λa in suspensions ($H=0.1$) of porous particles with various fixed charge density (Q_{fix}) at different κa. *Dashed lines*: $Pe_1=Pe_2=0$. (A) $\kappa a=0.1$; (B) $\kappa a=1$; (C) $\kappa a=10$.

double layer polarization is profound for highly permeable porous particles, or small λa. Its impact diminishes when λa increases due to the increasing extra hydrodynamic drag force, as we discussed earlier for a single porous particle situation. This concludes the discussion in this chapter.

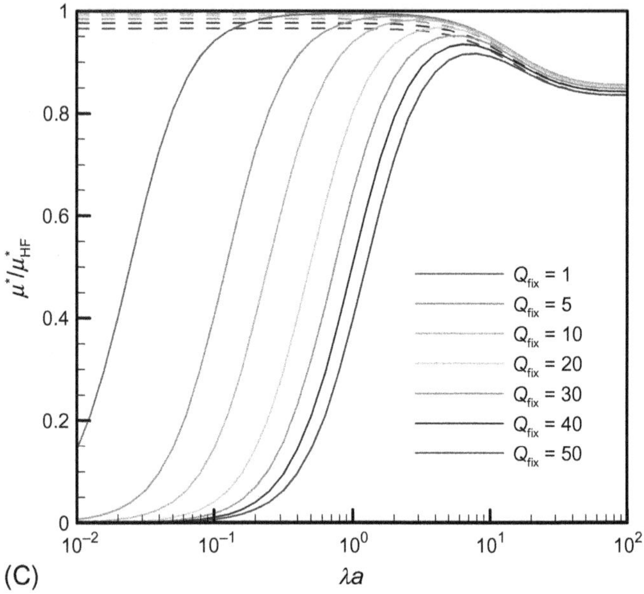

FIG. 7.15—CONT'D

REFERENCES

[1] Fuoss RM. Polyelectrolytes. Science 1948;108(2812):545–50.

[2] Kogej K, Skerjanz J, Radeva T. Surfactant science series. vol. 99. New York: Marcell Dekker Inc.; 2001.

[3] Oosawa F. Polyelectrolytes. In: Polyelectrolytes. New York: Marcel Dekker; 1971.

[4] Overbeek JTG. Polyelectrolytes, past, present and future. Pure Appl Chem 1976;46(2–4): 91–101.

[5] Mannion J, Craighead H. Nanofluidic structures for single biomolecule fluorescent detection. Biopolymers 2007;85(2):131–43.

[6] Bell D, Dunnill P. The influence of precipitation reactor configuration on the centrifugal recovery of isoelectric soya protein precipitate. Biotechnol Bioeng 1982;24(11):2319–36.

[7] Pont L, et al. Analysis of transthyretin in human serum by capillary zone electrophoresis electrospray ionization time-of-flight mass spectrometry. Application to familial amyloidotic polyneuropathy type I. Electrophoresis 2015;36(11–12):1265–73.

[8] Woolley AT, Mathies RA. Ultra-high-speed DNA sequencing using capillary electrophoresis chips. Anal Chem 1995;67(20):3676–80.

[9] Valaskovic GA, Kelleher NL, McLafferty FW. Attomole protein characterization by capillary electrophoresis—mass spectrometry. Science 1996;273(5279):1199.

[10] Woolley AT, Sensabaugh GF, Mathies RA. High-speed DNA genotyping using microfabricated capillary array electrophoresis chips. Anal Chem 1997;69(11):2181–6.

[11] Westermeier R. Electrophoresis in practice: a guide to methods and applications of DNA and protein separations. Hoboken, NJ: John Wiley & Sons; 2016.

[12] Manning GS. Limiting laws and counterion condensation in polyelectrolyte solutions I. Colligative properties. J Chem Phys 1969;51(3):924–33.

[13] Brinkman H. A calculation of the viscous force exerted by a flowing fluid on a dense swarm of particles. Appl Sci Res 1949;1(1):27–34.
[14] Hill RJ, Saville D, Russel W. Polarizability and complex conductivity of dilute suspensions of spherical colloidal particles with uncharged (neutral) polymer coatings. J Colloid Interface Sci 2003;268(1):230–45.
[15] Duval JF, et al. Humic substances are soft and permeable: evidence from their electrophoretic mobilities. Environ Sci Technol 2005;39(17):6435–45.
[16] Keh HJ, Liu CP. Electric conductivity and electrophoretic mobility in suspensions of charged porous spheres. J Phys Chem C 2010;114(50):22044–54.
[17] Yeh L-H, Hsu J-P. Effects of double-layer polarization and counterion condensation on the electrophoresis of polyelectrolytes. Soft Matter 2011;7(2):396–411.
[18] Fujita H. Notes on the calculation of electrophoresis of polyelectrolytes with partial free-drainage. J Phys Soc Jpn 1957;12(8):968–73.
[19] Hermans J. Sedimentation and electrophoresis of porous spheres. J Polym Sci 1955;18(90):527–34.
[20] Hermans J. Electrophoresis of charged polymer molecules with partial free drainage. Koninkl Ned Akad Wetenschap Proc 1955;58:182–7.
[21] He YY, Wu E, Lee E. Electrophoresis in suspensions of charged porous spheres in salt-free media. Chem Eng Sci 2010;65(20):5507–16.
[22] He YY, Lee E. Electrophoresis in concentrated dispersions of charged porous spheres. Chem Eng Sci 2008;63(23):5719–27.
[23] Tsai P, Huang CH, Lee E. Electrophoresis of a charged colloidal particle in porous media: boundary effect of a solid plane. Langmuir 2011;27(22):13481–8.
[24] Hsu HP, Lee E. Electrophoresis of a charged porous sphere normal to an air-water interface. Phys Chem Chem Phys 2012;14:15729–38.
[25] Huang CH, Hsu HP, Lee E. Electrophoretic motion of a charged porous sphere within micro- and nanochannels. Phys Chem Chem Phys 2012;14(2):657–67.
[26] Hsu H-P, Lee E. Electrophoresis of a single charged porous sphere in an infinite medium of electrolyte solution. J Colloid Interface Sci 2013;390(1):85–95.
[27] Fang W, Lee E. Diffusiophoretic motion of an isolated charged porous sphere. J Colloid Interface Sci 2015;459:273–83.
[28] Darcy H. Les fontaines publique de la ville de Dijon. Paris: Dalmont; 1856. p. 647.
[29] Debye P, Bueche AM. Intrinsic viscosity, diffusion, and sedimentation rate of polymers in solution. J Chem Phys 1948;16(6):573–9.
[30] Tchesskaya TY. Kinetics of semidilute and dense polymer solution: the time dependence of single-chain dynamics. J Mol Liq 2005;120(1–3):143–6.
[31] Einstein A. Ann Phys (Leipzig) 1906;19:289.
[32] Neale G, Epstein N. Creeping flow relative to permeable spheres. Chem Eng Sci 1973;28(10):1865–74.
[33] Dukhin SS, et al. On the applicability of the Brinkman equation in soft surface electrokinetics. J Colloid Interface Sci 2010;350(1):1–4.
[34] Phenrat T, et al. Adsorbed polyelectrolyte coatings decrease Fe0 nanoparticle reactivity with TCE in water: conceptual model and mechanisms. Environ Sci Technol 2009;43(5):1507–14.
[35] Ballauff M, Borisov O. Polyelectrolyte brushes. Curr Opin Colloid Interface Sci 2006;11(6):316–23.
[36] Balazs AC, et al. Theory of polymer chains tethered at interfaces. Prog Surf Sci 1997;55(3):181–269.

[37] Wittemann A, Ballauff M. Secondary structure analysis of proteins embedded in spherical polyelectrolyte brushes by FT-IR spectroscopy. Anal Chem 2004;76(10):2813–9.

[38] Galaev IY, Mattiasson B. "Smart" polymers and what they could do in biotechnology and medicine. Trends Biotechnol 1999;17(8):335–40.

[39] Mei Y, et al. High catalytic activity of platinum nanoparticles immobilized on spherical polyelectrolyte brushes. Langmuir 2005;21(26):12229–34.

[40] Sharma G, Ballauff M. Cationic spherical polyelectrolyte brushes as nanoreactors for the generation of gold particles. Macromol Rapid Commun 2004;25(4):547–52.

[41] Donath E, Pastushenko V. Electrophoretical study of cell surface properties. The influence of the surface coat on the electric potential distribution and on general electrokinetic properties of animal cells. J Electroanal Chem Interfacial Electrochem 1979;104:543–54.

[42] Wunderlich R. The effects of surface structure on the electrophoretic mobilities of large particles. J Colloid Interface Sci 1982;88(2):385–97.

[43] Levine S, et al. Theory of the electrokinetic behavior of human erythrocytes. Biophys J 1983;42(2):127–35.

[44] Sharp KA, Brooks DE. Calculation of the electrophoretic mobility of a particle bearing bound polyelectrolyte using the nonlinear Poisson-Boltzmann equation. Biophys J 1985;47(4):563–6.

Chapter 8

Electrophoretic Motion of a Porous Spherical Particle Normal to Planar Boundaries

8.1 INTRODUCTION

The configuration of a charged porous particle near a planar boundary is encountered very often in tissue engineering in biomedical and biochemical applications [1], where the polyelectrolyte molecules such as DNAs or proteins are often modeled as a porous sphere and the tissue structure as a solid plane [2, 3]. Other applications are also abundant and available in the literature [4–7]. Corresponding applications involving the air-water interface are very popular as well with relevant experimental works reported in various fields [8–11].

As for the theoretical treatment of these two electrokinetically similar planar configurations, many research endeavors have been conducted in the past few decades. Among them, Lee's group in particular explored both the hydrodynamic and the electrostatic boundary effects due to the presence of either a solid plane or an air-water interface, taking full account of the double layer polarization effect [12–19]. Bi-spherical coordinates coupled with pseudo-spectral method are adopted and numerical approach was undertaken there.

In comparison with the rigid particle situation treated in Chapters 3 and 4, the porous particle has a unique feature that most of the counterions reside within the particle due to its permeability to the electrolyte solution. This enhances the polarization, or uneven distribution of mobile space charges in the interior of the porous particle. As discussed in Chapter 7, unlike the double layer polarization exterior to a rigid particle which retards the particle motion in general, this corresponding polarization in the interior actually is the very mechanism to generate the electric driving force leading to the electrophoretic motion of a porous particle, the same role as played by the double layer polarization on the shear plane in the rigid particle situation. The polarization of the bulk double layer exterior to the porous particle, on the other hand, is motion-deterring as the rigid particle situation, where an induced electric field

Interface Science and Technology, Vol. 26. https://doi.org/10.1016/B978-0-08-100865-2.00008-4

against the particle motion is generated. And the higher the particle charge level is, the more profound this phenomenon is, as stronger electrostatic Coulomb attractive force is resulted by the higher amount of fixed charges contained in the porous particle and much more significant polarization of the interior space charge as well. As a result, the boundary confinement effect here is much more complicated due to the particle permeability, which turns out to enhance the interior double player polarization effect under some circumstances and results in an intriguing phenomenon that the closer the porous particle to the planar boundary is, the faster it may move. Moreover, the overall boundary effect in terms of the reduction of particle mobility is not so significant as the corresponding rigid particle situation. Not at all. The detailed behavior is discussed with figures, plots, and tables later in this chapter.

8.2 THEORY

8.2.1 Solid Plane

The system diagram is shown in Fig. 8.1 and the corresponding mesh diagram in calculation is shown in Fig. 8.2. Note that the system is similar to that in Chapter 3, except an extra domain within the spherical porous particle needs to be included, which has already been treated in Chapter 7 for a single porous particle. All we need to do here is to re-assemble them with some necessary conversions.

The governing equations and associated boundary conditions outside the porous particle are the same as those in Chapter 3, i.e., Eqs. (3.3)–(3.7) and Eqs. (3.16)–(3.26).

Within the porous particle, the governing equations are the same as those in Chapter 7, i.e., Eqs. (7.4) and (7.8) to (7.11). Note that they have to be written in bi-spherical coordinates in computation now to match with the governing system outside the particle.

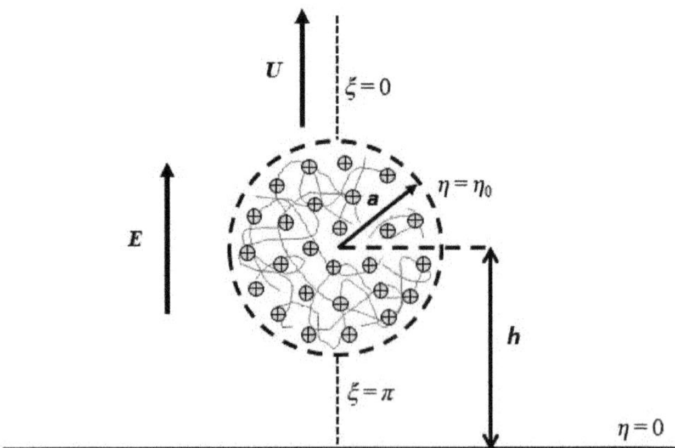

FIG. 8.1 System diagram for a solid plane.

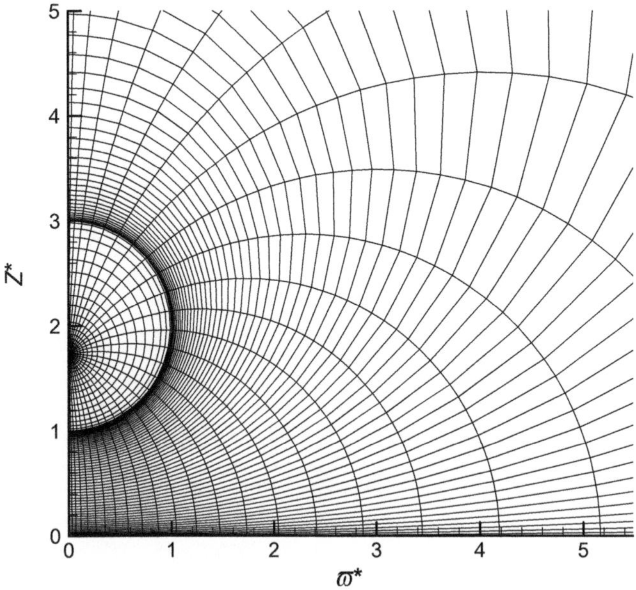

FIG. 8.2 Mesh diagram in computation.

The boundary conditions across the porous particle surface have to be written in bi-spherical coordinates as well. They are:

$$\left.\frac{\partial \phi_{\mathrm{e}}^{*}}{\partial \eta}\right|_{\eta=\eta_0^-}=\left.\frac{\partial \phi_{\mathrm{e}}^{*}}{\partial \eta}\right|_{\eta=\eta_0^+} \tag{8.1}$$

$$\left.\frac{\partial \delta\phi^{*}}{\partial \eta}\right|_{\eta=\eta_0^-}=\left.\frac{\partial \delta\phi^{*}}{\partial \eta}\right|_{\eta=\eta_0^+} \tag{8.2}$$

$$\left.\frac{\partial g_{\mathrm{j}}^{*}}{\partial \eta}\right|_{\eta=\eta_0^-}=\left.\frac{\partial g_{\mathrm{j}}^{*}}{\partial \eta}\right|_{\eta=\eta_0^+} \tag{8.3}$$

$$\left.\frac{\partial \Psi^{*}}{\partial \eta}\right|_{\eta=\eta_0^-}=\left.\frac{\partial \Psi^{*}}{\partial \eta}\right|_{\eta=\eta_0^+} \tag{8.4}$$

$$\left.\frac{\partial^2 \Psi^{*}}{\partial \eta^2}\right|_{\eta=\eta_0^-}=\left.\frac{\partial^2 \Psi^{*}}{\partial \eta^2}\right|_{\eta=\eta_0^+} \tag{8.5}$$

$$\left.\frac{\partial^3 \Psi^{*}}{\partial \eta^3}\right|_{\eta=\eta_0^-}=\left.\frac{\partial^3 \Psi^{*}}{\partial \eta^3}-(\lambda a)^2\frac{c^{*2}}{x^2}\left(\frac{\partial \Psi^{*}}{\partial \eta}-\frac{c^{*2}\sin^2\xi\sinh\eta}{x^3}U^{*}\right)\right|_{\eta=\eta_0^+} \tag{8.6}$$

Subsequent procedure of the particle mobility evaluation is the same as that used in Chapter 1, i.e., Eqs. (1.73)–(1.81), as elaborated in Chapter 7. The details can be found elsewhere [20]. And we are done.

8.2.2 Air-Water Interface

The system diagram is shown in Fig. 8.3, and the mesh diagram in computation is the same as Fig. 8.2.

Note that the system under consideration is essentially the same as the solid plane case just treated above. Only the hydrodynamic and electrostatic conditions at the air-water interface ($\eta=0$) need to be replaced now as follows:

$$\frac{\partial \phi_e^*}{\partial \eta}=0 \tag{8.7}$$

$$\frac{\partial \delta\phi^*}{\partial \eta}=-E^* \tag{8.8}$$

$$g_j^*=-\delta\phi^* \tag{8.9}$$

$$\frac{\partial \Psi^*}{\partial \xi}=0 \tag{8.10}$$

$$\frac{\partial \delta\phi^*}{\partial \eta}\frac{\partial \phi_e^*}{\partial \xi}-\frac{x}{c^* \sin\xi}\frac{\partial^2 \Psi^*}{\partial \eta^2}=0 \tag{8.11}$$

The physical essence behind Eqs. (8.7) and (8.11) is the same as that in Chapter 4. Subsequent treatment in order to obtain the particle mobility is exactly the same as well. The details can be found elsewhere [14]. And we are done.

8.3 RESULTS AND DISCUSSION

8.3.1 Solid Plane

Fig. 8.4A–C depicts the general Henry's charts for a highly permeable porous particle ($\lambda a=1$) at three representative charge levels ($Q_{fix}=1$, 10, and 30) at various separation distances. The mobility is determined by the balance of

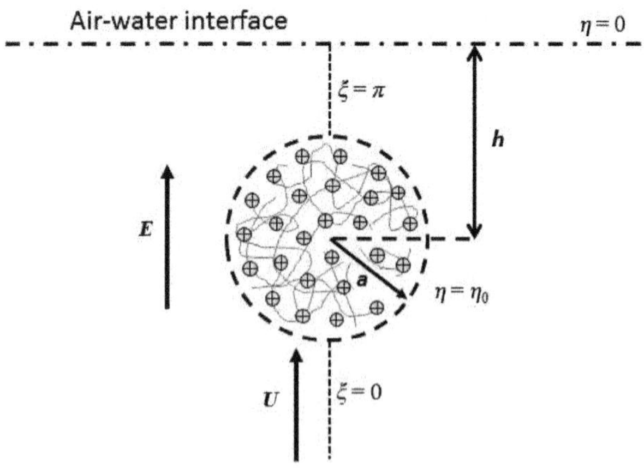

FIG. 8.3 System diagram for an air-water interface.

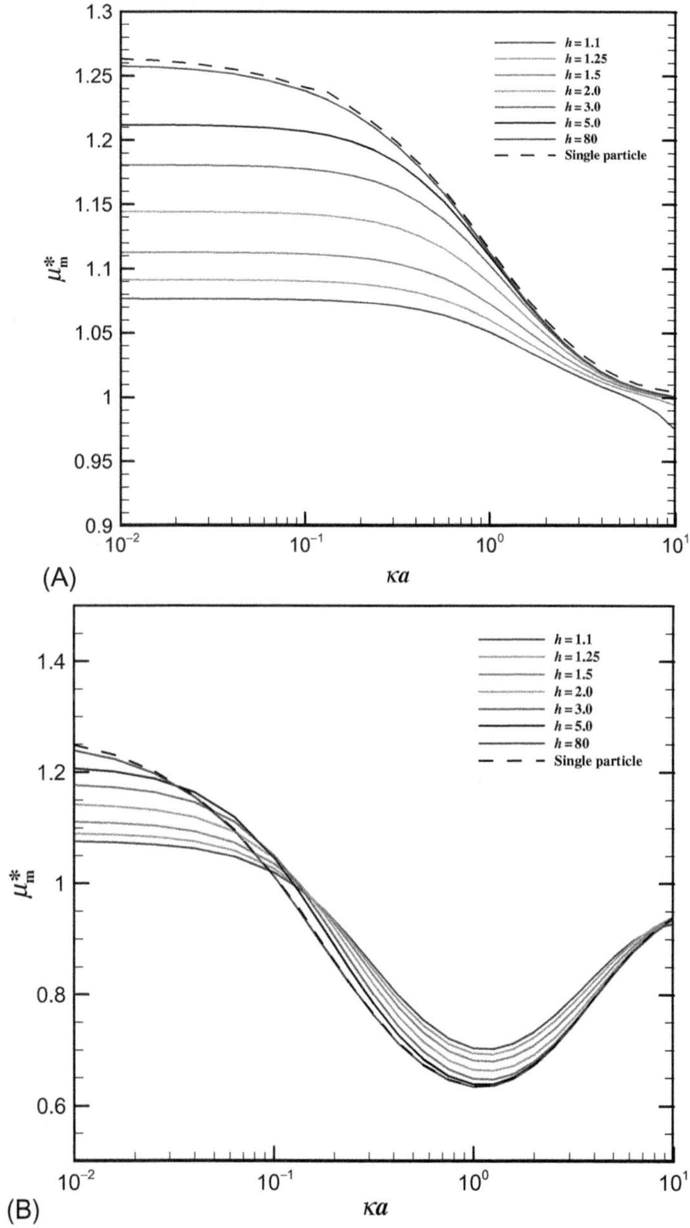

FIG. 8.4 Scaled mobility $\mu_m^*\left(=\frac{\mu^*}{Q_{fix}/(\lambda a)^2}\right)$ as a function of κa at different h for $\lambda a=1$. (A) $Q_{fix}=1$; (B) $Q_{fix}=10$; (C) $Q_{fix}=30$.

Continued

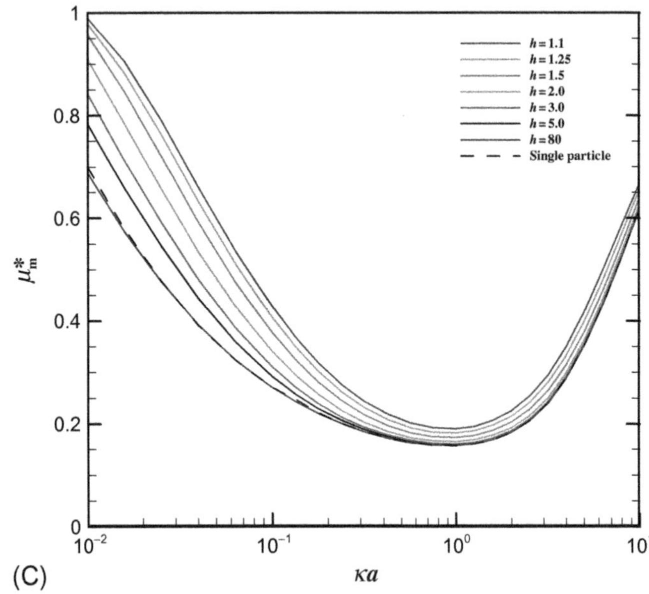

FIG. 8.4—CONT'D

electric driving force and hydrodynamic drag force. The former increases as the particle approaches the solid plane. This is due to the attractive image-charge force analog as explained in Chapter 3, which can be considered as proportional to a characteristic electric field established between the tip of the particle and the plane ahead: $E_c = \frac{\phi_r}{h-1}$, as the surface potential is held at zero on the plane surface. The latter, on the other hand, always increases as the particle approaches the plane. Obviously, the latter dominates in Fig. 8.4A as the mobility reduction is observed over the entire range of κa. This is understandable as the charge level is so low ($Q_{fix}=1$); hence, the electric driving force is too weak to make any difference. Moreover, the thicker the double layers is, the more profound this reduction is, as most of the counterions actually reside exterior to the particle for such a lowly charged porous particle; hence, the geometric boundary confinement effect still compresses them to a significant extent.

An intriguing phenomenon is observed in Fig. 8.4B that a porous particle may enhance its speed as it approaches the plane when κa is in the range from 0.1 to 10, roughly. Or equivalently, it may slow down leaving the plane, depending upon the direction of the applied electron field. This is against the reasoning based on the hydrodynamic retarding effect alone. Obviously, the electric driving force dominates here in this range, as $Q_{fix}=10$ is apparently high enough to respond to the increasing characteristic electric field in the gap (E_c) to generate a significant enhancement of electric driving force based on the electrostatic mechanism that more than compensates the increasing hydrodynamic drag force. Moreover, the boundary confinement effect also drives more counterions in the double layer originally in the rear region into the interior of the porous particle, which increases the local electric field strength and hence its electric driving force as well. This is why the increase of mobility as the particle approaches the plane is most profound around $\kappa a = 1$, where the double layer polarization is most significant. This effect diminishes as κa gets large ($\kappa a = 10$), where hardly any outside counterions can be driven into the interior of the particle anymore to increase its effect charge and hence driving force. Fig. 8.4C shows the dominance of the attractive image-charge force and the boundary confinement effect over the hydrodynamic drag force. The dominance there goes over all the κa range considered for a highly charged particle with $Q_{fix} = 30$.

Fig. 8.5A–D shows the contour plots of ρ_e^*, ϕ_e^*, $\delta\phi^*$, and $\delta\rho^*$ for a representative highly charged ($Q_{fix} = 30$) and highly permeable ($\lambda a = 1$) particle very close to the plane ($h = 1.1$). The boundary confinement is clearly demonstrated. Unlike the corresponding rigid particle treated in Chapter 3, the interior of the particle is affected as well now. Note in particular Fig. 8.5D, the boundary confinement drives essentially all the counterions into the lower hemisphere of the particle interior to increase its polarization there as well as the overall effective charges, as discussed earlier.

Corresponding flow fields are shown as well in Fig. 8.6A and B. Similar to those observed in Chapter 3, two separate vortex flows with opposite orientations appear in the neighborhood of the particle. The only difference is that the flow can penetrate into the interior of a porous particle now in this chapter.

Fig. 8.7 shows the mobility profile as a function of separation distance between the particle and the plane for three orders of double layer thickness

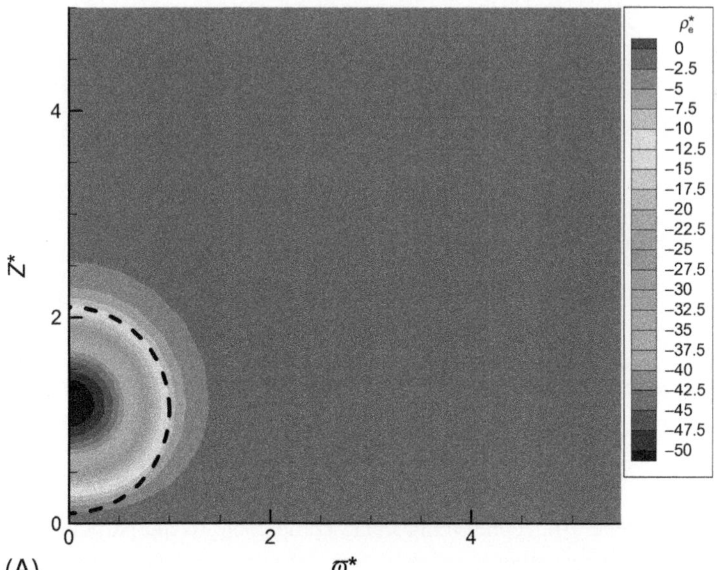

(A)

FIG. 8.5 Contour plots of various scaled variables for $\kappa a = 1$, $Q_{fix} = 30$, $h = 1.1$, and $\lambda a = 1$. (A) Scaled equilibrium charge density (ρ_e^*); (B) scaled equilibrium electric potential (ϕ_e^*); (C) scaled disturbed electric potential ($\delta\phi^*$); (D) scaled disturbed charge density ($\delta\rho^*$).

Continued

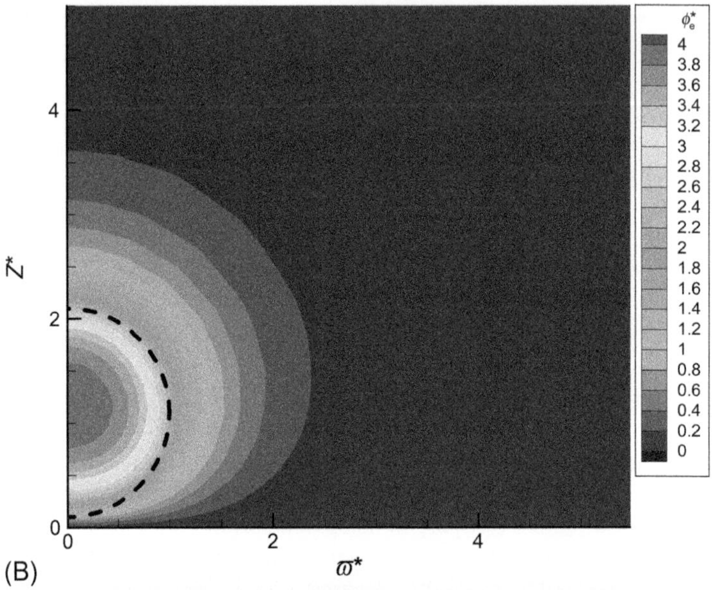

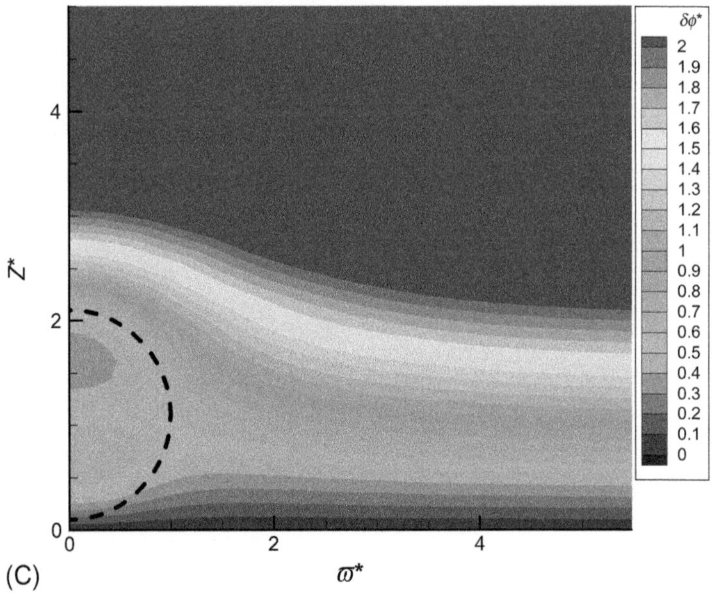

FIG. 8.5—CONT'D

($\kappa a = 0.1$, 1, and 10), representing thick, comparable to the particle radius, and thin double layer situations, respectively. The enhancement of particle speed is clearly shown as it approaches the plane. Note that the overall particle mobility or speed is the lowest when the double layer thickness is comparable to the particle radius, i.e., $\kappa a = 1$. This is due to the motion-deterring double layer polarization effect which is most profound for κa around unity.

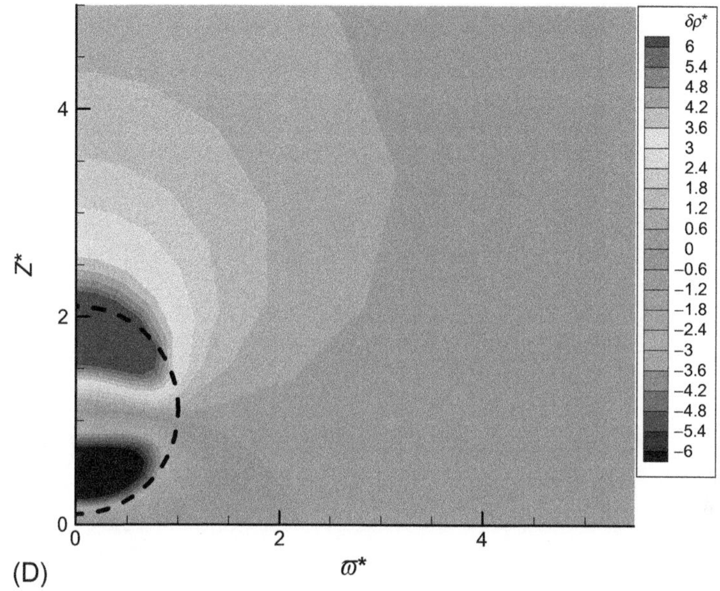

FIG. 8.5—CONT'D

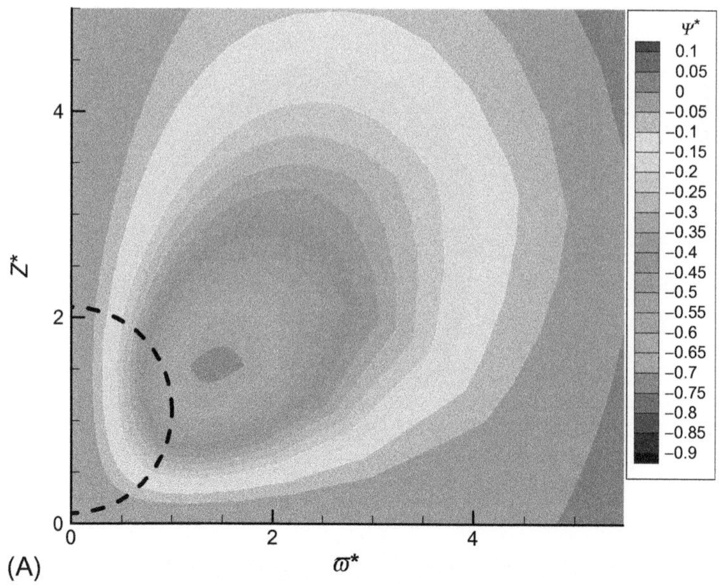

FIG. 8.6 Contour plot of the scaled stream function (Ψ^*) for $\kappa a = 1$, $Q_{\text{fix}} = 30$, $h = 1.1$, and $\lambda a = 1$. (A) Small scale (close-up view); (B) large scale.

Continued

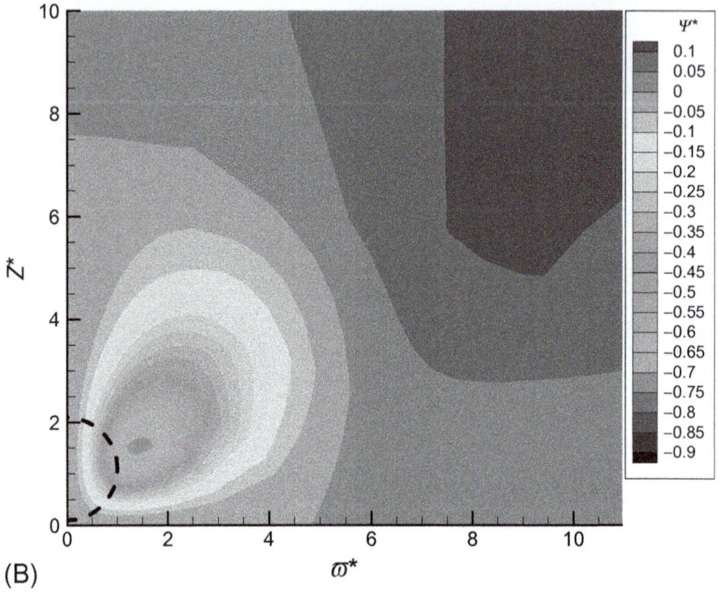

FIG. 8.6—CONT'D

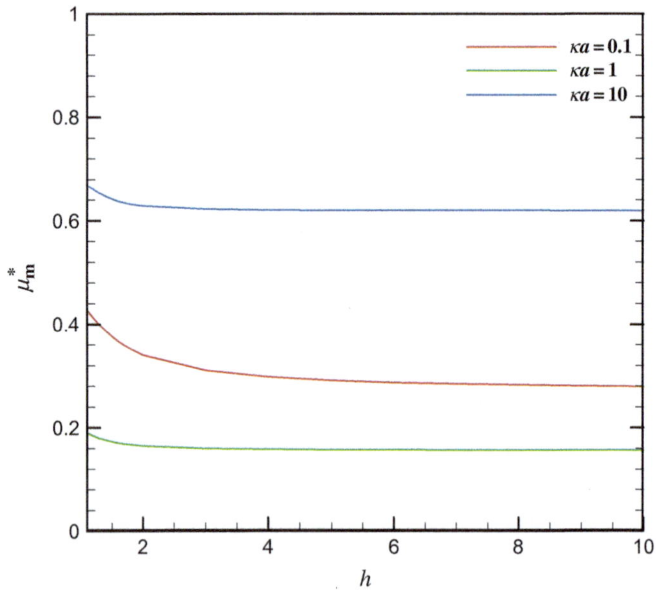

FIG. 8.7 Scaled mobility μ_m^* as a function of h at different κa for $Q_{fix} = 30$ and $\lambda a = 1$.

Corresponding table of correction factor as a function of separation distance (h) at various double layer thicknesses (κa) is also given in Table 8.1 to facilitate the possible use of interested readers.

TABLE 8.1 Correction Factor Γ for a Solid Plane With $Q_{fix} = 30$

	κa							
h	*0.01*	*0.1*	*0.5*	*1*	2	*5*	*7*	*10*
1.1	1.451	1.582	1.249	1.211	1.229	1.186	1.136	1.074
1.2	1.440	1.526	1.209	1.175	1.190	1.149	1.109	1.063
1.4	1.416	1.432	1.148	1.121	1.131	1.095	1.067	1.039
1.6	1.388	1.358	1.108	1.088	1.093	1.062	1.043	1.024
1.8	1.360	1.302	1.082	1.067	1.068	1.042	1.029	1.015
2	1.333	1.259	1.065	1.053	1.051	1.030	1.020	1.010
4	1.180	1.103	1.020	1.011	1.006	1.002	1.000	0.997
6	1.127	1.064	1.010	1.004	1.002	1.000	0.998	0.996
8	1.100	1.046	1.005	1.002	1.001	0.999	0.998	0.996
10	1.083	1.036	1.003	1.001	1.000	0.999	0.998	0.995
20	1.047	1.015	1.001	1.000	1.000	0.999	0.997	0.995
40	1.023	1.005	1.000	1.000	1.000	0.999	0.997	0.995
60	1.013	1.002	1.000	1.000	1.000	0.999	0.997	0.995
80	1.008	1.002	1.000	1.000	1.000	0.999	0.997	1.000
∞	1	1	1	1	1	1	1	1

8.3.2 Air-Water Interface

Corresponding general Henry's charts, as shown in Fig. 8.8A–C, are similar to those for the solid plane case except the electric driving is weaker now; hence,

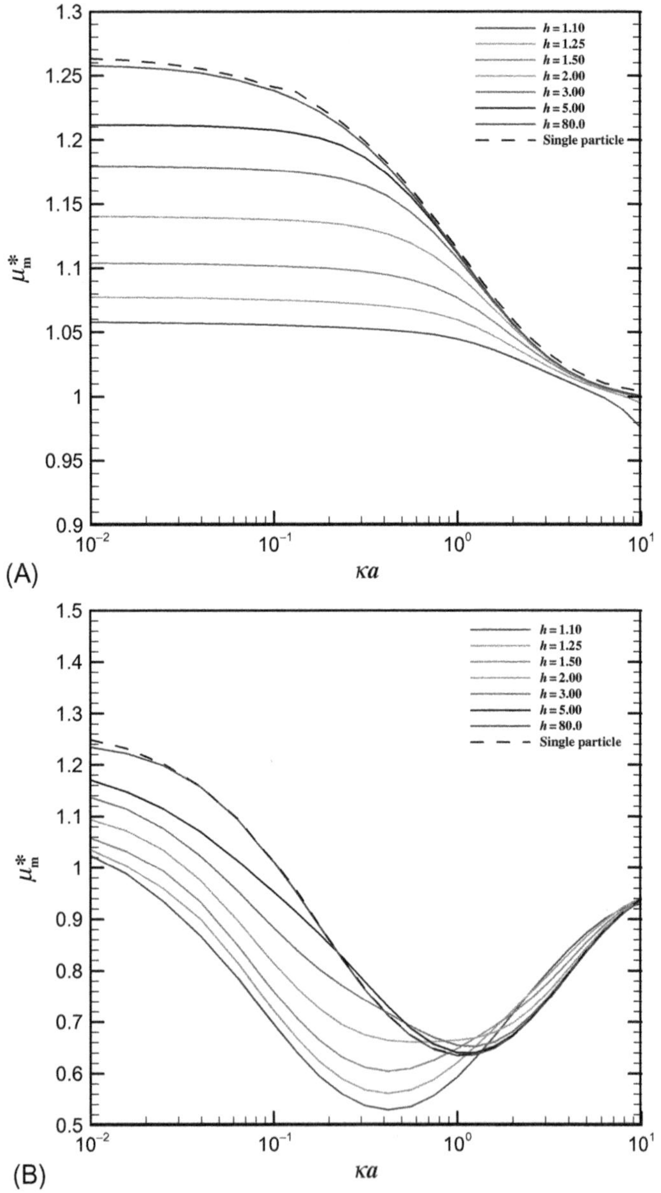

FIG. 8.8 Scaled mobility μ_m^* as a function of κa at different h for $\lambda a=1$. (A) $Q_{fix}=1$; (B) $Q_{fix}=10$; (C) $Q_{fix}=30$.

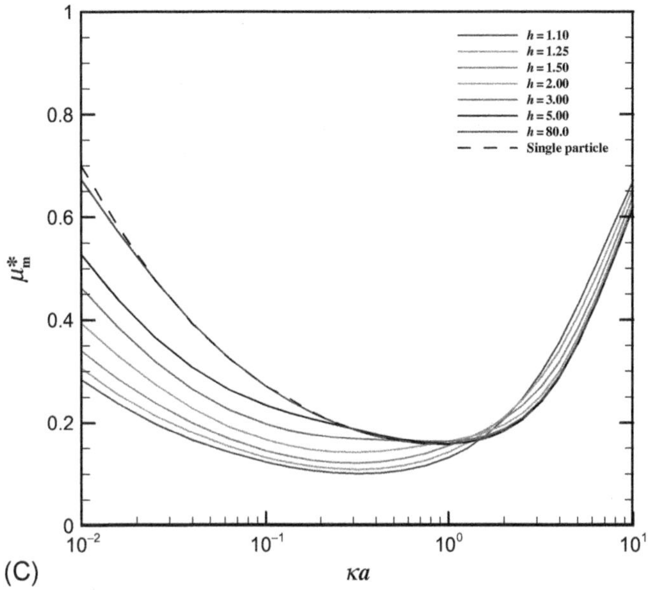

(C)

FIG. 8.8—CONT'D

the mobility reduction behavior is observed over a wider range of κa (Fig. 8.8B and C), if not entirely (Fig. 8.8A). This is due to the repulsive nature of the image-charge force analog which significantly reduces the electric driving force as the particle has to move against an extra electric field generated by it. Moreover, the fluid at the air-water interface is allowed to move horizontally, with the possible presence of coions in it. Thus, the backward flow to drive coions into the particle interior is weakened as well. As a result, the intriguing phenomenon that a porous particle may move faster as it approaches a planar boundary only happens when κa is relatively large, as shown in Fig. 8.8B and C.

Corresponding contour plots of ρ_e^*, ϕ_e^*, $\delta\phi^*$, and $\delta\rho^*$ are given in Fig. 8.9A–D for a particle both highly charged ($Q_{\text{fix}}=30$) and highly permeable ($\lambda a=1$) very close to the interface ($h=1.1$). The plots are similar in essence to the solid plane case just treated, except that there is no clearly visible polarization in the upper hemisphere interior region now. The repulsive image-charge force analog attracts many counterions to the gap area and penetrate into the interior of the porous particle there. Hence, the depletion of counterions is fulfilled again, leading to the apparent disappearance of polarization there. Nearly half of the electric driving force is lost as a result, thus the particle moves slower than a single one, as shown in Fig. 8.8C. Corresponding flow fields are shown as well in Fig. 8.10A and B for completeness, with

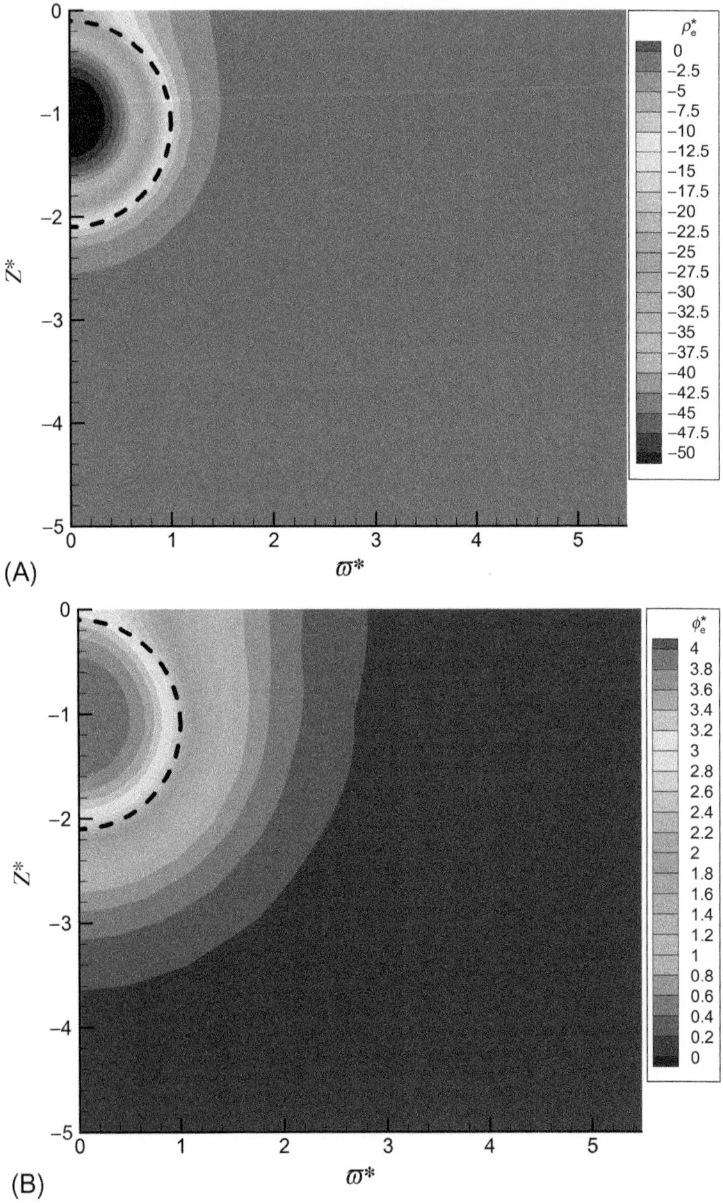

FIG. 8.9 Contour plots of various scaled variables for $\kappa a=1$, $Q_{\text{fix}}=30$, $h=1.1$, and $\lambda a=1$. (A) Scaled equilibrium charge density (ρ_e^*); (B) scaled equilibrium electric potential (ϕ_e^*); (C) scaled disturbed electric potential ($\delta\phi^*$); (D) scaled disturbed charge density ($\delta\rho^*$).

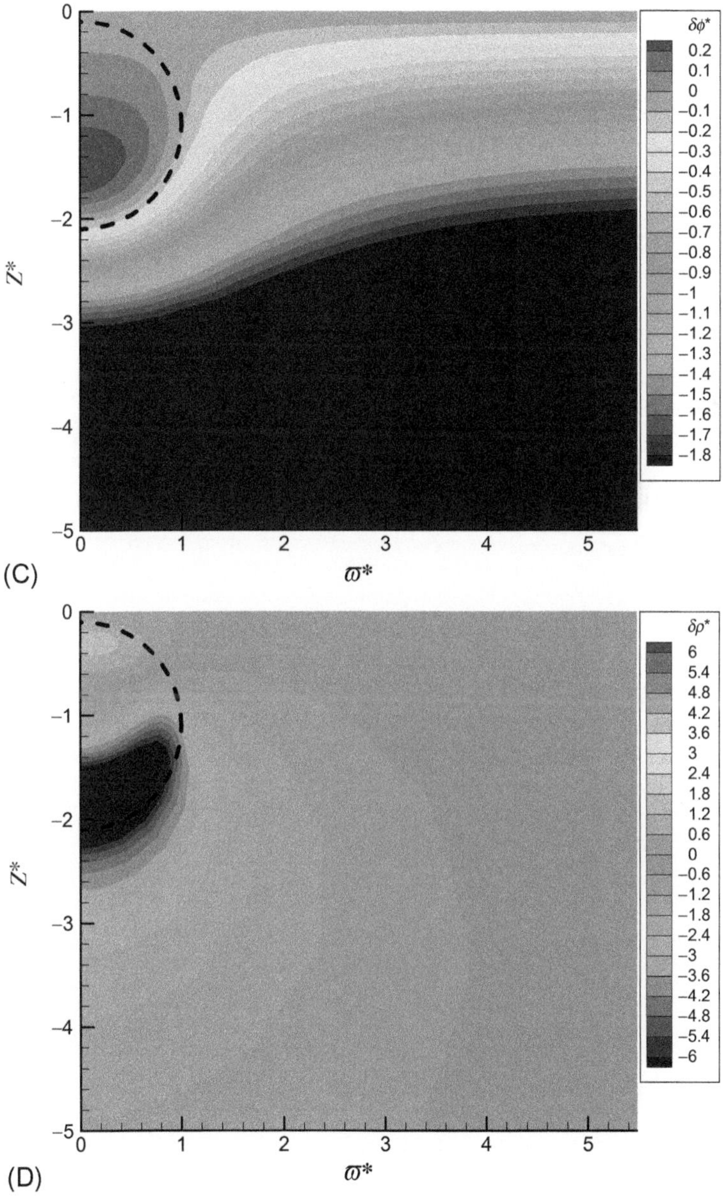
δϕ*
0.2
0.1
0
−0.1
−0.2
−0.3
−0.4
−0.5
−0.6
−0.7
−0.8
−0.9
−1
−1.1
−1.2
−1.3
−1.4
−1.5
−1.6
−1.7
−1.8
Z*
ϖ*
(C)
δρ*
6
5.4
4.8
4.2
3.6
3
2.4
1.8
1.2
0.6
0
−0.6
−1.2
−1.8
−2.4
−3
−3.6
−4.2
−4.8
−5.4
−6
Z*
ϖ*
(D)

FIG. 8.9—CONT'D

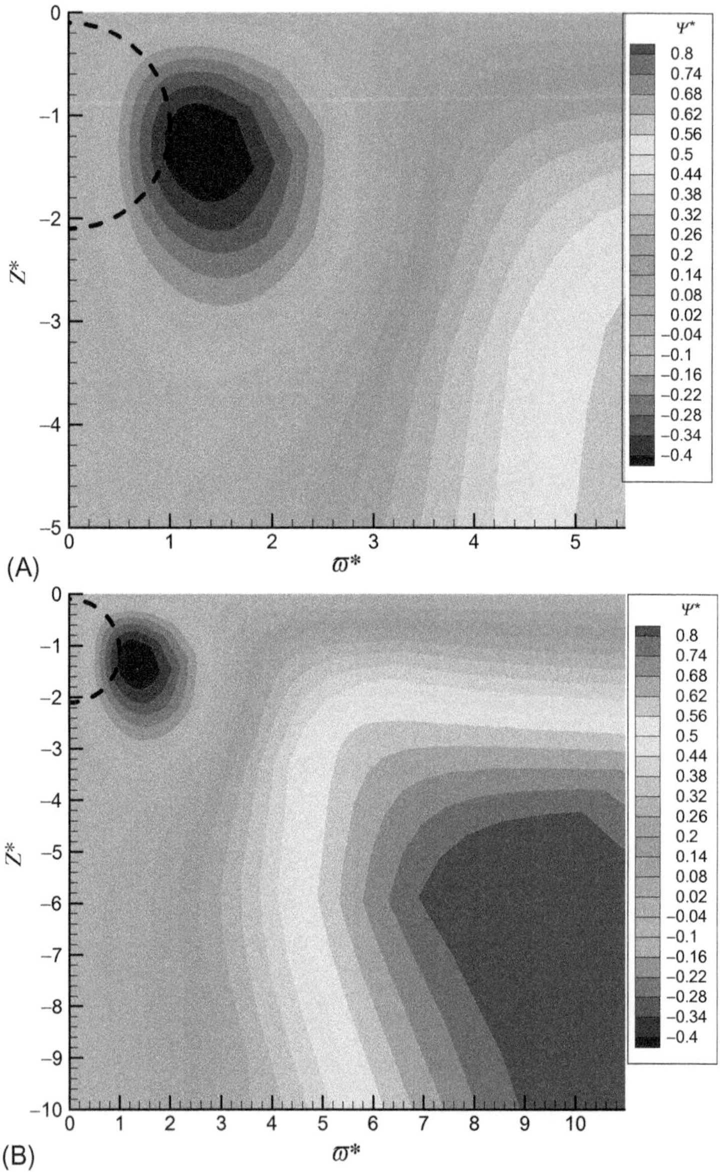

FIG. 8.10 Contour plot of the scaled stream function (Ψ^*) for $\kappa a = 1$, $Q_{\text{fix}} = 30$, $h = 1.1$, and $\lambda a = 1$. (A) Small scale (close-up view); (B) large scale.

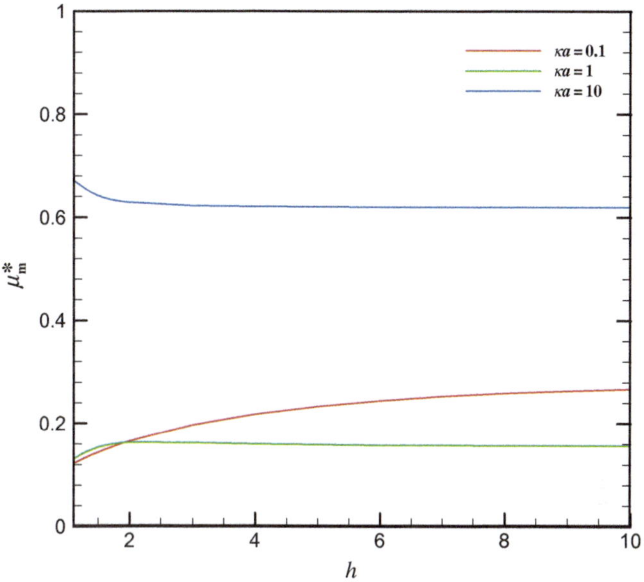

FIG. 8.11 Scaled mobility μ_m^* as a function of h at different κa for $Q_{fix} = 30$ and $\lambda a = 1$.

similar observation as those in Chapter 4 for a rigid particle. Note that the streamlines pass through the porous particle now due to its permeability.

Fig. 8.11 shows the behavior for three representative double layer thicknesses when a porous particle both highly charged ($Q_{fix} = 30$) and highly permeable ($\lambda a = 1$) moves toward the air-water interface from far away ($h = 10$). Only the profile corresponding to a very thin double layer ($\kappa a = 10$) exhibits the intriguing behavior observed earlier, as shown in Fig. 8.8C as well.

Corresponding table of correction factor as a function of separation distance (h) at various double layer thicknesses (κa) is also given in Table 8.2 to facilitate the possible use of interested readers.

TABLES OF BOUNDARY EFFECT EXPRESSED AS A CORRECTION FACTOR Γ

($\Gamma = \mu_{m,h}/\mu_{m,\infty}$, where $\mu_{m,h}$ refers to the particle mobility at a separation distance h, whereas $\mu_{m,\infty}$ refers to the mobility in an infinite medium.)

TABLE 8.2 Correction Factor Γ for the Air-Water Interface With $Q_{fix}=30$

	κa							
h	*0.01*	*0.1*	*0.5*	*1*	2	5	7	*10*
1.1	0.417	0.454	0.624	0.835	1.126	1.211	1.150	1.079
1.2	0.439	0.476	0.661	0.883	1.147	1.170	1.118	1.066
1.4	0.480	0.517	0.729	0.954	1.142	1.105	1.071	1.040
1.6	0.517	0.553	0.789	1.000	1.117	1.067	1.045	1.025
1.8	0.549	0.585	0.840	1.027	1.091	1.045	1.030	1.016
2	0.578	0.614	0.882	1.041	1.071	1.032	1.021	1.010
4	0.737	0.807	1.022	1.021	1.009	1.003	1.000	0.997
6	0.800	0.904	1.019	1.007	1.003	1.000	0.998	0.996
8	0.835	0.958	1.011	1.003	1.001	0.999	0.998	0.996
10	0.858	0.988	1.006	1.002	1.001	0.999	0.998	0.995
20	0.915	1.016	1.001	1.000	1.000	0.999	0.997	0.995
40	0.958	1.006	1.000	1.000	1.000	0.999	0.997	0.995
60	0.977	1.002	1.000	1.000	1.000	0.999	0.997	0.995
80	0.987	1.002	1.000	1.000	1.000	0.999	0.997	0.995
∞	1	1	1	1	1	1	1	1

REFERENCES

[1] Johnston AP, Read ES, Caruso F. DNA multilayer films on planar and colloidal supports: sequential assembly of like-charged polyelectrolytes. Nano Lett 2005;5(5):953–6.

[2] Klenchin V, et al. Electrically induced DNA uptake by cells is a fast process involving DNA electrophoresis. Biophys J 1991;60(4):804–11.

[3] Viovy J-L. Electrophoresis of DNA and other polyelectrolytes: physical mechanisms. Rev Mod Phys 2000;72(3):813.

[4] Rouzina I, Bloomfield VA. Competitive electrostatic binding of charged ligands to polyelectrolytes: planar and cylindrical geometries. J Phys Chem 1996;100(10):4292–304.

[5] Bremmell KE, Jameson G, Biggs S. Polyelectrolyte adsorption at the solid/liquid interface: interaction forces and stability. Colloids Surf A Physicochem Eng Asp 1998;139(2): 199–211.

[6] Borkovec M, Papastavrou G. Interactions between solid surfaces with adsorbed polyelectrolytes of opposite charge. Curr Opin Colloid Interface Sci 2008;13(6):429–37.

[7] Norde W. Adsorption of proteins from solution at the solid-liquid interface. Adv Colloid Interf Sci 1986;25:267–340.

[8] Ruths J, et al. Polyelectrolytes I: polyanion/polycation multilayers at the air/monolayer/water interface as elements for quantitative polymer adsorption studies and preparation of heterosuperlattices on solid surfaces. Langmuir 2000;16(23):8871–8.

[9] Monteux C, et al. Adsorption of oppositely charged polyelectrolyte/surfactant complexes at the air/water interface: formation of interfacial gels. Langmuir 2004;20(1):57–63.

[10] Ahrens H, et al. Polyelectrolyte adsorption onto insoluble monolayers at the air/water interface. Macromolecules 2001;34(13):4504–12.

[11] Noskov B, et al. Dynamic surface properties of polyelectrolyte/surfactant adsorption films at the air/water interface: poly (diallyldimethylammonium chloride) and sodium dodecylsulfate. Langmuir 2007;23(19):9641–51.

[12] Cheng W-L, He Y-Y, Lee E. Electrophoresis of a soft particle normal to a plane. J Colloid Interface Sci 2009;335(1):130–9.

[13] Chih M-H, Lee E, Hsu J-P. Electrophoresis of a sphere normal to a plane at arbitrary electrical potential and double layer thickness. J Colloid Interface Sci 2002;248(2):383–8.

[14] Hsu H-P, Lee E. Electrophoresis of a charged porous sphere normal to an air–water interface. Phys Chem Chem Phys 2012;14(45):15729–38.

[15] Lee E, Chen C-T, Hsu J-P. Electrophoresis of a rigid sphere in a Carreau fluid normal to a planar surface. J Colloid Interface Sci 2005;285(2):857–64.

[16] Liao C, et al. Electrophoretic motion of a liquid droplet and a bubble normal to an air–water interface. Colloids Surf A Physicochem Eng Asp 2016;507:124–33.

[17] Tsai P, Fang H, Lee E. Electrophoresis of a charge-regulated sphere normal to an air–water interface. J Phys Chem B 2011;115(20):6484–94.

[18] Tsai P, Huang C-H, Lee E. Electrophoresis of a charged colloidal particle in porous media: boundary effect of a solid plane. Langmuir 2011;27(22):13481–8.

[19] Tsai P, et al. Electrophoresis of a spherical particle normal to an air–water interface. Electrophoresis 2010;31(20):3363–71.

[20] O'Brien RW, White LR. Electrophoretic mobility of a spherical colloidal particle. J Chem Soc, Faraday Trans 2 Mol Chem Phys 1978;74:1607–26.

Chapter 9

Electrophoretic Motion of a Porous Spherical Particle in a Cylindrical Pore

9.1 INTRODUCTION

As discussed in Chapter 5 for rigid particles, electrophoresis of porous particles like polyelectrolytes, such as DNAs and proteins, is often the target samples in capillary electrophoresis operations as well [1–4]; DNA sequencing and protein separation for further clinic diagnosis in biochips, for instance, are direct applications in biochemical and biomedical fields [2, 4–12]. Hence, the knowledge of the boundary effect of a cylindrical pore is very important.

Many theoretical investigations have been conducted regarding this issue [4, 13–22], including Lee's group who solved the complete general electrokinetic equations taking full account of the double layer polarization effect for both charged porous and soft particles [18, 19]. The boundary effect is similar to that discussed in Chapter 5 for a rigid particle in general, except that the permeability of a porous particle makes the particle move much faster, and as a result, much stronger double layer polarization effect. Moreover, the permeability also results in a phenomenon that most counterions may actually reside within the sphere, especially when the particle is highly charged, as we have discussed in Chapter 7. This makes the boundary confinement effect upon the double layer much weaker, as there is not much counterions left outside the particle anyway, leading to a complicated and sometimes intriguing behavior of the electrophoretic motion. The details will be discussed later.

9.2 THEORY

The system diagram is shown in Fig. 9.1.

The geometric configuration is similar to the one in Chapter 5, where a rigid particle in a cylindrical pore is considered. The only difference is

Interface Science and Technology, Vol. 26. https://doi.org/10.1016/B978-0-08-100865-2.00009-6

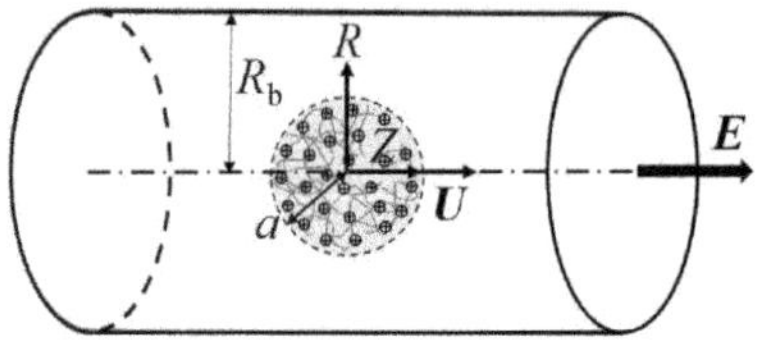

FIG. 9.1 System diagram.

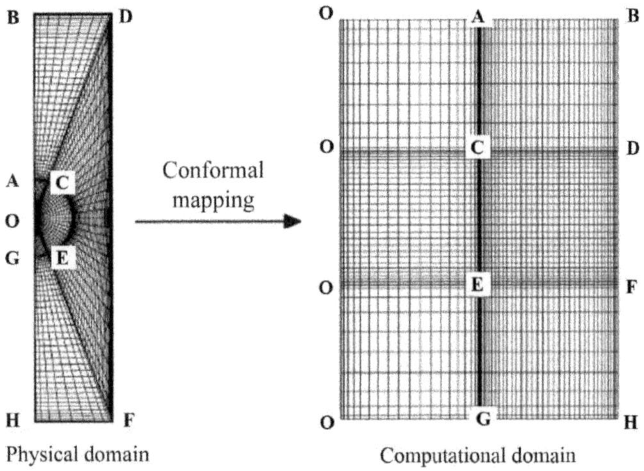

FIG. 9.2 Conformal mapping and mesh diagram in computation.

that an additional spherical porous region has to be treated now. Moreover, corresponding extra domain decomposition and subsequent conformal mapping within the porous sphere region are needed as well. Three more subregions are generated to represent the region within the porous sphere, as shown in Fig. 9.2. The mathematical details can be found elsewhere [18].

As spherical coordinates are used here, the governing equations for the fluid inside the porous particle are exactly the same as those in Chapter 7, i.e., Eq. (7.4) and Eqs. (7.8) to (7.11), with the Brinkman model assumed here. Corresponding boundary conditions on the particle surface and its center are the same as those in Chapter 7 as well, i.e., Eqs. (7.5) to (7.7), Eqs. (7.12) to (7.21), and so on.

The governing equation outside the porous particle as well as their associated boundary conditions on the pore wall and upstream and downstream fully developed profiles is exactly the same as those treated in Chapters 1 and 5 for a rigid particle. They are Eqs. (1.29)–(1.33) and Eqs. (5.1)–(5.10), respectively. Subsequent treatments are exactly the same as those in Chapter 5, and the particle mobility is obtained in the end.

9.3 RESULTS AND DISCUSSION

9.3.1 General Henry's Charts in a Narrow Cylindrical Pore With $R_b^* = 1.2$

Fig. 9.3 shows the scaled dimensionless mobility μ_m^*, defined as $\mu_m^* = \frac{\mu^*}{(Q_{fix}/\lambda a^2)}$, profiles at various levels of particle charges. Polarization effect of the double layer, indicated by the local minima, remains essentially intact even when the particle is so close to the pore wall here with $R_b^* = 1.2$. This is due to the fact that most of the counterions reside within the porous particle, and the remaining ones outside cluster mainly in the longitudinal direction. Thus, the lateral confinement effect of the cylindrical pore does not change the distribution of ions much, as there is not much outside there to be confined anyway. The reduction of particle mobility is clearly demonstrated in comparison with the single particle situation. Note that the extent of reduction is not so drastic even when the double layer is thick ($\kappa a = 0.01$) as shown in Fig. 9.3, due to the same reasoning as elaborated above.

9.3.2 Effect of Particle Permeability, λa

Fig. 9.4 shows the corresponding mobility profile μ^* for a representative highly charged particle ($Q_{fix} = 30$ here) with various levels of permeability

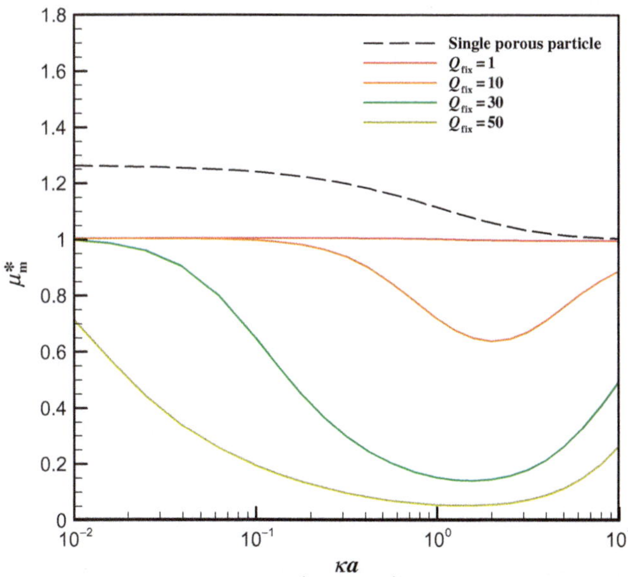

FIG. 9.3 Scaled dimensionless mobility $\mu_m^* \left(= \frac{\mu^*}{(Q_{fix}/\lambda a^2)}\right)$ versus κa respectively for $\lambda a = 1$, $\phi_w = 0$, and $R_b^* = 1.2$ with $Q_{fix} = 1$, 10, 30, and 50.

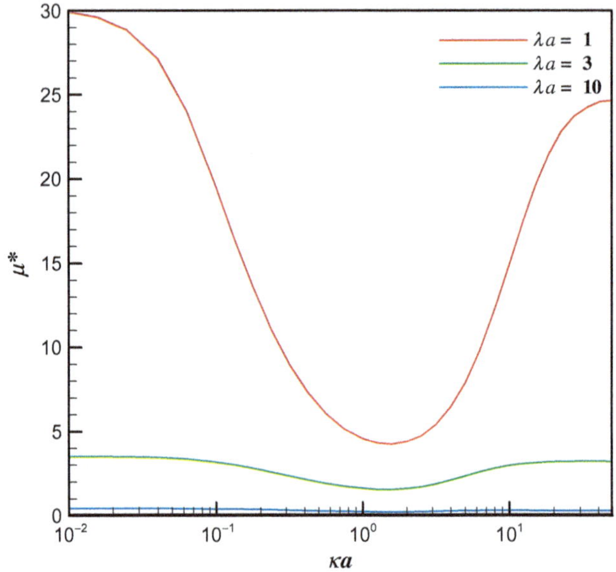

FIG. 9.4 Dimensionless mobility $\mu^* \left(=\frac{U^*}{E^*}\right)$ versus κa respectively for $Q_{\text{fix}}=30$, $\phi_{\text{w}}=0$, and $R_{\text{b}}^*=1.2$ with $\lambda a=1$, 3, and 10.

($\lambda a=1$, 3, and 10, respectively). Highly permeable particle ($\lambda a=1$) moves much faster than the rest two and with much significant double layer polarization, similar to the situation of a single porous particle. Note that the actual mobility reflects the true particle velocity here with $\mu^*=U^*/E^*$, hence μ^* is proportional to U^* or U. This is mainly due to the drastic reduction of hydrodynamic drag force, which is proportional to the square of particle permeability λa as indicated in Eq. (7.16): $F_{\text{D}}=\iiint_{r^*\leq 1}\gamma \mathbf{v}\cdot\mathbf{i}_z dV$, or $F_D^* = -\frac{8}{3}\pi(\lambda a)^2\Psi^*\Big|_{r^*=1}$ in dimensionless form. Hence, the drag force for $\lambda a=1$ is only 1% of the drag force for $\lambda a=10$, should the velocity distribution is the same within the porous particle. Even if the velocity fields will not be exactly the same in general due to the involvement of electric charges, either fixed or mobile, order-of-magnitude difference in particle velocity is still anticipated here as the ratio of $\lambda a=1$:3:10 corresponds to $(\lambda a)^2=1$:9:100, which crosses three orders of magnitude and particle velocity is proportional to the hydrodynamic drag force according to the Stokes' law: $F_{\text{D}}=6\pi\eta_f Ua$. The relative magnitude of particle mobility in Fig. 9.4 still reflects this reasoning, even in such a narrow cylindrical pore with $R_{\text{b}}^*=1.2$. Note that the profile corresponding to $\lambda a=10$ is almost invisible due to its small magnitude in comparison with $\lambda a=1$.

9.3.3 Lateral Boundary Confinement Effect

Fig. 9.5A and B shows the equilibrium distribution of the charge density ρ_{e}^* and corresponding electrostatic potential ϕ_{e}^* in a narrow pore. Lateral boundary

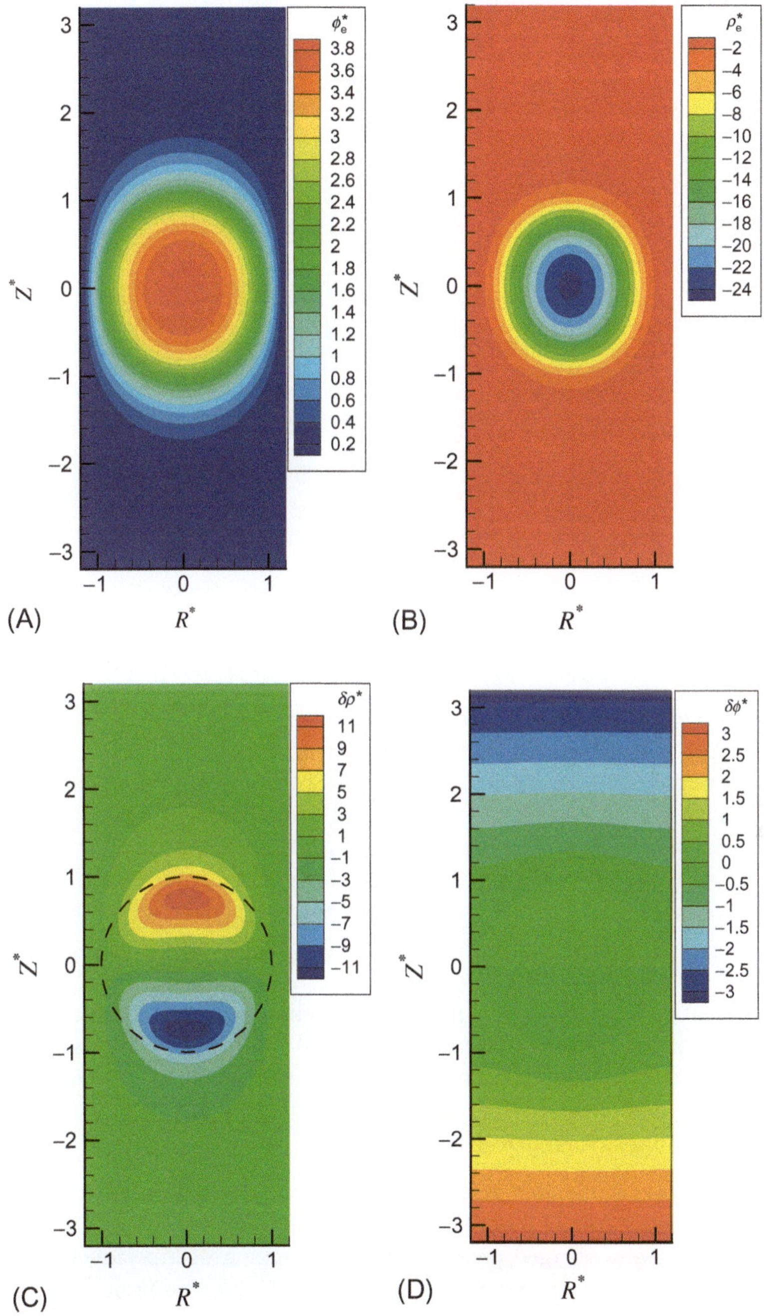

FIG. 9.5 Contour plots of different scaled variables at $\kappa a=1$, $Q_{\text{fix}}=30$, $R_b^*=1.2$, and $\phi_w=0$. (A) Scaled equilibrium electric potential (ϕ_e^*); (B) scaled equilibrium charge density (ρ_e^*); (C) scaled disturbed charge density ($\delta\rho^*$); (D) scaled disturbed electric potential ($\delta\phi^*$); (E) scaled stream function (Ψ^*).

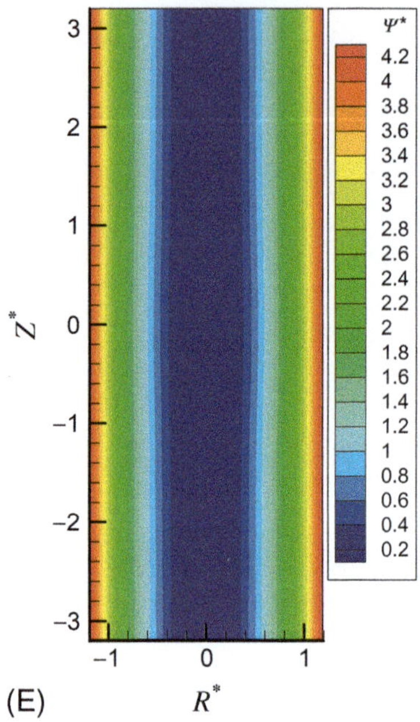

FIG. 9.5—CONT'D

confinement effect squeezes the original concentric shape to the observed elliptic shape. The polarization of the diffuse layer, however, is still similar to the corresponding single particle situation, as shown in Fig. 9.5C. This is due to the fact, as mentioned above, that most of the counterions within the double layer reside in the particle interior anyway. Corresponding plot for the disturbance of the electric potential $\delta\phi^*$ is shown in Fig. 9.5D as well. Note that it is this information that is used directly in the evaluation of electric driving force of the particle: $F_E = \iiint_{r^* \le 1} \rho_{fix}(-\nabla\delta\phi)\cdot \mathbf{i}_z dV$, or $F_E^* = -\frac{4}{3}\pi Q_{fix}\delta\phi^*\Big|_{r^*=1}$ in dimensionless form (Eq. 7.17). Corresponding flow field is also shown in Fig. 9.5E where the streamline plot of the scaled stream function Ψ^* is presented. Nearly straight ahead flow pattern is observed across the porous particle.

9.3.4 Impact of Electroosmosis Flow

Fig. 9.6 shows the impact of the electroosmosis flow of a charged cylindrical pore. The observation is consistent with that of a rigid particle in Chapter 5 in that a negatively charged pore enhances the motion of a positively charged particle, whereas a positively charged pore deters its motion in general. The reasoning behind is the same as that in Chapter 5.

An intriguing behavior is noted in Fig. 9.6 as well that a single porous particle may actually move faster in a chargeless ($\phi_w = 0$) narrow channel ($R_b^* = 1.2$), especially when the double layer is thick. Apparently, the lateral

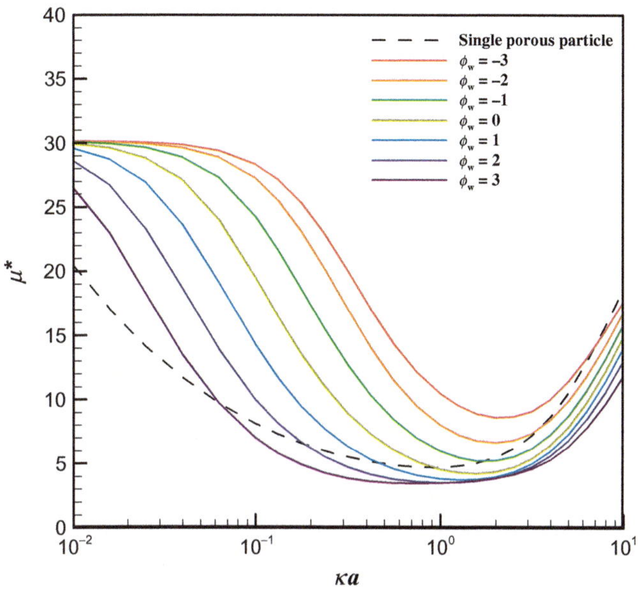

FIG. 9.6 Dimensionless mobility versus κa for $Q_{\text{fix}}=30$, $\lambda a=1$, and $R_b^*=1.2$ with $\phi_w=-3$ to 3, respectively.

boundary confinement effect is severe in this situation, and the squeezed-in extra counterions, together with the possible nearby coions, strengthen the uneven antisymmetric distribution between hemispheres within the particle further, which leads to an enhancement of particle motion. This electrostatic confinement effect diminishes when the double layer gets thin, and the hydrodynamic boundary effect sets in and tells the story thereafter.

9.3.5 Effect of Pore Radius

Fig. 9.7 exhibits the effect of the pore radius upon the particle mobility. Again the intriguing behavior when the double layer is thick is observed: When the double layer is relatively thick, the narrower the cylindrical pore is, the faster the particle moves! The single porous sphere serves as the consistent limiting case. The mechanism behind is again the lateral boundary confinement effect, which gets more significant with narrower pore.

9.3.6 Contour Plots of Electrokinetic Parameters

For completeness, Figs. 9.8A–9.9E present the contour plots of electrokinetic parameters of interest for both the positively and the negatively charged pore wall with $\phi_w=+3$ and $\phi_w=-3$, respectively. An insightful overview of the whole picture can thus be obtained by these plots, which shed the light into this complicated electrokinetic system.

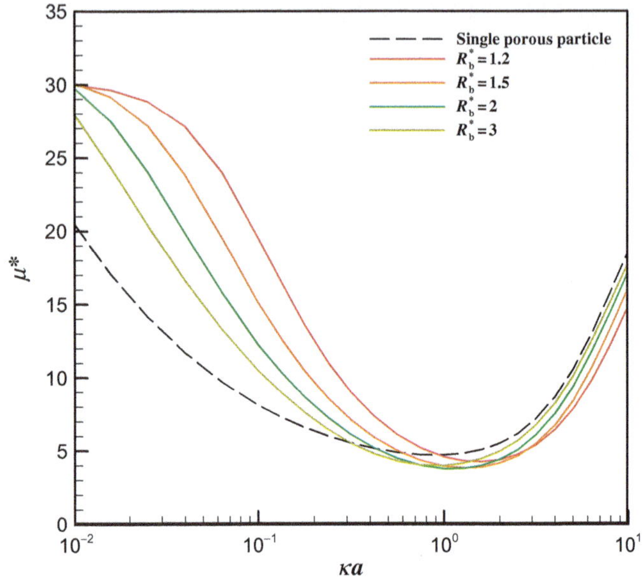

FIG. 9.7 Dimensionless mobility versus κa for $\phi_w=0$, $Q_{fix}=30$, and $\lambda a=1$ with $R_b^*=1.2$, 1.5, 2, and 3, respectively.

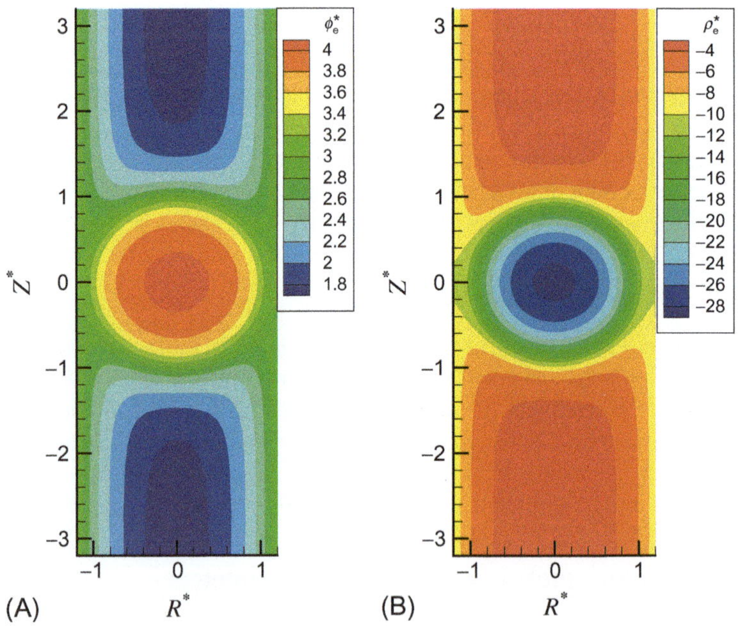

FIG. 9.8 Contour plots of different scaled variables at $\kappa a=1$, $\phi_w=3$, $R_b^*=1.2$, and $Q_{fix}=30$. (A) Scaled equilibrium electric potential (ϕ_e^*); (B) scaled equilibrium charge density (ρ_e^*); (C) scaled disturbed charge density ($\delta\rho^*$); (D) scaled disturbed electric potential ($\delta\phi^*$); (E) scaled stream function (Ψ^*).

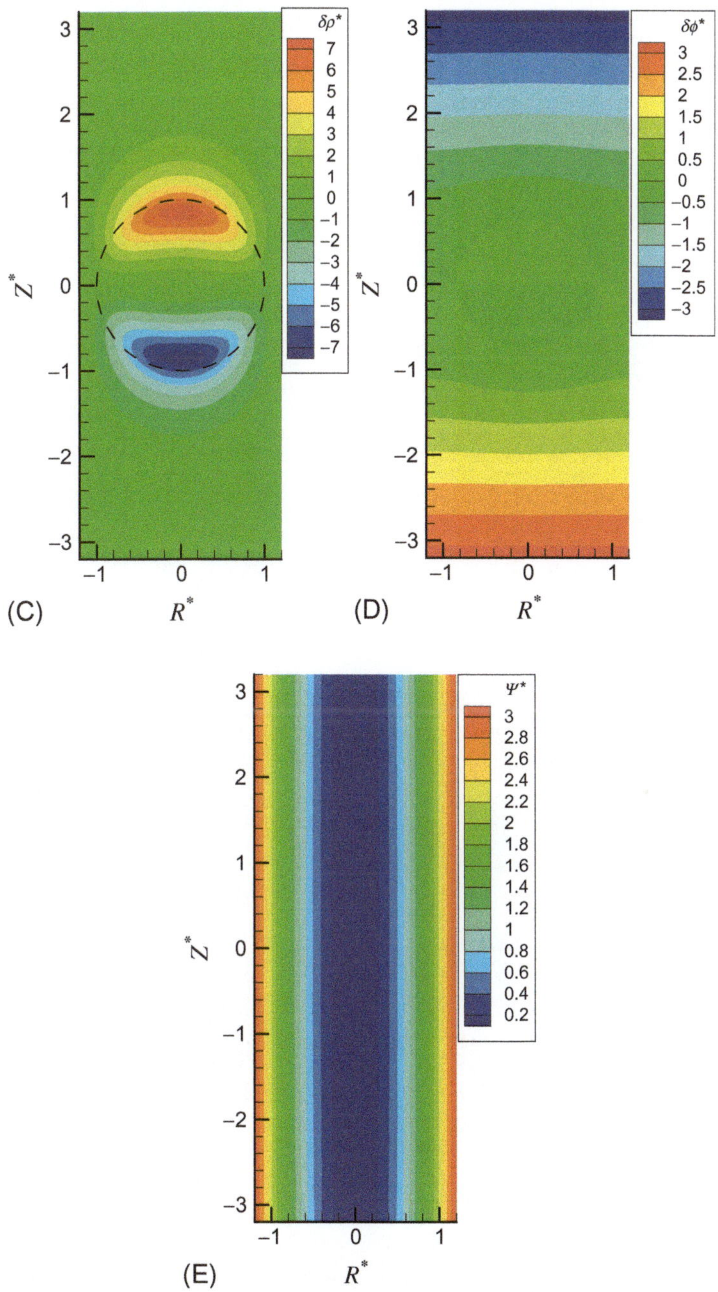
δρ*
7
6
5
4
3
2
1
0
−1
−2
−3
−4
−5
−6
−7
Z*
R*
(C)
δφ*
3
2.5
2
1.5
1
0.5
0
−0.5
−1
−1.5
−2
−2.5
−3
Z*
R*
(D)
Ψ*
3
2.8
2.6
2.4
2.2
2
1.8
1.6
1.4
1.2
1
0.8
0.6
0.4
0.2
Z*
R*
(E)

FIG. 9.8—CONT'D

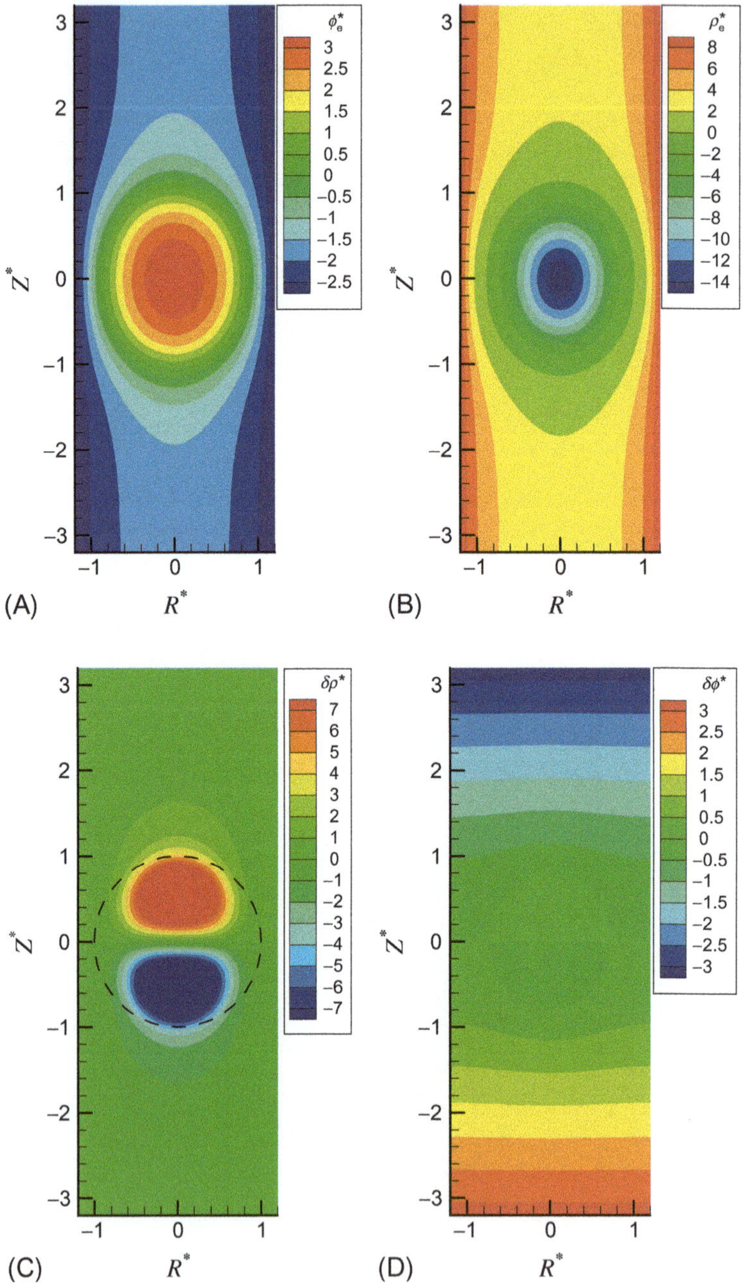

FIG. 9.9 Contour plots of different scaled variables at $\kappa a = 1$, $\phi_w = -3$, $R_b^* = 1.2$, and $Q_{fix} = 30$. (A) Scaled equilibrium electric potential (ϕ_e^*); (B) scaled equilibrium charge density (ρ_e^*); (C) scaled disturbed charge density ($\delta\rho^*$); (D) scaled disturbed electric potential ($\delta\phi^*$); (E) scaled stream function (Ψ^*).

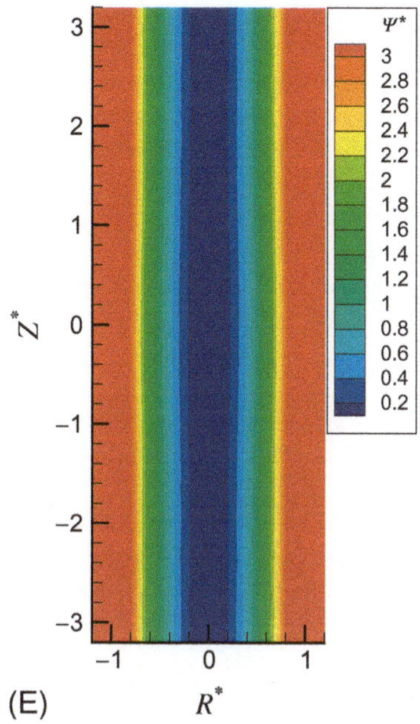

FIG. 9.9—CONT'D

REFERENCES

[1] Chang H, Yeo L. Electrokinetically driven microfluidics and nanofluidics. New York: Cambridge University Press; 2010.

[2] Kostal V, Arriaga EA. Recent advances in the analysis of biological particles by capillary electrophoresis. Electrophoresis 2008;29(12):2578–86.

[3] Lagerqvist J, Zwolak M, Di Ventra M. Fast DNA sequencing via transverse electronic transport. Nano Lett 2006;6(4):779–82.

[4] Kang Y, Li D. Electrokinetic motion of particles and cells in microchannels. Microfluid Nanofluid 2009;6(4):431–60.

[5] Ren J, Fang N, Wu D. Inverse-flow derivatization for capillary electrophoresis of DNA fragments with laser-induced fluorescence detection. Anal Chim Acta 2002;470(2):129–35.

[6] Lowe AB, McCormick CL. Polyelectrolytes and polyzwitterions: synthesis, properties, and applications. Washington, DC: ACS Publications; 2006.

[7] Heller C. Capillary electrophoresis of proteins and nucleic acids in gels and entangled polymer solutions. J Chromatogr A 1995;698(1–2):19–31.

[8] Kao Y-Y, et al. Analysis of amino acids and biogenic amines in breast cancer cells by capillary electrophoresis using polymer solutions containing sodium dodecyl sulfate. J Chromatogr A 2010;1217(4):582–7.

[9] Li Z, et al. Electrophoresis of periodontal pathogens in poly (ethyleneoxide) solutions with uncoated capillary. Anal Biochem 2015;471:70–2.

[10] Carrilho E, et al. Rapid DNA sequencing of more than 1000 bases per run by capillary electrophoresis using replaceable linear polyacrylamide solutions. Anal Chem 1996;68(19):3305–13.

[11] Kim Y, Yeung ES. Separation of DNA sequencing fragments up to 1000 bases by using poly (ethylene oxide)-filled capillary electrophoresis. J Chromatogr A 1997;781(1–2):315–25.

[12] Shimomura M, Sawadaishi T. Bottom-up strategy of materials fabrication: a new trend in nanotechnology of soft materials. Curr Opin Colloid Interface Sci 2001;6(1):11–6.

[13] Keh H-J, Anderson J. Boundary effects on electrophoretic motion of colloidal spheres. J Fluid Mech 1985;153:417–39.

[14] Keh HJ, Chiou JY. Electrophoresis of a colloidal sphere in a circular cylindrical pore. AICHE J 1996;42(5):1397–406.

[15] Shugai AA, Carnie SL. Electrophoretic motion of a spherical particle with a thick double layer in bounded flows. J Colloid Interface Sci 1999;213(2):298–315.

[16] Teubner M. The motion of charged colloidal particles in electric fields. J Chem Phys 1982;76(11):5564–73.

[17] Chao K, Chen B, Wu J. Numerical analysis of field-modulated electroosmotic flows in microchannels with arbitrary numbers and configurations of discrete electrodes. Biomed Microdevices 2010;12(6):959–66.

[18] Huang CH, Hsu HP, Lee E. Electrophoretic motion of a charged porous sphere within micro- and nanochannels. Phys Chem Chem Phys 2012;14(2):657–67.

[19] Huang CH, et al. Electrophoresis of a soft particle within a cylindrical pore: polarization effect with the nonlinear Poisson-Boltzmann equation. J Phys Chem B 2010;114(31): 10114–25.

[20] Hermans J. Sedimentation and electrophoresis of porous spheres. J Polym Sci 1955;18(90): 527–34.

[21] Hermans J. Electrophoresis of charged polymer molecules with partial free drainage. Koninkl Ned Akad Wetenschap Proc 1955;58:182–7.

[22] Imai N, Iwasa K. Theory of electrophoresis of polyelectrolytes. Israel J Chem 1973;11(2–3): 223–33.

Chapter 10

Dynamic Electrophoresis of Porous Spherical Particles

10.1 INTRODUCTION

The general theory underlying the dynamic electrophoresis and the observed phenomena were discussed in Chapter 6 for the rigid particle situation. For the porous particle systems, the fundamental theory is essentially the same except that the electric driving force is from the fixed charges within the particle instead of the surface charges, which makes the total charges much larger and hence leads to a much stronger electric driving force in response to the applied AC electric field in general. Moreover, the hydrodynamic motion-deterring drag force is much weaker due to the permeability of a porous particle, especially when the particle is highly permeable. Combining these two factors of stronger electric driving force and weaker hydrodynamic drag force, the dynamic mobility of a porous particle is normally much larger than the corresponding rigid particle.

On the other hand, the distribution of counterions is fundamentally different from that of a rigid particle. Due to the permeability of a porous particle and strong Coulomb attraction force when the particle is highly charged, most of the counterions may reside in the interior of the particle leading to the "counterion condensation effect" as discussed earlier in Chapter 7. The particle fixed charges, hence, are screened by these condensed ions there, leading to an effective particle charges much less than the nominal one represented by Q_{fix}. When an AC electric field is applied, however, those "condensed" counterions can still move back and forth accordingly in the particle interior in response to the field due to its mobile nature, which is also emphasized by the theoretical treatment regarding them as represented by homogeneously charged fluid. In other words, the convection of fluid flow in a porous particle has significant impact upon the motion of electrolyte ions; hence, the corresponding double layer polarization, or single diffuse layer polarization to be precise, as discussed in Chapter 7.

Interface Science and Technology, Vol. 26. https://doi.org/10.1016/B978-0-08-100865-2.00010-2

Note that the polarization within the particle actually determines the particle velocity via the incremental change of the local electric potential there, which can lead to a significant local electric field ($\nabla \delta\phi$), although it assumes a small value itself. With this picture in mind, we realize that the dynamic mobility for a porous particle would be significantly different from the rigid particle case discussed earlier in Chapter 6, both qualitatively and quantitatively, as we shall discuss later.

As mentioned earlier in Chapter 7, the charged porous particle is an excellent model for polyelectrolytes like DNAs and proteins; hence, the dynamic electrophoresis finds many applications in biochemical and biomedical applications involving them [1–14]. Moreover, the so-called "soft particle" with an inner rigid sphere covered by a concentric outer permeable layer of porous nature is also adopted to model such important systems like the "spherical polyelectrolyte brush" (SPB), which has a wide variety of applications as well [15, 16]. Note that its unique properties and functions are mainly from the outer porous layer of polymers, such as the load and release of drugs or other therapeutic agents with biomedical applications [16].

Theoretical studies on the porous particle, in particular, have seldom been reported in the literature, if any. On the other hand, quite a few were presented for the corresponding soft particles [17–23]. Issues like stress jump across the interface between the porous layer and the solution, the hydrodynamic drag force of the inner rigid particle, among other things, were discussed there. In particular, giant dynamic mobility for an SPB soft particle system was observed experimentally also [16], as we have predicted at the beginning, indicating the unique and facilitating feature pertinent to the porous particle system. Obviously, this observation has nothing to do with the extra drag force from the inner rigid sphere, which can only slow down the particle motion. Definitely, it is from the outer porous layer, especially the permeability there.

We focus here on a uniformly charged porous particle here in this chapter, as the presence of an inner rigid sphere has no role to play in general, except inducing a motion-deterring drag force. Moreover, inclusion of an extra parameter representing the size of the inner rigid particle often just muddies the water and prevents any meaningful and universally applicable conclusion to be drawn, at least from the viewpoint of theoretical analysis.

10.2 THEORY

10.2.1 Single Porous Particle

The system diagrams are shown in Fig. 10.1A and B for a single porous particle and the suspension of porous particles, respectively. The governing

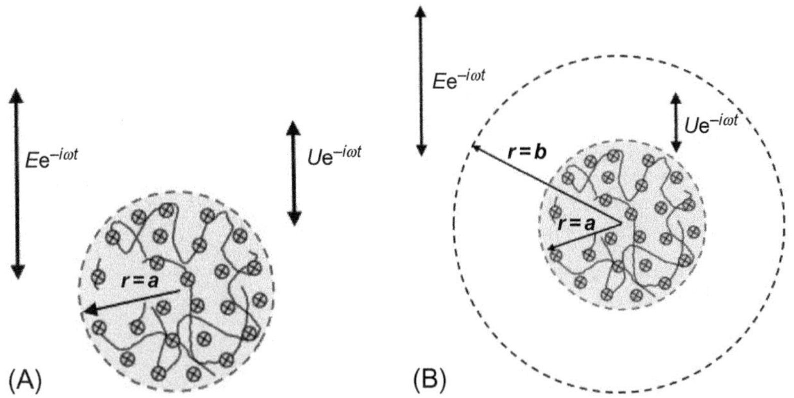

FIG. 10.1 System diagram. (A) Single porous particle; (B) porous particles suspensions.

equations for the dynamic electrophoresis of a single porous particle as well as suspensions are similar to those in Chapter 6 where corresponding rigid particle situations were treated, with an additional porous region within the particle considered. This region, however, is well-analyzed in Chapter 7 where corresponding DC electrophoresis is considered. Therefore, the task here is to combine these two sets of treatments together, which can be done by either way. We decide to start with the governing system outside the particle as in Chapter 6 and simply add an additional domain inside the porous sphere and an extra time-dependent sinusoidal factor, $e^{-i\omega t}$, to all of the variables there that will be affected by the AC electric field applied. Following the same analysis and methodology adopted in Chapter 6, the final one-dimensional governing equations in a binary electrolyte solution system within the porous particle are:

$$\nabla^{*2}\phi_e{}^* = -\frac{(\kappa a)^2}{1+\alpha}[\exp(-\phi_e{}^*) - \exp(\alpha\phi_e{}^*)] - Q_{\text{fix}} \quad 0 \le r^* \le 1 \tag{10.1}$$

$$D^{*4}\psi_R - (\lambda a)^2 D^{*2}\psi_R - \omega^* D^{*2}\psi_I = -\frac{(\kappa a)^2}{1+\alpha}\left[G_{1R}\exp\left(-\phi_e^*\right) + \alpha G_{2R}\exp\left(\alpha\phi_e^*\right)\right] \frac{d\phi_e^*}{dr^*}, \quad 0 \le r^* \le 1 \tag{10.2}$$

$$D^{*4}\psi_I - (\lambda a)^2 D^{*2}\psi_I + \omega^* D^{*2}\psi_R = -\frac{(\kappa a)^2}{1+\alpha}\left[G_{1I}\exp\left(-\phi_e^*\right) + \alpha G_{2I}\exp\left(\alpha\phi_e^*\right)\right] \frac{d\phi_e^*}{dr^*}, \quad 0 \le r^* \le 1 \tag{10.3}$$

Note both the real and imaginary parts are presented explicitly.

And the boundary conditions across the surface of the porous particles are the same as those in Chapter 7 as well, but now of dynamic nature with an extra time-dependent factor $e^{-i\omega t}$ associated with each variable. They, however, cancel each other in corresponding boundary conditions and hence, yield the same form as in Chapter 7 for the static DC electrophoresis situation.

The governing equations outside the particle as well as the boundary conditions far away from the particle are the same as those in Chapter 6 for a single rigid particle, as we have mentioned. Subsequent numerical treatment and final evaluation of particle mobility are the same as well.

10.2.2 Suspensions

Cell model is applied and the governing equations as well as the associated boundary conditions are the same as above except that an outer virtual surface is incorporated into the system instead. The boundary conditions there are the same as those in Chapter 6. The rest of the treatment is also the same as there.

10.3 RESULTS AND DISCUSSION

10.3.1 Single Porous Particle

The electroacoustic spectrometry, the measurement of dynamic mobility as a function of frequency for a single porous particle, is presented in Figs. 10.2–10.10 with various amount of fix charges (Q_{fix}), ranging from 1 to 100, and various permeability (λa), ranging from 1 to 10, respectively. These situations cover extensively the range of highly permeable ($\lambda a = 1$) to moderately and poorly permeable ($\lambda a = 3$ and 10), as well as highly charged ($Q_{\text{fix}} = 30$, 50 and 100, respectively) and lowly charged ($Q_{\text{fix}} = 1$ and 10) porous particles. The electrolyte strength of the suspending medium, κa, or the double layer thickness, ranges from thick ($\kappa a = 0.1$), comparable to the particle size ($\kappa a = 1$) and thin ($\kappa a = 10$). Essentially, all of the information that can be possibly provided by the dynamic electrophoresis are contained in these diagrams completely, which we will discuss in details as follows.

As we elaborated earlier in the introduction section, the much larger amount of carried charges coupled with much smaller hydrodynamic drag force due to the particle permeability makes a porous particle move much faster than the corresponding rigid particle in general, or much larger electrophoretic mobility. As shown in Figs. 10.2–10.10, the magnitude of particle mobility for a high permeable one ($\lambda a = 1$) is order-of-magnitude larger than the less permeable ones ($\lambda a = 3$ and 10). Note that in order to highlight the much faster particle speed, dimensionless particle mobility $\mu^* = U^*/E^*$ is used here, which is directly proportional to the actual particle speed. Sharp peaks,

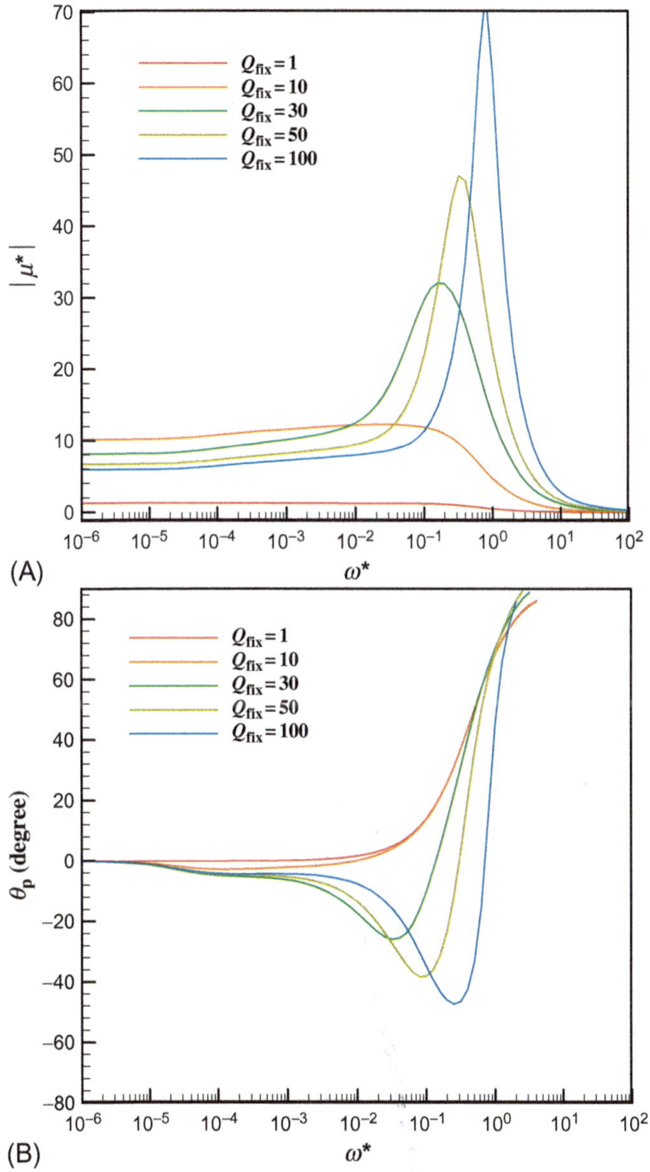

FIG. 10.2 (A) Magnitude of the dimensionless mobility $|\mu^*|$ versus the scaled frequency ω^* at various Q_{fix} with $\lambda a=1$ and $\kappa a=0.1$. (B) Phase angle of the dimensionless mobility θ_p versus the scaled frequency ω^* at various Q_{fix} with $\lambda a=1$ and $\kappa a=0.1$.

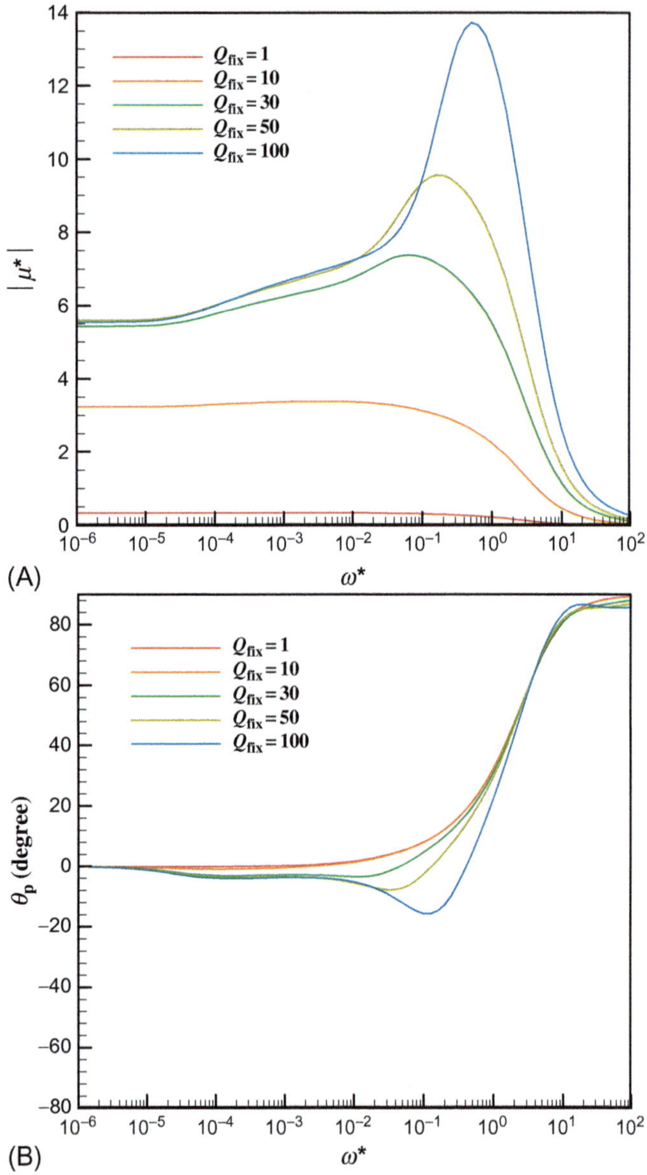

FIG. 10.3 (A) Magnitude of the dimensionless mobility $|\mu^*|$ versus the scaled frequency ω^* at various Q_{fix} with $\lambda a=3$ and $\kappa a=0.1$. (B) Phase angle of the dimensionless mobility θ_p versus the scaled frequency ω^* at various Q_{fix} with $\lambda a=3$ and $\kappa a=0.1$.

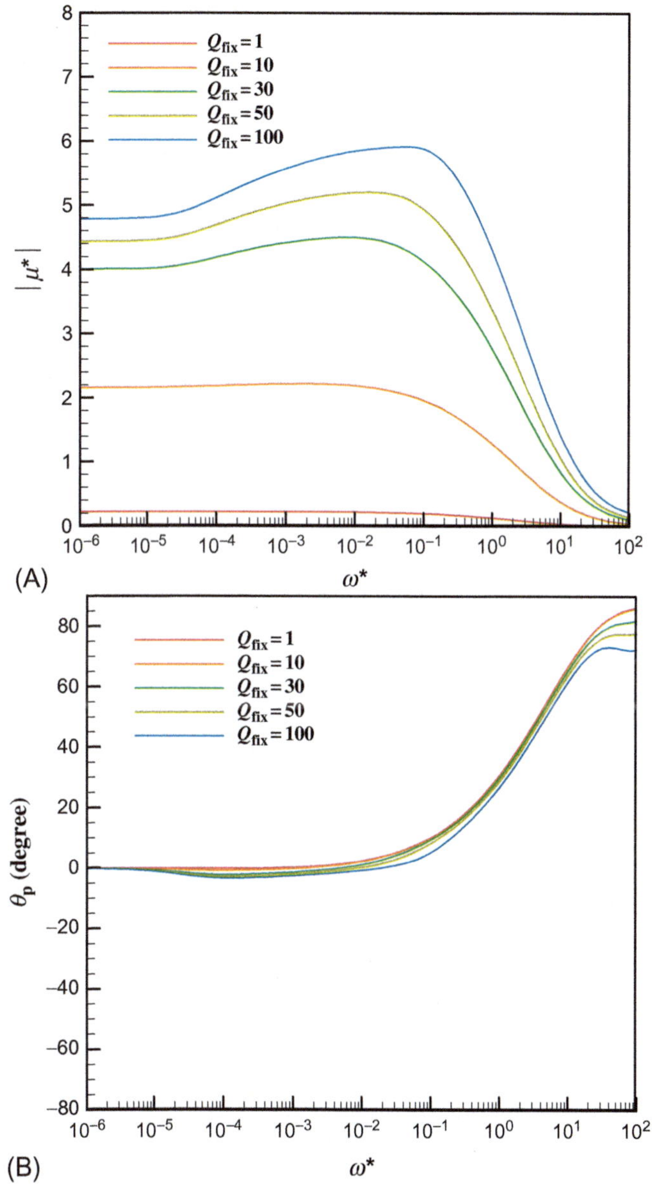

FIG. 10.4 (A) Magnitude of the dimensionless mobility $|\mu^*|$ versus the scaled frequency ω^* at various Q_{fix} with $\lambda a = 10$ and $\kappa a = 0.1$. (B) Phase angle of the dimensionless mobility θ_p versus the scaled frequency ω^* at various Q_{fix} with $\lambda a = 10$ and $\kappa a = 0.1$.

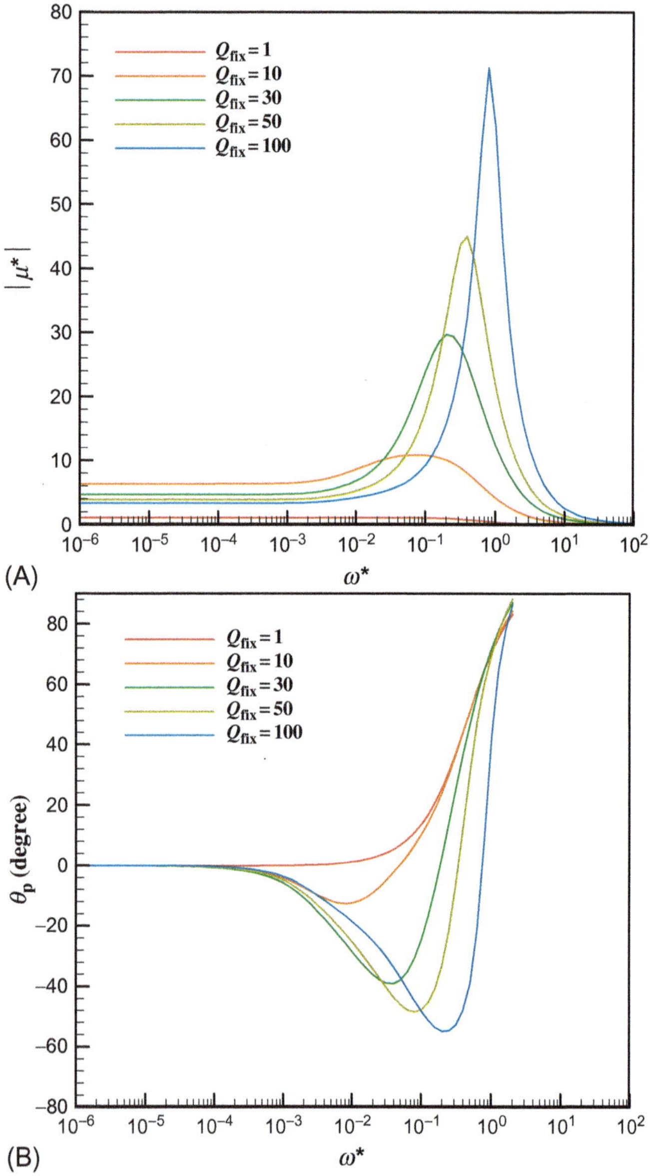

FIG. 10.5 (A) Magnitude of the dimensionless mobility $|\mu^*|$ versus the scaled frequency ω^* at various Q_{fix} with $\lambda a = 1$ and $\kappa a = 1$. (B) Phase angle of the dimensionless mobility θ_p versus the scaled frequency ω^* at various Q_{fix} with $\lambda a = 1$ and $\kappa a = 1$.

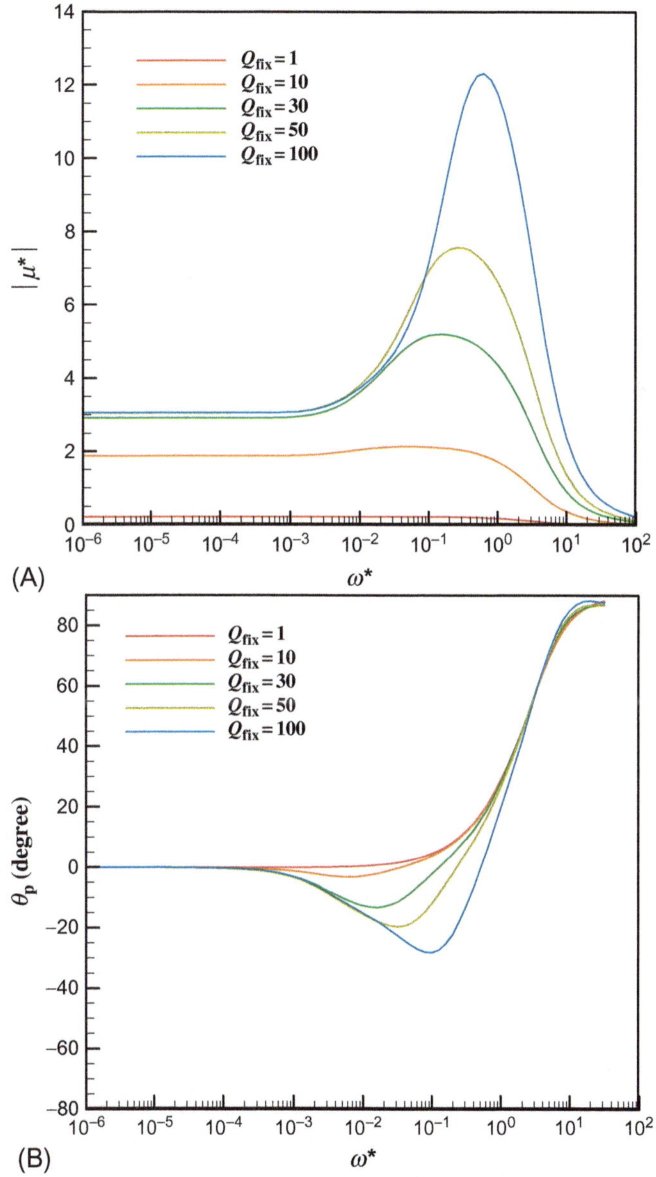

FIG. 10.6 (A) Magnitude of the dimensionless mobility $|\mu^*|$ versus the scaled frequency ω^* at various Q_{fix} with $\lambda a = 3$ and $\kappa a = 1$. (B) Phase angle of the dimensionless mobility θ_p versus the scaled frequency ω^* at various Q_{fix} with $\lambda a = 3$ and $\kappa a = 1$.

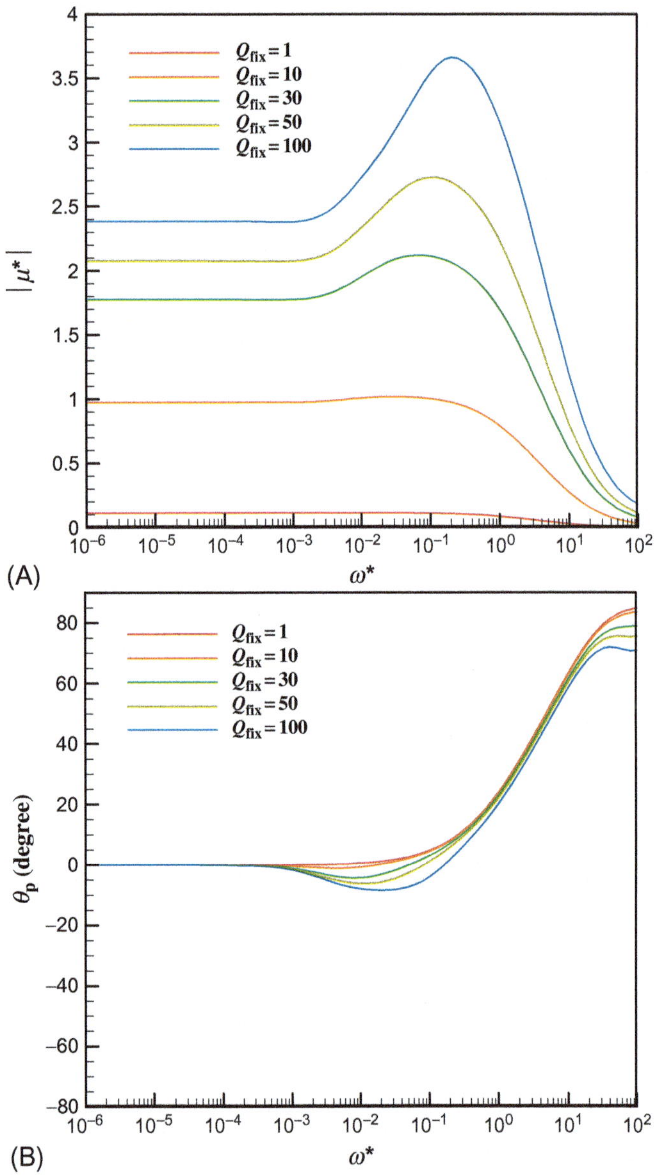

FIG. 10.7 (A) Magnitude of the dimensionless mobility $|\mu^*|$ versus the scaled frequency ω^* at various Q_{fix} with $\lambda a = 10$ and $\kappa a = 1$. (B) Phase angle of the dimensionless mobility θ_{p} versus the scaled frequency ω^* at various Q_{fix} with $\lambda a = 10$ and $\kappa a = 1$.

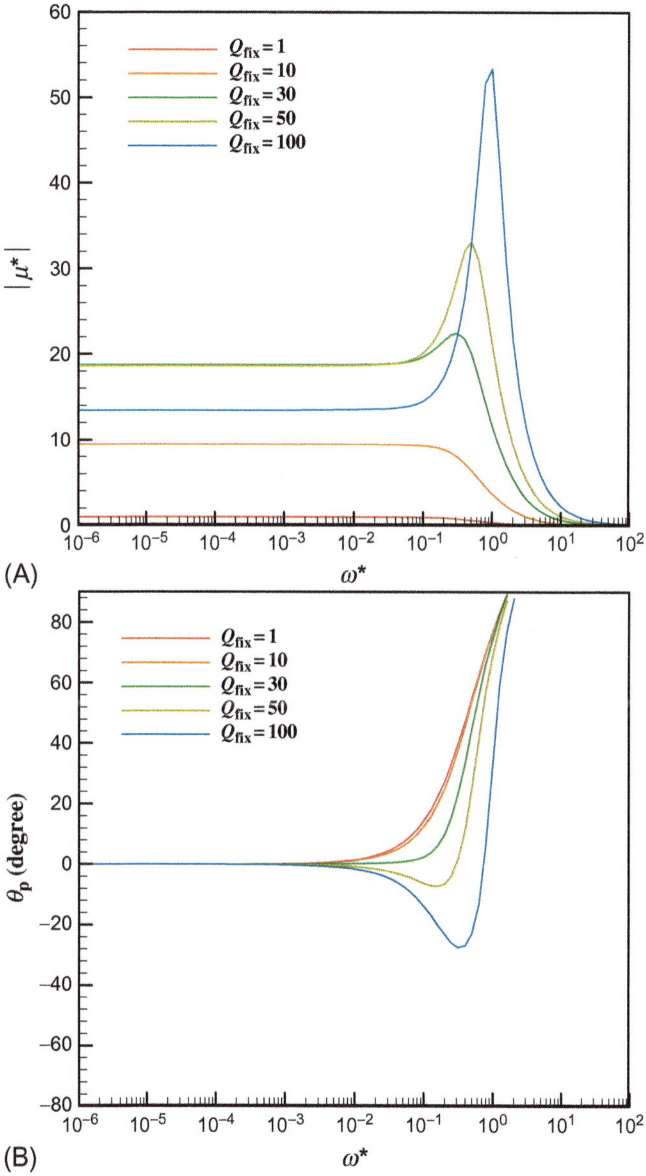

FIG. 10.8 (A) Magnitude of the dimensionless mobility $|\mu^*|$ versus the scaled frequency ω^* at various Q_{fix} with $\lambda a = 1$ and $\kappa a = 10$. (B) Phase angle of the dimensionless mobility θ_p versus the scaled frequency ω^* at various Q_{fix} with $\lambda a = 1$ and $\kappa a = 10$.

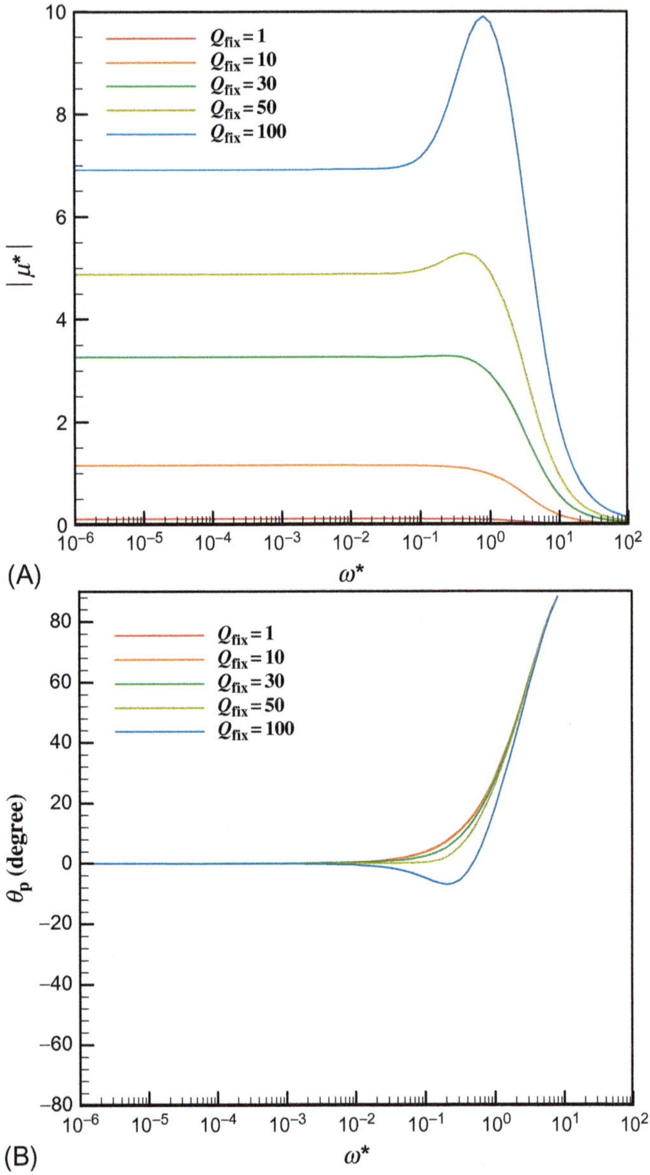

FIG. 10.9 (A) Magnitude of the dimensionless mobility $|\mu^*|$ versus the scaled frequency ω^* at various Q_{fix} with $\lambda a=3$ and $\kappa a=10$. (B) Phase angle of the dimensionless mobility θ_p versus the scaled frequency ω^* at various Q_{fix} with $\lambda a=3$ and $\kappa a=10$.

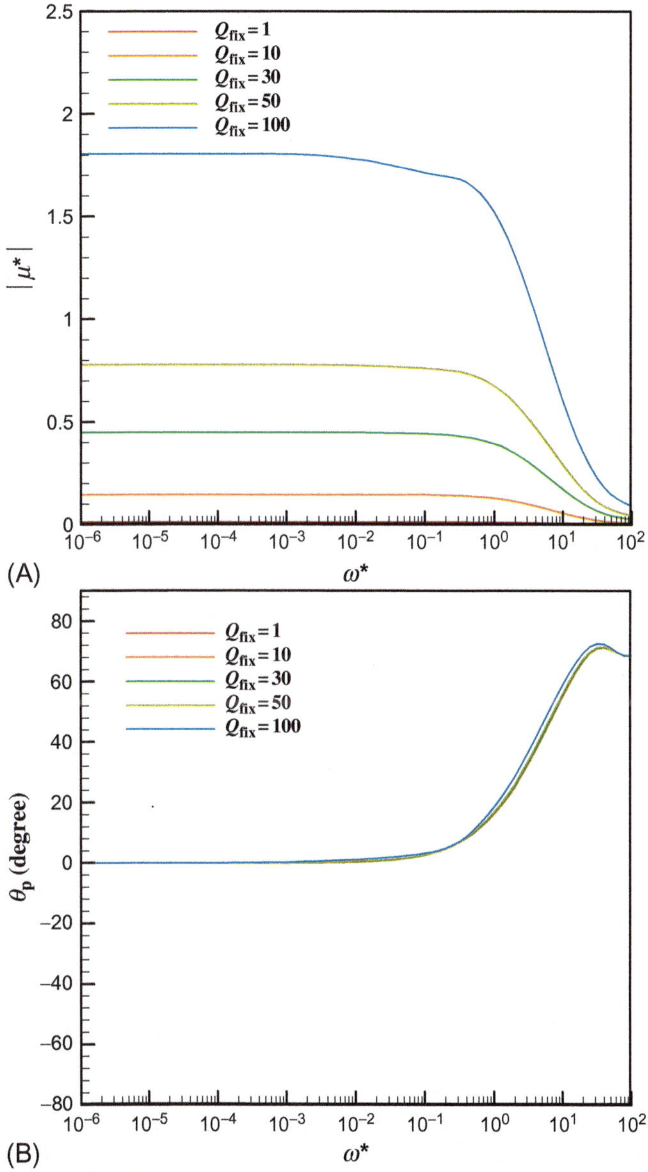

FIG. 10.10 (A) Magnitude of the dimensionless mobility $|\mu^*|$ versus the scaled frequency ω^* at various Q_{fix} with $\lambda a = 10$ and $\kappa a = 10$. (B) Phase angle of the dimensionless mobility θ_p versus the scaled frequency ω^* at various Q_{fix} with $\lambda a = 10$ and $\kappa a = 10$.

hence, are observed in these figures for a porous particle both highly permeable and highly charged, especially when the double layer is very thick ($\kappa a = 0.1$). This is due to the enlarged extent of double layer polarization outside the particle when the motion-deterring diffusion of electrolyte ions cannot catch up with the increasing AC electric field frequency. The ions outside are swept far away from the highly mobile particle leading to a drastic increase of particle mobility, as discussed in Chapter 6. When the frequency further increases, however, these mobility profiles do not coincide with one another to yield a Smoluchowski type of behavior. Separate peaks are normally observed here. This is due to the fact that even when the outside counterions are motionless at ultra high frequency, the handsome amount of counterions essentially confined within the particle due to the counterion condensation phenomenon still move hand-in-hand with the porous particle itself with no delay whatsoever, kind of like a general bulk-type immobile Stern layer. Thus, the mobility behavior is still charge level- (Q_{fix}) dependent, as the extent of counterion condensation is charge level-dependent. And that's why complete separate peaks are observed in these figures here.

Moreover, it is observed that these peaks are about the same for $\kappa a = 0.1$, 1, and 10. This is because the counterions of the double layer are mostly within the particle itself in this situation, due to its permeability and counterion condensation effect of electrostatic nature discussed above. This fact is essentially universal regardless of the levels of κa. Hence, the behaviors of particles of the same permeability are similar here, unlike the corresponding rigid particle situation where all the counterions are outside the particle and thus the polarization is very sensitive of the κa value.

10.3.2 Suspensions

We now go on to examine the impact of neighboring particles, which consists of two aspects: double layer overlapping of electrostatic nature and hydrodynamic swarming effect of extra drag force. We focus on highly permeable particles when the double layer thickness is comparable to the particle radius, i.e., $\lambda a = 1$, to highlight the unique feature pertinent to porous particles. $Q_{\text{fix}} = 10$, 30, and 50 are examined, respectively, in Figs. 10.11–10.13.

An astonishing phenomenon is observed in general: The more concentrated the particle suspension is, the higher the particle mobility is!; at least in the low to moderate frequency range. This is consistent with the observation in the corresponding DC or static electrophoresis with similar reasoning behind it. Concentrated suspensions of porous particles are like fenceless beautiful gardens for kids to enter and play there freely. There is no difference whether they are from any specific porous particle. The electric driving force simply increases with more counterions driven in, due to the inherent nature of antisymmetric distribution of them and the unit cell model treatment here. The more ions are poured in, the stronger the electric driving force; hence the particle mobility. This concludes our discussion here for this chapter.

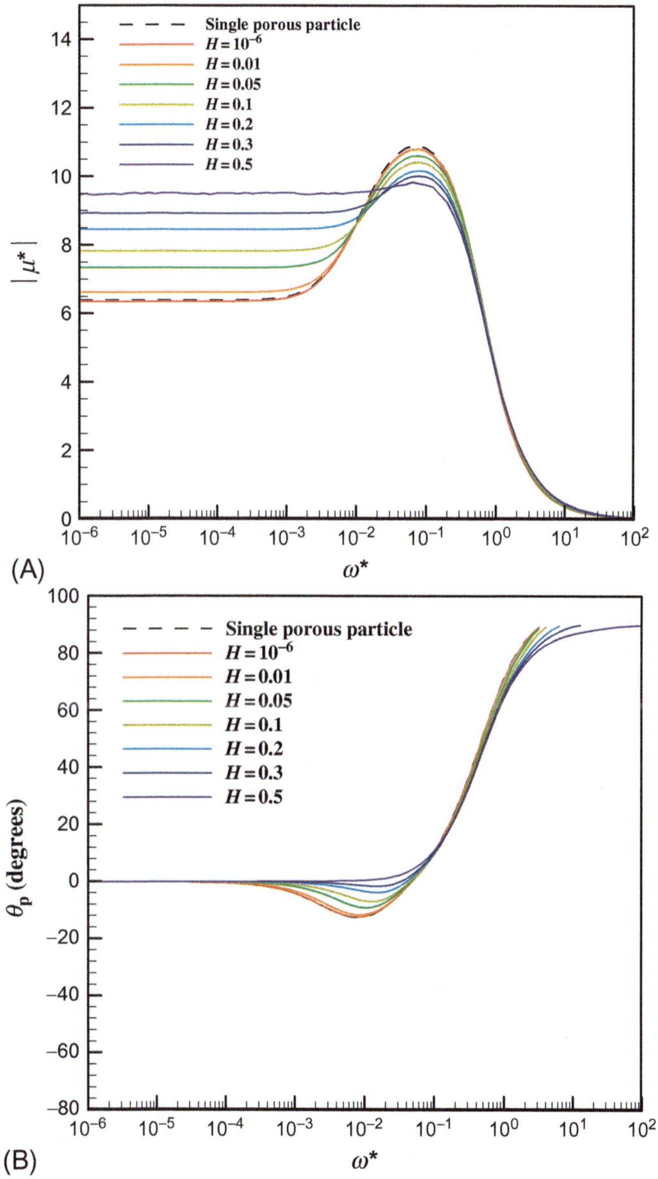

FIG. 10.11 (A) Magnitude of the dimensionless mobility $|\mu^*|$ versus the scaled frequency ω^* at various H with $Q_{\text{fix}}=10$, $\lambda a=1$, and $\kappa a=1$. (B) Phase angle of the dimensionless mobility θ_{p} versus the scaled frequency ω^* at various H with $Q_{\text{fix}}=10$, $\lambda a=1$, and $\kappa a=1$.

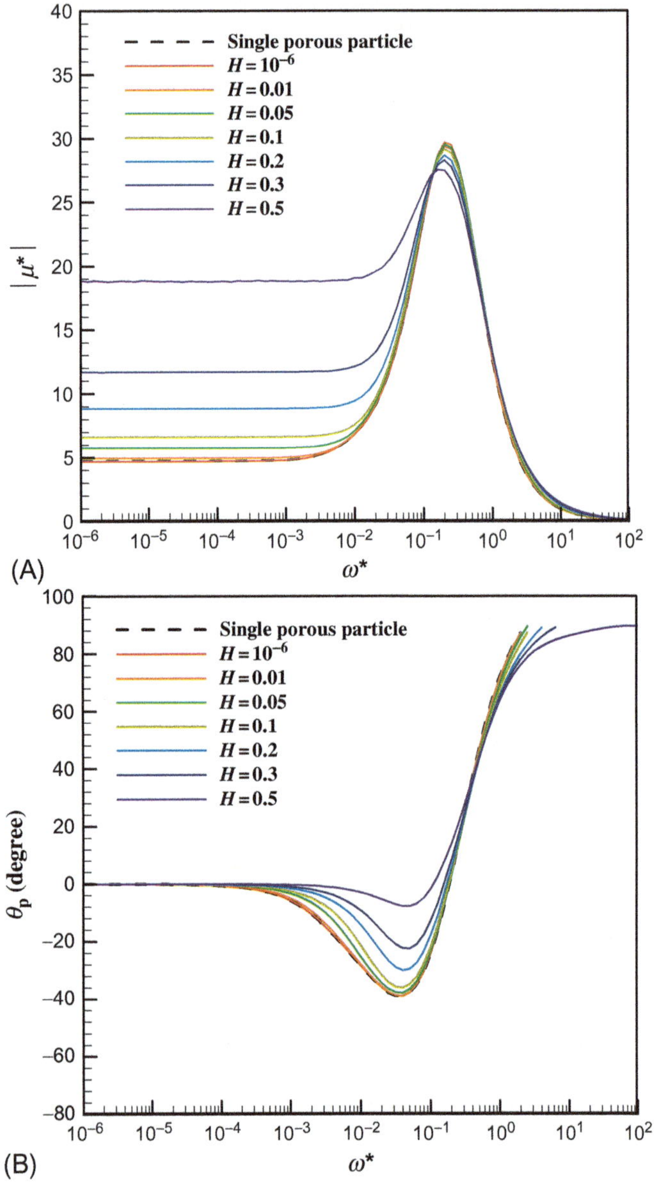

FIG. 10.12 (A) Magnitude of the dimensionless mobility $|\mu^*|$ versus the scaled frequency ω^* at various H with $Q_{\text{fix}}=30$, $\lambda a=1$, and $\kappa a=1$. (B) Phase angle of the dimensionless mobility θ_{p} versus the scaled frequency ω^* at various H with $Q_{\text{fix}}=30$, $\lambda a=1$, and $\kappa a=1$.

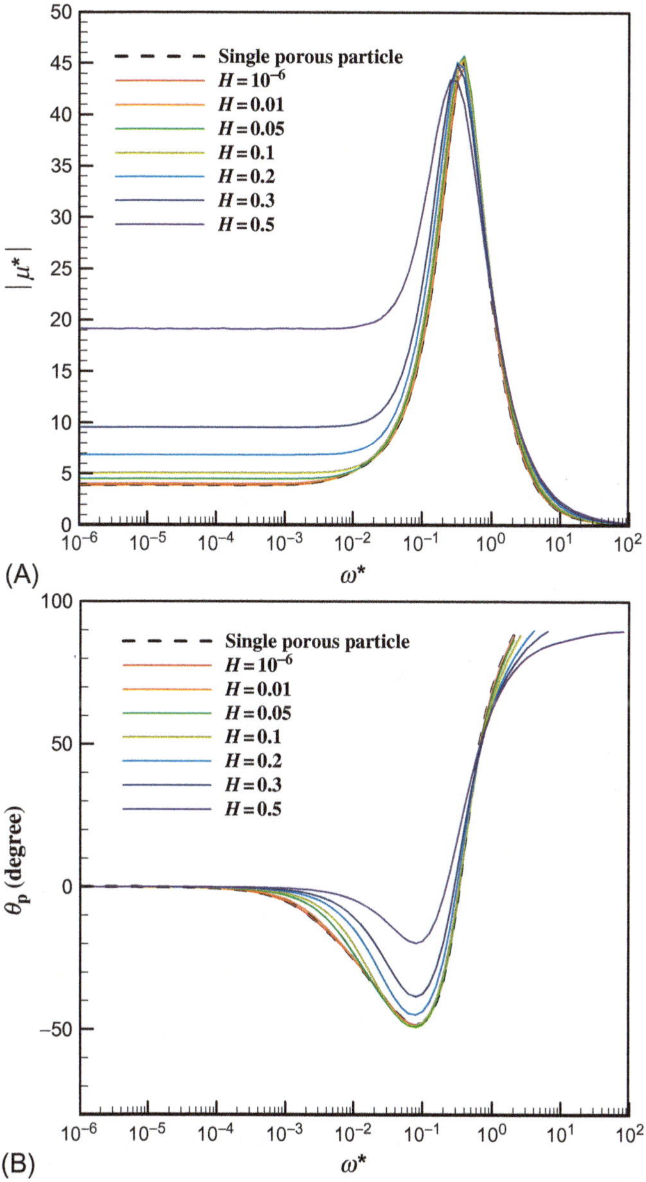

FIG. 10.13 (A) Magnitude of the dimensionless mobility $|\mu^*|$ versus the scaled frequency ω^* at various H with $Q_{\text{fix}}=50$, $\lambda a=1$, and $\kappa a=1$. (B) Phase angle of the dimensionless mobility θ_p versus the scaled frequency ω^* at various H with $Q_{\text{fix}}=50$, $\lambda a=1$, and $\kappa a=1$.

REFERENCES

[1] Galler K, Bräutigam K, Große C, Popp J, Neugebauer U. Making a big thing of a small cell—recent advances in single cell analysis. Analyst 2014;139(6):1237–73.

[2] El-Ali J, Sorger PK, Jensen KF. Cells on chips. Nature 2006;442(7101):403.

[3] Wei Z, Li X, Zhao D, Yan H, Hu Z, Liang Z, et al. Flow-through cell electroporation microchip integrating dielectrophoretic viable cell sorting. Anal Chem 2014;86(20):10215–22.

[4] Rothemund PW. Folding DNA to create nanoscale shapes and patterns. Nature 2006;440(7082):297.

[5] Mikkila J, Eskelinen A-P, Niemela EH, Linko V, Frilander MJ, Torma P, et al. Virus-encapsulated DNA origami nanostructures for cellular delivery. Nano Lett 2014;14(4):2196–200.

[6] Kuzyk A, Yurke B, Toppari JJ, Linko V, Törmä P. Dielectrophoretic trapping of DNA origami. Small 2008;4(4):447–50.

[7] Shen B, Linko V, Dietz H, Toppari JJ. Dielectrophoretic trapping of multilayer DNA origami nanostructures and DNA origami-induced local destruction of silicon dioxide. Electrophoresis 2015;36(2):255–62.

[8] Ahadian S, Yamada S, Ramon-Azcon J, Ino K, Shiku H, Khademhosseini A, et al. Rapid and high-throughput formation of 3D embryoid bodies in hydrogels using the dielectrophoresis technique. Lab Chip 2014;14(19):3690–4.

[9] Clevers H. The cancer stem cell: premises, promises and challenges. Nat Med 2011;17(3):313.

[10] Ong JM, da Cruz L. A review and update on the current status of stem cell therapy and the retina. Br Med Bull 2012;102:133–46.

[11] Bhasin A, Srivastava MP, Mohanty S, Bhatia R, Kumaran SS, Bose S. Stem cell therapy: a clinical trial of stroke. Clin Neurol Neurosurg 2013;115(7):1003–8.

[12] Hoppe PS, Coutu DL, Schroeder T. Single-cell technologies sharpen up mammalian stem cell research. Nat Cell Biol 2014;16(10):919.

[13] Adamczyk Z, Jamroży K, Batys P, Michna A. Influence of ionic strength on poly(diallyldimethylammonium chloride) macromolecule conformations in electrolyte solutions. J Colloid Interface Sci 2014;435:182–90.

[14] Walldal C, Åkerman B. Effect of ionic strength on the dynamic mobility of polyelectrolytes. Langmuir 1999;15(16):5237–43.

[15] Ballauff M. Spherical polyelectrolyte brushes. Prog Polym Sci 2007;32(10):1135–51.

[16] Jiménez ML, Delgado A, Ahualli S, Hoffmann M, Witteman A, Ballauff M. Giant permittivity and dynamic mobility observed for spherical polyelectrolyte brushes. Soft Matter 2011;7(8):3758–62.

[17] Ohshima H. Dynamic electrophoretic mobility of a soft particle. J Colloid Interface Sci 2001;233(1):142–52.

[18] Lopez-Garcia JJ, Grosse C, Horno J. Numerical study of colloidal suspensions of soft spherical particles using the network method—2. AC electrokinetic and dielectric properties. J Colloid Interface Sci 2003;265(2):341–50.

[19] Min WL, Lee E, Hsu JP. Dynamic electrophoresis of a spherical dispersion of soft particles subject to a stress-jump condition. J Colloid Interface Sci 2006;299(1):464–71.

[20] Lou J, Lee E. Dynamic electrophoresis of concentrated droplet dispersions at arbitrary surface potentials. Chem Eng Sci 2007;62(22):6234–45.

[21] Ahualli S, Jimenez ML, Carrique F, Delgado AV. AC electrokinetics of concentrated suspensions of soft particles. Langmuir 2009;25(4):1986–97.
[22] Ahualli S, Ballauff M, Arroyo FJ, Delgado AV, Jimenez ML. Electrophoresis and dielectric dispersion of spherical polyelectrolyte brushes. Langmuir 2012;28(47):16372–81.
[23] Merlin J, Duval JFL. Electrodynamics of soft multilayered particles dispersions: dielectric permittivity and dynamic mobility. Phys Chem Chem Phys 2014;16(29):15173–88.

Chapter 11

Gel-Electrophoresis of Rigid Spherical Particles in Suspension and in a Cylindrical Pore

11.1 INTRODUCTION

The electrophoretic separation of polyelectrolytes like DNAs and proteins is often conducted in a polymer gel-like matrix immersed in electrolyte solutions, referred to as the gel-electrophoresis in practice [1–6]. When the gel is contained in a capillary, it is called capillary gel-electrophoresis, or CGE, which has been the normal practice in biomedical or biochemical applications in terms of separation of DNAs or proteins. In comparison, the conventional electrophoresis in electrolyte solution is sometimes referred to as the "solution electrophoresis" or "free electrophoresis," and when conducted in a capillary, it is called capillary zone electrophoresis or CZE [7]. The reason for the introduction of gel material to the conventional electrophoresis is mainly due to the fact that constant charge-to-mass ratio happens for large macromolecules like DNAs or proteins once the molecular weight past over some threshold value [8]. The incremental increase of the electric driving force is offset by the corresponding increase of hydrodynamic drag force simultaneously. Thus, it becomes impossible to separate these particles based on the distinctive electrophoretic mobility alone [9]. Extra separation mechanism needs to be added to the system based on something else, such as the size difference [10–12]. Polymer gel in this regard is an excellent candidate due to its porous structure, which provides size-dependent molecular sieving function [13]. Moreover, polymer gel can also reduce joule-heating effect and corresponding convection flow to enhance the overall separation performance, among other merits [9].

Polymer gel used in gel-electrophoresis is normally a highly hydrophilic polymeric substance, or hydrogel, with three-dimensional cross-linking structure [13, 14–17]. Due to its highly hydrophilic nature, large amount of water

Interface Science and Technology, Vol. 26. https://doi.org/10.1016/B978-0-08-100865-2.00011-4

or electrolyte solution can be absorbed into the interior of the three-dimensional cross-linking structure of polymer chains and push them away to form a jungle-gym like morphology. This leads to a significant increase of the apparent volume (swelling) and creates abundant empty space in the porous medium for the particles to pass through. Two kinds of hydrogels are most often used: polyacrylamide and agarose [9,14,16], with the former good for relatively small particles, say globular proteins or DNAs smaller than 2000 base pairs [18–20]; and the latter for larger ones [21,22]. Their functions compensate for each other in general.

Note that, however, the role of the heterogeneous polymer gel in CGE has been replaced by homogeneous polymeric solution with linear polymers dissolved in aqueous electrolyte solution since nearly two decades ago [4,23]. The fixed porous structure in gel is replaced by the dynamic pore due to temporary entanglement of polymer chains in the solution, which forms a size-dependent mechanism to separate particles in motion [2,3,6,24]. It has been used extensively in practical applications of CGE, such as the separation of quantum dots [12].

Note also that the rheological response of a fluid passing through a heterogeneous porous structure is normally well-modeled by the so-called Brinkman equation, Eq. (7.2), where an extra drag force term proportional to the local fluid velocity is added to the momentum equation, or the modified Stokes' equation, to account for the extra hindrance effect from the solid components in the porous structure [25]. And for the polymeric solution consisting of linear polymers dissolved in a Newtonian fluid like water, Debye and Bueche [26] developed the corresponding equation describing the rheological response of this polymer solution which takes exactly the same form as the Brinkman equation, and together, they are often referred to as the Debye-Bueche-Brinkman equation, or simply DBB equation in fluid mechanics [27, 28]. Note that while the Brinkman equation in a heterogeneous porous structure is of semi-analytical nature, the Debye-Bueche equation in a homogeneous polymeric solution is theoretically exact based on the Einstein's equation for drag force [29].

Hence, the rheological response for a colloidal particle passing through either a gel or a polymeric solution can be fully taken account by the DBB equation. However, there is yet another factor to be reckoned with: the size-dependency of the chance whether a particle can successfully pass through either a static pore (gel) or a dynamic pore (polymeric solution). This is an issue of statistic nature as the orientation of the particle approaching a pore may not be the same if the particle is of irregular shape itself. Moreover, the distribution of pore size is normally not uniform as well.

In short, there are many studies of the particle migration in a porous medium based on the idea of hindered diffusion [30, 31]. The presence of a porous gel, or a polymeric solution, generates an extra drag force for one thing, and moreover, the variation of pore sizes leads to the variation of easiness for particles of different size to pass through as well. The former can be treated using the DBB model to account for the extra hydrodynamic drag force, the so-called long-range force [32–34]. The latter, however, is of probability nature and can only be properly treated with theories based on statistics, as those done by Ogston and others [14]. The latter is often referred to as the steric effect or the short-range effect [32–40]. These studies can be applied to the gel-electrophoresis analysis considered here as a correction factor.

Here we focus on the long-range effect by treating the gel as a Brinkman medium, or a Debye-Bueche-Brinkman (DBB) medium for a polymeric solution in general, taking the extra drag force from the heterogeneous polymer chains into account as well as the electric body force due to the electrolyte solution contained within the gel, or the polymeric electrolyte solution itself. The overall performance of the electrophoretic separation can be obtained by the combination of both the long-range effect treated here and the short-range effect treated elsewhere [41], such as the product of both factors, although it is still an open field in general [42–44].

11.2 THEORY

11.2.1 Single Rigid Spherical Particle in a DBB Polymeric Fluid

The system diagram is shown in Fig. 11.1A.

The governing equations and associated boundary conditions are exactly the same as those in Chapter 1, except that an extra term representing the rheological response in a DBB polymeric solution needs to be added to the hydrodynamic Stokes' equation as follows:

$$\eta_f \nabla^2 \boldsymbol{v} - \nabla p - \rho \nabla \phi - \gamma \boldsymbol{v} = 0 \tag{11.1}$$

Subsequent mathematical treatment is exactly the same as well, and Eq. (11.1) reduces to the dimensionless form in term of stream function.

$$E^{*4}\Psi^* - (\lambda a)^2 E^{*2}\Psi^* + \frac{(\kappa a)^2}{1+\alpha}\left(\frac{\partial \rho_e^*}{\partial r^*}\frac{\partial \delta\phi^*}{\partial \theta} - \frac{\partial \phi_e^*}{\partial r^*}\frac{\partial \delta\rho^*}{\partial \theta}\right)\sin\theta = 0 \tag{11.2}$$

Subsequent solution procedure of the electric and flow fields as well as the evaluation of particle mobility are the same as in Chapter 1.

11.2.2 Suspension of Rigid Spherical Particles in a Polymeric DBB Fluid

The system diagram is shown in Fig 11.1B.

Kuwabara's unit cell model coupled with Shilov and Zharkikh (S-Z) electric boundary conditions is adopted here. The system is essentially the same as that treated in Chapter 2 for a Newtonian fluid. Thus, simply replace the Newtonian counterparts with Eqs. (11.1) and (11.2) here and we are done.

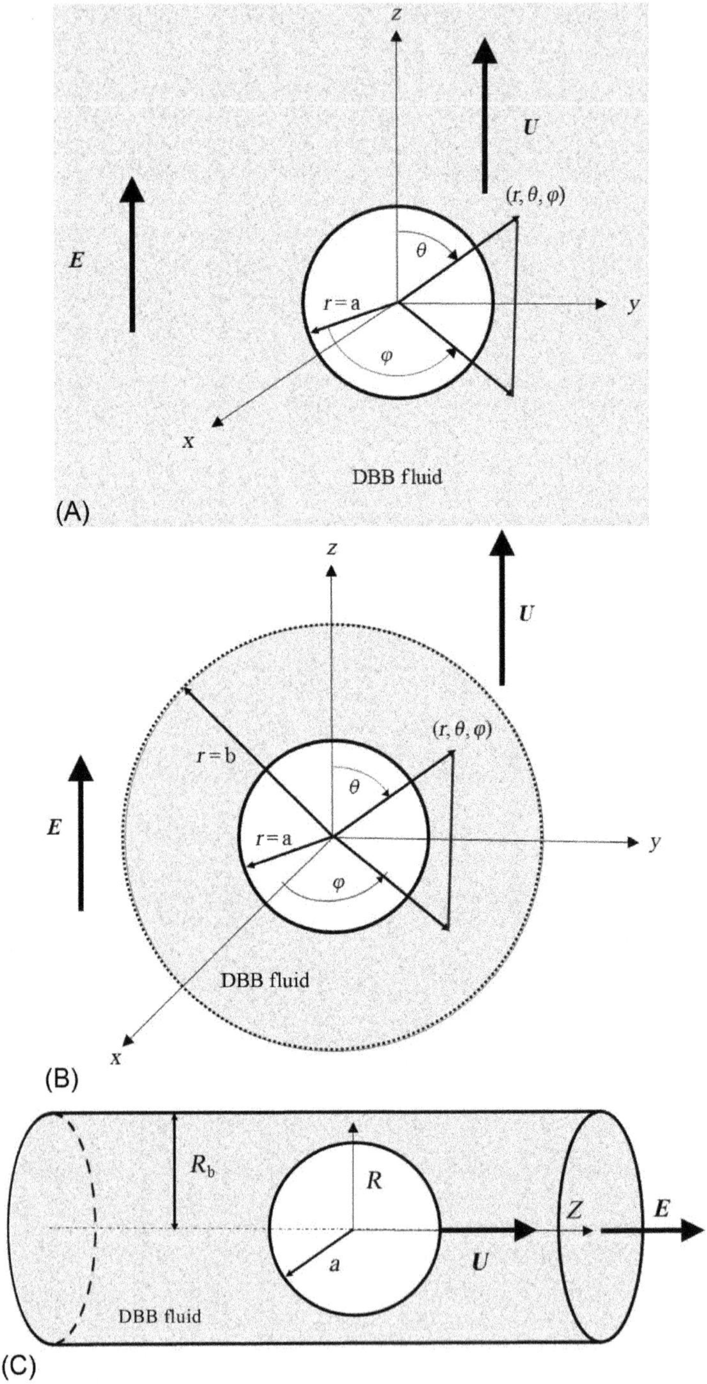

FIG. 11.1 System diagram in a DBB fluid. (A) Single rigid spherical particle; (B) suspension of rigid spherical particles; (C) a rigid spherical particle in a cylindrical pore.

11.2.3 A Rigid Spherical Particle in a Cylindrical Pore Filled With Polymeric DBB Fluid

The system diagram is shown in Fig. 11.1C.

The system is essentially identical to that in Chapter 5 for a Newtonian solution. So, again replace the Newtonian counterparts there with Eqs. (11.1) and (11.2) and we are done. The details can also be found elsewhere [45].

Note that in all the above cases, γ is further replaced by the so-called Debye-Bueche screening length, $\lambda^{-1}=(\eta_f/\gamma)^{1/2}$, which indicates the permeability of the fluid as discussed in Chapter 7. Larger λ implies less permeability and corresponding dimensionless number λa characterizes the fluid permeability considered here. Normally $\lambda a=1$ corresponds to a highly permeable fluid, whereas $\lambda a=10$ behaves almost like a rigid material. Note that $\lambda a=0$ reduces to the Newtonian fluid case with no involvement of polymer contribution.

11.3 RESULTS AND DISCUSSION

The general Henry's charts for a single spherical particle with various surface potential ($\phi_r=1$–5) in a DBB polymeric electrolyte solution with various permeability ($\lambda a=1$, 3, and 10) are shown in Fig. 11.2A–C.

The general behavior is similar to the corresponding Newtonian case in Chapter 1, but no local minima are observed anymore here for highly charged

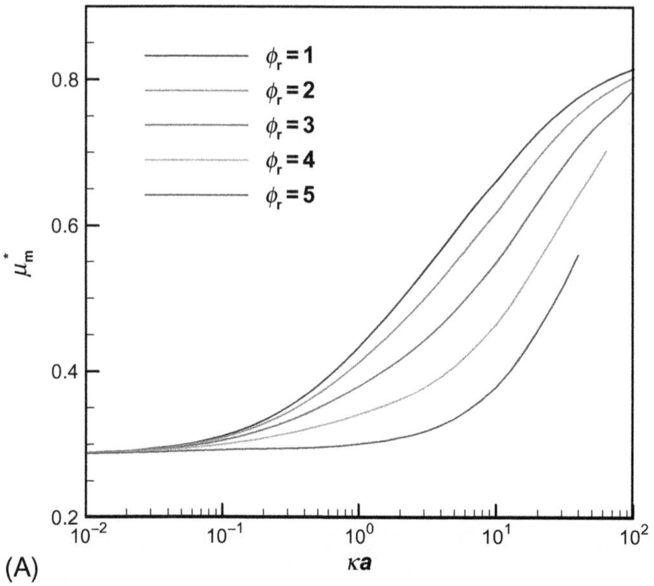

FIG. 11.2 Scaled mobility μ_m^* as a function of κa at various ϕ_r for a single rigid spherical particle in a DBB fluid. (A) $\lambda a=1$; (B) $\lambda a=3$; (C) $\lambda a=10$.

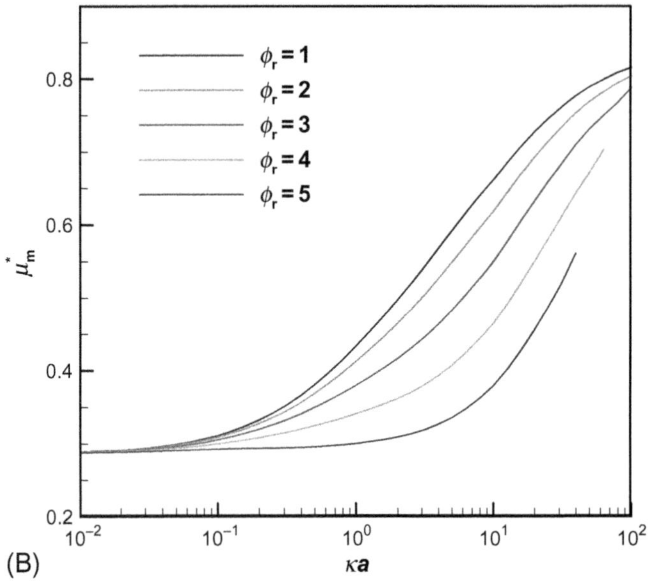

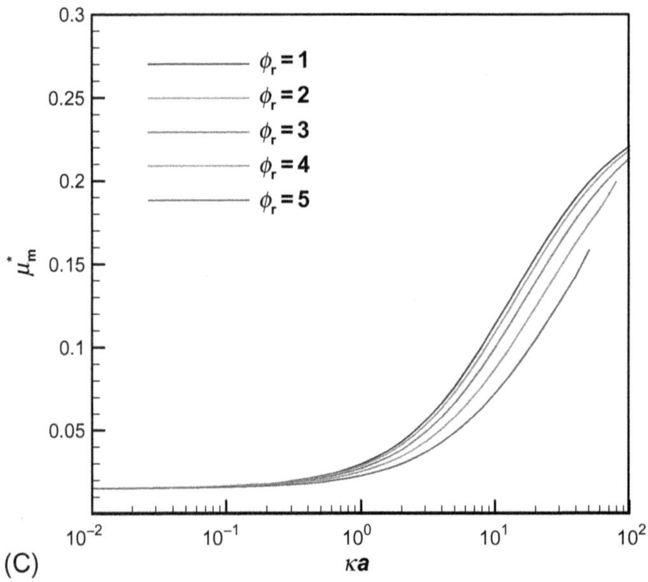

FIG. 11.2—CONT'D

particles. This indicates that the polarization effect is much weaker or negligible in a DBB solution due to the dominant motion-retarding effect of the extra hydrodynamic drag force. The larger λa is, the stronger this extra hydrodynamic drag force is, leading to a smaller particle mobility in general. Note that the magnitude of this force is proportional to the square of λa. Hence, $\lambda a = 1$, 3, and 10 stand for three orders of magnitude difference. The effect of λa upon particle mobility is shown in Fig. 11.3 for a highly charged particle with $\phi_r = 3$. Note that $\lambda a = 0$ reduces to the Newtonian case in Chapter 1. Again, no local minimum is observed in these mobility profiles in a DBB fluid, or the polarization effect in a DBB fluid is negligible in practice. Note that over four orders of permeability magnitude are examined in this chart.

General Henry's chart for a representative particle suspension is shown as well for highly charged particles ($\phi_r = 3$) in a highly permeable DBB fluid ($\lambda a = 1$) in Fig. 11.4. The general behavior is similar to that observed in Chapter 2, except that the polarization effect is so weak that no local minimum is observed. The particle mobility is smaller than the corresponding Newtonian fluid in general. Fig. 11.5 demonstrates the extent of double layer polarization, $\delta\rho^*$, for a representative highly charged particle ($\phi_r = 3$) in a highly permeable DBB fluid ($\lambda a = 1$) at $\kappa a = 1$. Note that the polarization extent is very similar to the corresponding Newtonian solution case presented in Chapter 1. This is because the electromigration and diffusion of electrolyte ions are essentially the same in both cases with the same diffusivity coefficient D_j used, justified by the assumption of dilute polymeric solution. And the convection mechanism there contributes to less than 15% of overall polarization anyway as discussed in Chapter 1. Here, the particle velocity in a DBB fluid is even slower, and hence, the convection contribution is even less.

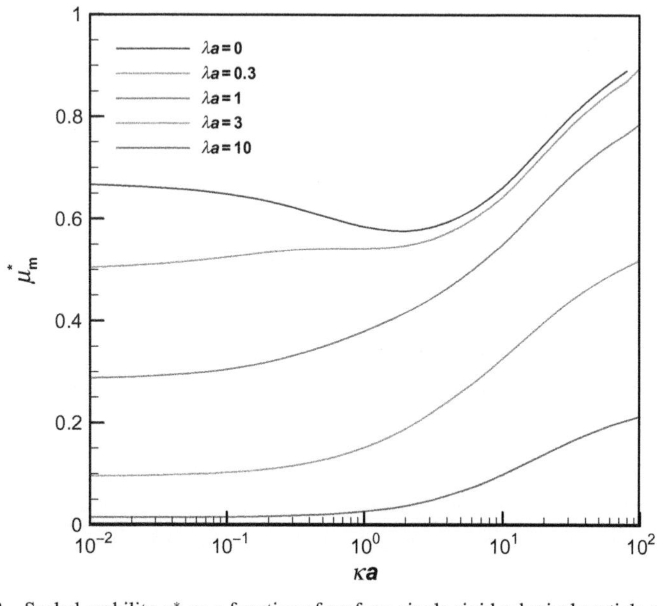

FIG. 11.3 Scaled mobility μ_m^* as a function of κa for a single rigid spherical particle with $\phi_r = 3$ in DBB fluids of various λa.

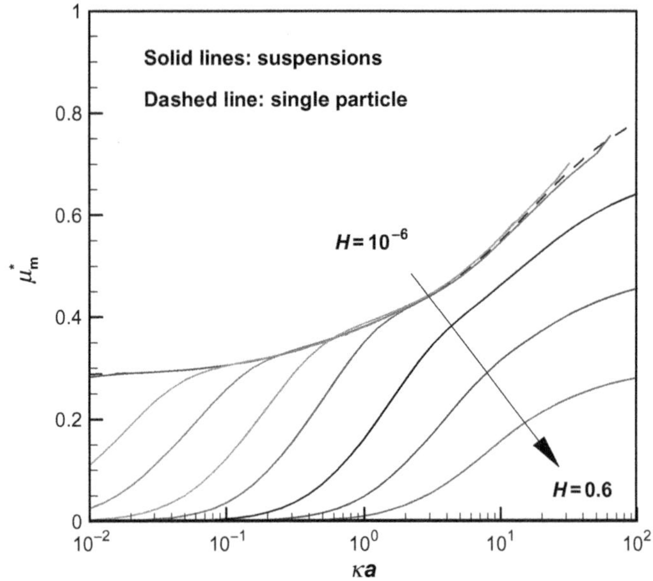

FIG. 11.4 Scaled mobility μ_m^* of suspended spherical particles ($\phi_r=3$) as a function of κa in a DBB fluid with $\lambda a=1$ at various levels of H ($H=10^{-6}$, 10^{-4}, 10^{-3}, 0.01, 0.05, 0.2, 0.4, and 0.6, respectively).

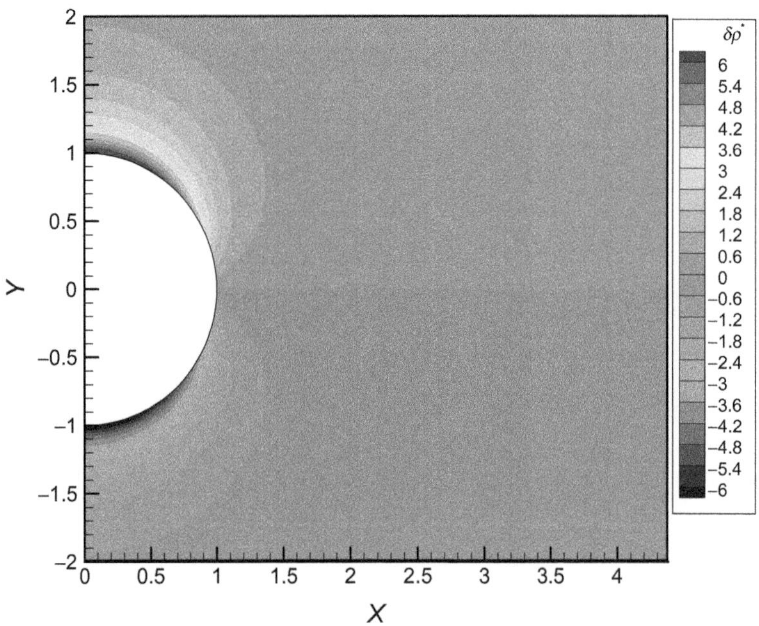

FIG. 11.5 Contour plot of the scaled disturbed charge density $\delta\rho^*$ a single rigid particle with $\phi_r=3$ and $\kappa a=1$ in a DBB fluid with $\lambda a=1$.

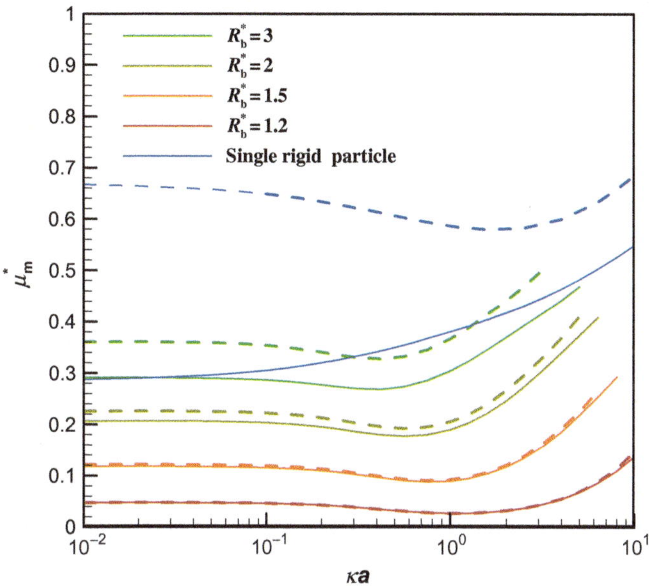

FIG. 11.6 Scaled mobility μ_m^* as a function of κa for a rigid spherical particle with $\phi_r = 3$ at various R_b^* in a cylindrical pore filled with a DBB fluid with $\lambda a = 1$. *Dashed lines*: results in corresponding Newtonian solutions ($\lambda a = 0$).

Corresponding phenomena in a cylindrical pore are shown in Figs. 11.6 and 11.7 for a representative system with $\phi_r = 3$ and $\lambda a = 1$, i.e., a highly charged rigid particle in a highly permeable DBB solution. As shown in Fig. 11.6, where the mobility profiles corresponding to various pore radii are presented, the boundary effect in terms of particle mobility reduction is observed in general. The narrower the pore is, the more significant this reduction is. It is interesting to note that the difference between a DBB fluid ($\lambda a = 1$) and a Newtonian fluid ($\lambda a = 0$) diminishes with narrower pore radius. For a very narrow pore with $R_b^* = 1.2$, the mobility profiles are essentially identical. This is due to the fact that the extra drag force in a DBB fluid is proportional to the local velocity field. In general, the presence of a particle in a cylindrical pore as well as the pore itself reduces the DBB fluid velocity there significantly, and the narrower the pore is, the slower the fluid moves, hence the smaller this local drag force term ($-\gamma \mathbf{v}$) is. As a result, rheological behavior of the DBB fluid reduces to the corresponding Newtonian solution when the pore gets narrower. This reasoning is also supported by the observation that particle mobility in Fig. 11.6 at $\kappa a = 0.01$ and $R_b^* = 1.2$ in the DBB fluid is only about 7% the mobility of a single particle in a Newtonian solution.

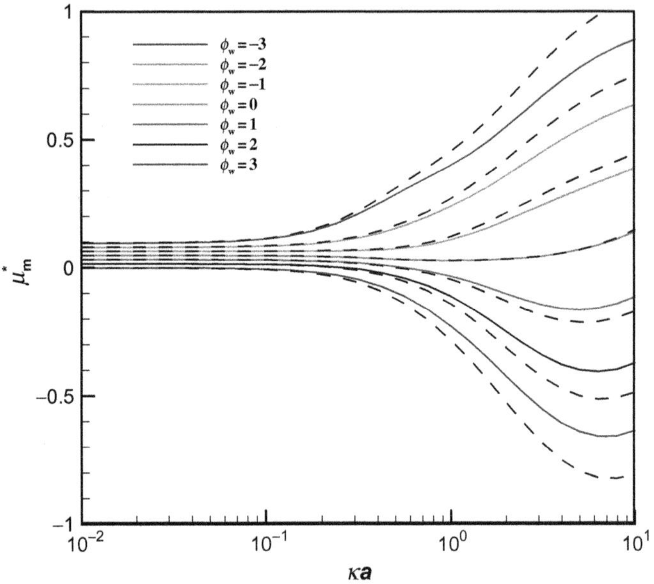

FIG. 11.7 Scaled mobility μ_m^* as a function of κa for a rigid spherical particle with $\phi_r = 3$ at various levels of ϕ_w in a cylindrical pore filled with a DBB fluid with $\lambda a = 1$. *Dashed lines*: results in corresponding Newtonian solutions ($\lambda a = 0$).

Note that the particle velocity is the same in magnitude as the corresponding fluid velocity far away from the particle.

Fig. 11.7 examines the impact of a charged wall for $\phi_w = -3$ to +3. An electroosmosis flow (EOF) is generated as a result. There is local minima in the mobility profiles in Fig. 11.7 for positive charged wall ($\phi_w > 0$), where the EOF slows down the positively charged particle motion. The apparent local minima are misleading. It is not due to motion-deterring polarization effect; it is the outcome of the stronger EOF as κa increases instead. Generally speaking, the higher the electrolyte strength (κ) is, the larger the EOF speed is, and hence, its effect upon the particle motion. The apparent local minima are the outcome of two corresponding mechanisms: the particle electrophoretic motion and the EOF from a charged wall. It is not from the motion-deterring double layer polarization effect.

REFERENCES

[1] Heller C. Capillary electrophoresis of proteins and nucleic acids in gels and entangled polymer solutions. J Chromatogr A 1995;698(1–2):19–31.

[2] Ren J, Fang N, Wu D. Inverse-flow derivatization for capillary electrophoresis of DNA fragments with laser-induced fluorescence detection. Anal Chim Acta 2002;470(2):129–35.

[3] Kao Y-Y, et al. Analysis of amino acids and biogenic amines in breast cancer cells by capillary electrophoresis using polymer solutions containing sodium dodecyl sulfate. J Chromatogr A 2010;1217(4):582–7.
[4] Chung M, Kim D, Herr AE. Polymer sieving matrices in microanalytical electrophoresis. Analyst 2014;139(22):5635–54.
[5] Li Z, et al. Electrophoresis of periodontal pathogens in poly (ethyleneoxide) solutions with uncoated capillary. Anal Biochem 2015;471:70–2.
[6] Song X, et al. Highly efficient size separation of CdTe quantum dots by capillary gel electrophoresis using polymer solution as sieving medium. Electrophoresis 2006;27(7):1341–6.
[7] Kuhn R, Hoffstetter-Kuhn S. Capillary electrophoresis: principles and practice. Berlin: Springer Science & Business Media; 2013.
[8] Hsu H-P. Counterion condensation and electrophoresis of soft matter. Ph.D. thesis. Taipei, Taiwan: Department of Chemical Engineering, National Taiwan University; 2012.
[9] Weinberger R. Practical capillary electrophoresis. Amsterdam: Elsevier; 2000.
[10] Deng Y, et al. A non-invasive genomic diagnostic method for bladder cancer using size-based filtration and microchip electrophoresis. Talanta 2015;136–44.
[11] Djerdjev AM, Beattie JK. Electroacoustic and ultrasonic attenuation measurements of droplet size and ζ-potential of alkane-in-water emulsions: effects of oil solubility and composition. Phys Chem Chem Phys 2008;10(32):4843–52.
[12] Li Y-Q, et al. A highly efficient capillary electrophoresis-based method for size determination of water-soluble CdSe/ZnS core–shell quantum dots. Anal Chim Acta 2009;647(2):219–25.
[13] De Gennes P-G, Gennes P-G. Scaling concepts in polymer physics. New York: Cornell University Press; 1979.
[14] Ogston AG. The spaces in a uniform random suspension of fibres. Trans Faraday Soc 1958;54:1754–7.
[15] Weinberger R. Size separations in capillary gels and polymer networks. In: Practical capillary electrophoresis. 2nd ed. San Diego: Academic Press; 2000. p. 245–92, Chapter 6.
[16] Stellwagen NC, Holmes DL. Resolution of a paradox in the electrophoresis of DNA in agarose gels. Electrophoresis 1990;11(8):649–52.
[17] Pernodet N, Maaloum M, Tinland B. Pore size of agarose gels by atomic force microscopy. Electrophoresis 1997;18(1):55–8.
[18] Lauer HH, McManigill D. Capillary zone electrophoresis of proteins in untreated fused silica tubing. Anal Chem 1986;58(1):166–70.
[19] Cohen A, Karger B. High-performance sodium dodecyl sulfate polyacrylamide gel capillary electrophoresis of peptides and proteins. J Chromatogr A 1987;397:409–17.
[20] Bushey MM, Jorgenson JW. Capillary electrophoresis of proteins in buffers containing high concentrations of zwitterionic salts. J Chromatogr A 1989;480:301–10.
[21] Woolley AT, Mathies RA. Ultra-high-speed DNA sequencing using capillary electrophoresis chips. Anal Chem 1995;67(20):3676–80.
[22] Woolley AT, Sensabaugh GF, Mathies RA. High-speed DNA genotyping using microfabricated capillary array electrophoresis chips. Anal Chem 1997;69(11):2181–6.
[23] Heiger DN, Cohen AS, Karger BL. Separation of DNA restriction fragments by high performance capillary electrophoresis with low and zero crosslinked polyacrylamide using continuous and pulsed electric fields. J Chromatogr A 1990;516(1):33–48.
[24] Radko SP, Chrambach A. Capillary zone electrophoresis of rigid submicron-sized particles in polyacrylamide solution: selectivity, peak spreading and resolution. J Chromatogr A 1999;848(1–2):443–55.

[25] Brinkman H. A calculation of the viscous force exerted by a flowing fluid on a dense swarm of particles. Flow Turbulence Combust 1949;1(1):27.
[26] Debye P, Bueche AM. Intrinsic viscosity, diffusion, and sedimentation rate of polymers in solution. J Chem Phys 1948;16(6):573–9.
[27] Ahlrichs P, Everaers R, Dünweg B. Screening of hydrodynamic interactions in semidilute polymer solutions: a computer simulation study. Phys Rev E 2001;64(4):040501.
[28] Tchesskaya T. A generalized Zimm model: hydrodynamic screening in polymer solutions. J Mol Liq 2009;150(1–3):77–80.
[29] Einstein A. Ann Phys (Leipzig) 1906;19:289.
[30] Viovy JL. Electrophoresis of DNA and other polyelectrolytes: physical mechanisms. Rev Mod Phys 2000;72(3):813–72.
[31] Salieb-Beugelaar GB, et al. Electrophoretic separation of DNA in gels and nanostructures. Lab Chip 2009;9(17):2508–23.
[32] Labrie J, Mercier JF, Slater GW. An exactly solvable Ogston model of gel electrophoresis V. Attractive gel-analyte interactions and their effects on the Ferguson plot. Electrophoresis 2000;21(5):823–33.
[33] Mercier JF, Slater GW. An exactly solvable ogston model of gel electrophoresis. 7. Diffusion and mobility of hard spherical particles in three-dimensional gels. Macromolecules 2001;34(10):3437–45.
[34] Mercier JF, Tessier F, Slater GW. An exactly solvable Ogston model of gel electrophoresis: VIII. Nonconducting gel fibers, curved field lines, and the Nernst-Einstein relation. Electrophoresis 2001;22(13):2631–8.
[35] Boileau J, Slater GW. An exactly solvable Ogston model of gel electrophoresis—VI. Towards a theory for macromolecules. Electrophoresis 2001;22(4):673–83.
[36] Slater GW, Guo HL. An exactly solvable Ogston model of gel electrophoresis: 1. The role of the symmetry and randomness of the gel structure. Electrophoresis 1996;17(6):977–88.
[37] Slater GW, Guo HL. An exactly solvable Ogston model of gel electrophoresis: 2. Sieving through periodic gels. Electrophoresis 1996;17(9):1407–15.
[38] Slater GW, Treurniet JR. Exactly solvable Ogston model of gel electrophoresis: 3. Percolation and sieving through two-dimensional gels. J Chromatogr A 1997;772(1–2):39–48.
[39] Mercier JF, Slater GW. An exactly solvable Ogston model of gel electrophoresis IV: sieving through periodic three-dimensional gels. Electrophoresis 1998;19(10):1560–5.
[40] Imanidis G, Luetolf P. An extended model based on the modified Nernst-Planck equation for describing transdermal iontophoresis of weak electrolytes. J Pharm Sci 2006;95(7):1434–47.
[41] Brady JF. Hindered diffusion. In: American Institute of Chemical Engineers Annual Meeting, San Francisco; 1994.
[42] Jackson GW, James DF. The permeability of fibrous porous media. Can J Chem Eng 1986;64(3):364–74.
[43] Ogston AG, et al. Transport of compact particles through solutions of chain-polymers. Proc R Soc Lond Ser A Math Phys Eng Sci 1973;333(1594):297–316.
[44] Johnson EM, et al. Hindered diffusion in agarose gels: test of effective medium model. Biophys J 1996;70(2):1017–23.
[45] Lee Y-F, et al. Electrophoretic and electroosmotic motion of a charged spherical particle within a cylindrical pore filled with Debye–Bueche–Brinkman polymeric solution. Langmuir 2016;32(49):13106–15.

Section C

Electrophoresis of Liquid Droplets and Gas Bubbles

Chapter 12

Electrophoresis of Liquid Droplets and Gas Bubbles

12.1 INTRODUCTION

Liquid droplet is encountered very often in daily life, such as milk, which is the liquid droplet of fat suspended in water [1]; or the emulsions used in cosmetic products with the effective element contained in droplets [2]. In colloidal industry, especially in food and pharmaceutical industries, the droplet has found numerous novel and important applications with a fast growing pace [3]. It can be used as a highly effective microreactor as well due to the fast rate of heat and mass transfer and nearly perfect mixing performance due to the tiny volume of the droplet inside and consequentially high specific surface area [4]. Moreover, droplet is an excellent candidate to model the behavior of a living cell, such as the human stem cells [5] or other mammalian cells [6]. It has also been used in oil drilling industry as the major horsepower for enhanced oil industry (EOR) [7]. Note that when the droplet size is smaller than one quarter of the wave length of visible light, which is about 100–175 nm, the droplet becomes transparent [8]. This property is highly desirable especially in food and cosmetic industries [9, 10].

Liquid droplet is often used to model an emulsion system as well with surfactant adsorbed onto the droplet surface, although droplet without the involvement of surfactant is also possible, such as the ones generated from the "Ouzo effect" [11–13]. From the electrokinetic point of view, as long as there is some kind of interface to prevent the penetrating of the outside suspending medium into the droplet interior, it can be treated somehow as a droplet. Some fundamental aspects of a droplet will be briefly introduced starting from the next paragraph.

12.1.1 Definition of a Droplet

A droplet is a volume of fluid within a thin film interface established between two immiscible fluids [14]. When the fluid inside the droplet is a gas, the

Interface Science and Technology, Vol. 26. https://doi.org/10.1016/B978-0-08-100865-2.00012-6

droplet reduces to a gas bubble, otherwise it is referred to as a liquid droplet, or simply droplet in short. The presence of a thin film is essential so that rigorous boundary conditions can be established there, as far as theoretical consideration is concerned. But for practical applications, this requirement does not need to be that rigid, so long as the interface can effectively separate the fluids inside and outside. Thus, the droplet can be viewed as a binary system of immiscible fluids whose interface has a unique structure and properties quite different from the bulk fluids [14].

12.1.2 Generation of Liquid Droplets or Gas Bubbles

12.1.2.1 Thermodynamic Consideration

The immiscibility of the two fluids involved is the key to the success of generating a droplet, which actually indicates that, thermodynamically, the droplet state in general is not favored, or simply unstable or metastable or kinetically stable in the sense that, given long enough time, the droplet system will collapse and go back to the original separate phases as two immiscible fluids. Hence, in principle, a certain amount of energy is needed to raise the free energy level from the original separate phases to the less stable droplet system. Thus, extra input of energy is needed in principle for the generation of droplets in general [15].

12.1.2.2 Surface Tension Consideration

Compared with the original separate phases, the formation of a droplet implies the creation of new surface area. Hence, the more number of droplets is produced, the more input of energy is required. And, for a given amount of total droplet volume, the tinier the droplets are, the more total surface area is generated. Considering the surface tension as energy contained within a unit area of surface, this means further more energy is required to generate tiny droplets.

The above two considerations form the basis of the droplet generation.

12.1.2.3 High Energy Methods Versus Low Energy Methods

There are two general ways of introducing energy needed to form droplets, which can be categorized in short as the "high energy methods" and the "low energy methods," respectively [16–18].

A liquid droplet can be easily produced via the input of mechanical energy, such as the mixing of two immiscible fluids within a traditional mixing tank [19, 20]. The blending motion of the mixing propellers there provides the physical mechanism needed to create new surface or interface area. It can also be produced via the introduction of ultrasonic wave [21]. A simple heating can generate droplet or a gas bubble as well [22]. Indeed, there are many ways to generate a droplet or a gas bubble, which we actually are pretty familiar with in our daily life, such as the decorative clam emitting gas bubbles in an aquarium tank. The key is the immiscibility nature of the two fluids. The rest is just the engineering ingenuity in the technology scope. The conventional introduction of mechanical energy, as mentioned earlier, is normally referred to as the "high energy methods" by some researchers [19, 20, 23], which has the shortcomings such as the droplet size which are normally widely spread and the control of the precise diameter is very difficult.

On the other hand, microfluidic operations have been proposed as an alternative way to generate droplets, with distinguished advantages such as a nearly mono-disperse distribution of droplet size and the precise size can be determined by careful design of the devices a priori [4, 24]. This type and other related operations are referred to as the "low energy methods" [18], which in fact has become a very promising and encouraging new trend to completely overhaul the traditional concept of production with unit operation processes in chemical engineering, in addition to an ideal way of droplet formation considered here [4].

12.1.3 Some Remarks on the Emulsion Systems

12.1.3.1 Emulsion in General

As mentioned earlier, a true droplet is essentially a binary system, with perhaps a tiny amount of impurities adsorbed on its surface, which cannot be completely eliminated in the real world. It is thermodynamically unstable in general, but can be kinetically stable if engineering ingenuity is introduced carefully. Most importantly, it has a well-defined interface of thin film nature which has some crucial implications on the physical boundary conditions there as we shall discuss later.

However, there is yet another closely related colloidal system involving two immiscible fluids as well with one suspended in the other called emulsion, which is a tertiary system involving a third component called "surfactant" [14]. Surfactant is a substance that can reduce the surface tension between the two originally immiscible fluids by forming an interfacing region in between, kind of like the interface concept in a true droplet system discussed earlier, but with a finite thickness which may not be considered as a thin film, especially when the droplet is very small [14]. Significant contribution to the total volume may result and typical percentage of 20% is not uncommon [14]. As a result, an emulsion system can be further decomposed to three regions: the outer suspending continuous phase, the inner dispersed liquid phase, and an interface region composed mainly of surfactants. The presence of this surfactant-rich interface raises the question of whether the outside momentum can be transferred across it, or it serves as a rigid shell making the inner fluid flow impossible. This is still an open question, in general, with no conclusive evidence demonstrated yet, and the answer could be actually system-dependent [25–28].

As true binary droplet systems like those from the "Ouzo effect" are very rare and impurities are bound to be present in a liquid system, and hence, adsorbed onto the droplet surface and functioning like surfactants there, the majority of droplet systems in the real world are in fact emulsions, or of emulsion nature with some surfactants on the droplet surface. Hence, we will discuss same basic contents of this "pseudo-droplet" type of system in more detail as follows. In general, the emulsion could be further categorized as macroemulsion, microemulsion, and nanoemulsion [26–28].

Macroemulsion

This is the conventional emulsion system with the size of micron-scale in general, say around 0.5 to several microns (10^{-6} m) [17]. The surfactant contribution to the total volume is typically around one to several percent; hence, the assumption of thin film and modeling as a liquid droplet is fully acceptable for a macroemulsion droplet. Size of this range makes significant scattering of light through; hence, they normally appear as milky latex suspensions, such as milk [29, 30]. Note that the emulsions encountered in microfluidic devices fall into this category as well, or even larger; hence, they can be treated and modeled as liquid droplets perfectly.

Microemulsion

Microemulsion is a thermodynamic stable, in that it can be formed spontaneously, tertiary system with added surfactants to two immiscible fluids. The size scale is well below a micron in general and it earns its name due to some historical reason with no implication of its actual size [31, 32]. In practice, the size that of below 100 nm is of major interest, as it shows very weak light scattering in this range and thus makes the emulsion nearly translucent. This property makes them highly attractive and desirable in cosmetic and other important food and pharmaceutical industries [33], as mentioned earlier. For the size this tiny, however, the involvement of the surfactant layer could be very large in terms of hydrodynamic drag force. The momentum transfer to the inner droplet fluid could be hindered to some degree, which so far has not been concluded for sure yet. The actual hydrodynamic behavior of a microemulsion droplet may stand somewhere between a perfect liquid droplet without the involvement of surfactant and a rigid particle with no permission of momentum transfer across the interphase at all. This dilemma can be resolved, in principle, by adopting an equivalent or effective droplet radius somewhere between the outer and inner boundaries of the surfactant layer, for example. The layer itself can also be regarded as a porous layer, like the one discussed in Chapter 7 for a spherical polyelectrolyte brush (SPB), except that it is on a droplet surface now instead of on a rigid particle.

Nanoemulsion

This is a system quite similar to the above-mentioned microemulsion, except that it is only kinetically stable in that mechanical energy needs to be added into the system, just like the situation in a true binary liquid droplet [26]. Theoretically speaking, no surfactant is needed and thus the interface is of the thin film nature and can be treated as a perfect liquid droplet. In practice, however, some surfactants are very often added, nonetheless, in order to facilitate the stability of the emulsion system over a reasonably long enough storage time or shelf time. And, that is why it is classified as a member of the general emulsion systems.

12.1.4 The Role of a Surfactant in General

The immiscibility of two fluids normally takes place between a polar fluid, such as water, and a nonpolar one, such as a hydrocarbon. As a result, emulsions are divided into either oil in water (O/W) or water in oil (W/O) situations, where oil refers to the hydrophobic nonpolar liquid immiscible with water, which is by far the most important polar solvent [23].

Surfactant is a substance that has a hydrophilic head group plus a hydrophobic tail group, thus it can form a buffer zone between the two immiscible liquids and reduce the total thermodynamic free energy there to form a stable emulsion droplet [14, 34].

In practical applications such as the encapsulation of nutrition or drugs into a droplet carrier for biomedical purposes [35], it is usually in the form of O/W, as the human blood is certainly an aqueous solution with many electrolytes within it. Thus, we will focus on the O/W droplet situation here. Note that no electrolytes exist in a nonpolar hydrophobic oil phase liquid in general, which simplifies the governing equations within the droplet a great deal, as only the purely hydrodynamic Stokes equation needs to be considered.

12.1.5 Gas Bubbles

When the interior of a droplet is filled with a gas such as air, a gas bubble or simply a bubble is formed, which is inviscid inside in that the viscosity there can be regarded as zero based on the order of magnitude analysis. In practice, the viscosity ratio of the droplet fluid to that of the outside suspending medium, σ_H, is in the order of 0.01 for a gas bubble. The rigid sphere, on the other hand, can be regarded as a droplet with infinitely large viscosity. In practice, the viscosity ratio is of the order of 100. Electrophoretic motion of bubble systems certainly is of interest as well and can be treated as a special case here with $\sigma_H = 0.01$ [36, 37].

After reviewing the above basic background of liquid droplets, we go on to investigate the electrophoretic behavior of a single liquid droplet or a gas bubble as well as their suspension systems in this chapter. There have been many theoretical analyses regarding the electrophoretic motion of a liquid droplet [38–53]. The approximating analytical formula valid for lowly charged droplet obtained by Ohshima [43, 44] is particularly interesting, which is shown below after conversion to the terminology defined here:

$$\mu_{\mathrm{m}}^{*} = \frac{2}{3}\left[\left(\frac{1}{3\sigma_{\mathrm{H}}+2}\right)\kappa a + \frac{3\sigma_{\mathrm{H}}+3}{3\sigma_{\mathrm{H}}+2}\right]\left[1+\frac{1}{2\left(1+\dfrac{1.86}{\kappa a}\right)^{3}}\right] \tag{12.1}$$

Very fast droplet motion when κa is large is indicated by this analytical formula. In fact, the mobility becomes infinitely large when κa approaches infinity. In other words, there is no corresponding Smoluchowski-type formula, or a modified version with a sole viscosity-dependent correction factor to it as one might think. Lee's group, on the other hand, considers the general situation of various droplet systems, either highly or lowly charged and with arbitrary double layer thickness, thus the double layer polarization effect is fully taken into account [45–53]. And they found exactly the same phenomenon as κa gets large regardless the charge levels of the droplet. Thus, the underlying mechanism must lie somewhere else, for instance, the fluidity of the droplet, which we will give a closer examination as follows.

Unlike the corresponding rigid particle situation, the fluid inside a moving droplet flows simultaneously, including the droplet surface. Thus, the hydrodynamic shear stress is reduced accordingly. For a fluid droplet, a modified Stokes law with a viscosity-dependent correction factor is available as follows [54]:

$$F_{\mathrm{D}} = 6\pi\eta_{\mathrm{f}} a U \frac{3\sigma_{\mathrm{H}}+2}{3\sigma_{\mathrm{H}}+3} \tag{12.2}$$

where F_{D} is the dimensional form of the hydrodynamic drag force, η_{f} is the viscosity of the suspending outside fluid, a is the droplet radius, U is the droplet velocity, and σ_{H} is the viscosity ratio of the fluid inside to the outside suspending fluid.

The hydrodynamic drag force decreases with decreasing viscosity ratio, σ_{H}, according to the above equation. For the limiting case of an inviscid interior fluid, or a gas bubble, the correction factor reduces to the lowest possible value of 2/3. Hence in general, the drag force for a droplet is always smaller than the corresponding rigid particle ($\sigma_{\mathrm{H}} = \infty$) due to the fluidity inside the droplet. Note that, however, this reduction is not much: only one-third at most, if the droplet velocity is the same. Thus, one might come up with a misleading conclusion that the electrophoretic mobility of a droplet is at most 50% larger than a corresponding rigid particle. However, in reality, the electrophoretic mobility of a droplet is orders of magnitude larger than a rigid sphere when κa is very large [55]. This obviously is not consistent with the hydrodynamic drag force reduction predicted by Eq. (12.2) alone.

The truth is the modified Stokes law, Eq. (12.2), is obtained for a purely hydrodynamic problem only with the fluid outside chargeless. Here, for the electrophoretic motion of a liquid droplet, however, the fluid outside contains electrolyte ions; hence, yields an extra electric body force upon the droplet when an external electric field is applied. In other words, mathematically the corresponding hydrodynamic drag force now contains an extra term related to the local space charge density in the vicinity of the droplet surface. Although the flow field inside the droplet here takes the same form as indicated by Eq. (12.2), the flow field outside is completely different with the involvement of an extra electric body force generated by the applied electric field. Thus, the droplet velocity U is also a function of κa now and turns out to be a very strong function of κa when κa is very large, which is not revealing at first glance of Eq. (12.2).

The key factor behind this intriguing phenomenon is the mobile droplet surface. Compared with the rigid particle situation as discussed in Chapter 1, where no-slip hydrodynamic boundary condition is assumed, a stagnant layer with compact immobile ions within it is thus resulted physically. This is referred to as the Stern layer in electrokinetics [11]. Here, the droplet surface is actually mobile, and thus, there is no stagnant Stern layer at all for a liquid droplet! The concept of a shear plane does not apply, or needed, either. We actually have only one single layer, the outer diffuse layer only. Thus, the fundamental of the entire double layer structure shatters. This is why there is no Smoluchowski-type behavior as for a rigid particle. But the terminology of the double layer can still be borrowed from there as the surface charges are strongly bounded to the droplet surface anyway. Even though it moves along with the mobile droplet surface, it assumes a uniform value over the entire droplet surface. Hence, the surface charge density can still be evaluated with the two-dimensional Gauss law as discussed in Chapter 1, and so is the corresponding electric driving force. Here, we just point out that due to the mobile droplet surface, which is fundamentally inconsistent with the basic assumption in the rigid particle situation, extra physical mechanisms may occur. For instance, when κa is large, double layer suppression happens and a large amount of counterions are driven toward the vicinity of the mobile droplet surface, and hence, move along with the convection flow there in the opposite direction of the droplet motion. This is equivalent to have an induced electroosmosis flow (EOF) against the droplet motion, with the coordinates set at the center of the moving droplet. The droplet motion thus is enhanced, especially when κa is large when the space charge density in the vicinity of the droplet is large as well. This extra EOF factor reduces the hydrodynamic drag force significantly with the extent proportional to the apparent space charge density there.

Note also that as the droplet velocity is determined once the hydrodynamic drag force is equal to the electric driving force, this reduction of hydrodynamic drag force can also be viewed, or analogized at least, as an increase of electric driving force. The movement of counterions against the droplet can be analogized as equivalent to the movement of coions of the same amount, but along the direction of the droplet motion instead. This is equivalent to have a strong excess surface conductivity added to the droplet; hence, a significant enhancement of electric driving force when κa is large. The key is the mobile surface of the fluid droplet, which physically makes possible the reasoning either by motion-enhancing EOF or equivalently by an effective excess surface conductivity. In short, the focus is not on the fluidity inside the droplet, it is on its mobile surface. Mathematically speaking, the local electric body force term, $-\rho\nabla\phi$, is involved both explicitly in the hydrodynamic drag force evaluation and implicitly in the electric driving force expression. Thus, either analog is okay. It is kind of like Janus in Greek mythology who has two faces.

The details will be illustrated in Section 12.3 with pertinent figures and plots. Here, we just point out the unique feature of a droplet conducting electrophoretic motion. The name of the game for the analysis of a droplet is not in the sheer reduction of purely hydrodynamic drag force given by Eq. (12.2), as one might think at first, not at all. It may not be too far away from the reality when κa is small, but definitely the one to watch in the show is the drastic increase of droplet mobility originated from the mobile surface when κa is large.

12.2 THEORY

12.2.1 Single Liquid Droplet or Gas Bubble

The system diagrams for a single droplet or gas bubble are shown in Fig. 12.1A. The governing equations outside the droplet or bubble are the same as those treated in Chapter 1 for a rigid particle, and so are the associated boundary conditions at infinity. On the droplet surface, the electric boundary conditions are the same as those in Chapter 1 as well, since a nonconducting nonpolar hydrophobic droplet containing no electrolytes inside is considered here.

As for the hydrodynamic boundary conditions there, the Rybczynski and Hadamard condition, or simply Hadamard condition, is adopted here, which assumes the continuity of both the shear stress and the tangential velocity across the droplet or bubble surface [56, 57]. They are, in spherical coordinates and in terms of dimensionless stream function Ψ^*,

$$\sigma_H\left[\frac{\partial^2\Psi^*}{\partial r^{*2}}-\frac{2}{r^*}\frac{\partial\Psi^*}{\partial r^*}-\frac{1}{r^{*2}}\frac{\partial^2\Psi^*}{\partial\theta^2}+\frac{\cos\theta}{r^{*2}\sin\theta}\frac{\partial\Psi^*}{\partial\theta}\right]_{r^*=1^-}$$
$$=\left[\frac{\partial^2\Psi^*}{\partial r^{*2}}-\frac{2}{r^*}\frac{\partial\Psi^*}{\partial r^*}-\frac{1}{r^{*2}}\frac{\partial^2\Psi^*}{\partial\theta^2}+\frac{\cos\theta}{r^{*2}\sin\theta}\frac{\partial\Psi^*}{\partial\theta}\right]_{r^*=1^+} \quad (12.3)$$

$$\left[\frac{\partial\Psi^*}{\partial r^*}\right]_{r^*=1^-}=\left[\frac{\partial\Psi^*}{\partial r^*}\right]_{r^*=1^+} \quad (12.4)$$

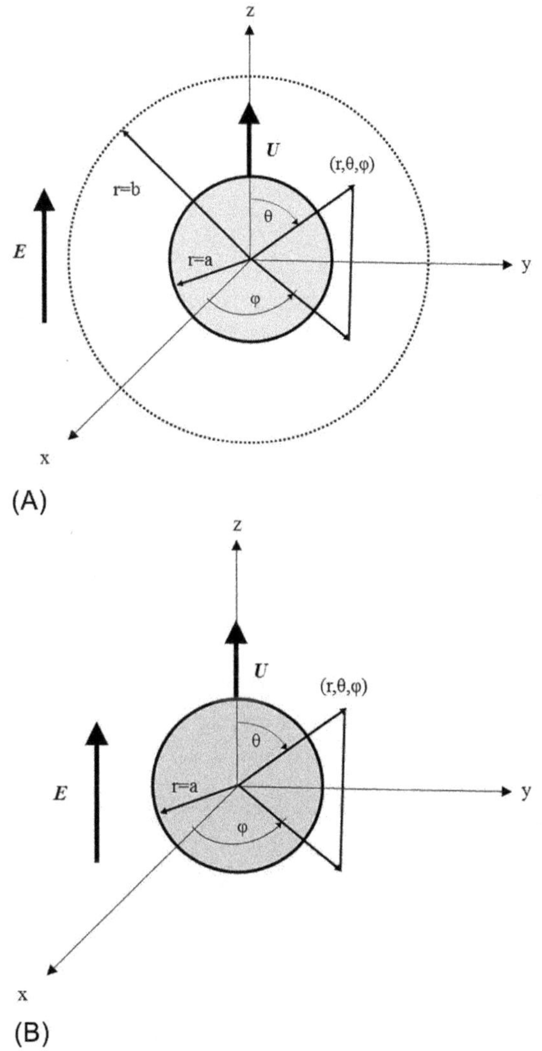

FIG. 12.1 System diagram. (A) A single droplet or bubble; (B) suspensions of droplets or bubbles.

where σ_H is the viscosity ratio of the fluid inside of the droplet or bubble to the fluid outside, and

$$v_r^* = -\frac{1}{r^{*2}\sin\theta}\frac{\partial\Psi^*}{\partial\theta} \tag{12.5}$$

$$v_\theta^* = \frac{1}{r^*\sin\theta}\frac{\partial\Psi^*}{\partial r^*} \tag{12.6}$$

The impermeability of the droplet or bubble surface yields the null normal velocity v_r^* there, i.e.,

$$\left[\frac{\partial \Psi^*}{\partial \theta}\right]_{r^*=1^+} = 0 \tag{12.7}$$

$$\left[\frac{\partial \Psi^*}{\partial \theta}\right]_{r^*=1^-} = 0 \tag{12.8}$$

Note that the droplet fluid treated here is a hydrophobic nonpolar one containing no electrolyte ions. Hence, the governing equation inside the droplet is simplified as

$$E^{*4}\Psi^* = 0 \tag{12.9}$$

As treated in Chapter 1, the insight that Ψ^* is separable as $\Psi^*(r^*, \theta) = \psi^*(r^*)\sin^2\theta$ leads to the one-dimensional form of Eq. (12.9):

$$D^{*4}\psi^* = 0 \tag{12.10}$$

And an analytical solution for ψ^* is available as follows [54]:

$$\psi^* = \frac{c_1}{r^*} + c_2 r^* + c_3 r^{*2} + c_4 r^{*4}, \; r^* \leq 1 \tag{12.11}$$

The finiteness and nullity of velocity at the droplet center yield $c_1 = c_2 = 0$ automatically. On the droplet surface, Eq. (12.8) yields

$$c_3 = -c_4 \tag{12.12}$$

Moreover, the continuity of tangent velocity across the droplet surface, Eq. (12.2), yields

$$2c_3 + 4c_4 = \left[\frac{d\psi^*}{dr^*}\right]_{r^*=1^+} \tag{12.13}$$

Thus, c_3 and c_4 can be obtained as

$$c_3 = -c_4 = -\frac{1}{2}\left[\frac{d\psi^*}{dr^*}\right]_{r^*=1^+} \tag{12.14}$$

Note that Eq. (12.14) implies that the entire flow field within the droplet is contingent upon the flow field outside the droplet, as it should be, since physically it is the electrokinetic motion outside the droplet that drives the purely hydrodynamic motion inside the droplet. Hence, Eq. (12.1), in its one-dimensional form, can be simplified as follows:

$$\left[\frac{d^2\psi^*}{dr^{*2}} - (3\sigma_H + 2)\frac{d\psi^*}{dr^*} + \frac{2}{r^{*2}}\psi^*\right]_{r^*=1^+} = 0 \tag{12.15}$$

This, together with the one-dimensional form of Eq. (12.7), $[\psi^*]_{r^*=1^+} = 0$, are the two hydrodynamic boundary conditions needed on the droplet surface.

Note that the flow field inside the droplet is explicitly decoupled from that of the outside electrolyte solution now. It has already been implicitly incorporated into the outside flow field via Eq. (12.15), the continuity of shear stress, however. Indeed, the forehander is smashed back to its original half court right above the tennis net. The details can be found elsewhere [58]. The rest of the treatment is exactly the same as that in Chapter 1.

12.2.2 Liquid Droplet or Gas Bubble Suspension

The system diagram is shown in Fig. 12.1B. The governing equations and associated boundary conditions are the same as above, except Kuwabara's hydrodynamic boundary conditions as well as the Shilov and Zharkikh's (S-Z) electric boundary conditions are applied on the outer virtual surface, which are exactly the same as those used in Chapter 2. Subsequent treatments are the same as well.

12.3 RESULTS AND DISCUSSION

Fig. 12.2A–C show the general Henry's charts for a sample liquid droplet ($\sigma_H = 0.5$) with various droplet surface potentials ranging from $\phi_r = 1$ to $\phi_r = 5$ in a KCl electrolyte solution. Three different scales of ordinates are used due to the large droplet mobility when κa is large. Motion-deterring double layer polarization effect is clearly demonstrated for highly charged droplets where local minima of mobility profiles of $\phi_r = 3$–5 are observed around $\kappa a = 1$–10 in Fig. 12.2A. The results predicted by Ohshima's analytical formula valid for lowly charged droplet, Eq. (12.1), are shown as well in dashed lines for comparison purpose [43]. No local minimum is observed for this analytical formula and it corresponds very closely to the $\phi_r = 1$ situation in practice.

Compared with the prediction by Hückel's formula, which yields $\mu_m^* = 2/3$ when κa is very small, say $\kappa a = 0.01$, the corresponding mobility for the droplet considered in Fig. 12.2A is about 0.85, regardless the surface charge level. This is consistent with the reduction of hydrodynamic drag force given by Eq. (12.1) for a purely hydrodynamic liquid droplet, as the electrolyte ions are sparse when κa is small. When κa increases past around unity, however, the mobility profiles start to increase so drastically that the frame of Fig. 12.2A can no longer contain them. This behavior is fundamentally different from the behavior of a rigid particle, as shown by the dotted lines there, where the mobility profile approaches asymptotically to the Smoluchowski's prediction of unity when κa is very large, say $\kappa a = 100$ in Fig. 12.2A.

Fig. 12.2B presents the identical mobility profiles in Fig. 12.2A, but with a larger ordinate scale to show the overall picture of the global behavior when κa is large. Much higher values at $\kappa a = 100$ are observed in comparison with the corresponding unity value of a rigid particle. The finiteness of these values

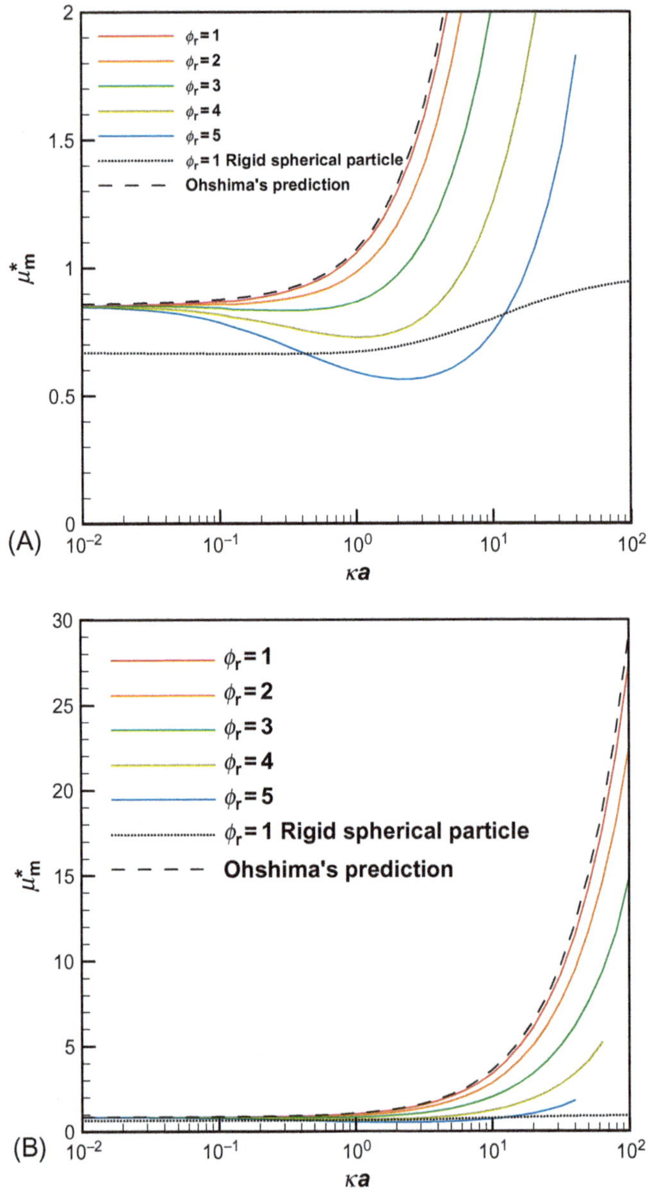

FIG. 12.2 Scaled mobility μ_m^* as a function of κa at different ϕ_r for a liquid droplet with $\sigma_H = 0.5$. (A) Small scale ordinate; (B) large scale ordinate; (C) log scale ordinate.

(Continued)

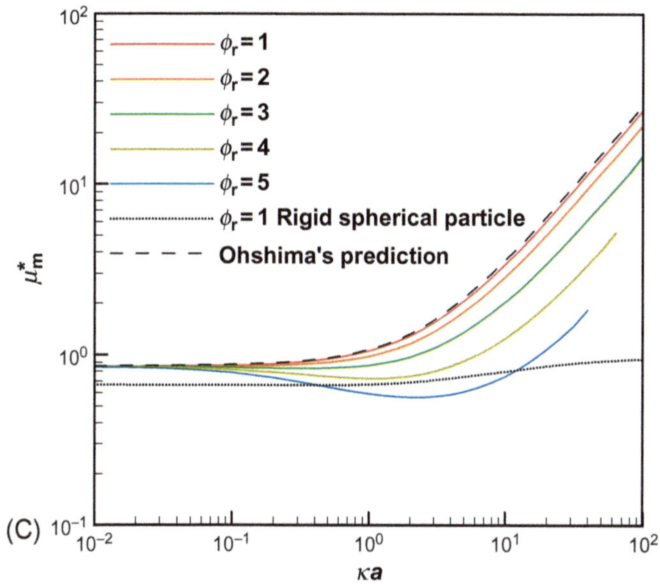

FIG. 12.2—CONT'D

TABLE 12.1 Ratio of the Scaled Disturbed Space Charge Density $\delta\rho^*$ of a Highly Charged Liquid Droplet ($\phi_r = 5$ and $\sigma_H = 0.5$) or a Gas Bubble ($\phi_r = 5$ and $\sigma_H = 0.01$) to the Corresponding Rigid Particle Situation

	κa				
σ_H	*1*	*10*	*20*	*50*	*100*
0.5	1.17	1.65	2.28	4.40	8.28
0.01	1.27	1.86	2.61	5.16	9.89

when κa is large also eliminates the possible concern on the numerical convergence of these sky high profiles in the smaller frame of Fig. 12.2A. Note that Ohshima's analytical formula also assumes the sky high behavior and approaches a large value at $\kappa a = 100$.

The reason behind this unexpected behavior at first glance has been discussed earlier in the introduction section, hence will not be repeated here again. As shown in Table 12.1, the scaled disturbed space charge density ($\delta\rho^*$) at the north pole of a droplet once an external electric field is applied could be many times larger than the corresponding rigid particle situation, either for a liquid droplet ($\sigma_H=0.5$) or a gas bubble ($\sigma_H=0.01$). Up to nearly ten times larger $\delta\rho^*$ is observed when κa is 100. Note that $\delta\rho^*$ here stands for the double layer polarization near the droplet surface, and hence, is the overall index of the initial driving mechanism of the droplet or bubble motion. The increase of this ratio indicates the increasing extent of this polarization, which in turn leads to the induced EOF or effective excess surface conductivity observed in droplet or bubble systems. Indeed, the mobile droplet surface drastically changes the electrostatic environment near the droplet; hence, its hydrodynamic behavior. Fig. 12.2C shows the corresponding mobility profiles in log scale. The overall polarization effect upon the entire range of κa examined is better demonstrated this way.

Fig. 12.3A–C present the corresponding general Henry's charts for a gas bubble ($\sigma_H=0.01$) at various mobility scales. Note that, in comparison with the liquid droplet situation, where the viscosity impact can be very profound thus fluid-dependent, all gas bubbles can actually be treated as an inviscid fluid as gas viscosity is generally much smaller than a liquid such as water. It turns out $\sigma_H=0.01$ stands as a perfect sample case as far as the calculation of bubble mobility is concerned.

To facilitate the possible use of readers who are interested in obtaining the accurate mobility prediction of a highly charged liquid droplet ($\sigma_H=0.5$) or a gas bubble ($\sigma_H=0.01$), correction factors based on the Ohshima's analytical formula [43] are provided in Fig. 12.4A and B for a liquid droplet and a gas bubble, respectively. Note that $\sigma_H=0.5$ is pretty close to a group of droplet systems of corn oil and ethyl acetate, which yields σ_H around 0.48–0.52 [59]. These nanoemulsions may be used as delivery systems in the food, pharmaceutical, and cosmetic industries [29]. On the other hand, $\sigma_H=1.0$ is close to the decane droplet system ($\sigma_H=0.99$) [60], and $\sigma_H=1.72$ to the mercury drop system [61], to name a few droplet systems frequently encountered in practical applications. Note that all the above viscosity measurements are based on the temperature of 25°C.

Note that the corresponding contour plots of equilibrium state variables such as ρ_e^* and ϕ_e^* are identical to the rigid particle situation treated in Chapters 1 and 2 as both systems are impermeable to the outside electrolyte solution. Fig. 12.5A–E show the double layer polarization plot $\delta\rho^*$ for a highly charged ($\phi_r=3$) liquid droplet ($\sigma_H=0.5$) with various double layer thickness ($\kappa a=1$, 10, 20, 50, and 100). The bulk polarization shape, in general, is similar to the rigid particle situation, but at a greater extent.

The flow fields are shown in Fig. 12.6A and B for a droplet ($\sigma_H=0.5$) and a bubble ($\sigma_H=0.01$), respectively. Inner vortex flow are observed in each case, with the viewpoint put at the center of the droplet or the bubble, where the velocity is set to zero. The vortex flow in a gas bubble is much stronger, with a large area motionless though (Fig. 12.6B). Nearly inviscid gas makes it difficult to transfer the momentum from the bubble surface to the inner region.

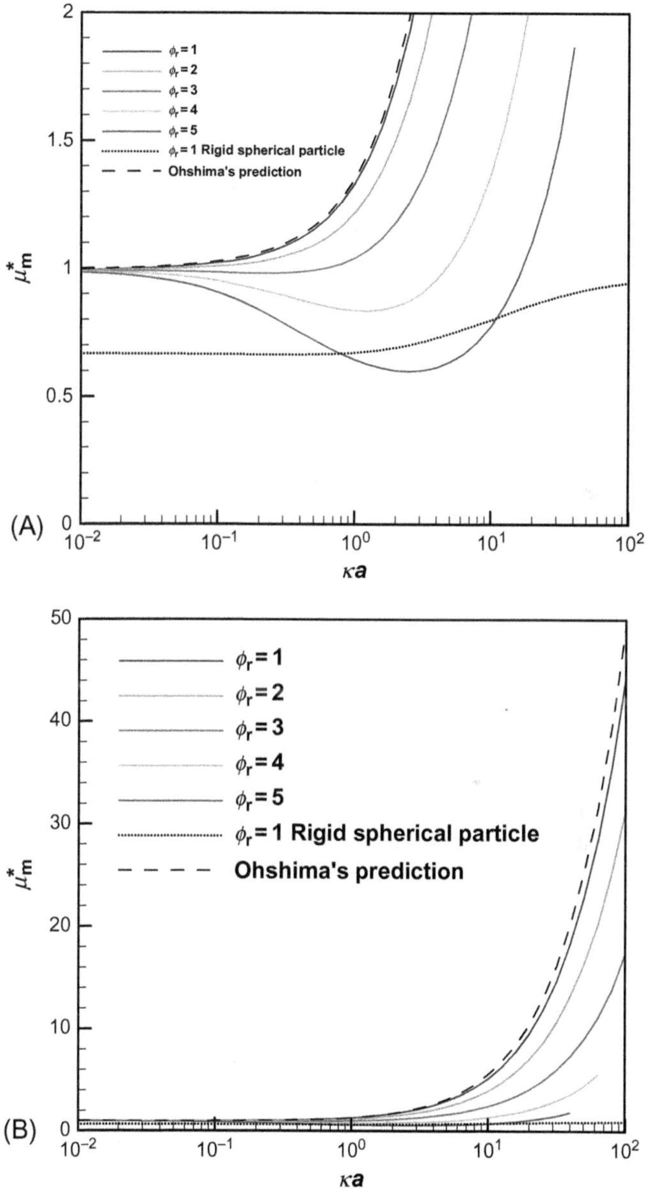

FIG. 12.3 Scaled mobility μ_m^* as a function of κa at different ϕ_r for a gas bubble with $\sigma_H = 0.01$. (A) Small scale ordinate; (B) large scale ordinate; (C) log scale ordinate.

(Continued)

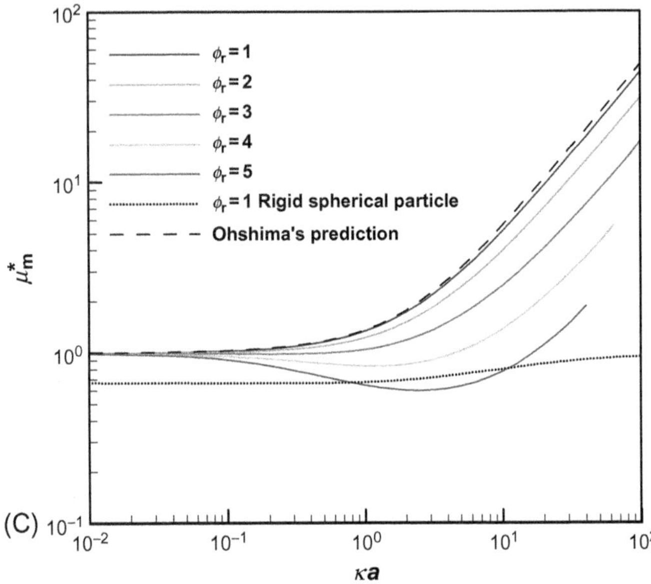

FIG. 12.3—CONT'D

The impact of droplet viscosity is shown in Fig. 12.7A–C, again in three different scales of ordinates. For the range of κa smaller than unity roughly, the hydrodynamic effect dominates and the reduction of drag force as viscosity ratio decreases is clearly observed in general. More importantly, the quantitative prediction is quite close to the one obtained by the analytical formula for a fluid droplet based on a purely hydrodynamic consideration, i.e., Eq. (12.1). When $\kappa a > 1$ roughly, however, the electric feature dominates and a huge enhancement of droplet mobility is observed.

As for the corresponding suspension systems, general Henry's charts are shown for a droplet ($\sigma_H = 0.5$) and a gas bubble ($\sigma_H = 0.01$) in Figs. 12.8 and 12.9, respectively, with different scales of ordinates. When double layer is relatively thick ($\kappa a < 5$, roughly), the phenomenon of effective excess surface conductivity has not happened yet and the behavior is similar to the one observed in Chapter 2 for the rigid particle suspension. Interactions between neighboring particles like motion-deterring double layer overlapping effect and hydrodynamic swarming effect determine the ultimate droplet mobility. When κa is even larger, these factors diminish and the dominant effect of an induced EOF or excess surface conductivity takes over which actually yields roughly a coincident mobility profile independent of the particle concentration in the suspension; indeed, a very interesting behavior. For a rigid particle suspension when κa is large, the double layer overlapping effect diminishes, and hence, the other motion-deterring hydrodynamic swarming effect dominates and the mobility decreases with increasing volume fraction of particles. For a droplet suspension, however, it is the induced EOF near

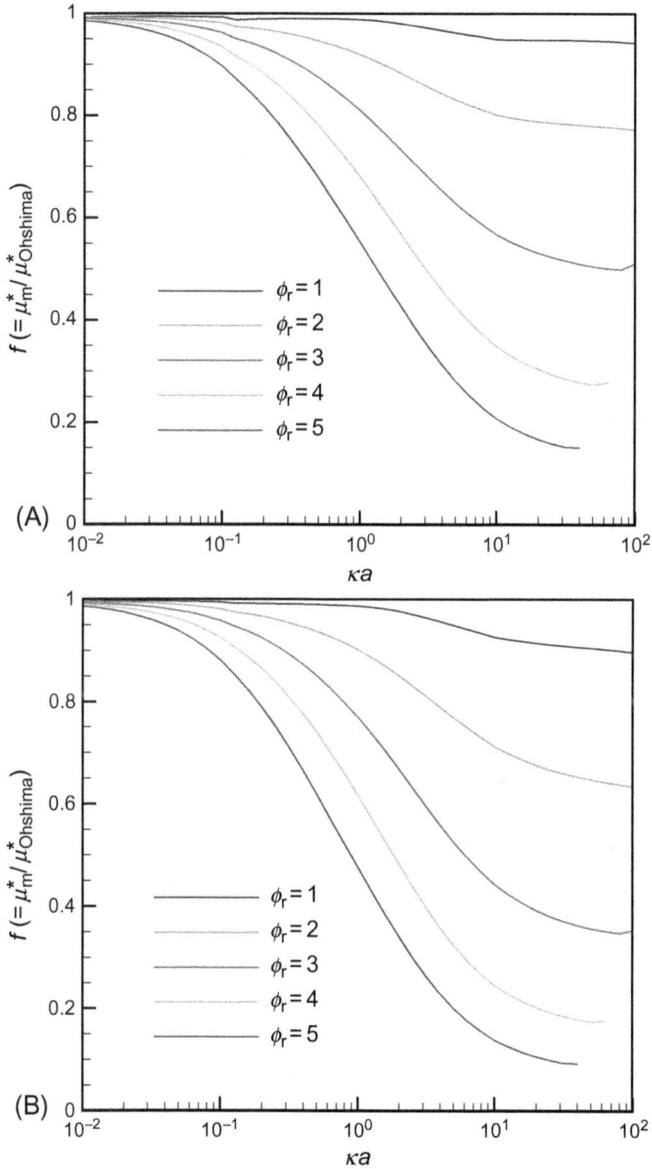

FIG. 12.4 Correction factor to Ohshima's analytical prediction as a function of κa at different ϕ_r. (A) liquid droplet with $\sigma_H = 0.5$; (B) gas bubble with $\sigma_H = 0.01$.

the droplet surface instead of the viscous drag that determines the ultimate droplet mobility. Hence, the presence of neighboring droplets almost has no hydrodynamic impact at all upon the droplet mobility. This explains the observed volume fraction-independent behavior of the droplet mobility.

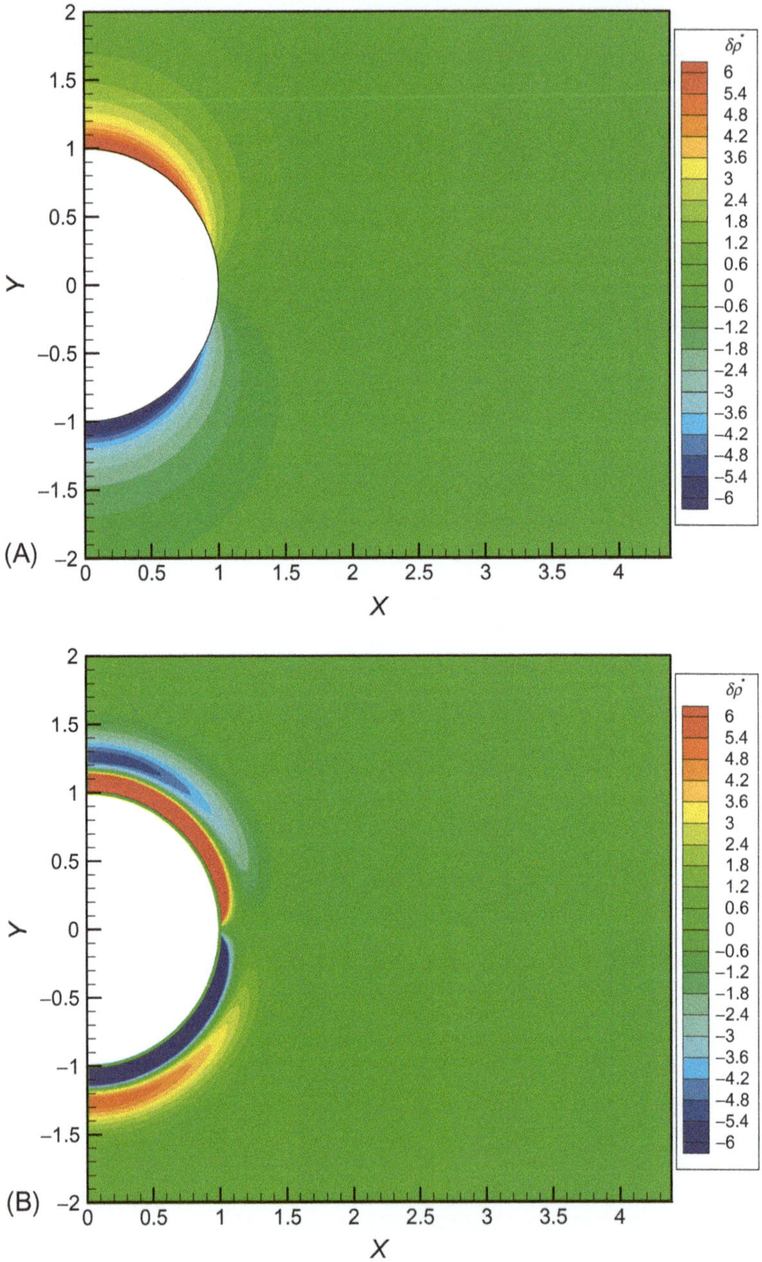

FIG. 12.5 Contour plots of disturbed charge concentration ($\delta\rho^*$) at various κa for a liquid droplet with $\phi_r = 3$ and $\sigma_H = 0.5$. (A) $\kappa a = 1$; (B) $\kappa a = 10$; (C) $\kappa a = 20$; (D) $\kappa a = 50$; (E) $\kappa a = 100$.

(Continued)

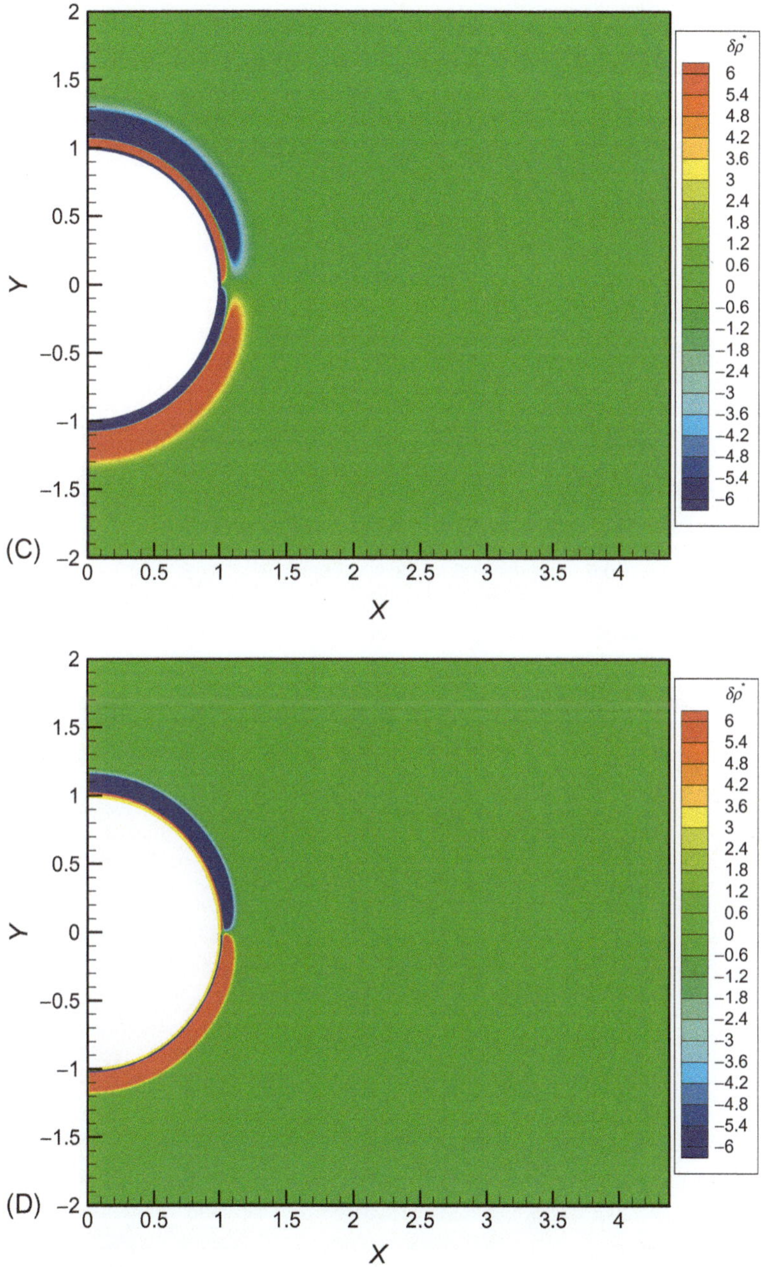

FIG. 12.5—CONT'D

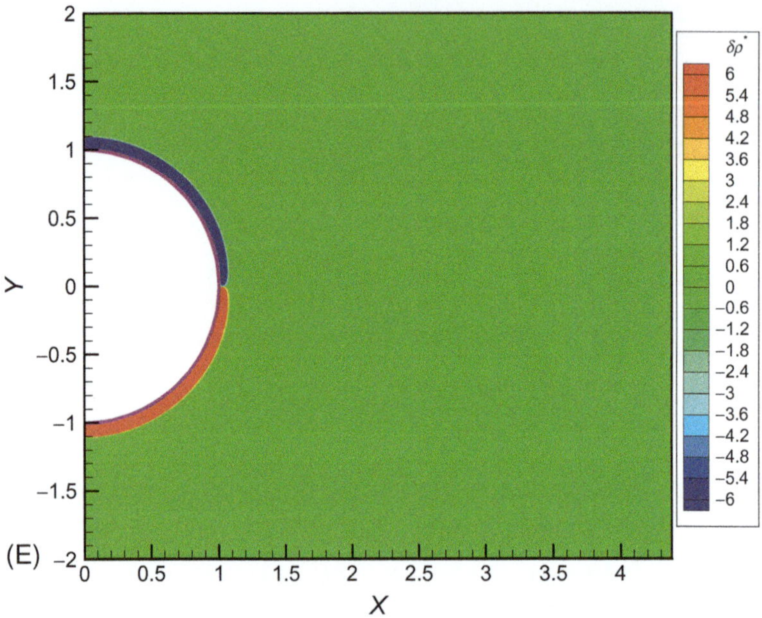

FIG. 12.5—CONT'D

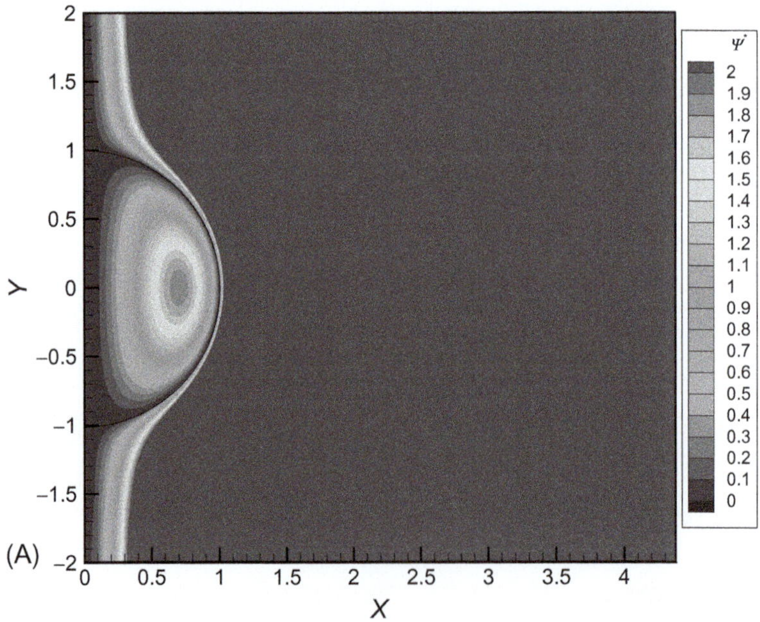

FIG. 12.6 Contour plot of scaled stream function (Ψ^*) at $\kappa a = 100$ for a fluid droplet with $\phi_r = 3$. (A) Liquid droplet with $\sigma_H = 0.5$; (B) gas bubble with $\sigma_H = 0.01$.

(Continued)

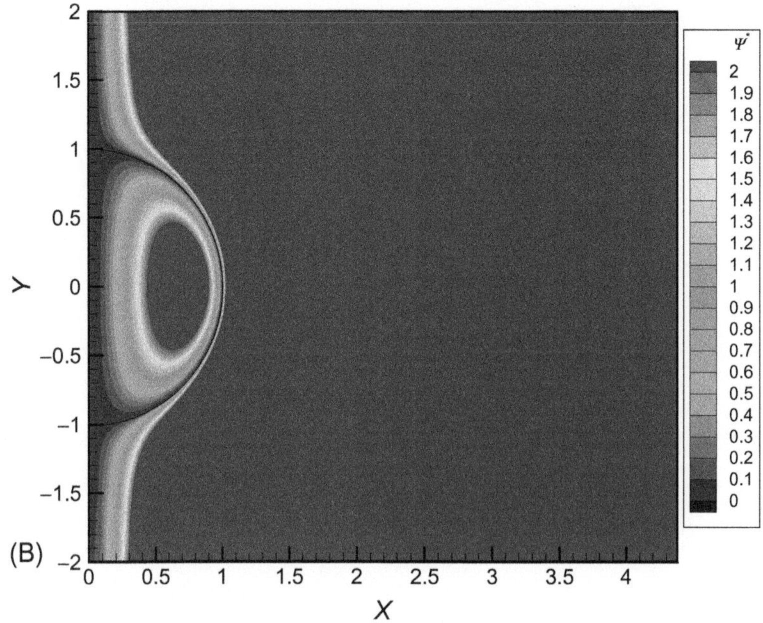

FIG. 12.6—CONT'D

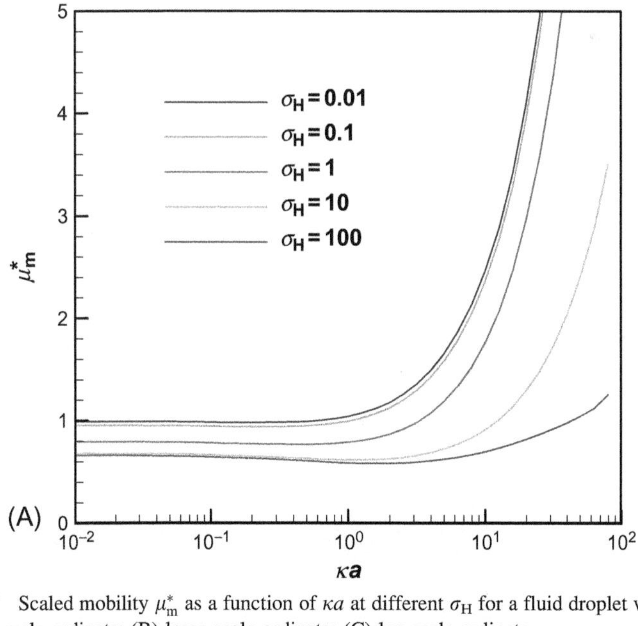

FIG. 12.7 Scaled mobility μ_m^* as a function of κa at different σ_H for a fluid droplet with $\phi_r = 3$. (A) Small scale ordinate; (B) large scale ordinate; (C) log scale ordinate.

(Continued)

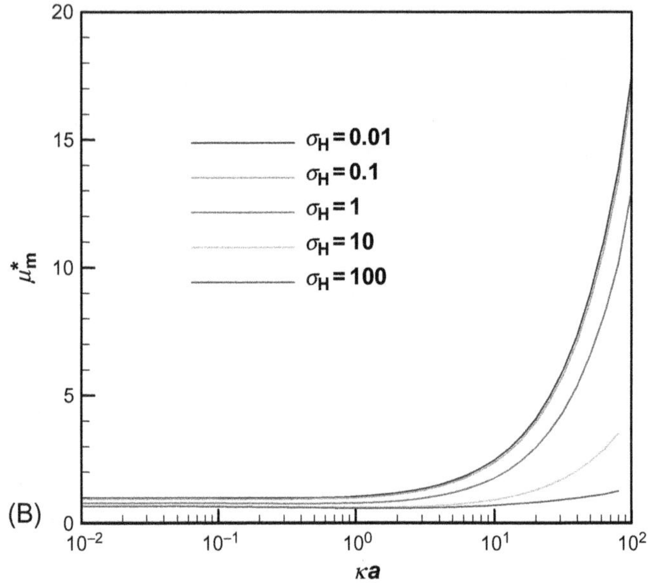

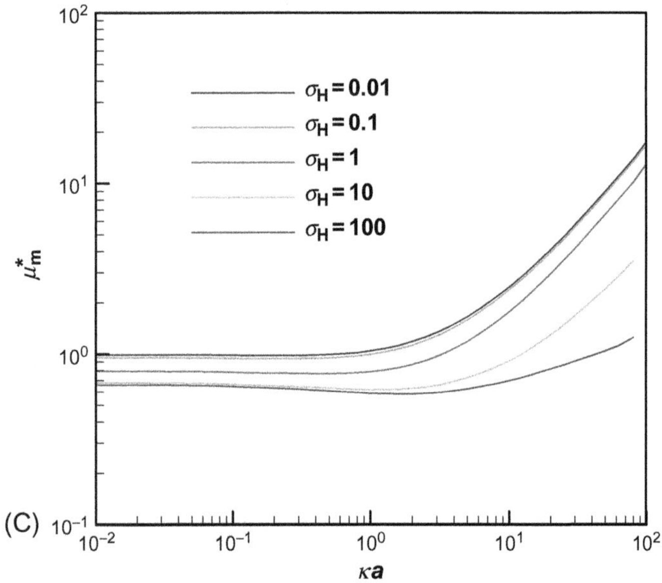

FIG. 12.7—CONT'D

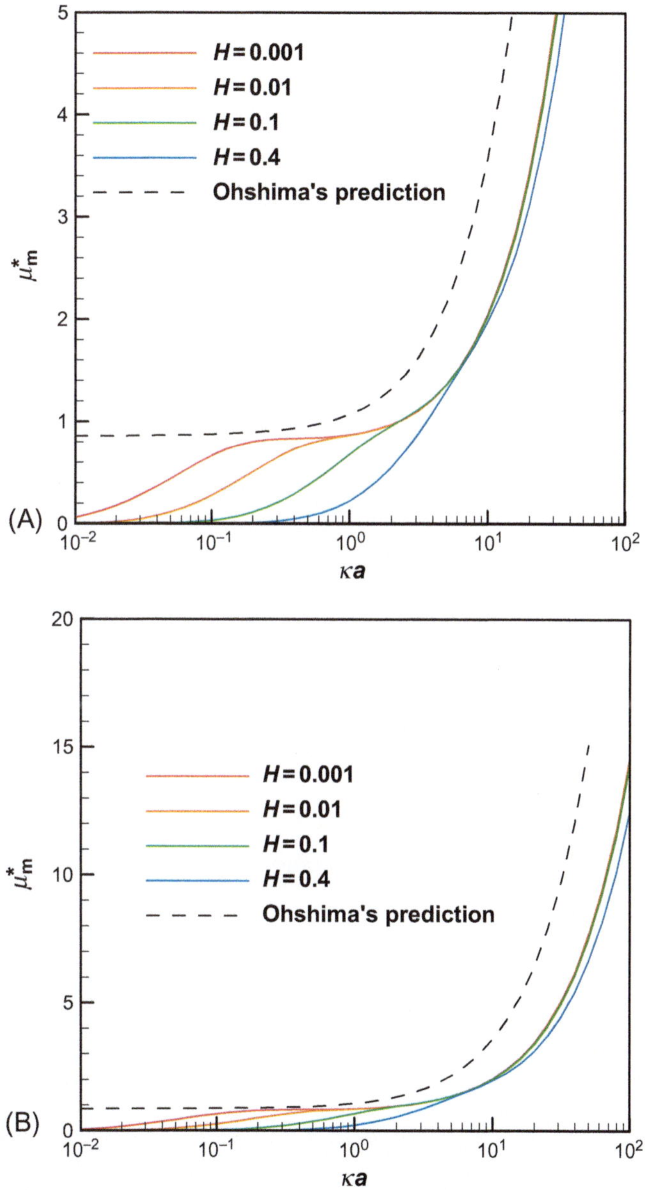

FIG. 12.8 Scaled mobility μ_m^* as a function of κa at different H for a liquid droplet with $\phi_r=3$ and $\sigma_H=0.5$. (A) Small scale ordinate; (B) large scale ordinate; (C) log scale ordinate.

(Continued)

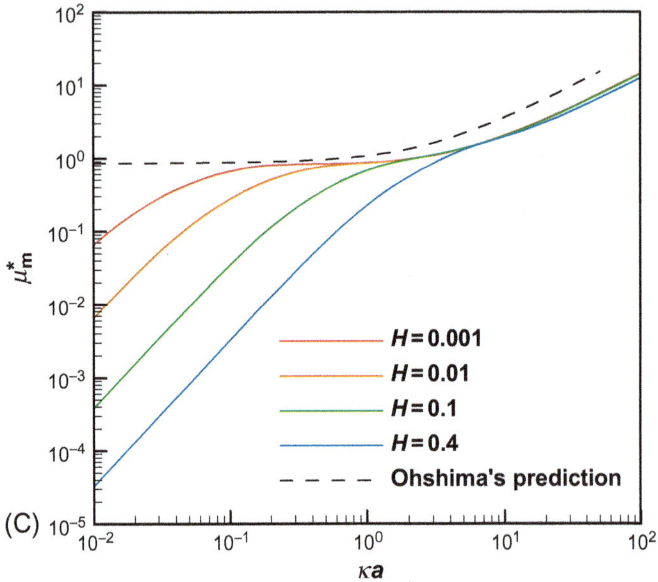

FIG. 12.8—CONT'D

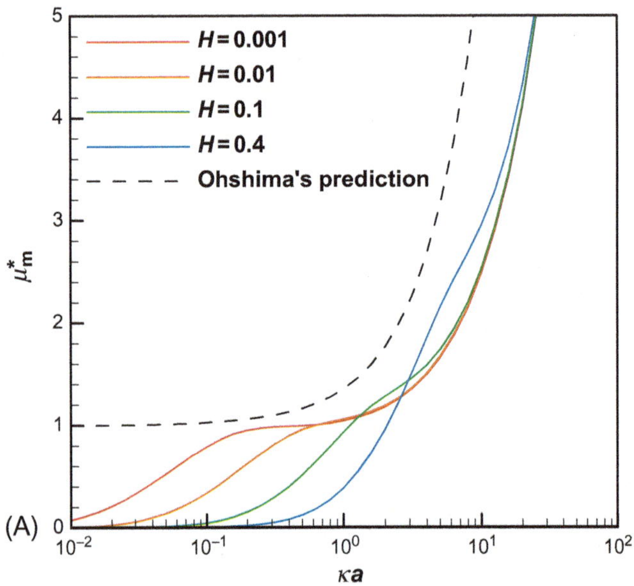

FIG. 12.9 Scaled mobility μ_m^* as a function of κa at different H for a gas bubble with $\phi_r = 3$ and $\sigma_H = 0.01$. (A) Small scale ordinate; (B) large scale ordinate; (C) log scale ordinate.

(Continued)

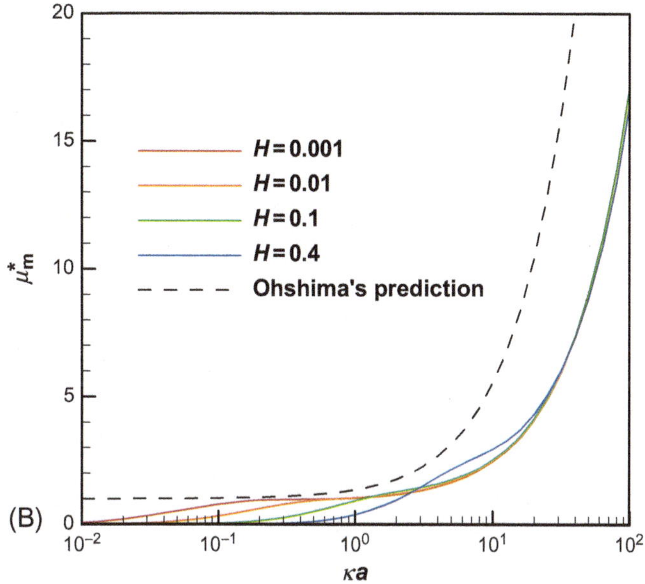

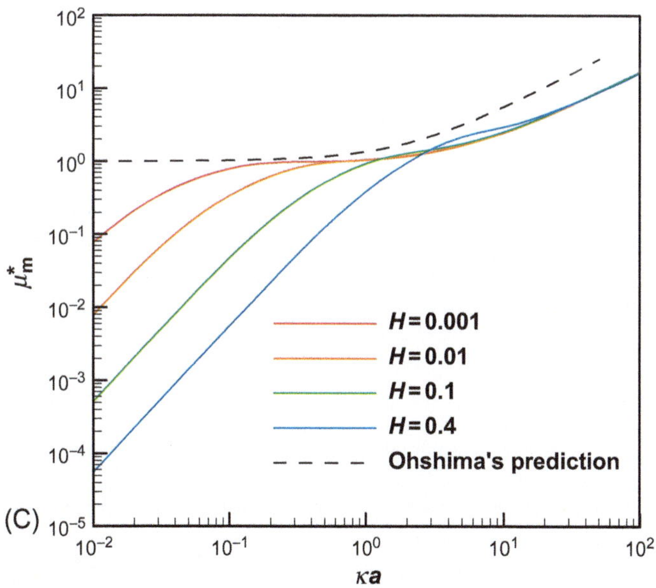

FIG. 12.9—CONT'D

REFERENCES

[1] Dickinson E. Proteins at interfaces and in emulsions stability, rheology and interactions. J Chem Soc Faraday Trans 1998;94(12):1657–69.

[2] Shaw DJ. Introduction to colloid and surface chemistry. London: Butterworths; 1966.

[3] Park K. Controlled drug delivery: challenges and strategies. Washington, DC: American Chemical Society; 1997.

[4] Shui L, Eijkel JC, van den Berg A. Multiphase flow in microfluidic systems—control and applications of droplets and interfaces. Adv Colloid Interf Sci 2007;133(1):35–49.

[5] Lee C-P, Chang H-C, Wei Z-H. Charged droplet transportation under direct current electric fields as a cell carrier. Appl Phys Lett 2012;101(1):014103.

[6] Im DJ, et al. Influences of electric field on living cells in a charged water-in-oil droplet under electrophoretic actuation. Biomicrofluidics 2011;5(4):044112.

[7] Stubenrauch C. Microemulsions background, new concepts, applications, perspectives. Wiley-Blackwell; 2009.

[8] Remington JP. Remington: the science and practice of pharmacy. vol. 1. Philadelphia: Lippincott Williams & Wilkins; 2006.

[9] Kumar P, Mittal KL. Handbook of microemulsion science and technology. Boca Raton, FL: CRC Press; 1999.

[10] Hamouda T, et al. A novel surfactant nanoemulsion with broad-spectrum sporicidal activity against Bacillus species. J Infect Dis 1999;180(6):1939–49.

[11] Tehrani-Bagha AR, et al. An Ouzo emulsion of toluene in water characterized by NMR diffusometry and static multiple light scattering. Colloids Surf A Physicochem Eng Asp 2016;494:81–6.

[12] Vitale SA, Katz JL. Liquid droplet dispersions formed by homogeneous liquid–liquid nucleation: "The Ouzo effect". Langmuir 2003;19(10):4105–10.

[13] Lu Z, Peng S, Zhang X. Influence of solution composition on the formation of surface nanodroplets by solvent exchange. Langmuir 2016;32(7):1700–6.

[14] Hunter RJ. Foundations of colloid science. Oxford: Oxford University Press; 1989.

[15] Boyson T, Pashley R. A study of oil droplet coalescence. J Colloid Interface Sci 2007;316(1):59–65.

[16] Acosta E. Bioavailability of nanoparticles in nutrient and nutraceutical delivery. Curr Opin Colloid Interface Sci 2009;14(1):3–15.

[17] Porter CJ, et al. Enhancing intestinal drug solubilisation using lipid-based delivery systems. Adv Drug Deliv Rev 2008;60(6):673–91.

[18] Tadros T, et al. Formation and stability of nano-emulsions. Adv Colloid Interf Sci 2004;108:303–18.

[19] Bouchemal K, et al. Nano-emulsion formulation using spontaneous emulsification: solvent, oil and surfactant optimisation. Int J Pharm 2004;280(1–2):241–51.

[20] Yin L-J, et al. Performance of selected emulsifiers and their combinations in the preparation of β-carotene nanodispersions. Food Hydrocoll 2009;23(6):1617–22.

[21] Kentish S, et al. The use of ultrasonics for nanoemulsion preparation. Innovative Food Sci Emerg Technol 2008;9(2):170–5.

[22] Rao J, McClements DJ. Stabilization of phase inversion temperature nanoemulsions by surfactant displacement. J Agric Food Chem 2010;58(11):7059–66.

[23] McClements DJ, Rao J. Food-grade nanoemulsions: formulation, fabrication, properties, performance, biological fate, and potential toxicity. Crit Rev Food Sci Nutr 2011; 51(4):285–330.

[24] Anna SL, Bontoux N, Stone HA. Formation of dispersions using "flow focusing" in microchannels. Appl Phys Lett 2003;82(3):364–6.
[25] Rees GD, Robinson BH. Microemulsions and organogels: properties and novel applications. Adv Mater 1993;5(9):608–19.
[26] McClements DJ. Nanoemulsions versus microemulsions: terminology, differences, and similarities. Soft Matter 2012;8(6):1719–29.
[27] Lee HS, et al. Cryogenic electron microscopy study of nanoemulsion formation from microemulsions. Langmuir 2014;30(36):10826–33.
[28] Tong K, Zhao C, Sun D. Formation of nanoemulsion with long chain oil by W/O microemulsion dilution method. Colloids Surf A Physicochem Eng Asp 2016;497:101–8.
[29] Prince LM. Microemulsions versus micelles. J Colloid Interface Sci 1975;52(1):182–8.
[30] Schulman JH, Stoeckenius W, Prince LM. Mechanism of formation and structure of micro emulsions by electron microscopy. J Phys Chem 1959;63(10):1677–80.
[31] Lissant KJ. Emulsions and emulsion technology. Soil Sci 1975;120(2):160.
[32] Overbeek JTG. The first rideal lecture. Microemulsions, a field at the border between lyophobic and lyophilic colloids. Faraday Discuss Chem Soc 1978;65:7–19.
[33] Schoch RB, Han J, Renaud P. Transport phenomena in nanofluidics. Rev Mod Phys 2008;80(3):839.
[34] Mitchell DJ, Ninham BW. Micelles, vesicles and microemulsions. J Chem Soc Faraday Trans 2: Mol Chem Phys 1981;77(4):601–29.
[35] Rhodes CT. Drug development and industrial pharmacy. Drug Dev Ind Pharm 1995;21(20): 2263–85.
[36] Schnitzer O, Frankel I, Yariv E. Electrophoresis of bubbles. J Fluid Mech 2014;753:49–79.
[37] Calgaroto S, Wilberg K, Rubio J. On the nanobubbles interfacial properties and future applications in flotation. Miner Eng 2014;60:33–40.
[38] Baygents J, Saville D. Electrophoresis of small particles and fluid globules in weak electrolytes. J Colloid Interface Sci 1991;146(1):9–37.
[39] Booth F. The cataphoresis of spherical fluid droplets in electrolytes. J Chem Phys 1951;19(11):1331–6.
[40] Craxford S, Gatty O, McKay H. CI. The theory of electrocapillarity—Part VI. A note on electrophoresis. London Edinburgh Dublin Phil Mag J Sci 1937;23(158):1079–82.
[41] Levich VG. Physicochemical hydrodynamics. Englewood Cliffs, NJ: Prentice Hall; 1962.
[42] Levine S, O'Brien R. A theory of electrophoresis of charged mercury drops in aqueous electrolyte solution. J Colloid Interface Sci 1973;43(3):616–29.
[43] Ohshima H. Electrokinetic phenomena in a concentrated dispersion of charged mercury drops. J Colloid Interface Sci 1999;218(2):535–44.
[44] Ohshima H. A simple expression for the electrophoretic mobility of charged mercury drops. J Colloid Interface Sci 1997;189(2):376–8.
[45] Hsu JP, Min WL, Lee E. Dynamic electrophoresis of droplet dispersions at low surface potentials. J Colloid Interface Sci 2007;306(2):421–7.
[46] Huang CH, Lee E. Electrophoretic motion of a liquid droplet in a cylindrical pore. J Phys Chem C 2012;116(28):15058–67.
[47] Lee E, Min WL, Hsu JP. Dynamic electrophoresis of a droplet in a spherical cavity. Langmuir 2006;22(8):3920–8.
[48] Lou J, Lee E. Dynamic electrophoresis of concentrated droplet dispersions at arbitrary surface potentials. Chem Eng Sci 2007;62(22):6234–45.
[49] Chiang CP, He YY, Lee E. Sedimentation velocity and potential in a concentrated suspension of charged liquid drops. Langmuir 2008;24(20):11361–9.

[50] Lee E, Chang CJ, Hsu JP. Electrophoresis of a concentrated aqueous dispersion of non-Newtonian drops. J Colloid Interface Sci 2005;282(2):486–92.
[51] Lee E, Chiang CP, Hsu JP. Electrophoresis of Carreau-drop dispersions with a charge-regulated surface. Langmuir 2006;22(4):1911–8.
[52] Lee E, Hu JK, Hsu JP. Electrophoresis of concentrated mercury drops. J Colloid Interface Sci 2003;257(2):250–7.
[53] Lee E, Lou SH, Hsu JP. Electrophoresis of a non-conducting Newtonian drop of low electrical potential normal to a plane. Chem Eng Sci 2006;61(14):4550–7.
[54] Happel J, Brenner H. Low Reynolds number hydrodynamics: with special applications to particulate media. vol. 1. New York: Springer Science & Business Media; 2012.
[55] Ohshima H. Theory of colloid and interfacial electric phenomena. vol. 12. Amsterdam: Elsevier; 2006.
[56] Hadamard J. Motion of liquid drops (viscous). Comp Rend Acad Sci Paris 1911;154:1735–55.
[57] Rybczynski W. On the translatory motion of a fluid sphere in a viscous medium. Bul Acad Sci Cracow Ser A 1911;40:40–6.
[58] Lin C-L. Dielectrophoretic phenomena of a liquid droplet in an electrolyte solution. Chemical Engineering Institute, National Taiwan University; 20151–133.
[59] Wooster TJ, Golding M, Sanguansri P. Impact of oil type on nanoemulsion formation and Ostwald ripening stability. Langmuir 2008;24(22):12758–65.
[60] Nespolo SA, et al. Hydrodynamic and electrokinetic properties of decane droplets in aqueous sodium dodecyl sulfate solutions. Langmuir 2001;17(23):7210–8.
[61] Weast RC, Astle MJ, Beyer WH. CRC handbook of chemistry and physics. vol. 1990. Boca Raton, FL: CRC Press; 1989.

Chapter 13

Electrophoretic Motion of a Liquid Droplet or a Gas Bubble Normal to Planar Boundaries

13.1 INTRODUCTION

Motion of a liquid droplet normal to a solid plane is encountered in many important fields such as biomedical or biochemical applications [1, 2]. For instance, a human cell or a drug-carrying capsule approaching a tissue can be modeled with this configuration [3]. The fouling of the marine vessel due to the adhesion of bacteria can be treated in this way as well [4].

On the other hand, the corresponding configuration with a droplet moving toward an air-water interface, in particular, finds many important applications as well [5, 6]. More specifically, it is deeply involved in the enhanced oil recovery (EOR) process to separate the oil phase in droplet form from the surfactant-containing solution phase, which is compressed into the void space in the porous rocks rich in oil, but cannot be extracted by the conventional oil draining equipment [7–9]. This large scale potential application of electrophoresis is vital to the efficient and reliable energy supply nowadays.

Indeed, these planar boundaries, either a solid plane or a free interface like an air-water interface, are classic boundaries which have been treated in the corresponding purely hydrodynamic field. There has been many literatures regarding this issue focusing on the creeping flow regime in particular, as summarized nicely by Happel and Brenner in their classic book [10]. The presence of a nearby planar boundary is found to reduce the particle speed in general, and the extent of this reduction can be expressed as a correction factor to the famous Stokes law of a fluid droplet [11].

As for the electrophoretic motion of a liquid droplet normal to a planar boundary, theoretical studies have not been very abundant [12, 13] and mainly done by Lee's group [14, 15]. Here in this chapter, we treat this interesting

Interface Science and Technology, Vol. 26. https://doi.org/10.1016/B978-0-08-100865-2.00013-8

system which is similar to Chapters 3 and 4, except an extra domain within the droplet has to be taken into account.

The droplet under consideration here is a nonpolar hydrophobic one containing no electrolyte ions within it, such as those encountered in typical oil-in-water (O/W) emulsions [16–18]. The droplet surface can still carry charges, however, via the adsorption mechanism in an aqueous electrolyte solution, for instance. Only one parameter, the viscosity ratio of the fluid inside to the outside, σ_{H}, is needed to fully characterize such a fluid droplet. Hence, a gas bubble can be modeled as a fluid droplet with trivial viscosity ratio, say 0.01.

The bi-spherical coordinates are used here as well, with a proper treatment of continuity across the droplet surface for both the velocity and the shear stress there, the so-called Rybczynski and Hadamard condition [10]. Results are presented with similar patterns like those in Chapters 3 and 4 for a representative liquid droplet with viscosity ratio $\sigma_{\mathrm{H}}=0.5$ in contrast to the suspending medium. Moreover, corresponding results for a gas bubble are also presented as it is a classic fluid in itself.

13.2 THEORY

13.2.1 Solid Plane

The system diagram is shown in Fig. 13.1A, with the mesh diagram in computation the same as Fig. 8.2.

Note that the system under consideration is essentially the same as that in Chapter 8, where a porous sphere is treated. All we need to do now is to replace the porous sphere with a liquid droplet. Note that as the droplet is a nonpolar hydrophobic one with no electrolyte ions in it, only the flow field needs to be resolved there as follows:

$$E^{*4}\Psi^{*}=0,\quad \eta>\eta_0 \tag{13.1}$$

where

$E^{*4}=E^{*2}E^{*2}$, and

$$E^{*2}=\frac{x^2}{c^{*2}}\left[\frac{\partial^2}{\partial\eta^2}+\frac{\partial^2}{\partial\xi^2}+\frac{\sinh\eta}{x}\frac{\partial}{\partial\eta}+\frac{1-\cos\xi\cosh\eta}{x\sin\xi}\frac{\partial}{\partial\xi}\right] \tag{13.2}$$

Note that Eq. (13.1) is the conventional Stokes equation written in bi-spherical coordinates and in dimensionless form. Associated boundary conditions at the droplet surface ($\eta=\eta_0$) are the same as those in Chapter 12, but in bi-spherical coordinates here as follows:

$$\phi_e^{*}=\phi_r \tag{13.3}$$

$$\frac{\partial\delta\phi^{*}}{\partial\eta}=0 \tag{13.4}$$

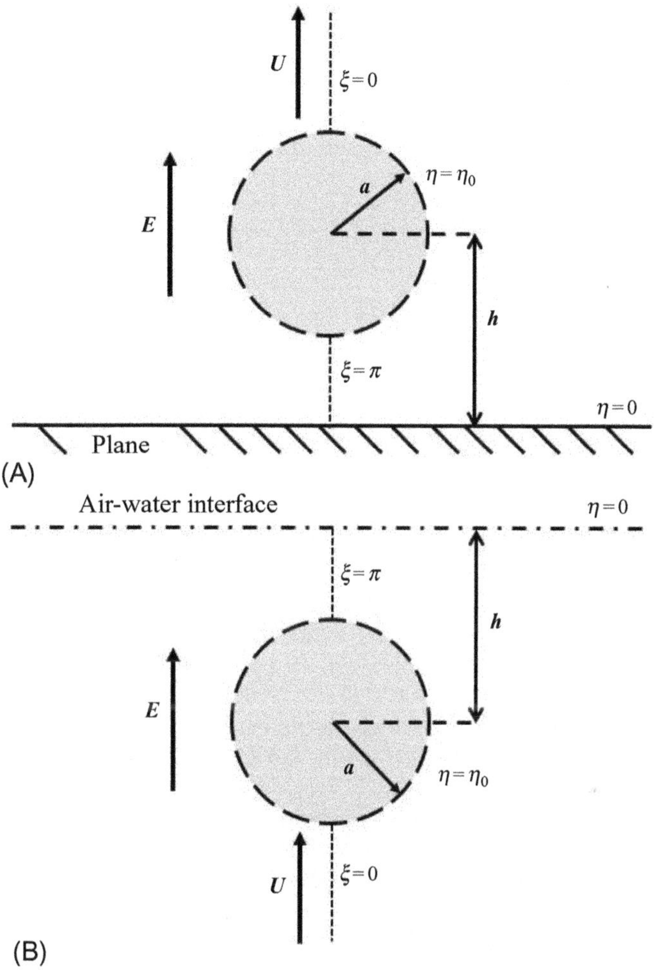

FIG. 13.1 System diagram. (A) Solid plane; (B) air-water interface.

$$\frac{\partial g_j^*}{\partial \eta}=0 \tag{13.5}$$

$$\Psi^*=-\frac{1}{2}\left(\frac{c^*\sin\xi}{x}\right)^2 U^* \tag{13.6}$$

$$\left.\frac{\partial \Psi^*}{\partial \eta}\right|_{\eta=\eta_0^-}=\left.\frac{\partial \Psi^*}{\partial \eta}\right|_{\eta=\eta_0^+} \tag{13.7}$$

and

$$\left\{\begin{array}{l} -\sin\xi\left(x^2\dfrac{\partial^2\Psi^{*(o)}}{\partial\eta^2}-2x\sinh\eta\dfrac{\partial\Psi^{*(o)}}{\partial\eta}-x\cosh\eta\Psi^{*(o)}+2\sinh^2\eta\Psi^{*(o)}\right) \\ +\sin\xi\left(x^2\dfrac{\partial^2\Psi^{*(o)}}{\partial\xi^2}-2x\sin\xi\dfrac{\partial\Psi^{*(o)}}{\partial\xi}-x\cos\xi\Psi^{*(o)}+2\sin^2\xi\Psi^{*(o)}\right) \\ +(1+2\sin^2\xi-\cos\xi\cosh\eta)\left(x\dfrac{\partial\Psi^{*(o)}}{\partial\xi}-\sin\xi\Psi^{*(o)}\right)-3\sin\xi\sinh\eta\left(x\dfrac{\partial\Psi^{*(o)}}{\partial\eta}-\sinh\eta\Psi^{*(o)}\right) \end{array}\right\}$$
$$=\sigma_H\left\{\begin{array}{l} -\sin\xi\left(x^2\dfrac{\partial^2\Psi^{*(i)}}{\partial\eta^2}-2x\sinh\eta\dfrac{\partial\Psi^{*(i)}}{\partial\eta}-x\cosh\eta\Psi^{*(i)}+2\sinh^2\eta\Psi^{*(i)}\right) \\ +\sin\xi\left(x^2\dfrac{\partial^2\Psi^{*(i)}}{\partial\xi^2}-2x\sin\xi\dfrac{\partial\Psi^{*(i)}}{\partial\xi}-x\cos\xi\Psi^{*(i)}+2\sin^2\xi\Psi^{*(i)}\right) \\ +(1+2\sin^2\xi-\cos\xi\cosh\eta)\left(x\dfrac{\partial\Psi^{*(i)}}{\partial\xi}-\sin\xi\Psi^{*(i)}\right)-3\sin\xi\sinh\eta\left(x\dfrac{\partial\Psi^{*(i)}}{\partial\eta}-\sinh\eta\Psi^{*(i)}\right) \end{array}\right\} \tag{13.8}$$

where in Eq. (13.8) $\dfrac{\partial\Psi^{*(o)}}{\partial\eta}$ refers to $\left.\dfrac{\partial\Psi^{*}}{\partial\eta}\right|_{\eta=\eta_0^-}$ and $\dfrac{\partial\Psi^{*(i)}}{\partial\eta}$ refers to $\left.\dfrac{\partial\Psi^{*}}{\partial\eta}\right|_{\eta=\eta_0^+}$ with the superscripts o standing for the exterior fluid side and i for the interior fluid side. Note that $\eta > \eta_0$ within the droplet. Note that Eq. (13.8) is the explicit form of shear stress continuity in bi-spherical coordinates, and σ_H is the viscosity ratio of the fluid within the droplet to the outside solution. Definitions of variables are essentially the same as those in Chapters 3 and 4, which can also be found in the List of Symbols at the end of the book.

The governing equations outside the droplet and associated boundary conditions on the solid plane are exactly the same as those in Chapter 3, that is, Eqs. (3.3)–(3.7) and Eqs. (3.16)–(3.26). The subsequent treatments and evaluation of droplet mobility are essentially the same as well; hence, will not be repeated here.

13.2.2 Air-Water Interface

The system diagram is shown in Fig. 13.1B, and the mesh diagram in computation is the same as Fig. 8.2. Note that the system is identical to the corresponding solid plane case above except that the hydrodynamic boundary conditions at the air-water interface, i.e., Eqs. (4.3)–(4.9), are used here instead. The rest is the same and we are done.

13.3 RESULTS AND DISCUSSION

13.3.1 Solid Plane

Note that the electrostatic feature of a droplet is exactly the same as that of a rigid sphere when the droplet is motionless before the application of an external electric field. Both colloidal entities are impenetrable to the outside suspending solution; hence, all the counterions reside outside and so are their corresponding double layer polarization. Hence, the equilibrium space charge density (ρ_e^*) and the electric potential (ϕ_e^*) are exactly the same as those presented in Chapter 3, and certainly there is no need to repeat them here again.

Typical general Henry's charts are shown in Fig. 13.2A–F to examine the boundary effect at various separation distances. An extra parameter of the fluid viscosity inside the droplet needs to be considered. Here, $\sigma_H = 0.5$ (a liquid droplet) and $\sigma_H = 0.01$ (a gas bubble) are selected as the two representative model cases. Note that far away from the solid plane, say $h = 80$, the particle mobility reduces to the corresponding single particle situation in these charts. Moreover, the analytical predictions by Ohshima [19] for a lowly charged droplet or a gas bubble are also recovered successfully when $\phi_r = 1$, indicating the reliability and accuracy of these charts.

The boundary effect in terms of mobility reduction is observed across the entire range of κa examined. The closer the particle to the solid plane is, the slower it moves. No peculiar phenomenon against intuition is observed as noticed in Chapter 8 where a porous particle may move faster as it gets close to the plane. The hydrodynamic factor apparently dominates here for a fluid droplet, as it does in the rigid particle case. The porous particle, on the other hand, is one of a kind. Qualitatively speaking, the motion of a droplet or a gas bubble is very similar to a rigid particle, which can be considered as a droplet with infinite viscosity.

The boundary confinement effect further results in a gradual disappearance of local minima of mobility profiles for highly charged droplets or bubbles when they approach the plane, as shown in Fig. 13.3A and B for $h = 2$ and $h = 1.1$, which corresponds to the situation when the double layer just touches the plane and when it is severely compressed, respectively. Similar behavior is also observed in Chapter 3 for a rigid particle, and the underlying mechanism and reasoning have already been discussed there. A liquid droplet or a gas bubble here simply moves faster, especially when the double layer is very thin, or κa is very large.

Fig. 13.4 depicts the severe confinement effect for a highly charged droplet with $\phi_r = 5$ and $\kappa a = 1$. Note that the polarization of double layer, although heavily compressed, still manages to take place somewhat in the narrow gap region. This is due to the hydrodynamic no-slip condition on the plane surface, which results in a much stronger confinement effect than the corresponding air-water interface, as shown later in Fig. 13.6, where the polarization in the narrow gap area is completely wiped out by the horizontal flow at the air-water interface. This is the fundamental difference of the two planar boundaries. The boundary confinement for a solid plane is stronger than an air-water interface.

Flow field and other contour plots are in fact quite similar to those of the rigid particle situation in Chapter 3, hence will not be presented here.

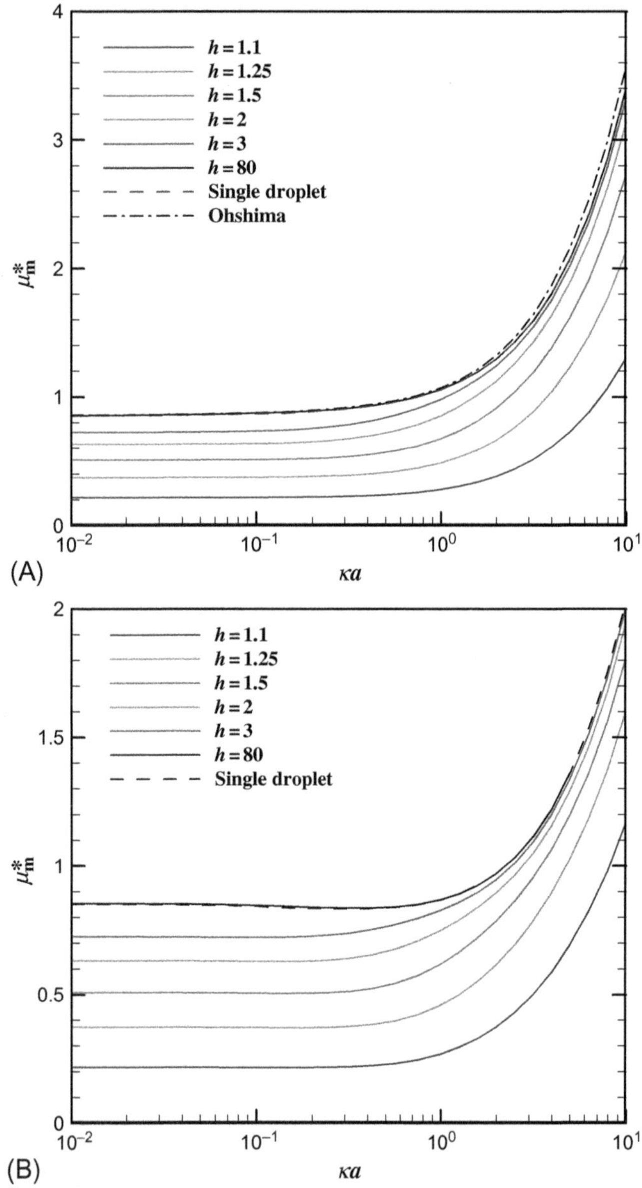

FIG. 13.2 Scaled mobility μ_{m}^{*} as a function of κa at different h. (A) $\phi_{\mathrm{r}}=1$ and $\sigma=0.5$; (B) $\phi_{\mathrm{r}}=3$ and $\sigma_{\mathrm{H}}=0.5$; (C) $\phi_{\mathrm{r}}=5$ and $\sigma_{\mathrm{H}}=0.5$; (D) $\phi_{\mathrm{r}}=1$ and $\sigma_{\mathrm{H}}=0.01$; (E) $\phi_{\mathrm{r}}=3$ and $\sigma_{\mathrm{H}}=0.01$; (F) $\phi_{\mathrm{r}}=5$ and $\sigma_{\mathrm{H}}=0.01$.

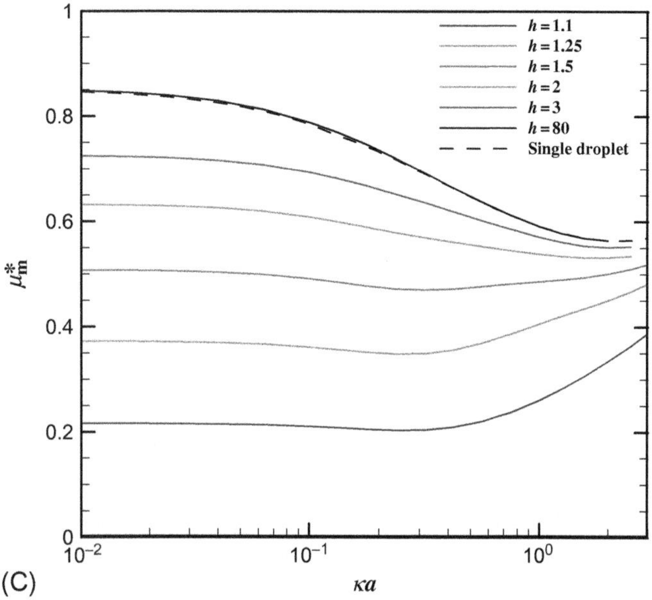

(C)

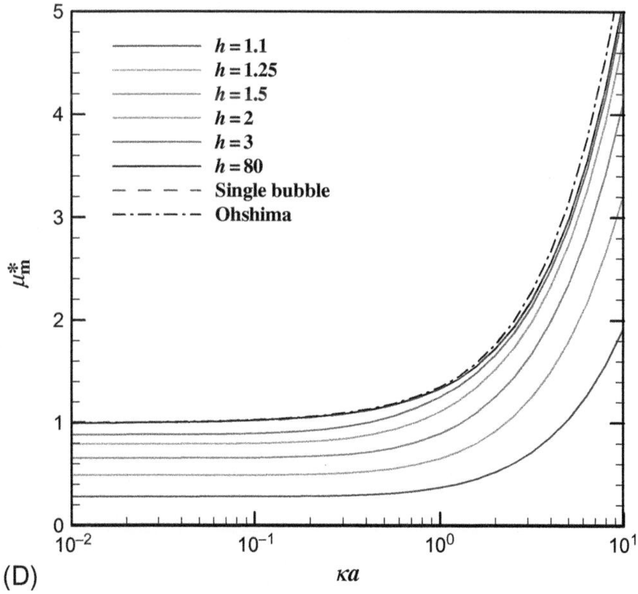

(D)

FIG. 13.2—CONT'D

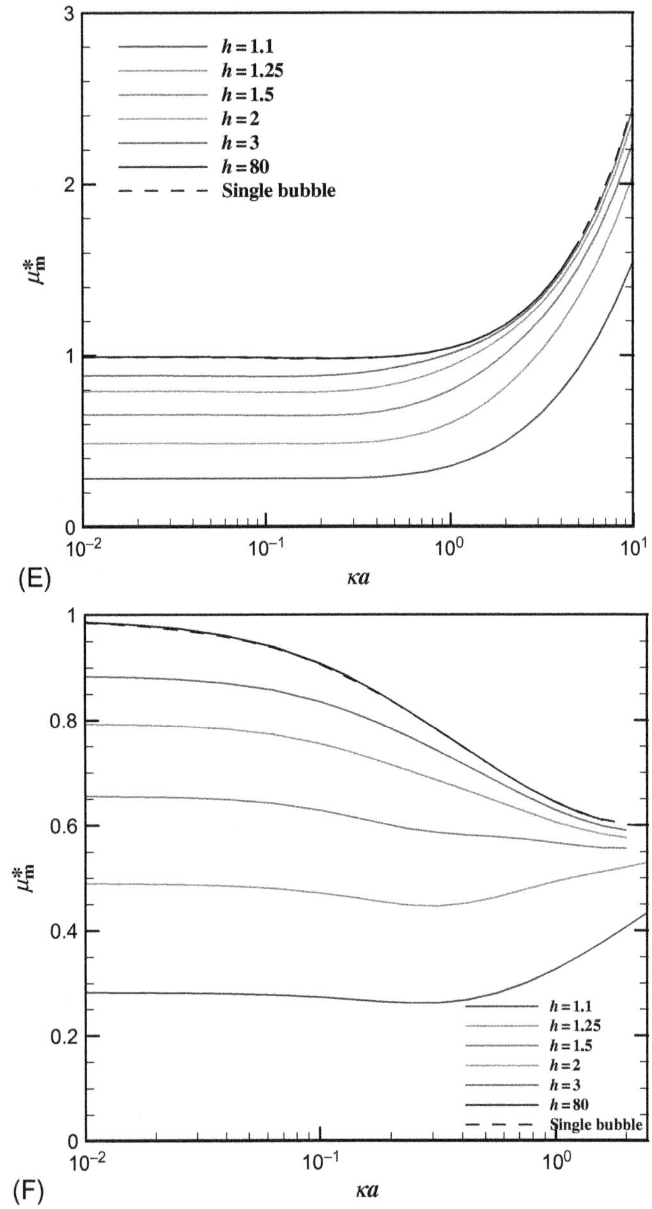

FIG. 13.2—CONT'D

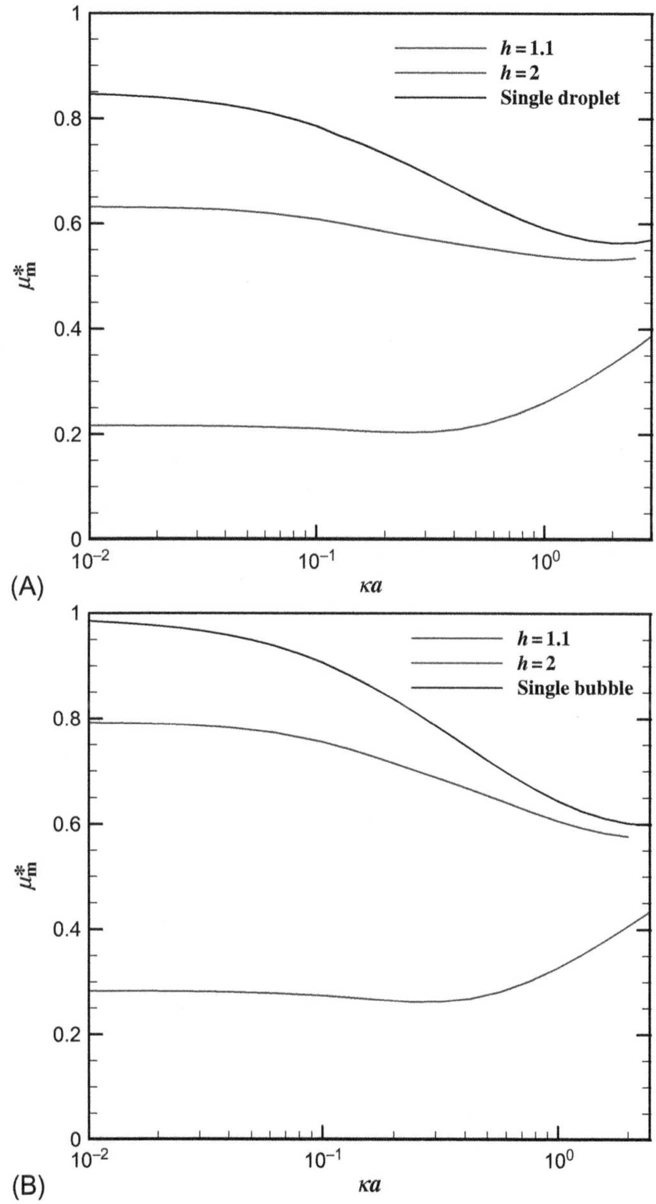

FIG. 13.3 Scaled mobility μ_m^* as a function of κa at $\phi_r = 5$ for $h = 2$, 1.1 and in comparison with a single droplet. (A) $\sigma_H = 0.5$; (B) $\sigma_H = 0.01$.

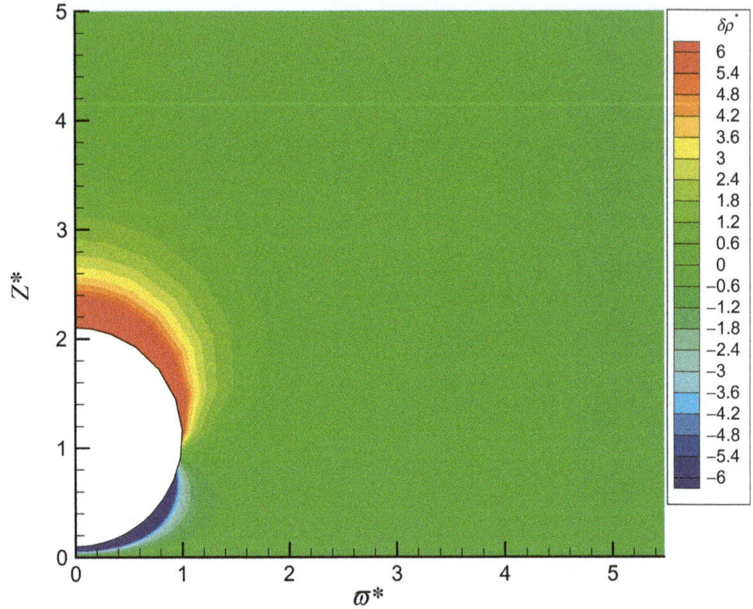

FIG. 13.4 Contour plot of the scaled disturbed charge density ($\delta\rho^*$) for $\kappa a=1$, $\phi_r=5$, $h=1.1$, and $\sigma_H=0.5$.

Tables of correction factor as a function of separation distance are presented in Tables 13.1 and 13.2 for a droplet ($\sigma_H=0.5$) and a gas bubble ($\sigma_H=0.01$), respectively.

13.3.2 Air-Water Interface

General Henry's charts are shown in Fig. 13.5A–F for a liquid droplet ($\sigma_H=0.5$) and a gas bubble ($\sigma_H=0.01$), respectively. Again the mobility reduction is observed across the entire range of κa and separation distance examined. No peculiar behavior is observed as that in Chapter 8 for a porous particle. Fig. 13.6 shows the severe boundary confinement effect upon the double layer polarization as discussed earlier.

Corresponding tables of correction factors as a function of separation distance are shown in Tables 13.3 and 13.4 for a liquid droplet ($\sigma_H=0.5$) and a gas bubble ($\sigma_H=0.01$), respectively.

TABLES OF BOUNDARY EFFECT EXPRESSED AS A CORRECTION FACTOR Γ

($\Gamma=\mu_{m,h}/\mu_{m,\infty}$, where $\mu_{m,h}$ refers to the particle mobility at a separation distance h, whereas $\mu_{m,\infty}$ refers to the mobility in an infinite medium.)

TABLE 13.1 Correction Factor With $\phi_r = 3$ and $\sigma_H = 0.5$ (Liquid Droplet) for a Solid Plane

	κa							
h	*0.01*	*0.1*	*0.5*	*1*	2	*5*	*7*	*10*
1.1	0.255	0.256	0.271	0.310	0.386	0.510	0.547	0.577
1.2	0.389	0.391	0.414	0.471	0.573	0.702	0.729	0.746
1.4	0.546	0.548	0.583	0.656	0.760	0.848	0.858	0.864
1.6	0.638	0.641	0.683	0.759	0.847	0.903	0.909	0.914
1.8	0.700	0.702	0.749	0.822	0.894	0.933	0.937	0.942
2	0.743	0.746	0.797	0.864	0.923	0.950	0.954	0.959
4	0.898	0.903	0.954	0.979	0.989	0.993	0.995	0.999
6	0.938	0.945	0.985	0.994	0.997	0.998	0.999	–
8	0.957	0.965	0.994	0.998	0.998	0.999	–	–
10	0.967	0.976	0.998	0.999	0.999	0.999	–	–
20	0.988	0.997	–	–	–	–	–	–
40	0.998	–	–	–	–	–	–	–
60	–	–	–	–	–	–	–	–
80	–	–	–	–	–	–	–	–
∞	1	1	1	1	1	1	1	1

Key: "–" refers to the fact that Γ approaches unity.

TABLE 13.2 Correction Factor With $\phi_r = 3$ and $\sigma_H = 0.01$ (Gas Bubble) for a Solid Plane

	κa							
h	*0.01*	*0.1*	*0.5*	*1*	*2*	*5*	*7*	*10*
1.1	0.286	0.286	0.299	0.340	0.423	0.558	0.599	0.631
1.2	0.440	0.440	0.458	0.518	0.626	0.757	0.783	0.796
1.4	0.610	0.609	0.636	0.709	0.811	0.887	0.892	0.892
1.6	0.703	0.701	0.734	0.807	0.888	0.929	0.930	0.931
1.8	0.761	0.759	0.796	0.864	0.926	0.950	0.951	0.953
2	0.801	0.799	0.838	0.899	0.948	0.963	0.964	0.967
4	0.927	0.927	0.967	0.986	0.993	0.994	0.995	0.999
6	0.956	0.958	0.990	0.996	0.997	0.997	0.999	–
8	0.969	0.973	0.996	0.998	0.998	0.998	0.999	–
10	0.977	0.982	0.998	0.999	0.999	0.999	–	–
20	0.991	0.997	–	–	–	–	–	–
40	0.997	–	–	–	–	–	–	–
60	–	–	–	–	–	–	–	–
80	–	–	–	–	–	–	–	–
∞	1	1	1	1	1	1	1	1

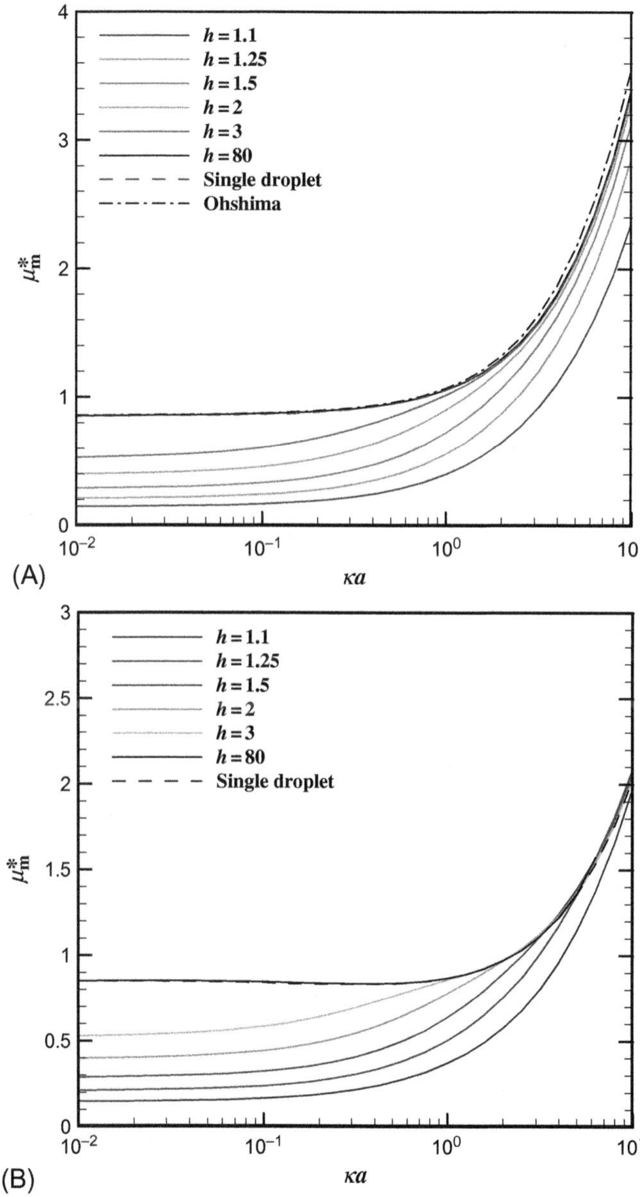

FIG. 13.5 Scaled mobility μ_m^* as a function of κa at different h. (A) $\phi_r=1$ and $\sigma_H=0.5$; (B) $\phi_r=3$ and $\sigma_H=0.5$; (C) $\phi_r=5$ and $\sigma_H=0.5$; (D) $\phi_r=1$ and $\sigma_H=0.01$; (E) $\phi_r=3$ and $\sigma_H=0.01$; (F) $\phi_r=5$ and $\sigma_H=0.01$.

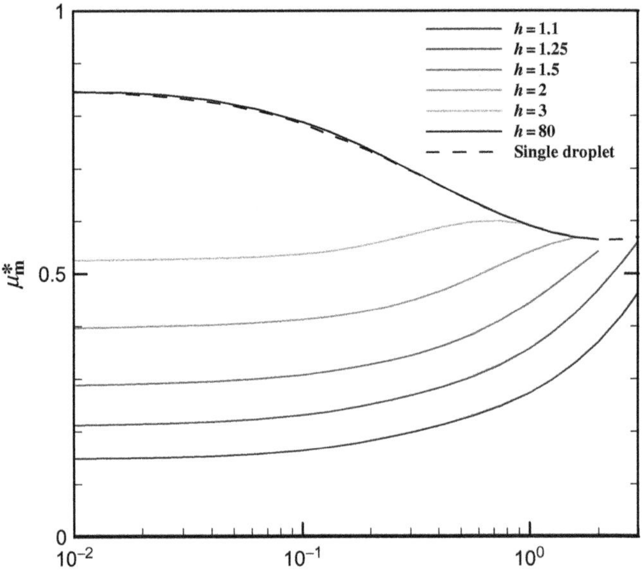

(C)

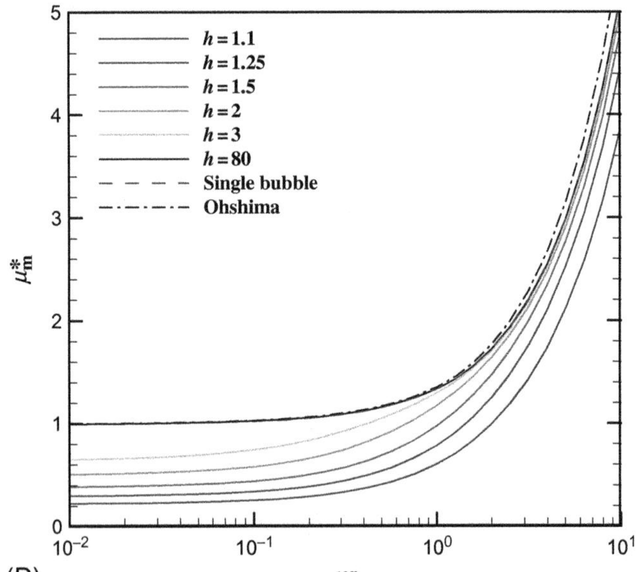

(D)

FIG. 13.5—CONT'D

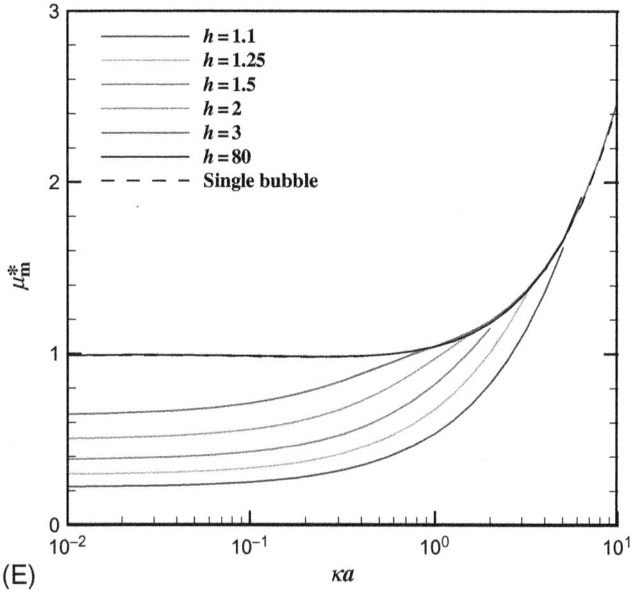

(E)

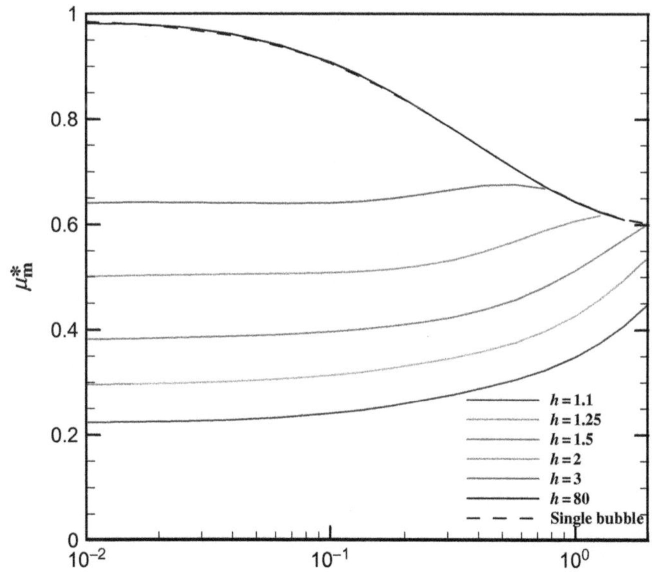

(F)

FIG. 13.5—CONT'D

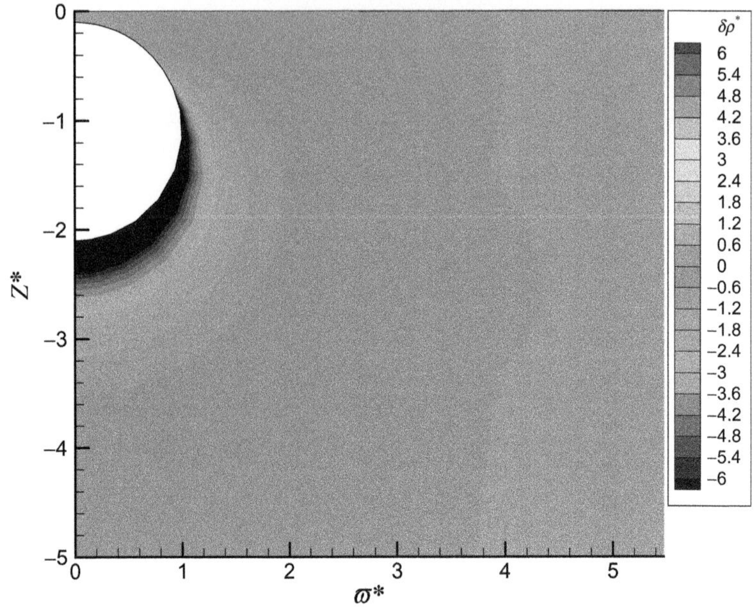

FIG. 13.6 Contour plot of the scaled disturbed charge density ($\delta\rho^*$) for $\kappa a=1$, $\phi_r=5$, $h=1.1$, and $\sigma_H=0.5$.

TABLE 13.3 Correction Factor With $\phi_r=3$ and $\sigma_H=0.5$ (Liquid Droplet) for an Air-Water Interface

	κa					
h	*0.01*	*0.1*	*0.5*	*1*	2	5
1.1	0.175	0.198	0.312	0.429	0.595	0.844
1.2	0.229	0.259	0.400	0.540	0.730	0.970
1.4	0.308	0.348	0.522	0.684	0.879	–
1.6	0.371	0.418	0.613	0.782	0.950	–
1.8	0.424	0.477	0.686	0.851	0.984	–
2	0.469	0.527	0.745	0.900	0.998	–
4	0.710	0.790	0.970	–	–	–
6	0.805	0.889	0.999	–	–	–
8	0.856	0.937	–	–	–	–
10	0.887	0.963	–	–	–	–
20	0.952	0.999	–	–	–	–
40	0.985	–	–	–	–	–
60	0.995	–	–	–	–	–
80	–	–	–	–	–	–
∞	1	1	1	1	1	1

TABLE 13.4 Correction Factor With $\phi_r = 3$ and $\sigma_H = 0.01$ (Gas Bubble) for an Air-Water Interface

	κa					
h	*0.01*	*0.1*	*0.5*	*1*	*2*	*5*
1.1	0.227	0.254	0.384	0.514	0.699	0.976
1.2	0.279	0.312	0.463	0.613	0.815	–
1.4	0.356	0.398	0.576	0.742	0.941	–
1.6	0.417	0.465	0.660	0.830	0.996	–
1.8	0.467	0.520	0.727	0.892	–	–
2	0.510	0.567	0.781	0.933	–	–
4	0.734	0.809	0.981	–	–	–
6	0.821	0.899	–	–	–	–
8	0.867	0.943	–	–	–	–
10	0.895	0.967	–	–	–	–
20	0.954	0.999	–	–	–	–
40	0.984	–	–	–	–	–
60	0.994	–	–	–	–	–
80	0.998	–	–	–	–	–
∞	1	1	1	1	1	1

REFERENCES

[1] Göpel W, Heiduschka P. Interface analysis in biosensor design. Biosens Bioelectron 1995;10(9–10):853–83.

[2] Feng J, Ganatos P, Weinbaum S. Motion of a sphere near planar confining boundaries in a Brinkman medium. J Fluid Mech 1998;375:265–96.

[3] Barry B. Drug delivery routes in skin: a novel approach. Adv Drug Deliv Rev 2002;54: S31–40.

[4] Brown DG, Hong Y. Impact of the charge-regulated nature of the bacterial cell surface on the activity of adhered cells. J Adhes Sci Technol 2011;25(17):2199–218.

[5] Jiang P, et al. Single-crystal colloidal multilayers of controlled thickness. Chem Mater 1999;11(8):2132–40.

[6] Liang Y-E, Tsao H-K, Sheng Y-J. Drops on hydrophilic conical fibers: gravity effect and coexistent states. Langmuir 2015;31(5):1704–10.

[7] Hunter RJ. Foundations of colloid science. Oxford: Oxford University Press; 2001.

[8] Sottmann T, Stubenrauch C, Stubenrauch C. Microemulsions: background, new concepts, applications, perspectives. Wiley-Blackwell; 2009.

[9] Raffa P, Broekhuis AA, Picchioni F. Polymeric surfactants for enhanced oil recovery: a review. J Pet Sci Eng 2016;145:723–33.
[10] Happel J, Brenner H. Low Reynolds number hydrodynamics: with special applications to particulate media. vol. 1. New York: Springer Science & Business Media; 2012.
[11] Wacholder E, Weihs D. Slow motion of a fluid sphere in the vicinity of another sphere or a plane boundary. Chem Eng Sci 1972;27(10):1817–28.
[12] Grashchenkov S. The effect of slip on the motion of two droplets and of a droplet close to a plane surface of a liquid. Aerosol Sci Technol 1996;25(2):101–12.
[13] Bart E. The slow unsteady settling of a fluid sphere toward a flat fluid interface. Chem Eng Sci 1968;23(3):193–210.
[14] Liao C, et al. Electrophoretic motion of a liquid droplet and a bubble normal to an air–water interface. Colloids Surf A Physicochem Eng Asp 2016;507:124–33.
[15] Lee E, Lou S-H, Hsu J-P. Electrophoresis of a non-conducting Newtonian drop of low electrical potential normal to a plane. Chem Eng Sci 2006;61(14):4550–7.
[16] Dickinson E. Proteins at interfaces and in emulsions stability, rheology and interactions. J Chem Soc Faraday Trans 1998;94(12):1657–69.
[17] Shaw DJ. Introduction to colloid and surface chemistry. London: Butterworths; 1966.
[18] Park K. Controlled drug delivery: challenges and strategies. Washington, DC: American Chemical Society; 1997.
[19] Ohshima H. A simple expression for the electrophoretic mobility of charged mercury drops. J Colloid Interface Sci 1997;189(2):376–8.

Chapter 14

Electrophoretic Motion of a Liquid Droplet and a Gas Bubble in a Cylindrical Pore

14.1 INTRODUCTION

Microfluidic and nanofluidic device are often used to manipulate the motion of a liquid droplet in biomedical analysis or biochemical operations [1–8]. Synchronized with the fast advancement of fabrication technology in microfluidic and nanofluidic applications, the state-of-the-art development in the field has been the lab-on-a-chip (LOC) system with all relevant operations integrated in a small and preferably portable device [9–11], as discussed in Chapters 5 and 9 where rigid and porous particles were treated. On the other hand, the study of a liquid droplet in a cylindrical pore is also closely related to the capillary electrophoresis (CE) of a single cell, as a human or mammalian cell is often modeled as a liquid droplet [12–16].

For the liquid droplet in particular, which has been proposed as an ideal microreactor with various merits such as perfect mixing and temperature control, a microfluidic system has proved to be an efficient and practical way for large scale commercial operations in a parallel way [5]. This resort is particularly attractive in biomedical or biochemical applications like in pharmaceutical industry [17], where how to manufacture high quality products efficiently with high economic added value is the name of the game [6]. Crystallization of proteins, for instance, has been carried out successfully with the manipulation of liquid droplet in a microfluidic channel [18]. The reason for its success is based on the fact that a droplet with uniform size can be easily generated within a microfluidic apparatus, as we have discussed in Chapter 12 [5]. Surface tension force is utilized in this type of maneuver and the droplet produced is of uniform size with very low cost of energy, in comparison with the conventional high energy methods, such as mixing with the input of mechanical energy or ultrasonic wave, where the droplet thus formed has a

Interface Science and Technology, Vol. 26. https://doi.org/10.1016/B978-0-08-100865-2.00014-X

very wide distribution of size usually. Hence, it is difficult, if not impossible, to harness it for any promising high quality further maneuvers in practical applications [19]. The incorporation of a liquid droplet into a microfluidic system, which in fact can be regarded as a microchemical plant, becomes a very natural selection with great potential in commercial applications. As a result, the boundary effect of the channel wall has to be investigated carefully in the design and operations of such systems, which is the focus of this chapter.

To manipulate the motion of a liquid droplet in a microfluidic device, such as in a cylindrical pore, there have been two types of physical mechanisms to be used in general: the electrowetting and the dielectrophoresis [20, 21]. The former modifies the wetting properties of a surface with an applied electric field, whereas the latter is based on the driving force of a dielectric colloidal entity in response to an uneven electric field [20]. The electrowetting, which is also referred to as the electrowetting on a dielectric (EWOD), often results in surface contamination of the liquid droplet. The dielectrophoresis (DEP), on the other hand, often requires a higher voltage and the resulted motion for a liquid droplet is not as fast as the EWOD approach [21]. In contrast to the above two approaches, electrophoresis is not only free from surface contamination, but also simpler and straightforward in operations, among other merits [22]. Indeed, electrophoresis has great potential in the manipulation of liquid droplet in microfluidic and nanofluidic systems.

Experimental studies regarding this issue have been abundant in the literature [23–25]. Corresponding theoretical analysis, however, has been very sparse. Lee's group studied the electrophoretic motion of a liquid droplet in a cylindrical pore focusing on the boundary confinement effect as well as the impact of the electroosmosis flow (EOF) generated by a charged wall [26–28]. Here, in this chapter, the extension to a gas bubble as well as the overall impact of droplet viscosity is treated. As we have discussed in Chapter 12, the droplet mobility is order-of-magnitude larger than the corresponding rigid particle situation when the electrolyte strength is large ($\kappa a > 10$, roughly), which is mainly due to the mobile droplet surface. The corresponding boundary confinement effect upon a droplet thus will be different as well, especially when κa is large. This will be examined carefully from wide to narrow cylindrical pores. Relevant contour plots are demonstrated as well to explore the physical mechanism underlying the droplet or bubble behavior.

14.2 THEORY

The system diagram is shown in Fig. 14.1. The configuration is similar to Chapter 9 where a porous particle in a cylindrical pore is considered, and the governing equations outside the droplet as well as the associated boundary conditions at the wall surface are the same as there too. And so is the domain decomposition scheme as well as the mesh diagram in computation.

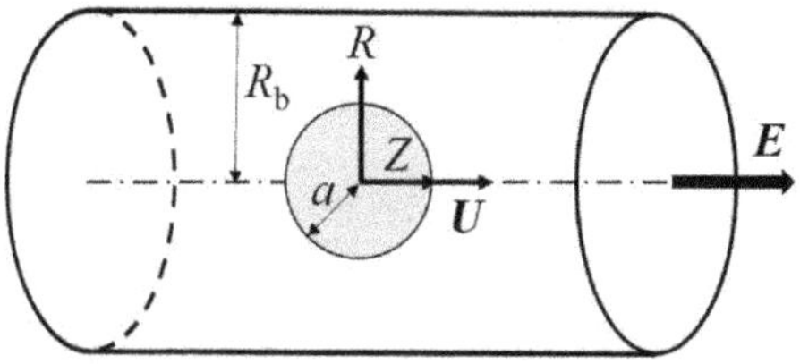

FIG. 14.1 System diagram.

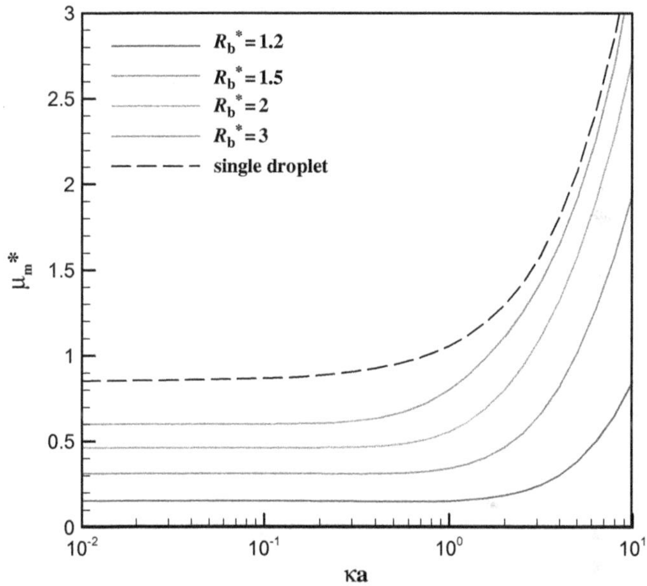

FIG. 14.2 Scaled dimensionless mobility μ_m^* versus κa at various R_b^* with $\phi_r=1$, $\phi_w=0$, and $\sigma_H=0.5$.

Within the droplet, which is assumed a nonpolar hydrophobic droplet containing no electrolyte ions, only the hydrodynamic equations in terms of dimensionless stream function Ψ^* are needed:

$$E^{*4}\Psi^*=0 \tag{14.1}$$

Boundary conditions on the droplet surface and the center, based on the same physical arguments as used in Chapter 12 for a single droplet, are the same as those adopted there too. The details can be found elsewhere [28].

14.3 RESULTS AND DISCUSSION

14.3.1 General Henry's Charts

Figs. 14.2–14.7 present the general Henry's charts for a liquid droplet ($\sigma_H=0.5$) and a gas bubble ($\sigma_H=0.01$), respectively, at various pore radii.

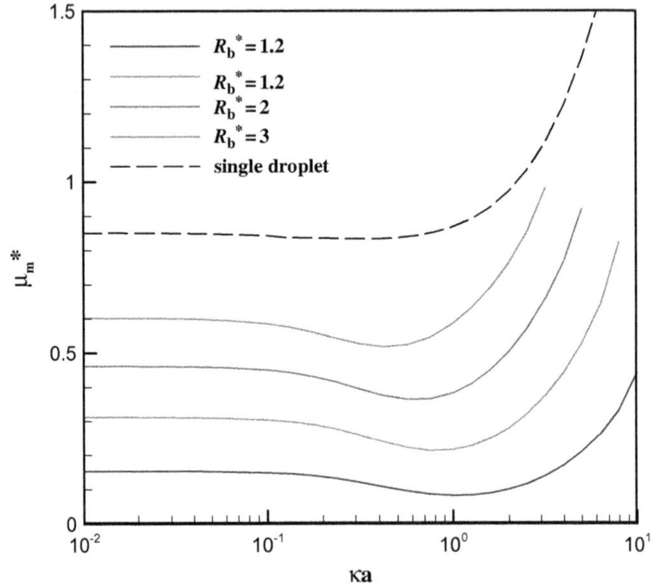

FIG. 14.3 Scaled dimensionless mobility μ_m^* versus κa at various R_b^* with $\phi_r = 3$, $\phi_w = 0$, and $\sigma_H = 0.5$.

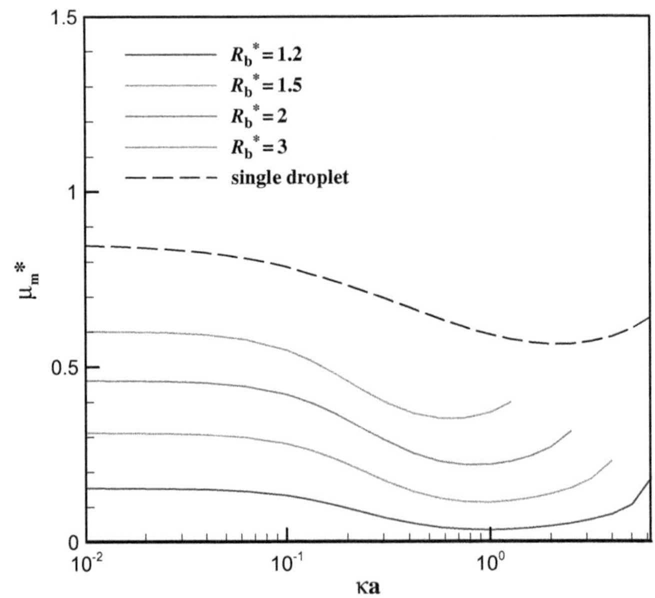

FIG. 14.4 Scaled dimensionless mobility μ_m^* versus κa at various R_b^* with $\phi_r = 5$, $\phi_w = 0$, and $\sigma_H = 0.5$.

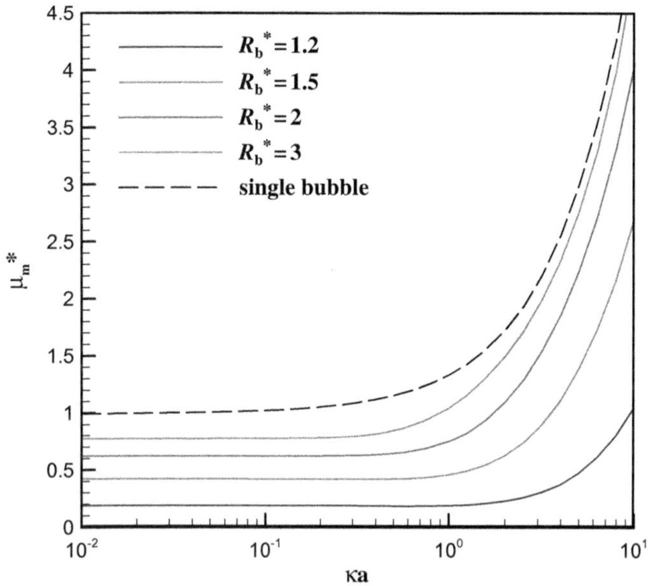

FIG. 14.5 Scaled dimensionless mobility μ_m^* versus κa at various R_b^* with $\phi_r=1$, $\phi_w=0$, and $\sigma_H=0.01$.

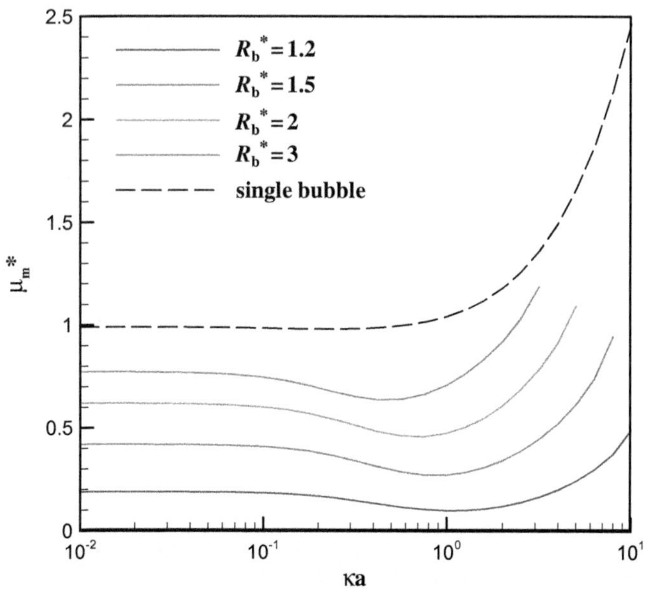

FIG. 14.6 Scaled dimensionless mobility μ_m^* versus κa at various R_b^* with $\phi_r=3$, $\phi_w=0$, and $\sigma_H=0.01$.

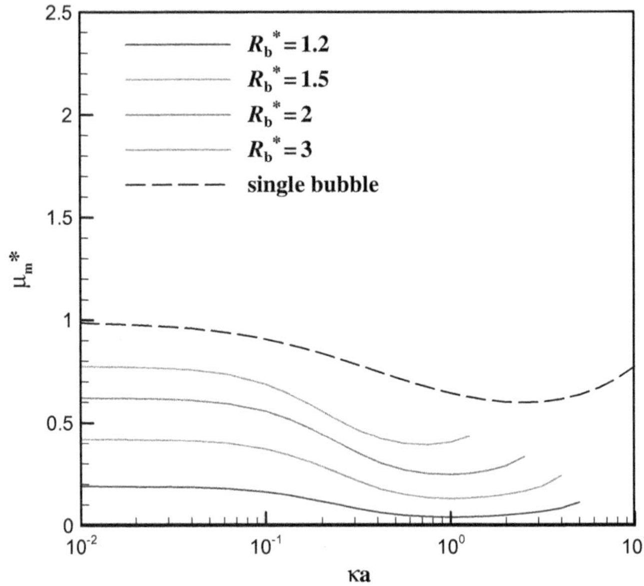

FIG. 14.7 Scaled dimensionless mobility μ_m^* versus κa at various R_b^* with $\phi_r=5$, $\phi_w=0$, and $\sigma_H=0.01$.

The reduction of particle mobility is very clear in comparison with the single droplet situation as shown. Polarization effect of the double layer for highly charged droplets or bubbles is essentially intact as the boundary effect is from the lateral direction where the polarization is hardly present anyway.

The mobility of a liquid droplet or a gas bubble certainly is larger than the corresponding rigid sphere when κa is large in general, as elaborated in Chapter 12. The presence of a nearby cylindrical pore does not affect this scenario much. The narrower the cylindrical pore is, the slower the particle moves. No weird behavior is observed as in the corresponding porous sphere situation as discussed in Chapter 9, where the particle may move faster in a narrower pore.

14.3.2 Contour Plots for a Moving Liquid Droplet or a Gas Bubble

It should be noted that no electrostatic results need to be presented here as they should be exactly the same as those in Chapter 5 for a rigid particle, since there is no difference between a motionless droplet/bubble and a motionless rigid particle, as both are impermeable and carry electric charges on their surfaces. Figs. 14.8 and 14.9, hence, show the contour plots of double layer polarization and corresponding flow field for a liquid droplet ($\sigma_H=0.5$) and a gas bubble ($\sigma_H=0.01$), respectively. Other than the difference in magnitude and the axisymmetric interior vortex flows within the droplet or bubble, no surprise is found here compared with the rigid particle situation discussed in Chapter 5.

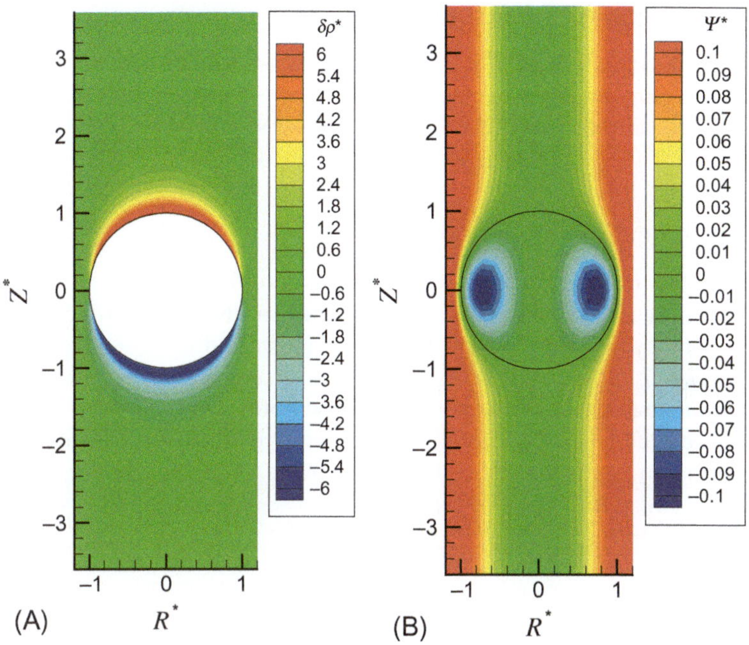

FIG. 14.8 Contour plots of different scaled variables at $\kappa a = 1$, $R_b^* = 1.2$, $\phi_r = 3$, $\phi_w = 0$, and $\sigma_H = 0.5$. (A) Scaled disturbed charge density ($\delta\rho^*$); (B) scaled stream function (Ψ^*).

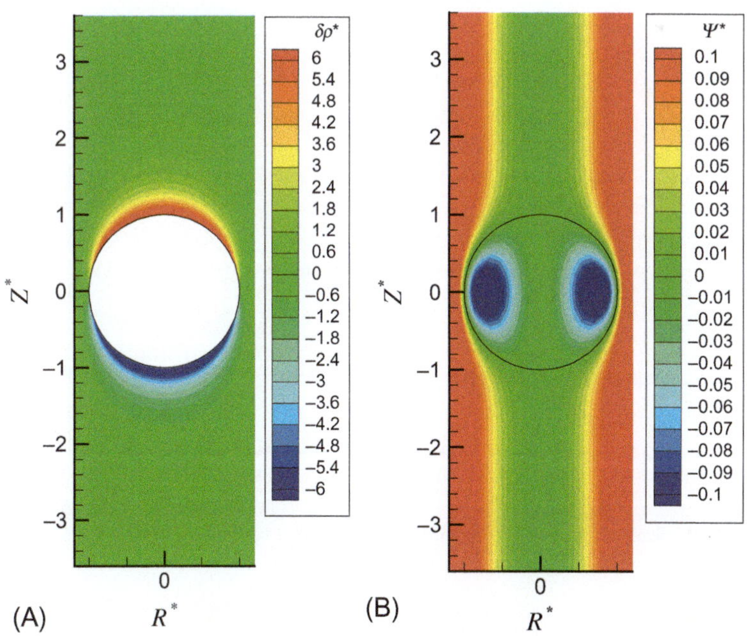

FIG. 14.9 Contour plots of different scaled variables at $\kappa a = 1$, $R_b^* = 1.2$, $\phi_r = 3$, $\phi_w = 0$, and $\sigma_H = 0.01$. (A) Scaled disturbed charge density ($\delta\rho^*$); (B) scaled stream function (Ψ^*).

14.3.3 Effect of EOF of a Charged Wall

Fig. 14.10 depicts the impact of EOF upon the liquid droplet ($\sigma_H = 0.5$) motion. Again, no surprise here. The EOF may enhance or retard the droplet motion depending on the sign of electric charges they carry, as discussed in Chapter 5. The overall scenario is also similar to the rigid particle situation.

14.3.4 Effect of Viscosity Ratio

Perhaps the most insightful information regarding the droplet motion is shown in Fig. 14.11 where the viscosity effect of the droplet fluid is examined. Note, $\sigma_H = 1.7$ there corresponds to a mercury drop. The more viscous the droplet fluid, the smaller its mobility is throughout the κa range examined. Note that when $\kappa a = 100$, the mobility profile reduces to that of a rigid particle discussed in Chapter 5. The presence of the pore wall retards its motion significantly in general and the double layer polarization effect is essentially intact in terms of local minima observed around $\kappa a = 1$ in those mobility profiles.

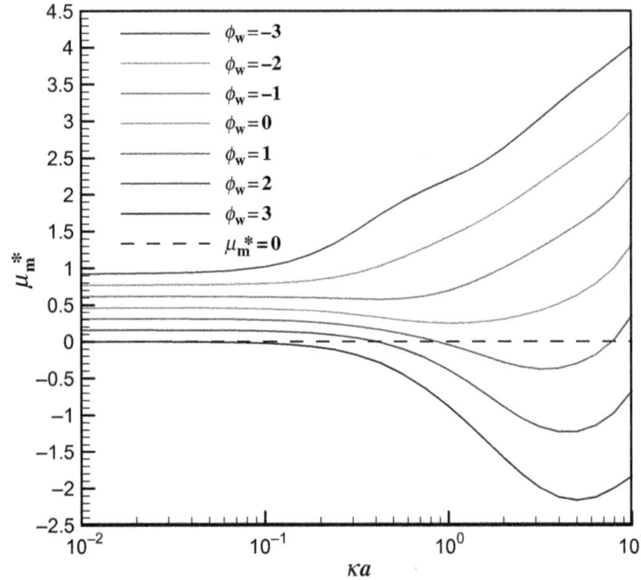

FIG. 14.10 Scaled dimensionless mobility μ_m^* versus κa at various ϕ_w with $R_b^* = 1.2$, $\phi_r = 3$, and $\sigma_H = 0.5$.

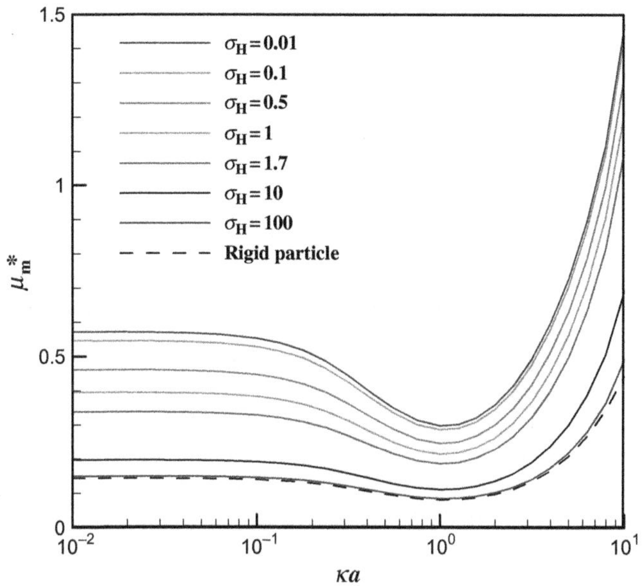

FIG. 14.11 Scaled dimensionless mobility μ_m^* versus κa at various σ_H with $R_b^* = 1.2$, $\phi_r = 3$, and $\phi_w = 0$.

REFERENCES

[1] Gu H, Duits MH, Mugele F. Droplets formation and merging in two-phase flow microfluidics. Int J Mol Sci 2011;12(4):2572–97.

[2] Günther A, Khan SA, Thalmann M, Trachsel F, Jensen KF. Transport and reaction in microscale segmented gas–liquid flow. Lab Chip 2004;4(4):278–86.

[3] Linder V, Sia SK, Whitesides GM. Reagent-loaded cartridges for valveless and automated fluid delivery in microfluidic devices. Anal Chem 2005;77(1):64–71.

[4] Sen AK, Sajeesh P. Electrokinetic transport and separation of droplets in a microchannel. Microfluid Nanofluid 2014;17(1):97–106.

[5] Shui L, Eijkel JC, van den Berg A. Multiphase flow in microfluidic systems—control and applications of droplets and interfaces. Adv Colloid Interf Sci 2007;133(1):35–49.

[6] Teh S-Y, Lin R, Hung L-H, Lee AP. Droplet microfluidics. Lab Chip 2008;8(2):198–220.

[7] Utada A, Lorenceau E, Link D, Kaplan P, Stone H, Weitz D. Monodisperse double emulsions generated from a microcapillary device. Science 2005;308(5721):537–41.

[8] Zhang M, Gong X, Wen W. Manipulation of microfluidic droplets by electrorheological fluid. Electrophoresis 2009;30(18):3116–23.

[9] van den Berg A, Lammerink TS. Micro total analysis systems: microfluidic aspects, integration concept and applications. In: Microsystem technology in chemistry and life science. Berlin, Heidelberg: Springer Verlag; 1998. p. 21–49.

[10] Haeberle S, Zengerle R. Microfluidic platforms for lab-on-a-chip applications. Lab Chip 2007;7(9):1094–110.

[11] McEvoy E, Marsh A, Altria K, Donegan S, Power J. Recent advances in the development and application of microemulsion EKC. Electrophoresis 2007;28(1–2):193–207.

[12] Redman EA, Mellors JS, Starkey JA, Ramsey JM. Characterization of intact antibody drug conjugate variants using microfluidic capillary electrophoresis–mass spectrometry. Anal Chem 2016;88(4):2220–6.

[13] Liu J-X, Aerts JT, Rubakhin SS, Zhang X-X, Sweedler JV. Analysis of endogenous nucleotides by single cell capillary electrophoresis-mass spectrometry. Analyst 2014;139(22):5835–42.

[14] Aerts JT, Louis KR, Crandall SR, Govindaiah G, Cox CL, Sweedler JV. Patch clamp electrophysiology and capillary electrophoresis–mass spectrometry metabolomics for single cell characterization. Anal Chem 2014;86(6):3203–8.

[15] Woods LA, Roddy TP, Ewing AG. Capillary electrophoresis of single mammalian cells. Electrophoresis 2004;25(9):1181–7.

[16] Alhusban AA, Gaudry AJ, Breadmore MC, Gueven N, Guijt RM. On-line sequential injection-capillary electrophoresis for near-real-time monitoring of extracellular lactate in cell culture flasks. J Chromatogr A 2014;1323:157–62.

[17] Wang L, Li X, Zhang G, Dong J, Eastoe J. Oil-in-water nanoemulsions for pesticide formulations. J Colloid Interface Sci 2007;314(1):230–5.

[18] Zheng B, Roach LS, Ismagilov RF. Screening of protein crystallization conditions on a microfluidic chip using nanoliter-size droplets. J Am Chem Soc 2003;125(37):11170–1.

[19] Christopher GF, Anna SL. Microfluidic methods for generating continuous droplet streams. J Phys D Appl Phys 2007;40(19):R319.

[20] Mukhopadhyay R. Diving into droplets. Washington, DC: ACS Publications; 2006.

[21] Hunt TP, Issadore D, Westervelt RM. Integrated circuit/microfluidic chip to programmably trap and move cells and droplets with dielectrophoresis. Lab Chip 2008;8(1):81–7.

[22] Im DJ, Noh J, Moon D, Kang IS. Electrophoresis of a charged droplet in a dielectric liquid for droplet actuation. Anal Chem 2011;83(13):5168–74.

[23] Arora A, Simone G, Salieb-Beugelaar GB, Kim JT, Manz A. Latest developments in micro total analysis systems. Anal Chem 2010;82(12):4830–47.

[24] Salieb-Beugelaar GB, Simone G, Arora A, Philippi A, Manz A. Latest developments in microfluidic cell biology and analysis systems. Anal Chem 2010;82(12):4848–64.

[25] Huebner A, Sharma S, Srisa-Art M, Hollfelder F, Edel JB. Microdroplets: a sea of applications? Lab Chip 2008;8(8):1244–54.

[26] Hsu J-P, Chen Z-S. Electrophoresis of a sphere along the axis of a cylindrical pore: effects of double-layer polarization and electroosmotic flow. Langmuir 2007;23(11):6198–204.

[27] Huang CH, Cheng WL, He YY, Lee E. Electrophoresis of a soft particle within a cylindrical pore: polarization effect with the nonlinear Poisson-Boltzmann equation. J Phys Chem B 2010;114(31):10114–25.

[28] Huang C-H, Lee E. Electrophoretic motion of a liquid droplet in a cylindrical pore. J Phys Chem C 2012;116(28):15058–67.

Chapter 15

Dynamic Electrophoresis of Liquid Droplets and Gas Bubbles

15.1 INTRODUCTION

The conventional static (DC) electrophoresis of droplet or bubble systems was discussed in Chapter 12. In practice, dynamic electrophoresis certainly can also be used in the droplet or bubble systems as well [1, 2]. As for the theoretical investigation, Lee's group explored the droplet suspension systems, either dilute or concentrated [3, 4]. Kuwabara's unit cell model was used there coupled with Shilov and Zharkikh (S-Z) boundary conditions [5, 6]. Double layer polarization effect was fully taken into account in these approaches.

In addition to the above direct analyses of classic droplet or bubble systems, dynamic electrophoresis has also been used in the dielectrophoresis (DEP) operations of living cells [7, 8]. The DEP operations are often carried out with AC electric field. Due to the intentionally designed uneven distribution of electric field around the dielectric colloidal entity, a net electric driving force is generated accordingly, which can be used in the separation or trapping of cells with corresponding ingenious arrangement of electrodes layout [9]. As cells such as human stem cells or general mammalian cells can often be modeled as liquid droplets [10–15], and the DEP/AC is actually the corresponding dynamic electrophoresis under uneven electric field, the analysis here in this chapter for the conventional dynamic electrophoresis will provide insight and relevant information there as well.

In general, due to the fluid flow inside the droplet, the drag force at the interface between the droplet and the solution is much less than that of a rigid particle. This suggests a higher dynamic electrophoretic mobility in liquid droplet systems. More importantly, as we elaborated in Chapter 12, the droplet surface is mobile, and thus, carries electric charges with it. This results in an extra convective migration of counterions near the droplet surface and in the opposite direction of particle motion; hence, an extra electroosmosis flow

Interface Science and Technology, Vol. 26. https://doi.org/10.1016/B978-0-08-100865-2.00015-1

is generated relative to the moving droplet. This mechanism greatly reduces the hydrodynamic drag force of the particle leading to a large particle speed in DC electrophoresis when κa is large, which implies the charge density is very high near the particle. And so is the scenario in corresponding droplet AC electrophoresis here under the same token. As a result, large mobility is expected here as well, like we observed in Chapter 10 for the porous particle systems. However, as the droplet is impermeable after all, the dynamic mobility is less than the porous particle situation due to the relatively higher hydrodynamic drag force here after all.

15.2 THEORY

The system diagrams are shown in Fig. 15.1A and B for a single liquid droplet or a gas bubble and their suspensions, respectively.

The treatment here is similar to that in Chapter 6 except with an additional domain inside the droplet to be reckoned with. Outside the droplet, both the governing equations and the associated boundary conditions, either at infinity or on the outer virtual surface in a cell model, are exactly the same as in

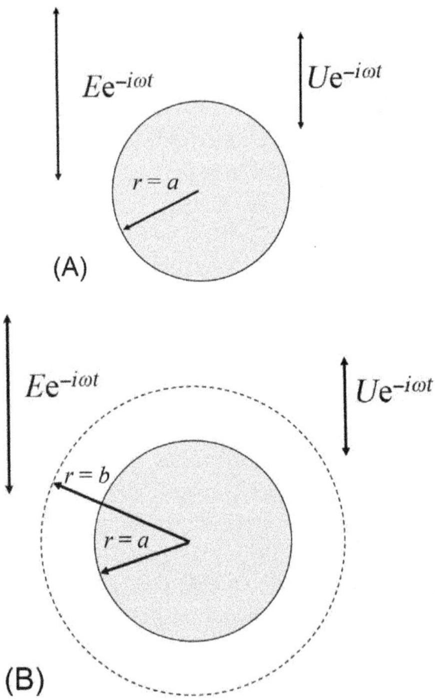

FIG. 15.1 System diagram. (A) Single liquid droplet or a gas bubble; (B) liquid droplet or gas bubble suspensions.

Chapter 6. Moreover, equilibrium state solutions are the same as those in Chapter 12 for static DC electrophoresis.

Note that the system here is also similar to that in Chapter 10 for a porous particle. Thus, all we need to do here is to replace the porous sphere treated there by a droplet. As the droplet treated here is a nonpolar hydrophobic one containing no electrolyte ions, only the hydrodynamic Stokes equation needs to be considered. It is, in dimensionless form,

$$E^{*4}\Psi^* = 0, \quad 0 \leq r^* \leq 1 \tag{15.1}$$

Associated boundary conditions on the droplet surface are based on the Rybczynski and Hadamard condition and assume the same form as those in Chapter 12 for a single droplet, after cancellation of the time-dependent factors $e^{-i\omega t}$ on both sides of the equations.

The subsequent procedure of particle mobility evaluation is the same as that in Chapter 6. And we are done.

15.3 RESULTS AND DISCUSSION

15.3.1 Single Liquid Droplet and Gas Bubble

Spectroscopy diagrams for a representative droplet ($\sigma_H = 0.5$) and a gas bubble ($\sigma_H = 0.01$) are presented for various levels of charge condition (ϕ_r) and double layer thickness (κa).

For thick double ($\kappa a = 0.1$), the magnitude of the dynamic mobility in dimensionless form is independent of the surface potential at high frequency of the AC current (Fig. 15.2). This Smoluchowski-type behavior is the most ideal situation for experimental operations aiming at determining the droplet surface potential, as we discussed in Chapter 6 for a rigid particle. For $\kappa a = 1$, similar behavior is observed as those considered in Chapters 6 and 10 for rigid and porous particles as well (Fig. 15.3). The magnitude of mobility gets even larger for a thin double layer as shown in Fig. 15.4 with $\kappa a = 10$. In general, the larger the κa value is, the faster the droplet moves, as discussed in Chapter 12 for the static DC electrophoresis. This scenario does not change in corresponding AC dynamic electrophoresis.

Corresponding diagrams for a gas bubble ($\sigma_H = 0.01$) exhibit similar behavior as the liquid droplet considered above, except even faster motion is observed in general due to the smaller drag force, as shown in Figs. 15.5–15.7. The impact of droplet viscosity is shown in Fig. 15.8, where the magnitude and the phase angle of the dynamic mobility profiles are shown at various levels of viscosity ratio (σ_H) for a representative highly charged droplet ($\phi_r = 3$) with double layer thickness comparable to the droplet surface ($\kappa a = 1$). In general, the less the ratio (σ_H) is, the faster the droplet moves due to the less significant hydrodynamic drag force. No surprising behavior is noticed here.

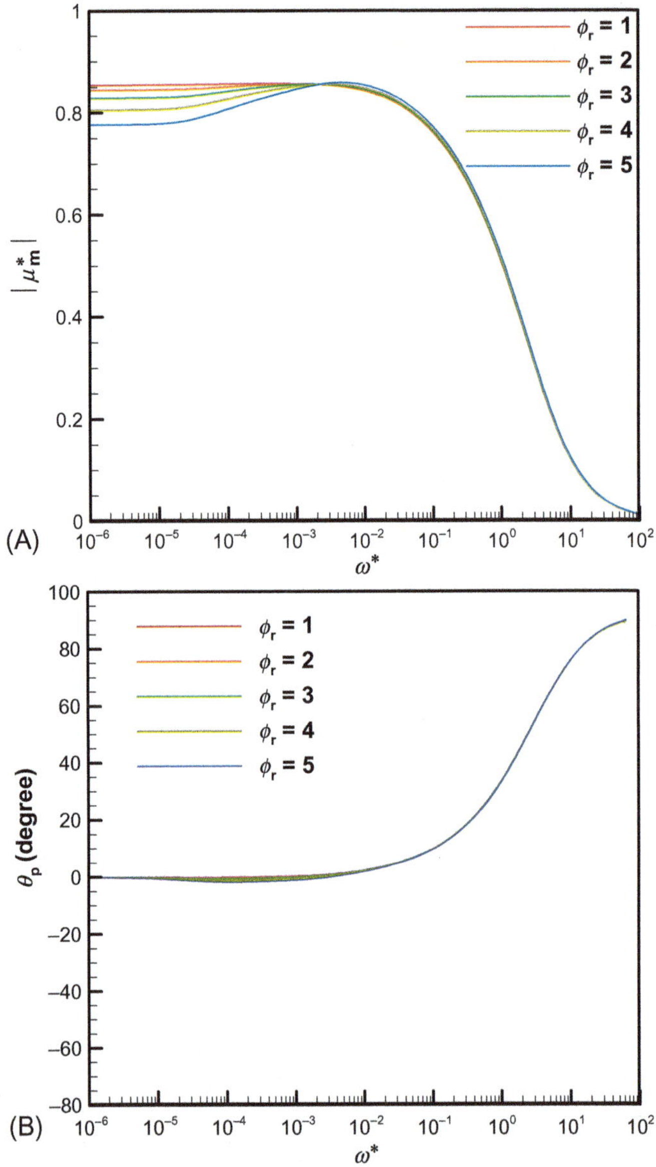

FIG. 15.2 (A) Magnitude of the scaled dimensionless mobility $|\mu_m^*|$ versus the scaled frequency ω^* at various ϕ_r with $\kappa a = 0.1$ and σ_H. (B) Phase angle of the scaled dimensionless mobility θ_p versus the scaled frequency ω^* at various ϕ_r with $\kappa a = 0.1$ and $\sigma_H = 0.5$.

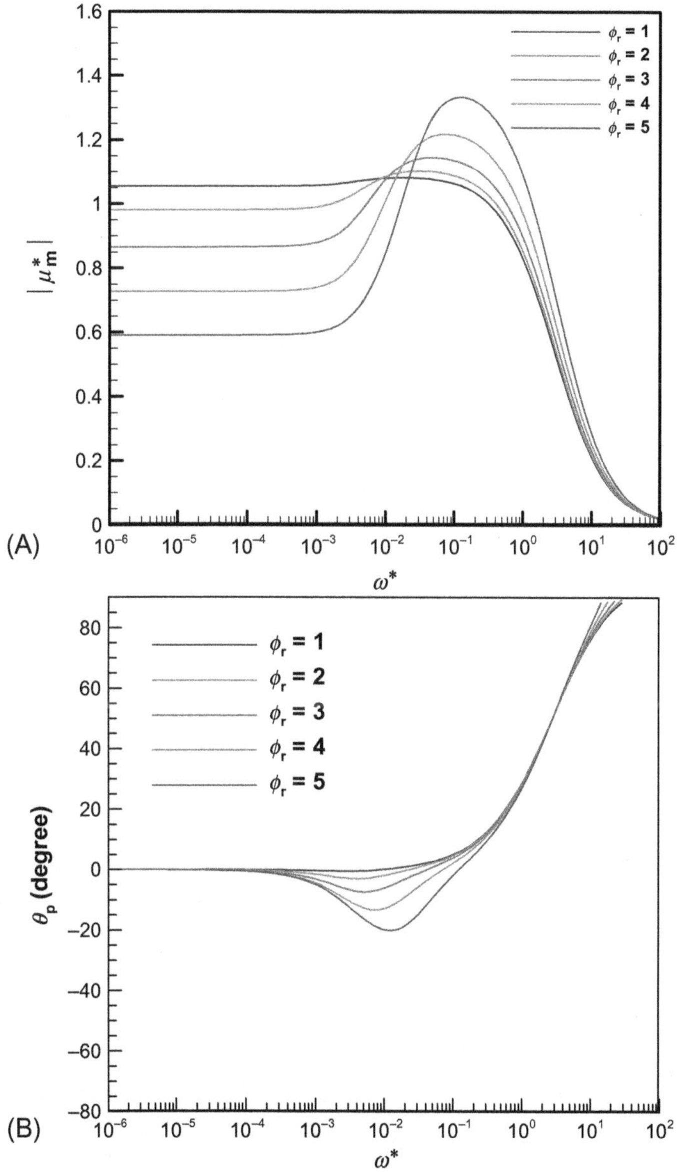

FIG. 15.3 (A) Magnitude of the scaled dimensionless mobility $|\mu_m^*|$ versus the scaled frequency ω^* at various ϕ_r with $\kappa a=1$ and $\sigma_H=0.5$. (B) Phase angle of the scaled dimensionless mobility θ_p versus the scaled frequency ω^* at various ϕ_r with $\kappa a=1$ and $\sigma_H=0.5$.

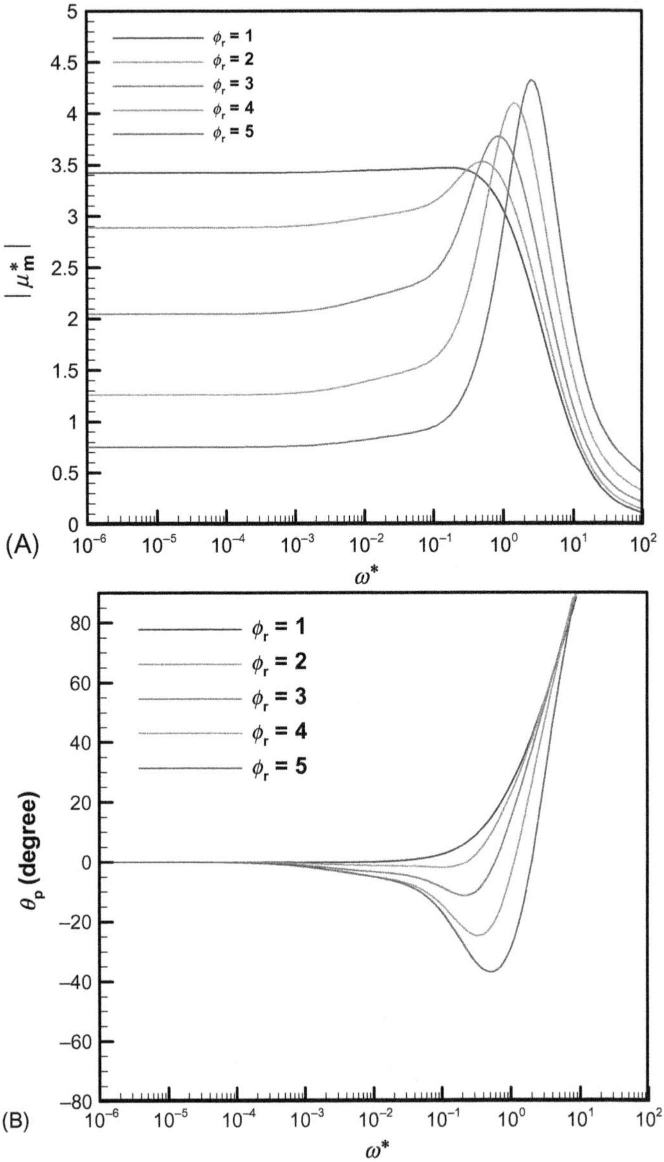

FIG. 15.4 (A) Magnitude of the scaled dimensionless mobility $|\mu_m^*|$ versus the scaled frequency ω^* at various ϕ_r with $\kappa a = 10$ and $\sigma_H = 0.5$. (B) Phase angle of the scaled dimensionless mobility θ_p versus the scaled frequency ω^* at various ϕ_r with $\kappa a = 10$ and $\sigma_H = 0.5$.

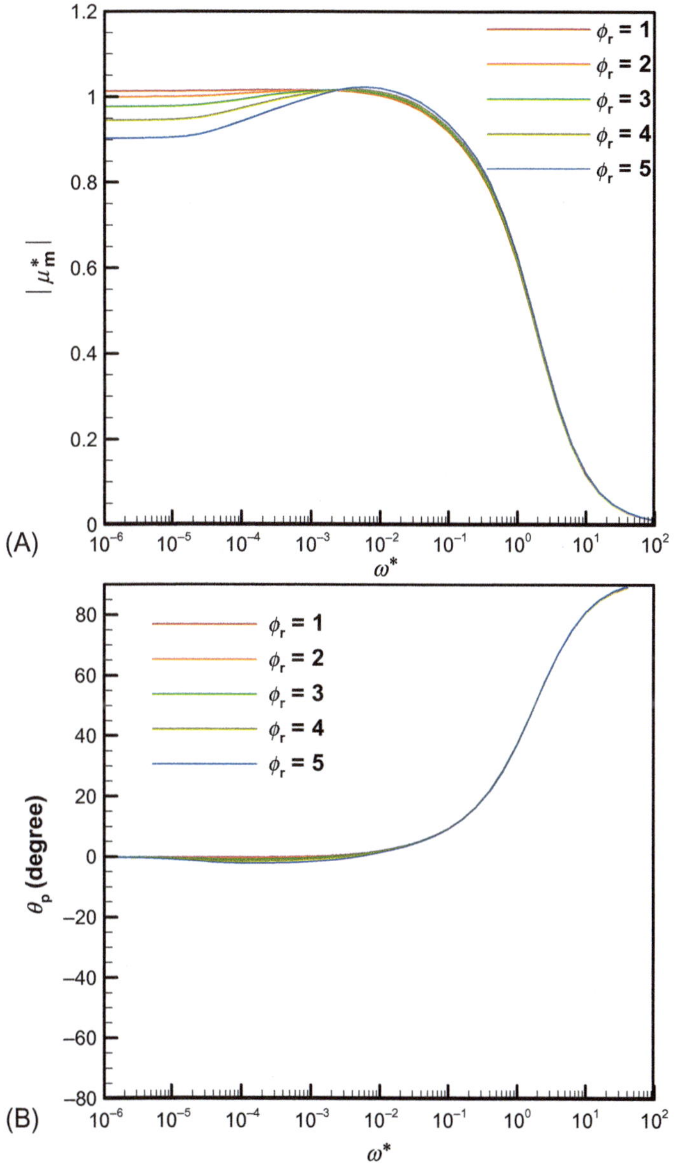

FIG. 15.5 (A) Magnitude of the scaled dimensionless mobility $|\mu_m^*|$ versus the scaled frequency ω^* at various ϕ_r with $\kappa a=0.1$ and $\sigma_H=0.01$. (B) Phase angle of the scaled dimensionless mobility θ_p versus the scaled frequency ω^* at various ϕ_r with $\kappa a=0.1$ and $\sigma_H=0.01$.

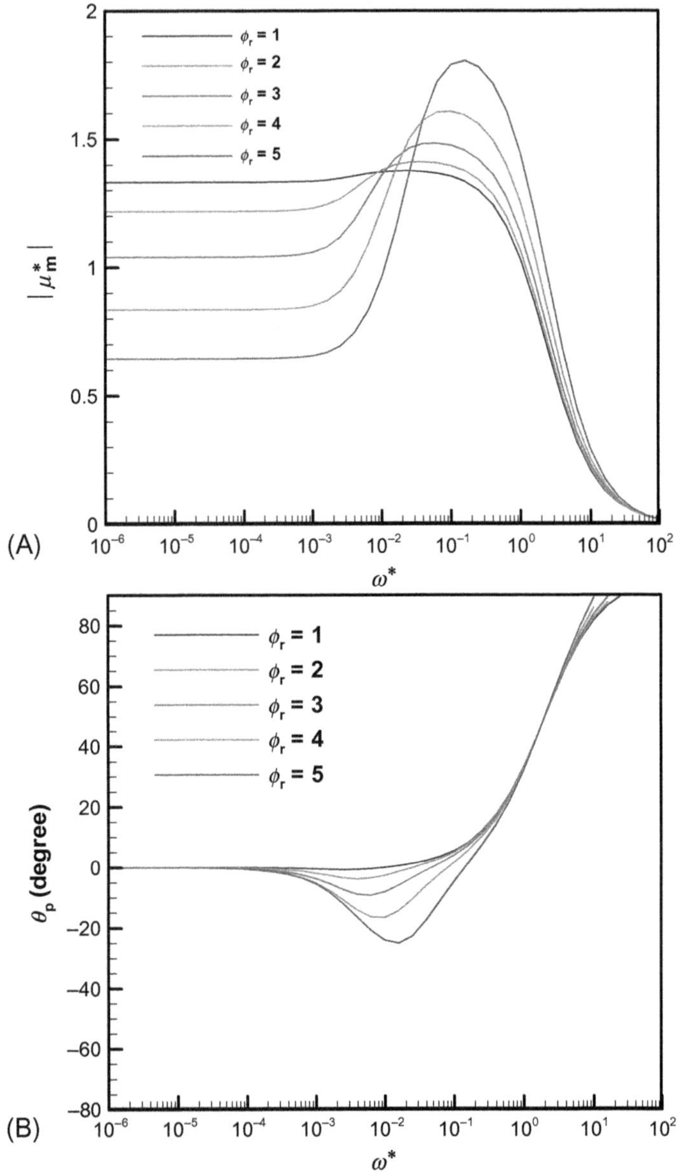

FIG. 15.6 (A) Magnitude of the scaled dimensionless mobility $|\mu_m^*|$ versus the scaled frequency ω^* at various ϕ_r with $\kappa a = 1$ and $\sigma_H = 0.01$. (B) Phase angle of the scaled dimensionless mobility θ_p versus the scaled frequency ω^* at various ϕ_r with $\kappa a = 1$ and $\sigma_H = 0.01$.

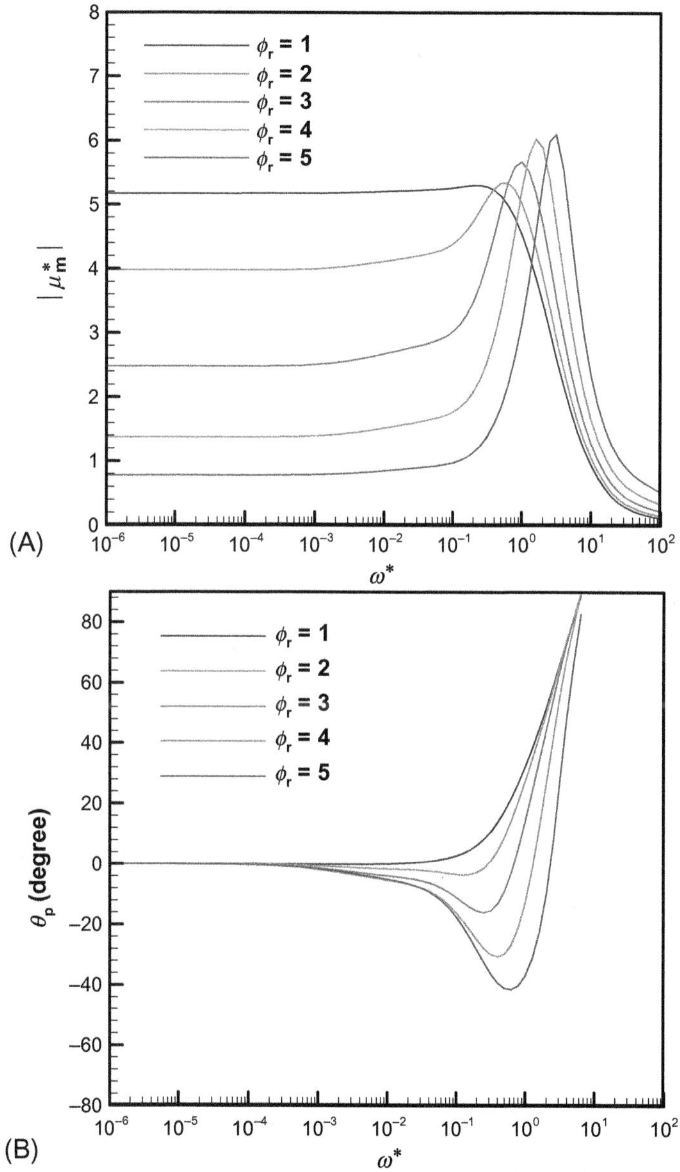

FIG. 15.7 (A) Magnitude of the scaled dimensionless mobility $|\mu_m^*|$ versus the scaled frequency ω^* at various ϕ_r with $\kappa a = 10$ and $\sigma_H = 0.01$. (B) Phase angle of the scaled dimensionless mobility θ_p versus the scaled frequency ω^* at various ϕ_r with $\kappa a = 10$ and $\sigma_H = 0.01$.

15.3.2 Suspensions of Liquid Droplets and Gas Bubbles

For droplet or bubble suspensions, the magnitude of the dynamic mobility decreases with increasing particle concentration level in general, as shown in Figs. 15.9–15.11 for a liquid droplet ($\sigma_H = 0.5$), and Figs. 15.12–15.14 for

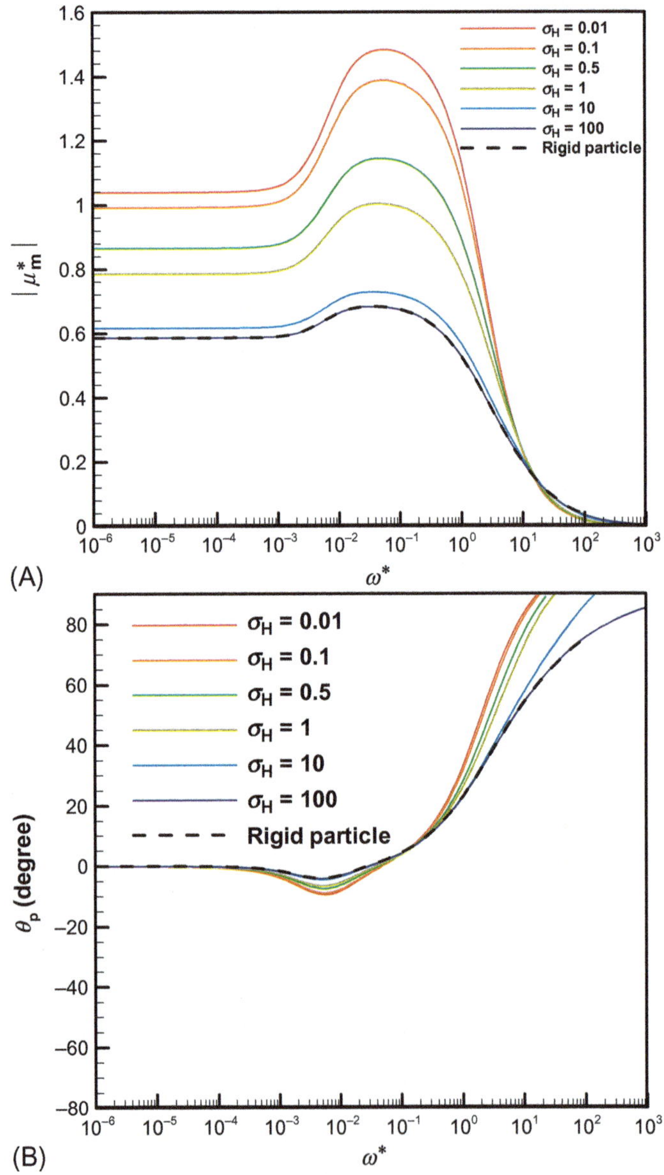

FIG. 15.8 (A) Magnitude of the scaled dimensionless mobility $|\mu_m^*|$ versus the scaled frequency ω^* at various σ_H with $\kappa a = 1$ and $\phi_r = 3$. (B) Phase angle of the scaled dimensionless mobility θ_p versus the scaled frequency ω^* at various σ_H with $\kappa a = 1$ and $\phi_r = 3$.

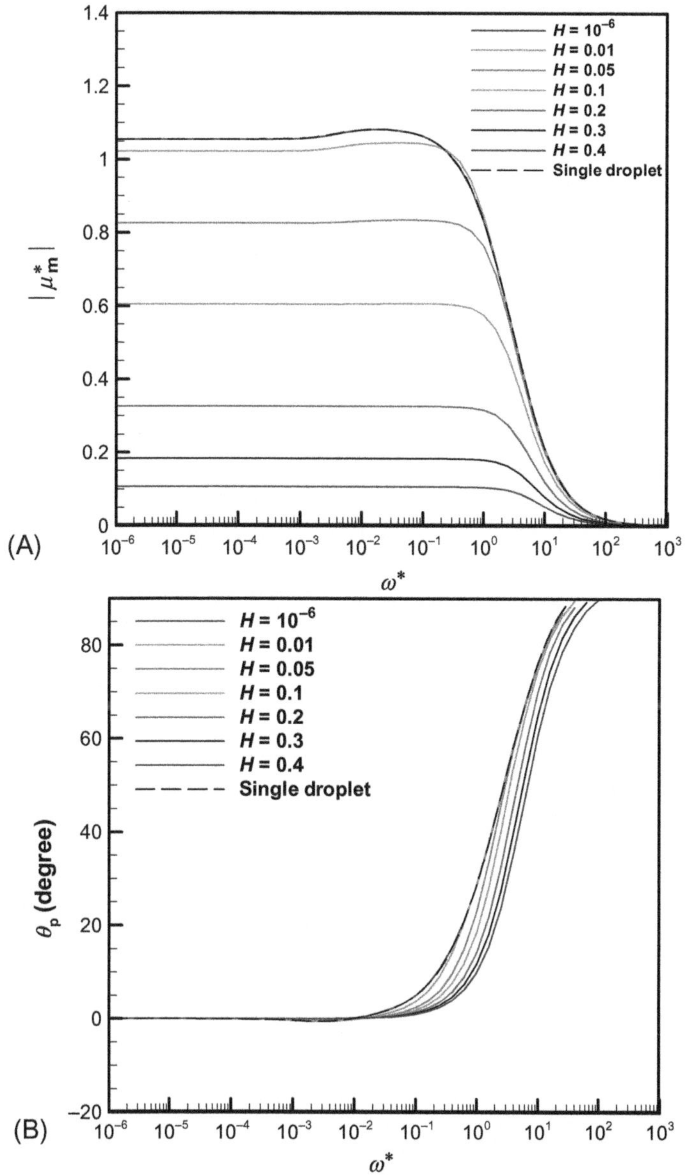

FIG. 15.9 (A) Magnitude of the scaled dimensionless mobility $|\mu_m^*|$ versus the scaled frequency ω^* at various volume fraction H with $\kappa a = 1$, $\sigma_H = 0.5$, and $\phi_r = 1$. (B) Phase angle of the scaled dimensionless mobility θ_p versus the scaled frequency ω^* at various volume fraction H with $\kappa a = 1$, $\sigma_H = 0.5$, and $\phi_r = 1$.

a gas bubble ($\sigma_H = 0.01$), respectively. The overall behavior is quite similar to the corresponding situation in Chapter 6, as both systems are impermeable to the fluid flow and the electrolyte ions. The mobility magnitude here is larger than the corresponding rigid particle situation in general.

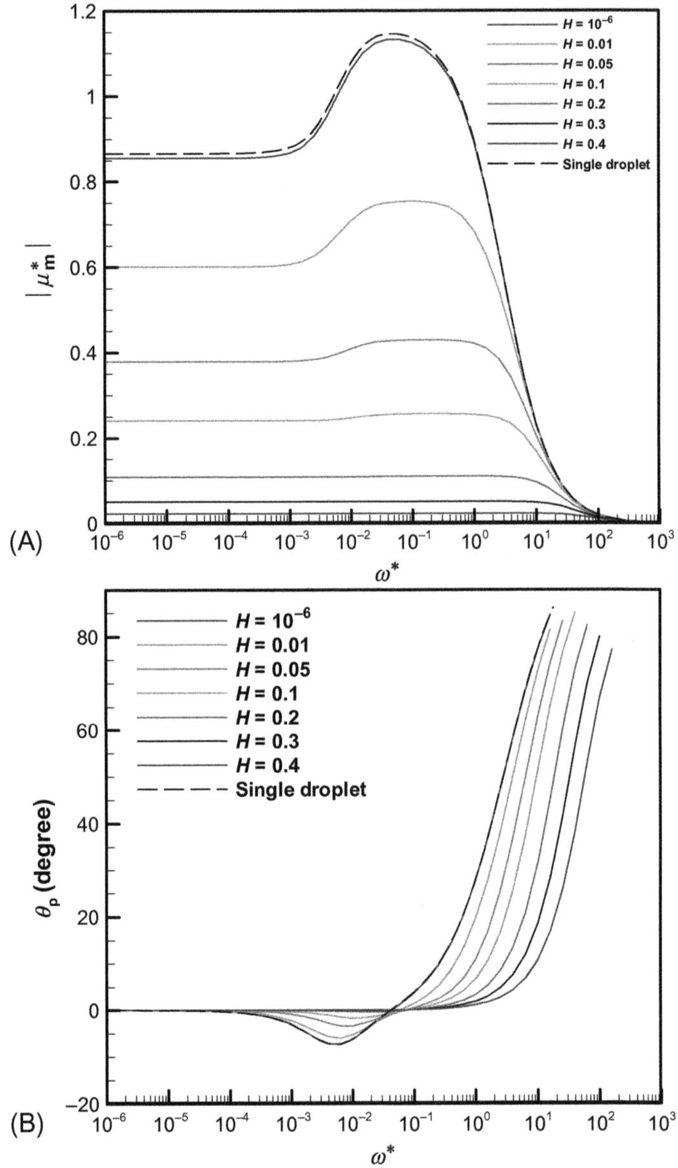

FIG. 15.10 (A) Magnitude of the scaled dimensionless mobility $|\mu_m^*|$ versus the scaled frequency ω^* at various volume fraction H with $\kappa a=1$, $\sigma_H=0.5$, and $\phi_r=3$. (B) Phase angle of the scaled dimensionless mobility θ_p versus the scaled frequency ω^* at various volume fraction H with $\kappa a=1$, $\sigma_H=0.5$, and $\phi_r=3$.

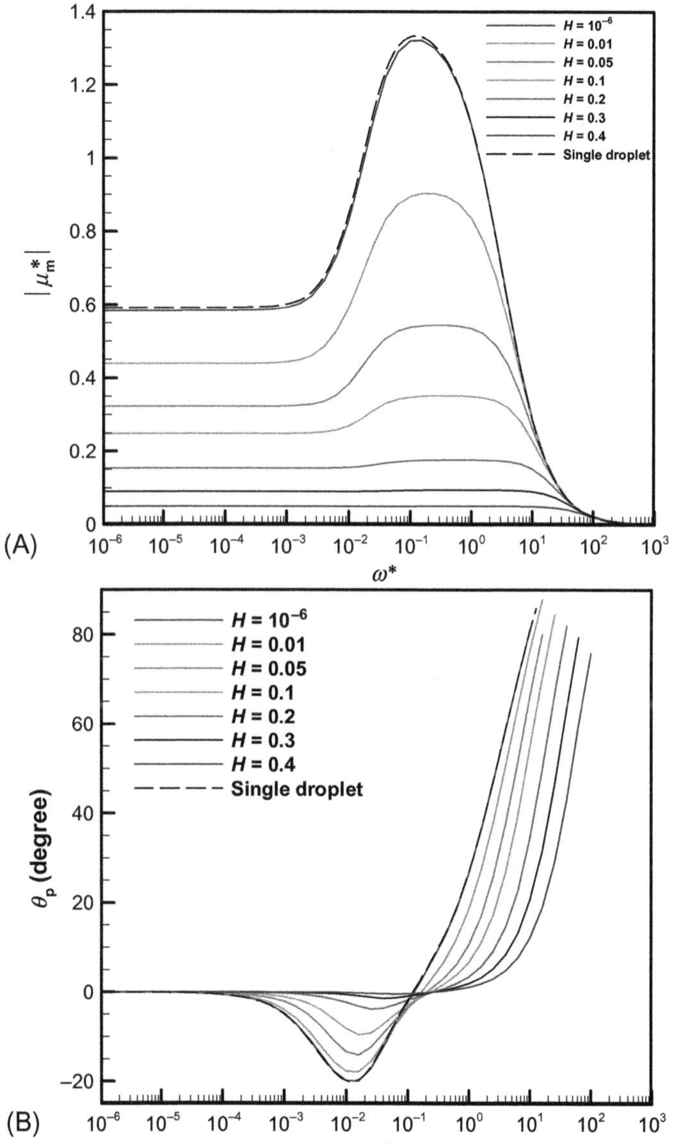

FIG. 15.11 (A) Magnitude of the scaled dimensionless mobility $|\mu_m^*|$ versus the scaled frequency ω^* at various volume fraction H with $\kappa a = 1$, $\sigma_H = 0.5$, and $\phi_r = 5$. (B) Phase angle of the scaled dimensionless mobility θ_p versus the scaled frequency ω^* at various volume fraction H with $\kappa a = 1$, $\sigma_H = 0.5$, and $\phi_r = 5$.

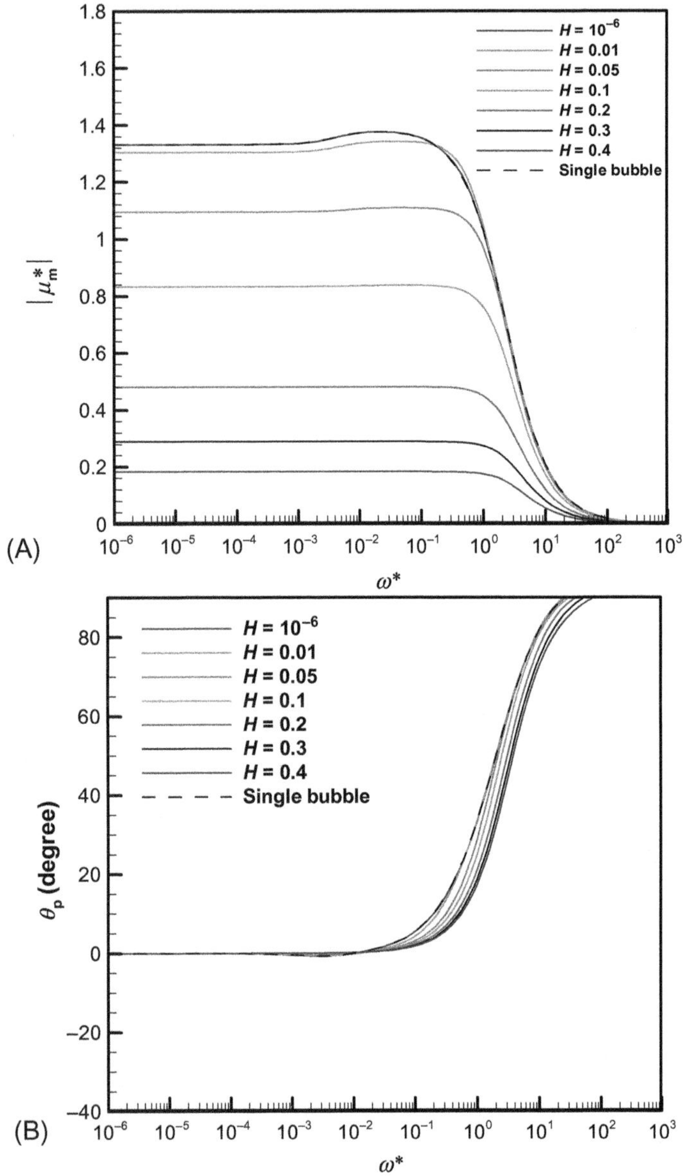

FIG. 15.12 (A) Magnitude of the scaled dimensionless mobility $|\mu_m^*|$ versus the scaled frequency ω^* at various volume fraction H with $\kappa a=1$, $\sigma_H=0.01$, and $\phi_r=1$. (B) Phase angle of the scaled dimensionless mobility θ_p versus the scaled frequency ω^* at various volume fraction H with $\kappa a=1$, $\sigma_H=0.01$, and $\phi_r=1$.

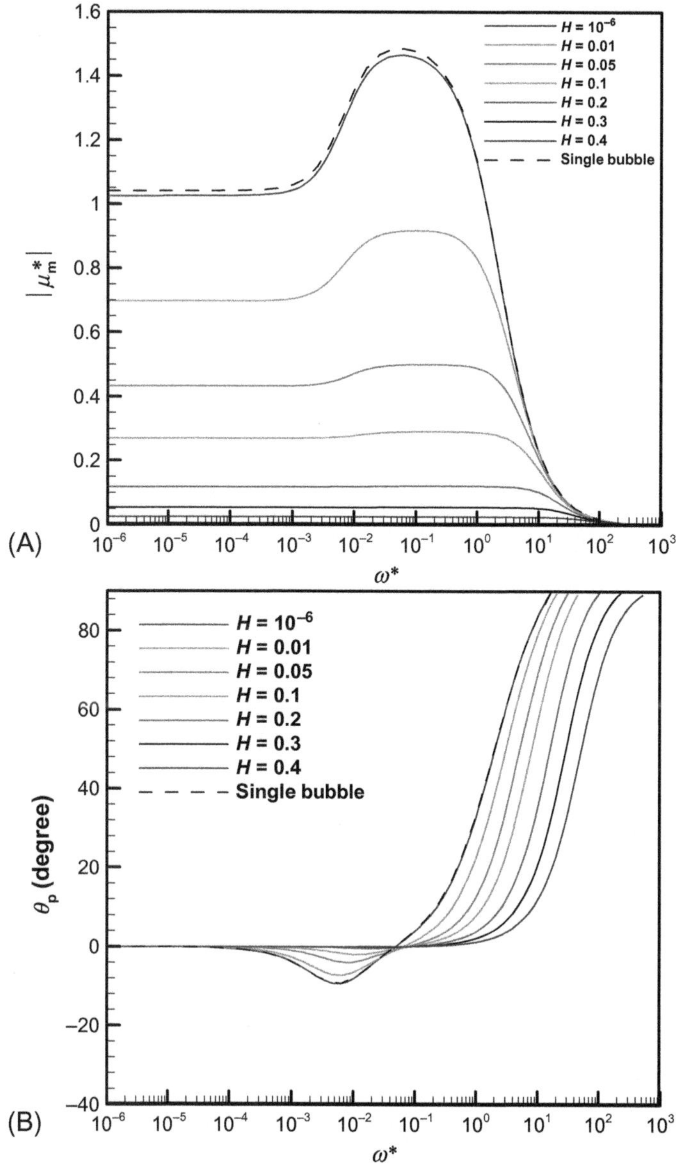

FIG. 15.13 (A) Magnitude of the scaled dimensionless mobility $|\mu_m^*|$ versus the scaled frequency ω^* at various volume fraction H with $\kappa a = 1$, $\sigma_H = 0.01$, and $\phi_r = 3$. (B) Phase angle of the scaled dimensionless mobility θ_p versus the scaled frequency ω^* at various volume fraction H with $\kappa a = 1$, $\sigma_H = 0.01$, and $\phi_r = 3$.

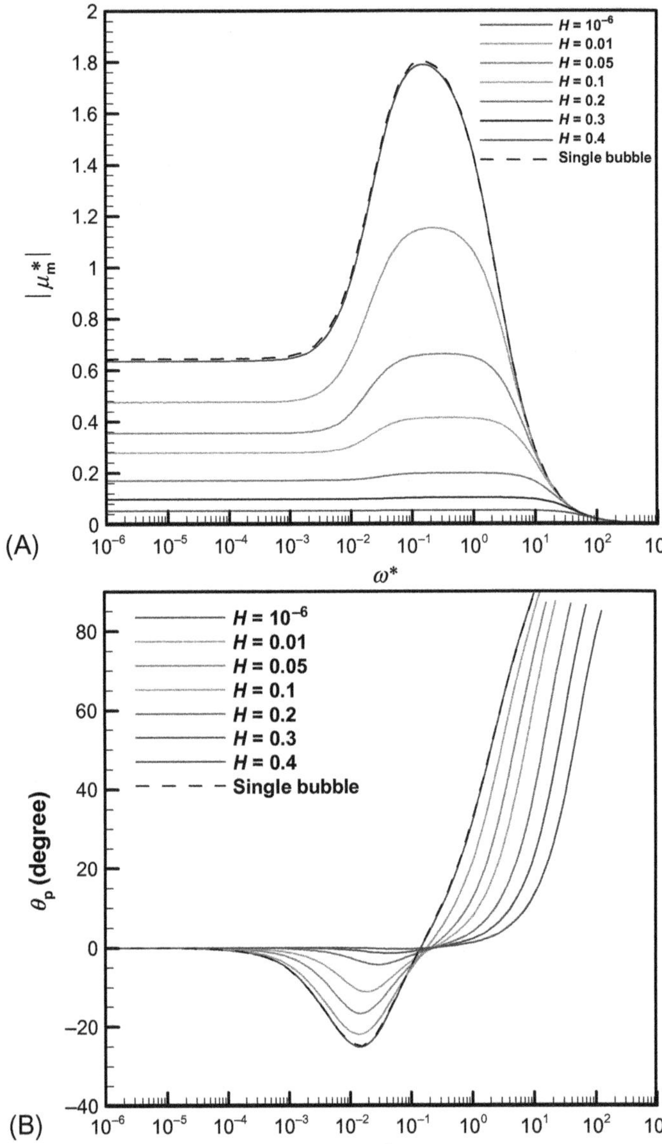

FIG. 15.14 (A) Magnitude of the scaled dimensionless mobility $|\mu_m^*|$ versus the scaled frequency ω^* at various volume fraction H with $\kappa a = 1$, $\sigma_H = 0.01$, and $\phi_r = 5$. (B) Phase angle of the scaled dimensionless mobility θ_p versus the scaled frequency ω^* at various volume fraction H with $\kappa a = 1$, $\sigma_H = 0.01$, and $\phi_r = 5$.

Overall, these figures shed light into the mysterious black box of the precise dynamic behavior of liquid droplet and gas bubble systems. They are valuable information in themselves.

REFERENCES

[1] Dukhin AS, Goetz PJ. Acoustic and electroacoustic spectroscopy for characterizing concentrated dispersions and emulsions. Adv Colloid Interf Sci 2001;92(1–3):73–132.

[2] Magual A, Horváth-Szabó G, Masliyah JH. Acoustic and electroacoustic spectroscopy of water-in-diluted-bitumen emulsions. Langmuir 2005;21(19):8649–57.

[3] Hsu J-P, Min W-L, Lee E. Dynamic electrophoresis of droplet dispersions at low surface potentials. J Colloid Interface Sci 2007;306(2):421–7.

[4] Lou J, Lee E. Dynamic electrophoresis of concentrated droplet dispersions at arbitrary surface potentials. Chem Eng Sci 2007;62(22):6234–45.

[5] Kuwabara S. The forces experienced by randomly distributed parallel circular cylinders or spheres in a viscous flow at small Reynolds numbers. J Phys Soc Jpn 1959;14(4):527–32.

[6] Shilov V, Zharkikh N, Borkovskaya YB. Theory of non-equilibrium electrosurface phenomena in concentrated disperse systems. 1. Application of non-equilibrium thermodynamics to cell model of concentrated dispersions. Colloid J USSR 1981;43(3):434–8.

[7] Galler K, Bräutigam K, Große C, Popp J, Neugebauer U. Making a big thing of a small cell–recent advances in single cell analysis. Analyst 2014;139(6):1237–73.

[8] El-Ali J, Sorger PK, Jensen KF. Cells on chips. Nature 2006;442(7101):403.

[9] Wei Z, Li X, Zhao D, Yan H, Hu Z, Liang Z, Li Z. Flow-through cell electroporation microchip integrating dielectrophoretic viable cell sorting. Anal Chem 2014;86(20):10215–22.

[10] Ahadian S, Yamada S, Ramon-Azcon J, Ino K, Shiku H, Khademhosseini A, Matsue T. Rapid and high-throughput formation of 3D embryoid bodies in hydrogels using the dielectrophoresis technique. Lab Chip 2014;14(19):3690–4.

[11] Clevers H. The cancer stem cell: premises, promises and challenges. Nat Med 2011;17(3):313.

[12] Ong JM, da Cruz L. A review and update on the current status of stem cell therapy and the retina. Br Med Bull 2012;102:133–46.

[13] Bhasin A, Srivastava MP, Mohanty S, Bhatia R, Kumaran SS, Bose S. Stem cell therapy: a clinical trial of stroke. Clin Neurol Neurosurg 2013;115(7):1003–8.

[14] Hoppe PS, Coutu DL, Schroeder T. Single-cell technologies sharpen up mammalian stem cell research. Nat Cell Biol 2014;16(10):919.

[15] Pethig R, Menachery A, Pells S, De Sousa P. Dielectrophoresis: a review of applications for stem cell research. J Biomed Biotechnol 2010;2010:182581.

Section D

Diffusiophoresis of Colloidal Particles

Chapter 16

Diffusiophoresis of Rigid Particles

16.1 INTRODUCTION

Traditionally diffusiophoresis refers to the particle motion in response to an applied concentration gradient of solute, electrolyte or nonelectrolyte, in the solution suspending the particle [1–5]. However, recently diffusiophoresis has found its new life in self-propelled Janus particles like microswimmers and nanoswimmers which does not need a superimposed concentration gradient in prior [6, 7]. Instead, the particles can convert chemical energy contained in the solution into mechanical energy by some catalytic reaction on part of its surface and drives themselves in motion [8–10]. This is achieved by the asymmetric distribution of catalyst on the particle surface. The reaction products, either electrolyte or nonelectrolyte, hence assert unbalanced force of osmosis nature in general upon the particle surface, which sets the particle in motion eventually [6, 7, 11, 12]. This has tremendous potential applications in biomedical fields as well as conventional processing industries [12, 13]. A fundamental revolution in technology actually is looming on the horizon [14–19].

Thus, diffusiophoresis nowadays can be further categorized as self-diffusiophoresis, as the Janus particles mentioned above, and the passive-diffusiophoresis where a concentration gradient is applied or established throughout the system in a macroscopic sense; in other words, the traditional perception of the diffusiophoresis [20–24]. To facilitate the fundamental understanding of the diffusiophoresis phenomenon, this book focuses on the passive-diffusiophoresis in the traditional scope.

As we have just mentioned at the beginning, the solution suspending the particle can contain either electrolyte or nonelectrolyte solutes. In the case of nonelectrolyte solutes, the particle motion is mainly due to the imbalance of the kinetic energy transport during collision with the particle by the solutes, thus is of osmotic pressure nature, plus perhaps the Van der Waal's attraction force or dipole force between them [3, 25]. The whole mechanism is kind of

Interface Science and Technology, Vol. 26. https://doi.org/10.1016/B978-0-08-100865-2.00016-3

like the gas kinetic theorem in fundamental chemistry, except that it happens mostly in liquid solution in practice. Actually, the very discovery of the diffusiophoresis phenomenon by Derjaguin is in an aerosol system [2]; the suspension of rigid particles in a gas medium with gas molecules treated as nonelectrolyte solutes. This type of motion in the traditional passive-diffusiophoresis, however, is quite slow in comparison with the corresponding situation of electrolyte solutes, and it has nothing to do with the electrokinetics, hence it will not be covered here. Theoretical analyses in this regard can be found in the following references [4, 26, 27].

However, in self-diffusiophoresis, strong driving force can be generated by nonelectrolyte solutes such as hydrogen molecules produced via the asymmetric catalytic reaction on a Janus particle coated by platinum on half of its surface and suspended in the peroxide solution [10]. Sometimes the particle motion is like a jet plane if a large amount of hydrogen is formed in gas phase [9, 11, 28]. Indeed, the self-diffusiophoresis is a horse of different color with great technical potential. Interested readers are suggested to read the following references [12, 15, 17–19].

For the electrolyte solutes, i.e., when the particle is suspendered in an electrolyte solution, the fundamental mechanisms for the particle motion in passive-diffusiophoresis are essentially the same as the electrophoresis phenomenon with exactly the same governing equations. The driving force now, however, is not due to an externally applied, or superimposed, electric field upon the colloidal system. Instead, it is due to the diffusion of electrolyte ions driven by a concentration gradient somehow established in the solution. The equilibrium distribution of the double layer is disturbed as a result once the ions diffuse across it, and a double layer polarization (deformation) around the particle is established accordingly which sets the particle in motion. After this point, the story is exactly like what happens in electrophoresis. Hence, diffusiophoresis is a member of the electrokinetics family, and the entire fundamental electrokinetic equations are applicable here as well. In particular, it is closely related to electrophoresis. The only difference is that electrophoresis is due to an applied external electric field, whereas diffusiophoresis is due to a concentration gradient of electrolytes. Thus, in terms of theoretical treatment, the difference is only the boundary conditions where the external driving force is applied.

In the bulk electrolyte solution, the solute concentration gradient favors the migration of electrolyte ions with higher diffusivity initially. However, these favored ions ahead instantly become the hindrance to the further migration of their kind due to the Coulomb repulsive force to slow down the motion of ions behind. Simultaneously, the migration of ions with lower diffusivity is enhanced by this Coulomb attractive force. Thus, at equilibrium, the migration rate of either electrolyte is the same, or no net electric current is generated by diffusion, to maintain the overall electroneutrality of the system. This can also be interpreted as a corresponding electric potential, often referred to as the "diffusion potential," generated within the electrolyte solution in the direction against the migration of the ions with higher diffusivity. Some researchers considered it as an "micro battery" built in the system in comparison with the external electric field applied in electrophoresis [8]. The double layer surrounding the charged particle responds to this diffusion potential exactly as a superimposed external electric field, thus the consequential particle motion is often referred to as simply "electrophoresis" by some researchers [4].

The key element in diffusiophoresis, however, is the extra mechanism of electrolyte migration due to the "selective allowance" of diffusion by the counterions across the borderline of the double layer, which denies the diffusion of the coions at the same time simply due to the Coulomb electrostatic force between the charged particle and the ions. Hence, within the double layer the counterions continue their motions under the diffusion mechanism just like in the bulk stream, with the extra involvement of particle charges of course, whereas the coions are basically absent there. The enrichment of counterions in the upstream and the equal amount of depletion in the downstream alter the ions' distribution there, leading to a double layer polarization. An electric driving force is thus generated, which normally drives the particle toward the region with higher solute concentration [29]. This mechanism is referred to as the "chemiphoresis" component in the sense that it is of osmotic pressure nature due to the uneven distribution of counterions within the double layer by this "selective allowance" of diffusion [4]. Chemi- refers to "chemical potential" here, which is simply a more rigorous term for concentration in thermodynamics [30]. Note that this chemiphoresis mechanism is absent in electrophoresis. There is no solute concentration gradient outside the borderline of the double layer, or the bulk electrolyte solution in electrophoresis, even though the ions there indeed are migrating in response to the external electric field. This extra element distinguishes diffusiophoresis from electrophoresis.

There has been many theoretical analyses regarding diffusiophoresis [1, 2, 4, 27, 29]. For a colloidal particle in a symmetric electric solution with very thin double layer, an analytical formula valid for arbitrary zeta potential is available as follows [4]:

$$U=\frac{\varepsilon}{\eta_{\mathrm{f}}}\left(\frac{kT}{ze}\right)\left\{\beta\zeta-2\left(\frac{kT}{ze}\right)\ln\left[1-\tanh^2\left(\frac{ze\zeta}{4kT}\right)\right]\right\}\left(\frac{\nabla C}{n_{10}}\right) \tag{16.1}$$

Or in dimensionless form:

$$\mu^*=\beta\phi_{\mathrm{r}}-2\ln\left[1-\tanh^2\left(\frac{\phi_{\mathrm{r}}}{4}\right)\right] \tag{16.2}$$

This can be regarded as the extended Smoluchowski formula, taking into account the chemiphoresis component, with the former part corresponding to the electrophoresis component of particle motion introduced by the diffusion potential measured by $\beta\phi_r$, and the latter part the chemiphoresis component [4, 8]. A dimensionless factor β defined as $\beta=(D_1-D_2)/(D_1+D_2)$ is introduced to account for the electrophoresis mechanism due to the distinct ion diffusivities. For instance, $\beta=0$, -0.2, and 0.64 for the aqueous electrolyte solutions of KCl, NaCl, and HCl, respectively. The larger the value of β, the more significant the electrophoresis component is. Note that the electrophoresis contribution is determined completely by the former part, whereas the chemiphoresis by the latter part. In other words, they are arithmetically addable under the condition of very thin double layer and symmetric electrolyte solution. For finite double layer thickness, these two components mingle with each other and cannot be separated. No analytical formula is available, at least not for arbitrary zeta potentials.

Various extensions were made later to consider situations like arbitrary double layer thickness, the double layer polarization effect, dilute or concentrated suspensions, particles of porous nature, liquid droplets, boundary effect, and so on [31–35]. Among them, Lee's group conducted a series of analyses on the highly charged colloidal entities in particular, taking full account of the double layer polarization effect [36–42].

Typically, the particle speed is about 1–10 μm/s [8], much smaller than the corresponding electrophoretic speed. This is because while the latter is proportional to the applied external electric field, which can be arbitrarily large in principle, the former is limited by the solute concentration gradient that can possibly be established upon the suspension system, as the saturation electrolyte concentration is the natural upper bound. For instance, no more than 6.15 g of NaCl (salt) can dissolve in one liter of water [43].

This sounds like a quite discouraging fact of the potential application of diffusiophoresis at first glance. However, this seemingly upper bound problem has been overcome with the fast advancement of microfluidic technology [44, 45]. Nowadays, a finite concentration difference can be applied across a very small distance to yield a very large concentration gradient in a designed apparatus [21, 46]. Moreover, the local concentration gradient near a surface of interest can be very large, although the overall gradient over a large distance in the bulk system is limited by the above-mentioned constraint [8].

Diffusiophoresis has some merits that electrophoresis lacks, for instance, no joule heating effect needs to be concerned for as no net electric current is passing through the suspension system, which is vital in operations involving living cells in particular [47]. For example, a raise of temperature more than 4°C is fatal for human cells [47]. Thus, it is particularly suitable to be used in human body in biomedical and biochemical applications.

Recently, many novel applications utilizing diffusiophoresis have been proposed. Among them, probably the most eye-catching demonstration of its potential application is the report that it can be used to detect the bone-crack place and guide the remedy medicine there to heal it, with the help of fluorescent quantum dots [48]. As the injured place will release specific chemicals to the nearby area, it establishes a concentration gradient in the neighborhood, which serves as a guiding signal as well as the driving force of the drug-carrying colloidal particles. The major handicap for the practical application of diffusiophoresis is the difficulty of harnessing it in a desired way, i.e., its poor maneuverability due to the task of establishing the desired solute concentration gradient, which by nature disturbs the system of interest greatly. The negative impact very often comes before any merit is achieved, or simply outweighing it. In biochemical or biomedical applications, however, as the bone-crack case mentioned above, the injury place in human body often releases some chemicals and corresponding concentration gradient is thus established naturally, which then serves as the driving force for the medicine carried in colloidal particles. No artificial establishment of solute concentration gradient is needed, and the human body certainly is a system full of all kinds of solutes in blood with concentration gradient everywhere. It can endure disturbances or has a self-adjusting mechanism to balance it. Indeed, life finds its own exit.

Here, in this chapter we treat the diffusiophoretic motion of a rigid particle theoretically. The results shed light on the mystery of this lesser known electrokinetic phenomenon which has a very promising potential in practical applications, especially in biochemical and biomedical fields [22, 49].

16.2 THEORY

16.2.1 Single Rigid Particle

The system diagram is shown in Fig. 16.1A, where a concentration gradient $\boldsymbol{\nabla} C$ is applied at infinity upon a single rigid spherical particle with constant surface potential ϕ_r in a binary z_1:z_2 electrolyte solution with $\alpha = -z_1/z_2$. Other than this difference in driving force at infinity, the governing equations are the same as the corresponding electrophoresis system treated in Chapter 1. Moreover, the boundary conditions at the particle surface are identical to those in Chapter 1 as well. The only difference here is the boundary conditions at infinity, where the oriented magnitude of the concentration field $\boldsymbol{\nabla}^* C^* = \boldsymbol{\nabla} n_{10}/\left(\frac{n_{10}}{a}\right)$ replaces the oriented magnitude of the dimensionless electric field $\boldsymbol{E}^*$ in Chapter 1 as follows:

$$\frac{\mathrm{d}\phi_e^*}{\mathrm{d}r^*} = -\frac{(\kappa a)r^* + 1}{r^*}\phi_e^* \tag{16.3}$$

$$\frac{\mathrm{d}G_1^*}{\mathrm{d}r^*} + \frac{2}{r^*}G_1^* = 3(\beta - 1)(\boldsymbol{\nabla}^* C^*)r^* \tag{16.4}$$

$$\frac{\mathrm{d}G_2^*}{\mathrm{d}r^*} + \frac{2}{r^*}G_2^* = 3\left(\beta + \frac{1}{\alpha}\right)(\boldsymbol{\nabla}^* C^*)r^* \tag{16.5}$$

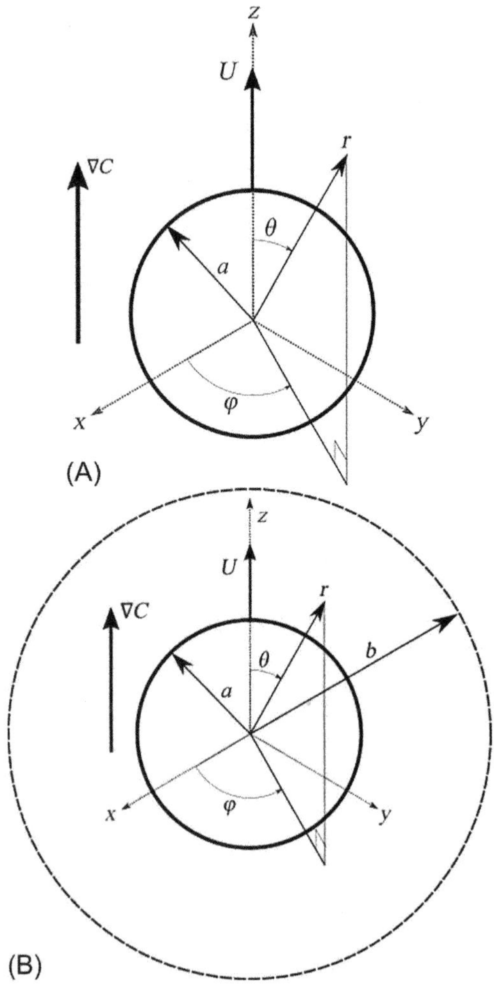

FIG. 16.1 System diagram. (A) Single rigid particle; (B) suspension of rigid particles.

$$\frac{\mathrm{d}\delta\Phi^*}{\mathrm{d}r^*}+\frac{2}{r^*}\delta\Phi^*=-3\beta(\nabla^*C^*)r^* \tag{16.6}$$

$$\frac{\mathrm{d}^2\psi^*}{\mathrm{d}r^{*2}}+\frac{1}{r^*}\frac{\mathrm{d}\psi^*}{\mathrm{d}r^*}-\frac{1}{r^{*2}}\psi^*=\frac{3}{2}U^* \tag{16.7}$$

$$\frac{\mathrm{d}^3\psi^*}{\mathrm{d}r^{*3}}-\frac{3}{r^{*2}}\frac{\mathrm{d}\psi^*}{\mathrm{d}r^*}+\frac{3}{r^{*3}}\psi^*=-\frac{3}{2r^*}U^* \tag{16.8}$$

These boundary conditions are derived following the process mentioned in Chapter 1. A large distance r_o^* is picked where the Debye-Hückel approximation can be applied. The details can be found elsewhere [42, 50]. Note that all the variables in Eqs. (16.3)–(16.8) are in one-dimensional form with the θ-dependence dropped off, as we did in Chapter 1. Note also that $\nabla^* C^*$ is the externally applied scalar constant bulk concentration gradient along the z-direction. The β factor above measures the magnitude of the diffusion potential:

$$\beta = \frac{D_1 - D_2}{D_1 + D_2} \tag{16.9}$$

where D_1 and D_2 are the diffusion coefficients of cations and anions, respectively. For example, $\beta = 0$ for KCl solution and $\beta = -0.2$ for NaCl solution. Note that Eq. (16.3) implies that $\phi_e^* = 0$ at infinity, or $n_{1e}^* = \alpha n_{2e}^*$ at infinity when the particle is at rest. In other words, the double layer never reaches there, which is the underlying precondition of Eqs. (16.4)–(16.6). Hence, if the double layer does reach there, as in suspension of particles treated separately below, Eqs. (16.4)–(16.6) fail to apply simultaneously as Eq. (16.3). Note also when $\beta = 0$ as in the sole chemiphoresis situation, Eqs. (16.4)–(16.6) still yield nontrivial boundary conditions. The general underlying physical background of Eqs. (16.4)–(16.6) will be discussed below for the suspension of rigid particles. Eqs. (16.7) and (16.8), on the other hand, simply imply that the coordinates are set at the center of the moving particle, thus a relative fluid velocity U against the moving direction of the particle is resulted far away from the particle, say at infinity.

Once the flow field and electric field are solved, following the same procedure as in Chapter 1, the diffusiophoretic mobility of the particle can be obtained as well, which is defined in dimensionless form as follows:

$$\mu^* = \frac{U^*}{\nabla^* C^*} = \frac{U/U_0}{\nabla C / \left(\frac{n_{10}}{a}\right)} \tag{16.10}$$

where U is the oriented particle diffusiophoretic velocity along the z-direction, U_0 is the characteristic velocity defined as $U_0 = \frac{\varepsilon}{\eta_f a}\left(\frac{kT}{z_1 e}\right)^2$, a is the particle radius, and n_{10} is the bulk concentration of cations. Note that ∇C used here is equivalent to ∇n_{10}, hence $\nabla^* C^*$ is equivalent to $\nabla^* n_{10}^*$. Note also that all the variables in Eq. (16.10) are scalars, including the two oriented magnitude of the concentration gradients, $\nabla^* C^*$ and $\nabla^* n_{10}^*$, as well as U^* and U.

16.2.2 Suspensions of Rigid Particles

The system diagram is shown in Fig. 16.2B, where Kuwabara's hydrodynamic unit cell model coupled with the Shilov and Zharkikh (S-Z) type of the electric boundary conditions is adopted to describe the electrokinetic response in a

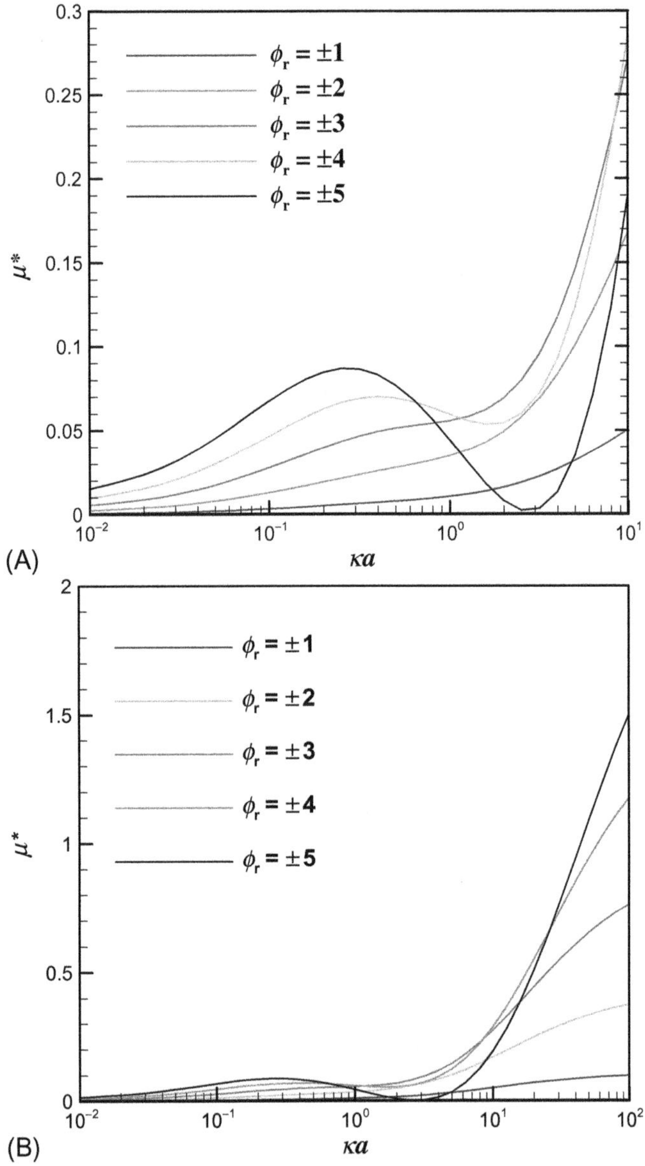

FIG. 16.2 Dimensionless mobility μ^* as a function of κa at various ϕ_r with $\beta = 0$. (A) Key: $\kappa a = 0.01$–10; (B) key: $\kappa a = 0.01$–100.

suspension of rigid spherical particles. The governing equations and boundary conditions on the particle surface are the same as the single particle situation discussed above. On the other hand, the boundary conditions at the outer virtual surface $\left(r^* = 1/\sqrt[3]{H}\right)$ are:

$$\frac{d\phi_e^*}{dr^*}=0 \tag{16.11}$$

$$G_1^*=-\delta\Phi^*-\frac{1}{\exp\left(-\phi_e^*\right)}(\nabla^*C^*)r^* \tag{16.12}$$

$$G_2^*=-\delta\Phi^*+\frac{1}{\alpha\exp\left(\alpha\phi_e^*\right)}(\nabla^*C^*)r^* \tag{16.13}$$

$$\delta\Phi^*=-\left[\frac{\dfrac{1}{Pe_1}-\dfrac{1}{Pe_2}}{\dfrac{1}{Pe_1}\exp\left(-\phi_e^*\right)+\dfrac{\alpha}{Pe_2}\exp\left(\alpha\phi_e^*\right)}\right](\nabla^*C^*)r^* \tag{16.14}$$

$$\psi^*=\frac{1}{2}U^*r^{*2} \tag{16.15}$$

$$\frac{d^2\psi^*}{dr^{*2}}-\frac{2}{r^{*2}}\psi^*=0 \tag{16.16}$$

Eq. (16.11) is the same as used in Chapter 2, where electrophoresis of suspensions of rigid particles is considered. No electric interaction takes place between adjacent unit cell from a local point of view and no external electric field is applied upon the system from a global point of view. The particle is at rest. Note that this electrostatic boundary condition allows the double layer overlapping and ϕ_e^* is not zero in general on the outer virtual surface.

Eqs. (16.12) and (16.13) reflect the initial driving force of the chemiphoresis component, the concentration gradient, exerted upon the virtual surface:

$$\delta n_j^*=\left(\nabla^*n_{j0}^*\right)r^*\cos\theta,\quad j=1,2 \tag{16.17}$$

Based on the linear perturbation analysis introduced in Chapter 1, it can be shown:

$$\delta n_j^*=-z_jn_{j0}^*\exp\left(-z_j\phi_e^*\right)\left(\delta\Phi^*+G_j^*\right)\cos\theta,\quad j=1,2 \tag{16.18}$$

Thus, Eq. (16.17) can be converted to the one-dimensional form:

$$-n_{10}^*\exp\left(-\phi_e^*\right)\left(\delta\Phi^*+G_1^*\right)=\left(\nabla^*n_{10}^*\right)r^* \tag{16.19}$$

$$\alpha n_{20}^*\exp\left(\alpha\phi_e^*\right)\left(\delta\Phi^*+G_2^*\right)=\left(\nabla^*n_{20}^*\right)r^* \tag{16.20}$$

Moreover, n_{20}^* can be further replaced by n_{10}^* via the following electroneutrality constraint:

$$n_{10}^*-\alpha n_{20}^*=0 \tag{16.21}$$

Hence, $\nabla^*n_{20}^*=\frac{1}{\alpha}\nabla^*n_{10}^*=\frac{1}{\alpha}\nabla^*C^*$. Moreover, as n_{10} is used as the characteristic concentration in the non-dimensionalization process, n_{10}^* is equal to unity and n_{20}^* is equal to $1/\alpha$. Note that $\nabla^*n_{10}^*=(\nabla n_{10})/\left(\frac{n_{10}}{a}\right)$ is nonzero, however, as ∇n_{10} is nonzero. So are $\nabla^*n_{20}^*$ and ∇^*C^*. They both are nonzero here. Substituting these identities into Eqs. (16.19) and (16.20) with some straightforward rearrangements, the two boundary conditions for G_j are obtained, i.e., Eqs. (16.12) and (16.13).

Eq. (16.14), on the other hand, is based on the following rationale. As there is no external electric field applied upon the system, the net flux of the electric charges, or the electric current, has to be zero on the outer virtual surface [51, 52]. According to the Nernst-Planck equation, we have:

$$z_1 D_1 \left(\nabla n_1 + \frac{z_1 e}{kT} n_1 \nabla \phi \right) + z_2 D_2 \left(\nabla n_2 + \frac{z_2 e}{kT} n_2 \nabla \phi \right) = 0 \tag{16.22}$$

or in dimensionless form:

$$\frac{1}{Pe_1}\left[\nabla^* n_1^* + n_1^* \nabla^* \phi^*\right] - \frac{\alpha}{Pe_2}\left[\nabla^* n_2^* - \alpha n_2^* \nabla^* \phi^*\right] = 0 \tag{16.23}$$

where Pe_1 and Pe_2 represent the dimensionless diffusivity coefficients for cations and anions, respectively, with $Pe_1 = \frac{U_0 a}{D_1}$ and $Pe_2 = \frac{U_0 a}{D_2}$.

When the particle is in motion under a small concentration gradient, Eq. (16.23) reduces to the following equation under linear perturbation analysis:

$$\frac{1}{Pe_1}\left[\nabla^* \delta n_1^* + n_{1e}^* \nabla^* \delta\phi^* + \delta n_1^* \nabla^* \phi_e^*\right] = \frac{\alpha}{Pe_2}\left[\nabla^* \delta n_2^* - \alpha n_{2e}^* \nabla^* \delta\phi^* - \alpha \delta n_2^* \nabla^* \phi_e^*\right] \tag{16.24}$$

Note that $\delta n_j^* \nabla^* \phi_e^*$ terms are all equal to zero by Eq. (16.11). Substituting Eq. (16.17) into Eq. (16.24) and cancelling the θ-dependence of δn_j and $\delta\phi$, the above equation can be converted to the one-dimensional form:

$$\frac{1}{Pe_1}\left[\nabla^* n_{10}^* + n_{10}^* \exp\left(-\phi_e^*\right) \nabla^* \delta\Phi^*\right] = \frac{\alpha}{Pe_2}\left[\nabla^* n_{20}^* - \alpha n_{20}^* \exp\left(\alpha\phi_e^*\right) \nabla^* \delta\Phi^*\right] \tag{16.25}$$

Note that the terms involving n_{10}^* and n_{20}^* can be replaced by unity and $1/\alpha$, respectively, as mentioned earlier. After these key simplifications and some straightforward rearrangements, Eq. (16.25) reduces to a Levine and Neale (L-N) type of boundary condition. Integration on both sides of the equal sign converts it to the Shilov and Zharkikh (S-Z) type of boundary condition used here, i.e., Eq. (16.14).

Note that Eqs. (16.12)–(16.14) reduce to Eqs. (16.4)–(16.6) for a single rigid particle if $\phi_e^* = 0$, i.e., if there is no double layer overlapping, or the suspension is very dilute.

Eq. (16.15) again reflects the fact that the coordinates are set at the center of the moving particle with velocity U. Eq. (16.16), on the other hand, is the null vorticity requirement imposed by the Kuwabara's unit cell model. For a two-dimensional axisymmetric flow field, a scalar equation indicating the magnitude of this one-dimensional vorticity vector in the φ-direction is resulted, as shown in Eq. (16.16).

16.3 RESULTS AND DISCUSSION

16.3.1 Single Rigid Particle

Fig. 16.2A presents the corresponding general Henry's chart in a symmetric KCl electrolyte solution with $\beta = 0$, hence only the chemiphoresis component is involved in the particle motion. In general, the particle mobility increases with increasing κa there except for $\phi_r = 4$ and 5, as shown in Fig. 16.2A. This is because as κa increases, more and more counterions are driven to the vicinity of the particle due to the double layer suppression effect. Hence, the established concentration gradient upon the double layer by diffusion will affect a large amount of ions there. This favors the particle motion via simple Coulomb's law, at least for ϕ_r ranging from ± 1 to ± 3. The behavior of highly charged ones, $\phi_r = \pm 4$ and ± 5, is peculiar. Local minima are observed for κa roughly between one to three, indicating the slowdown of the particle motion there. This is very different from the electrophoresis behavior where highly charged particles always move faster, although local minima are observed for profiles of the scaled mobility μ_m^* which has been further divided by ϕ_r. The reason lies in the fact that the double layer polarization here is generated by the concentration gradient initially on the double layer boundary, which is a cause; whereas in electrophoresis double layer polarization is the consequence of a superimposed electric field. The former affects the ions distribution within the double layer by diffusion from the boundary like a relay game; the latter, on the other hand, affects the entire ions distribution simultaneously.

Apparently, the highly charged particle hinders the migration of counterions once they enter the double layer. The strong Coulomb attraction force surpasses the initial diffusion-induced migration across the double layer via the Nernst-Planck ion flux equation. The downstream ions have a hard time leaving the double layer, and this back congestion in traffic hinders the further pouring in of ions in the upstream. The traffic jam at the exit of a parking lot prevents the further inflow of cars at the entrance waiting to get in. The double layer polarization is, thus, reduced significantly, and so is the eventual particle motion. This will be discussed in detail again later on.

Note that Fig. 16.2A is only a close-up view with κa limited to 10. Fig. 16.2B presents the whole picture with κa all the way up to 100. The thinner the double layer is (large κa), the faster the particle moves, with a fashion much steeper than the corresponding electrophoresis behavior. The correction of the mobility prediction for finite κa is drastic. This behavior is quite similar to the electrophoresis of a liquid droplet treated in Chapter 12.

Note that no matter whether the particle is positively or negatively charged, the particle always moves toward the higher concentration end of the electrolytes here in chemiphoresis, indicated by positive particle velocity (μ^*) against positive concentration gradient (∇C). This is because there is a net migration of counterions across the entire double layer driven by the established concentration gradient of electrolyte ions, ∇C_{10}. Hence, there is a buildup of counterions in the upstream double layer facing the oncoming diffusion of ions, whereas an equal amount of depletion of ions in the downstream double layer. An antisymmetric double layer polarization is thus established surrounding the particle, which generates an electric driving force normally toward the higher concentration end, which can be best perceived as due to the imbalance of the Coulomb attraction force between the charged particle and the counterions. Note that there is no net migration of counterions across the double layer in electrophoresis, it just redistributes from one hemisphere to the other. There is no actual migration of ions across any part of the double layer boundary in electrophoresis either. In diffusiophoresis, however, migration of ions, predominantly the counterions, across the double layer boundary takes place constantly, or in a dynamic way.

Fig. 16.3 shows the particle mobility as a function of particle surface potential at various double layer thickness. The symmetric profiles imply the particle always moves toward the higher concentration end of electrolytes, regardless it is positively charged (plus ϕ_r) or negatively charged (minus ϕ_r), as μ^* is always positive there. The reason for this has just been explained. Note that the Smoluchowski-type analytical prediction valid for very thin double layer ($\kappa a \gg 1$) overestimates the particle mobility in general, as shown by the dashed line in Fig. 16.3. The thicker the double layer is, the more severe the extent of this overestimation is in general, at least when $|\phi_r|$ is smaller

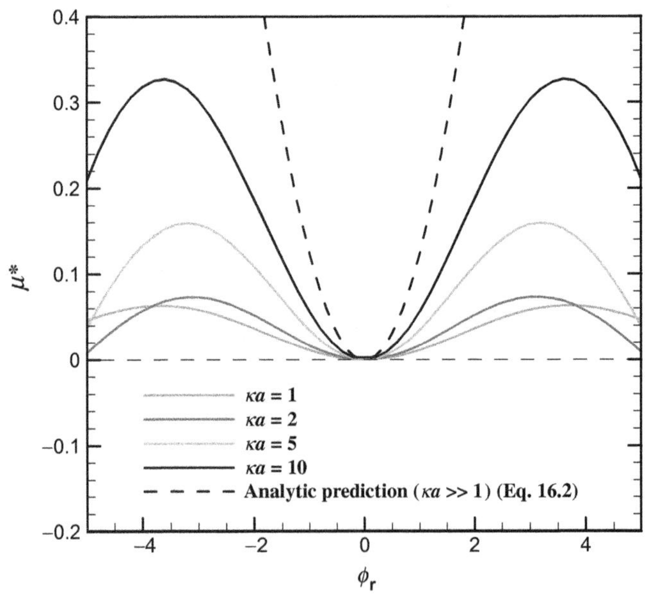

FIG. 16.3 Dimensionless mobility μ^* as a function of ϕ_r at various κa with $\beta = 0$.

than about 4. Thick double layer implies less counterions in the vicinity of the particle, and hence, the driving force there in general. Note also that there are local maxima in these mobility profiles around $|\phi_r| = 3$. This implies a highly charged particle may actually move slower than a lowly charged one in diffusiophoresis. Or once past a threshold value of ϕ_r, the higher the particle charge level is, the slower it moves. Indeed, an intriguing phenomenon, which has already been discussed above in Fig. 16.2A. Here, we elaborate it further with another analog. With this scenario in the backdrop, the migration of the diffusion-driven ions through the double layer surrounding a highly charged particle is like sailing a space ship past an enormous star with huge mass with the orbit very close to it. The strong gravity will pull the unwilling space ship toward it. Similarly, the strong Coulomb attraction force will suck in the passing-by counterions toward it at a much closer distance than the less charged ones, and very reluctantly to let it leave. The resulted double layer polarization, which is the very driving force in chemiphoresis, is significantly weakened, leading to a slower particle motion.

Corresponding contour plots of the dimensionless polarization $\delta\rho^*$ are shown in Fig. 16.4A for a very highly charged particle with $\phi_r = 5$. Much

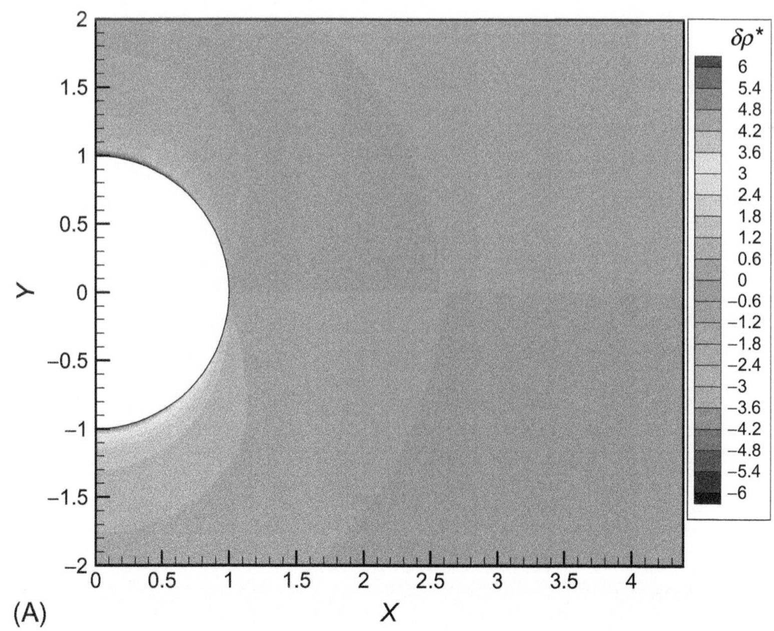

FIG. 16.4 Contour plot of the dimensionless disturbed space charge density $\delta\rho^*$ for a single rigid particle with $\beta = 0$ and $\kappa a = 1$. (A) $\phi_r = 5$; (B) $\phi_r = 3$; (C) $\phi_r = 1$.

Continued

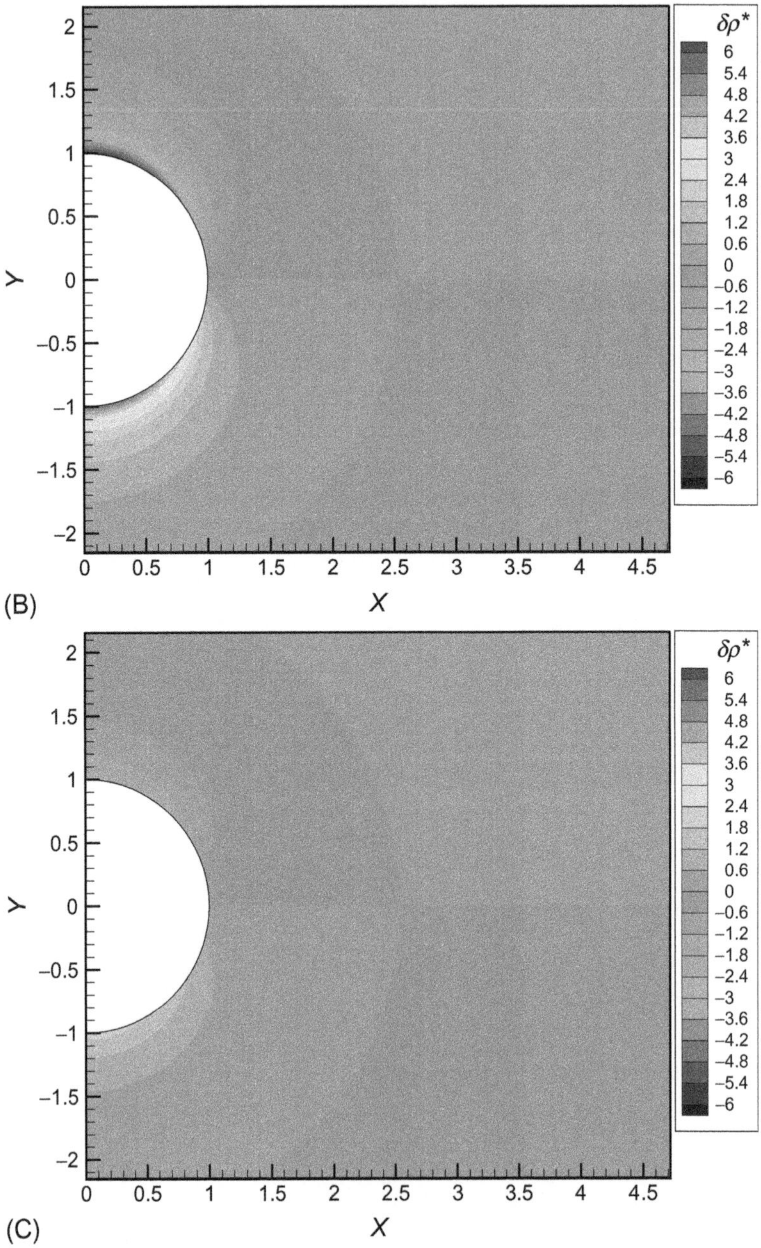

FIG. 16.4—CONT'D

weaker polarization of double layer is observed there, compared with the less charged situation ($\phi_r = 3$) shown in Fig 16.4B. Fig. 16.4C ($\phi_r = 1$) shows the physical aspect consistent with the elaboration just mentioned, which is a very nice and intuitive one. Of course, other factors such as the distribution of the perturbed local electric potential (in dimensionless form) $\delta\phi^*$ are important as well. The ultimate particle mobility is determined by the joint effect of various relevant electrokinetic factors, which can only be obtained via rigorous computation as done here.

The discussion on chemiphoresis is done at this point. Corresponding discussion on general diffusiophoresis is shown below, with the involvement of the electrophoresis component from the diffusion potential in an electrolyte solution with $\beta \neq 0$, e.g., NaCl, where $\beta = -0.2$.

Fig. 16.5A shows the general Henry's chart for positively charged rigid particles. Note that the electrophoresis component in diffusiophoresis is orientation. For instance, when β is negative, an induced diffusion potential against the motion of anions will be established, as $\beta = \frac{D_1 - D_2}{D_1 + D_2}$. Thus, the corresponding electric field is along the same direction of the ions diffusion. This means the resulted electric driving force will push back the motion of a positively charged particle toward the lower concentration end of electrolytes. A negative particle mobility is, thus, observed reflecting this fact, as

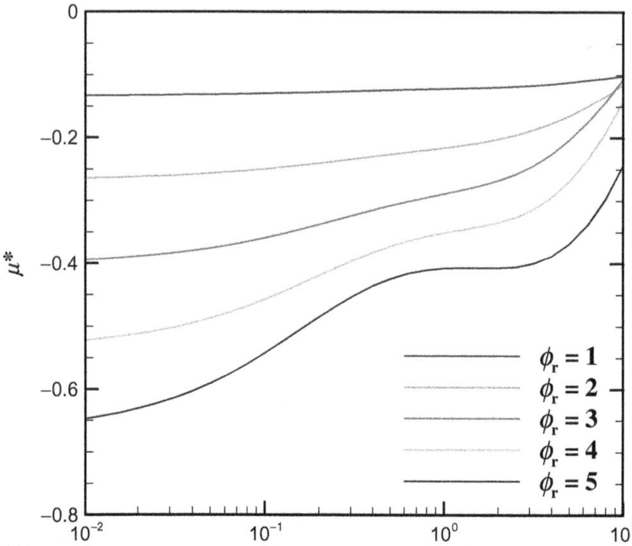

(A)

FIG. 16.5 Dimensionless mobility μ^* as a function of κa at various ϕ_r with $\beta = -0.2$. (A) Positive ϕ_r, key: $\kappa a = 0.01$–10; (B) positive ϕ_r, key: $\kappa a = 0.01$–100; (C) negative ϕ_r, key: $\kappa a = 0.01$–10; (D) negative ϕ_r, key: $\kappa a = 0.01$–100.

Continued

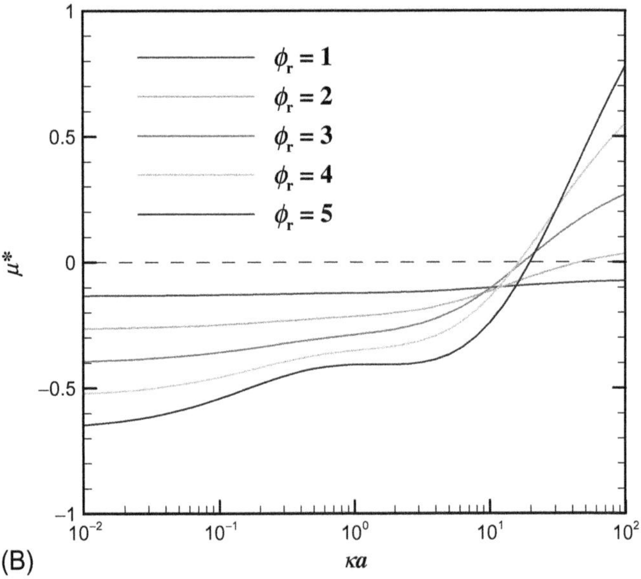

(B)

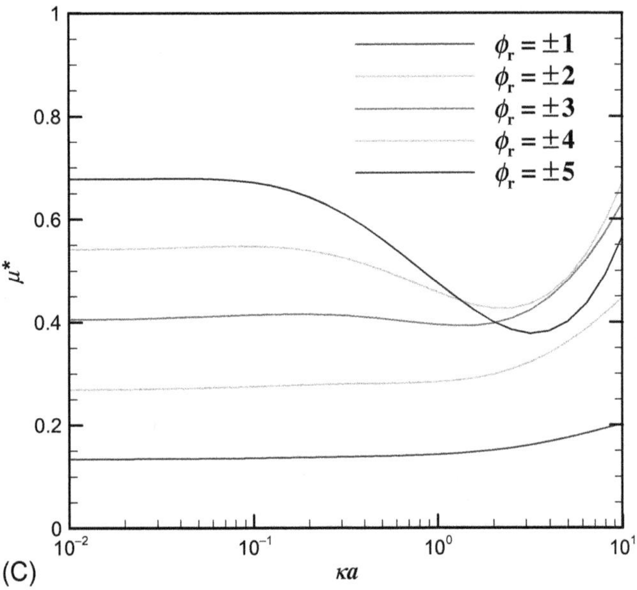

(C)

FIG. 16.5—CONT'D

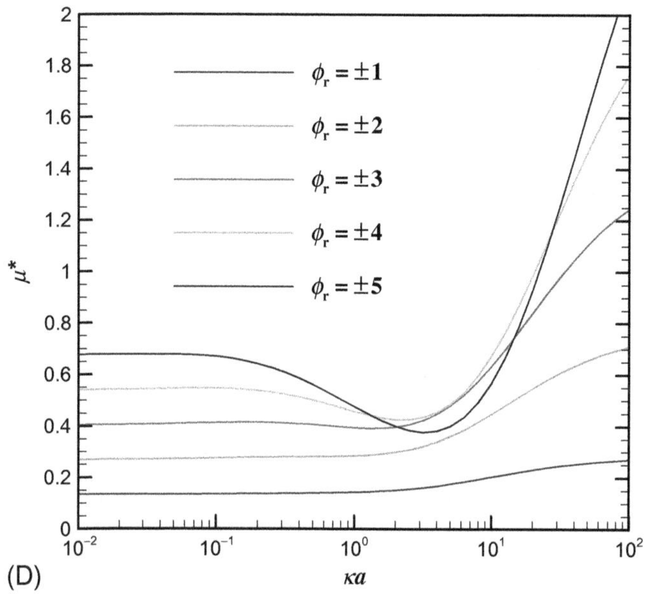

FIG. 16.5—CONT'D

shown in Fig. 16.5A. This is opposite to the direction of particle motion from the chemiphoresis component. Hence, they will compete with each other. As the electrophoresis component is normally larger than the chemiphoresis component, a positively charged particle normally moves toward the lower concentration end.

The particle speed decreases with increasing κa, as the motion-deterring chemiphoresis component increases accordingly. Note also that the largest particle speed appears at $\kappa a = 0.01$ in Fig. 16.5A, or the most dilute electrolyte solution, as the effect of chemiphoresis nearly disappears there.

Fig. 16.5B extends the κa range to 100. The chemiphoresis component becomes dominant eventually as κa further increases, as we discussed in Fig. 16.2B for $\beta = 0$. Hence, a reverse of the particle motion direction is observed. The particle moves toward the higher concentration end now at large κa.

Fig. 16.5C, on the other hand, presents the general Henry's chart for negatively charged particles in a NaCl solution. Now the particle moves toward the higher concentration end with positive mobility. Note that the local minima of the profiles for $\phi_r = -4$ and -5 are due to the motion-deterring double layer polarization effect. Again, the chemiphoresis component takes over when κa gets large, as shown in Fig.16.5D, where the range of κa is extended to 100. Corresponding mobility values at $\kappa a = 0.01$ in Fig. 16.5D stand for the purely electrophoresis component, on the other hand.

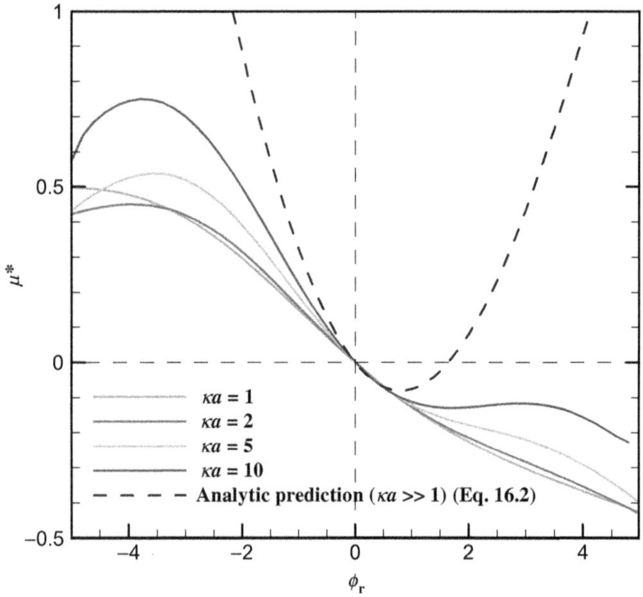

FIG. 16.6 Dimensionless mobility μ^* as a function of ϕ_r at various κa with $\beta = -0.2$.

Fig. 16.6 shows the mobility profile as a function of particle surface potential ϕ_r in NaCl solution. Now the profiles are not symmetric anymore, as there are two competing components with opposite direction of respective driving force, electrophoresis by the diffusion potential and chemiphoresis by the concentration gradient. Enhancement of particle mobility is observed for negatively charged particles, and retardation is observed for positively charged ones. The overall behavior otherwise is similar to what observed in Fig. 16.3 for sole chemiphoresis situation with $\beta = 0$.

Fig. 16.7A–C present the mobility profile as a function of ϕ_r for various β values. Six different electrolyte solutions with β ranging from -0.596 (NaOH) to 0.789 (H_2CO_3) are considered with $\kappa a = 0.1$, 1, and 10 to represent thick, comparable, and thin double layer thickness, respectively. As shown in Fig. 16.7A for a thick double layer ($\kappa a = 0.1$), the chemiphoresis component turns out to be very small or even negligible, thus the electrophoresis component dominates. Since $\mu^* = \beta\phi_r$ in electrophoresis, nearly straight lines with slopes equal to β is observed for these profiles. Note that they all pass through the origin as the mobility is null for a chargeless particle ($\phi_r = 0$). As shown in Fig. 16.7B, when the double layer thickness is comparable to the particle radius, the chemiphoresis apparently is getting noticeable, thus the mobility profiles cease to be nearly antisymmetric to the origin, and nonlinear effect at highly charged condition starts to exhibit with curvy shape there. The situation gets even more profound when the double layer gets further thinner, as

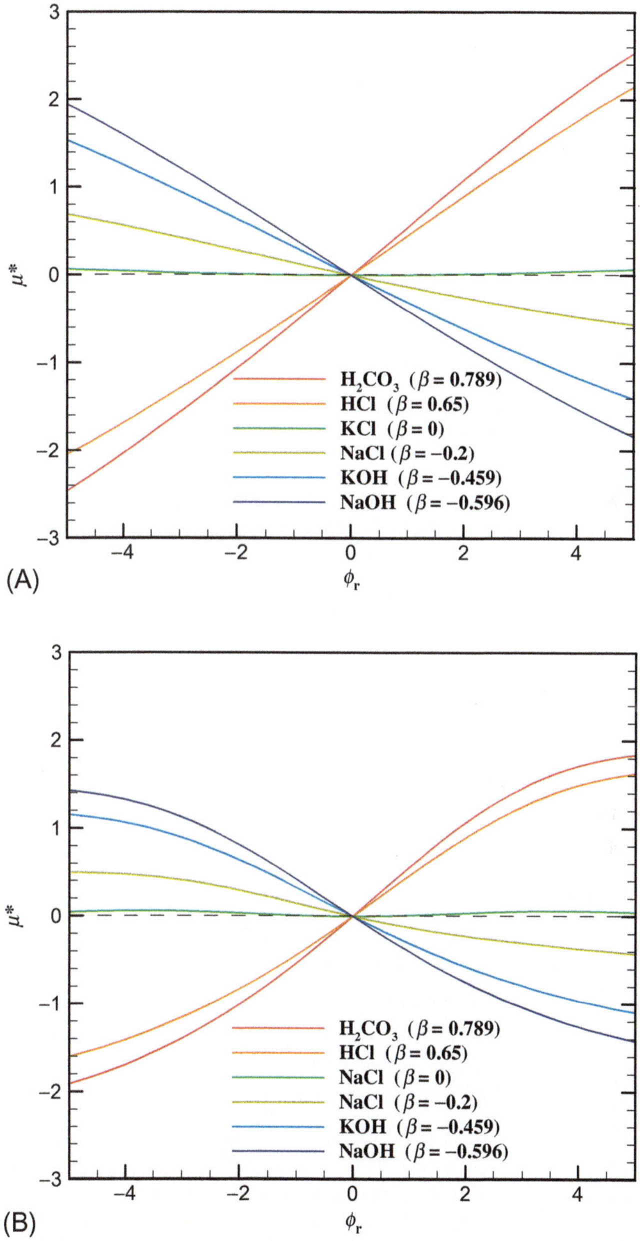

FIG. 16.7 Dimensionless mobility μ^* as a function of ϕ_r at various β. (A) $\kappa a = 0.1$; (B) $\kappa a = 1$; (C) $\kappa a = 10$.

Continued

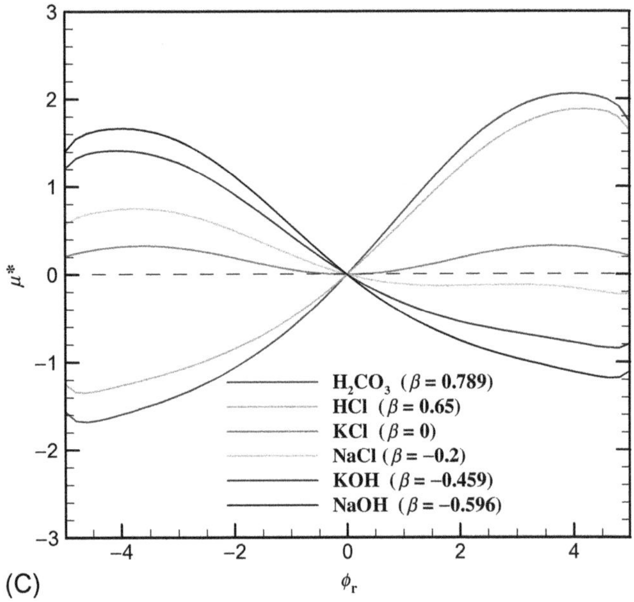

FIG. 16.7—CONT'D

shown in Fig. 16.7C where $\kappa a = 10$. The chemiphoresis component starts to be so significant that even local extrema are generally observed at highly charged condition. This concludes our discussion on the diffusiophoresis of a single rigid particle.

16.3.2 Suspension

We first check the chemiphoresis phenomenon ($\beta = 0$). Fig. 16.8A–C presents the dimensionless particle mobility μ^* as a function of κa at various particle volume fractions ($H = 10^{-6}$ to 0.4) for $\phi_r = 1$, 3, and 5, respectively. Note that this is similar to the general Henry's chart treated in Chapter 2 for the electrophoresis in suspensions. The general behavior is similar to the corresponding single particle situation. Higher surface potential yields larger particle mobility in general. When κa is small, the electrolyte ions are simply too sparse to yield any appreciable particle motion, especially when the suspension gets concentrated due to the hydrodynamic swarming effect. This motion-deterring effect continues its dominance in general and results in significant deviations of particle mobility for different volume fractions H at $\kappa a = 100$. The more concentrated the suspension is, the slower the particles moves as expected.

An intriguing phenomenon is observed, however, that the particles in concentrated suspension may actually move faster than a single particle over some κa range around unity. This is further demonstrated in Fig. 16.9A–C, where the mobility is expressed as a function of particle volume fraction H

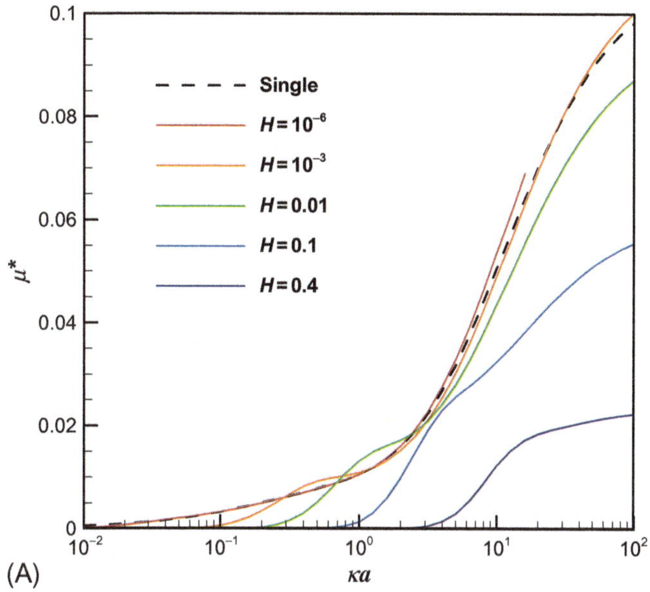

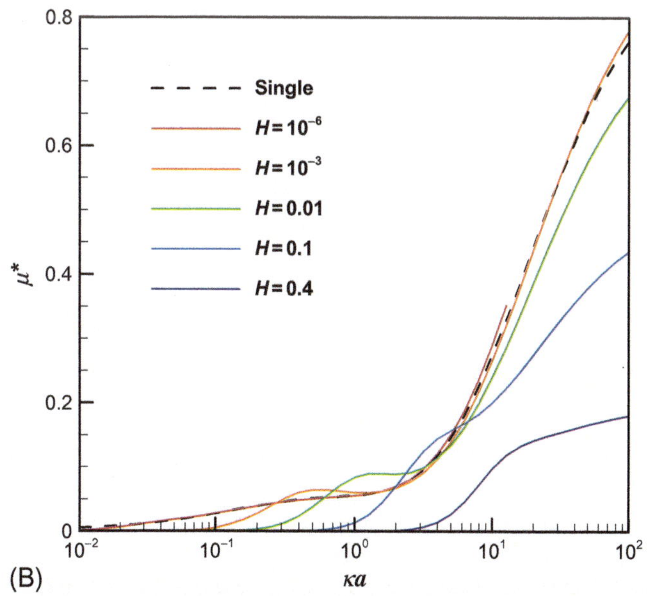

FIG. 16.8 Dimensionless mobility μ^* as a function of κa at various H with $\beta=0$. (A) $\phi_r=1$; (B) $\phi_r=3$; (C) $\phi_r=5$.

Continued

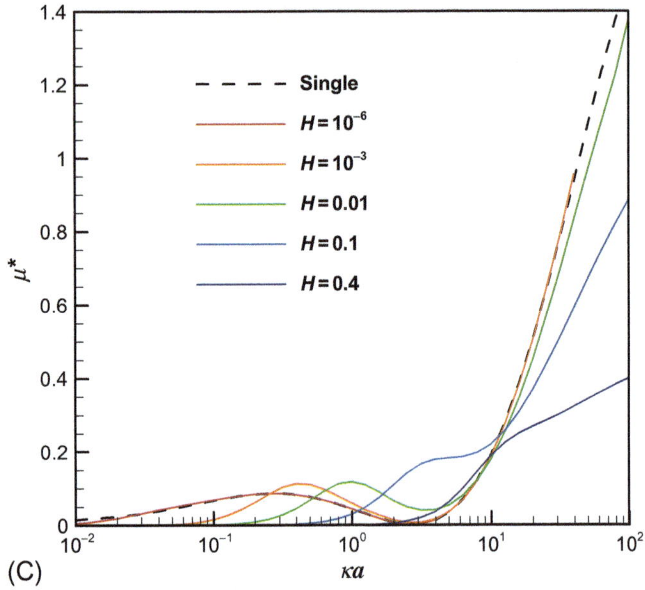

FIG. 16.8—CONT'D

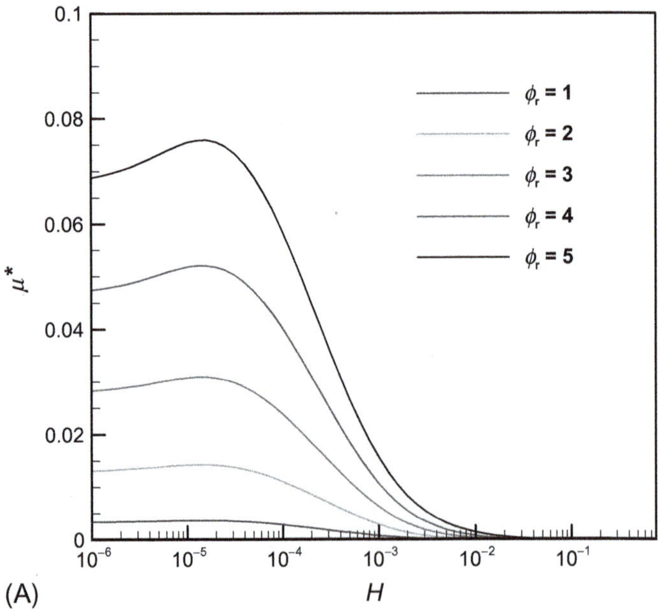

FIG. 16.9 Dimensionless mobility μ^* as a function of H at various ϕ_r with $\beta = 0$. (A) $\kappa a = 0.1$; (B) $\kappa a = 1$; (C) $\kappa a = 10$.

Continued

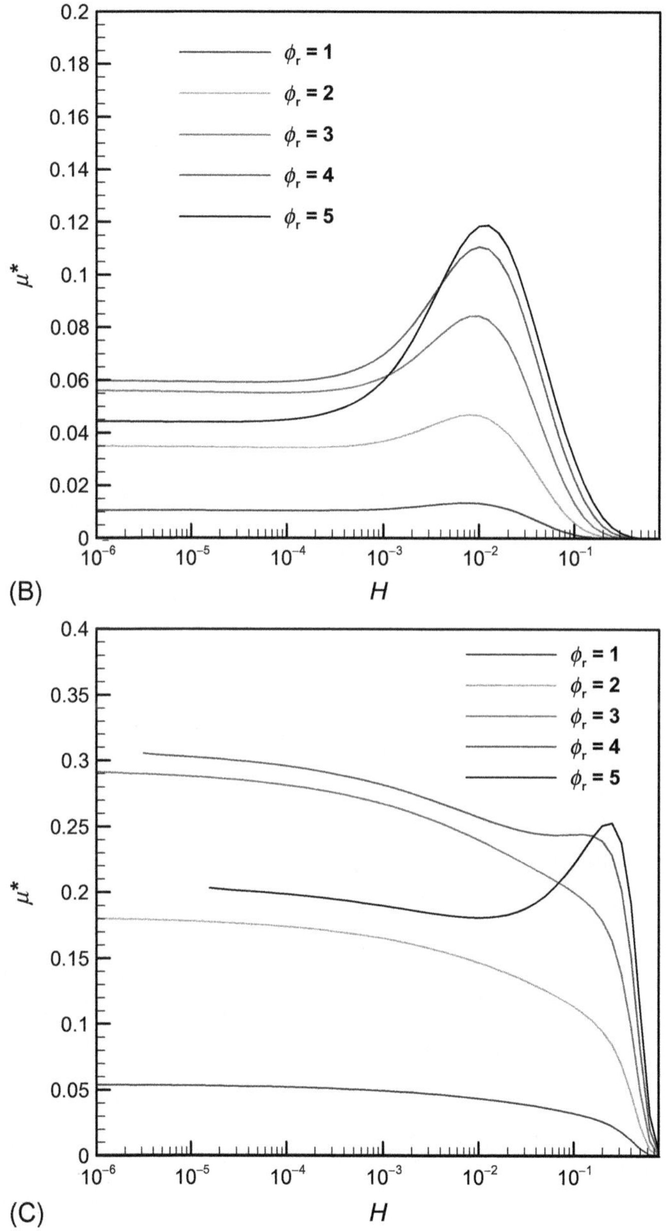

FIG. 16.9—CONT'D

at various surface potentials ($\phi_r = 1$–5) and κa values ($\kappa a = 0.1$, 1, and 10). Note that the profiles would be exactly the same for corresponding negatively charged particles here as $\beta = 0$. The mobility profiles of highly charged particles exhibit local maxima as shown. In other words, there is a range of volume fraction over which the more concentrated the suspension is, the faster the particles in it move.

This happens regardless of the charge level of the particle. But the higher the value of ϕ_r is, the more significant the extent of the local maximum in the corresponding mobility profile expressed as a function of H, as shown in Fig. 16.9A–C. Apparently, the suppression of counterions by the boundary confinement effect from the neighboring particles is the underlying physical mechanism, which increases with increasing volume fraction H. As diffusiophoresis is very sensitive to the ion concentrations within the double layer, or κa, as shown in Fig. 16.2A and B, the particle mobility increases with increasing suspension concentrations accordingly. Once past some threshold value of H, however, the hydrodynamic swarming effect dominates again and the mobility decreases with further increase of H. Eventually, the mobility reduces to nullity when H approaches unity.

Contour plots of the dimensionless double layer polarization $\delta\rho^*$ and the dimensionless disturbed electric potential $\delta\phi^*$ are shown in Figs. 16.10A–C and 16.11A–C, respectively, for a highly charged particle ($\phi_r = 3$) at $\kappa a = 1$

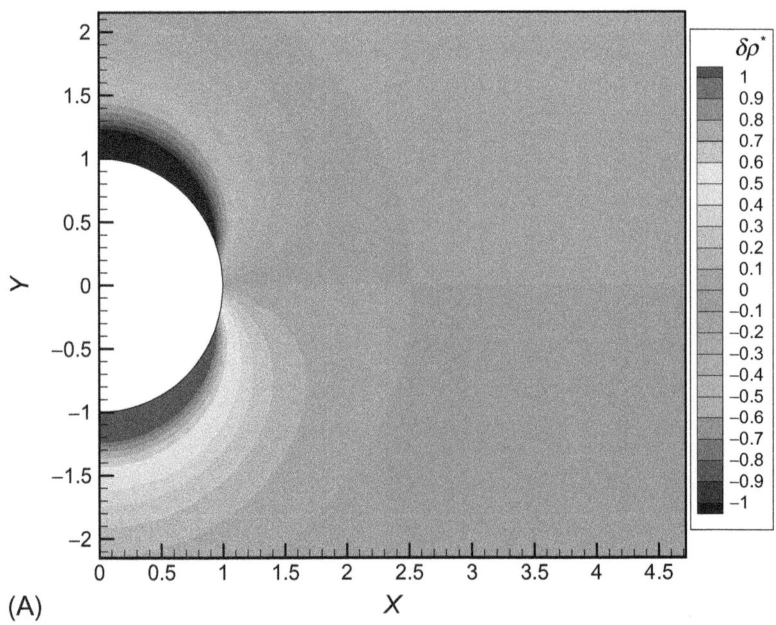

FIG. 16.10 Contour plot of the dimensionless disturbed space charge density $\delta\rho^*$ with $\phi_r = 3$, $\kappa a = 1$, and $\beta = 0$. (A) $H = 10^{-3}$; (B) $H = 0.1$; (C) $H = 0.4$.

Continued

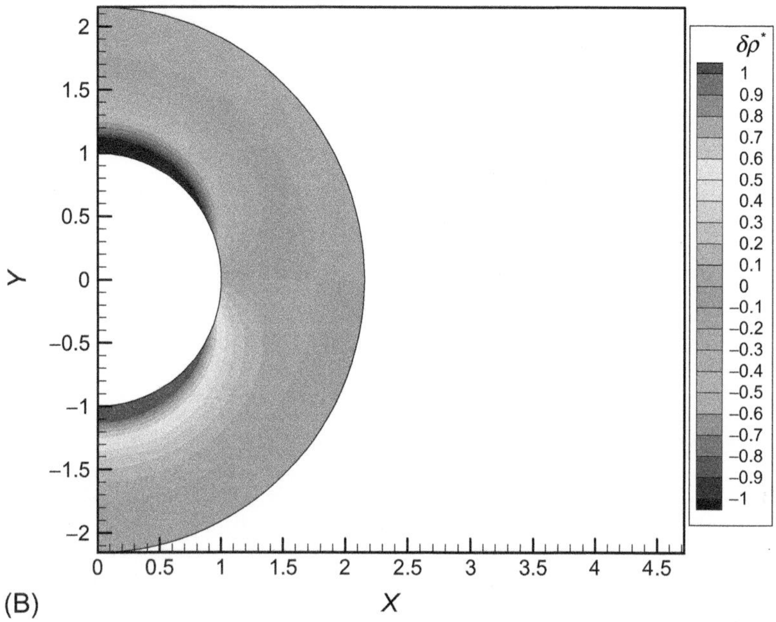

(B)

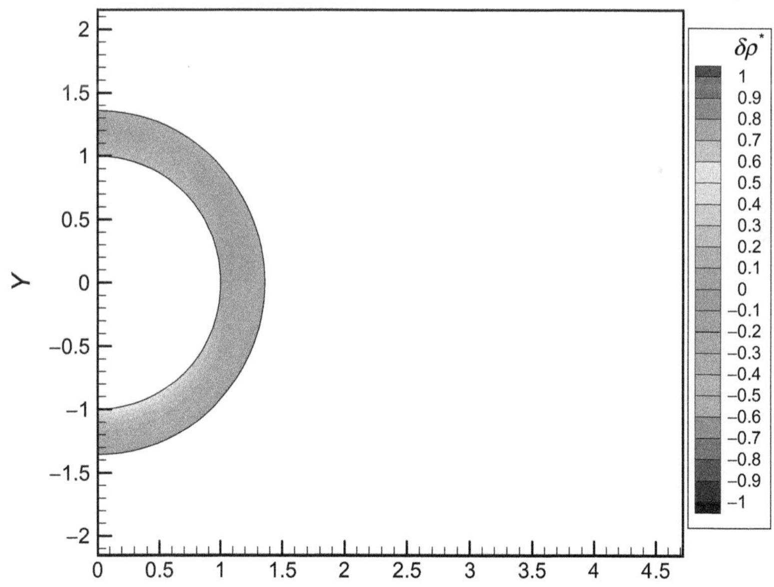

(C)

FIG. 16.10—CONT'D

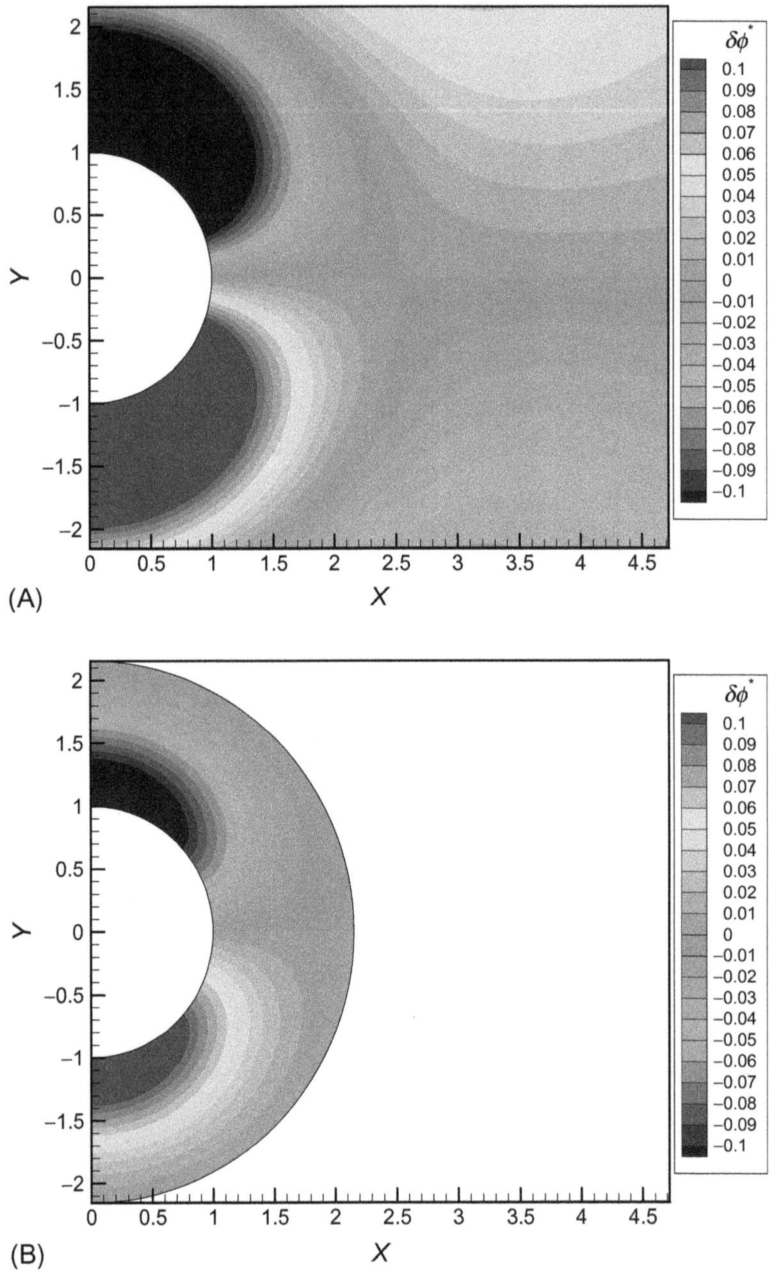

FIG. 16.11 Contour plot of the dimensionless disturbed potential $\delta\phi^*$ with $\phi_r=3$, $\kappa a=1$, and $\beta=0$. (A) $H=10^{-3}$; (B) $H=0.1$; (C) $H=0.4$.

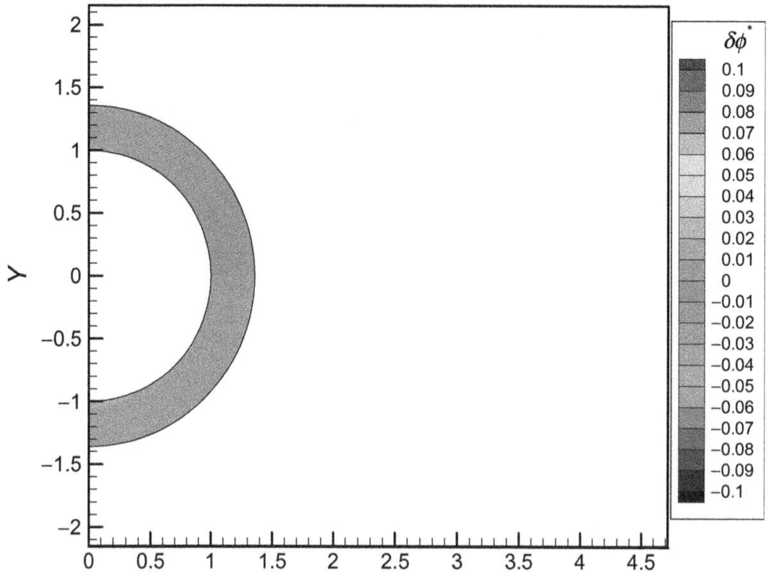

(C)

FIG. 16.11—CONT'D

in various suspension concentrations ($H = 10^{-3}$, 0.1, and 0.4). The behavior is similar to the electrophoresis treated in Chapter 2 in terms of double layer overlapping effect. Corresponding plots of $\phi_r = 5$ and $\kappa a = 10$ are shown in Fig. 16.12A–C for $\delta\rho^*$, and Fig. 16.13A–C for $\delta\phi^*$, respectively. Induced secondary polarization of ions generates an induced electric field against the particle motion at large κa ($\kappa a = 10$) here in general.

Corresponding general Henry's charts in NaCl solution $\beta = -0.2$ are shown in Fig. 16.14A–C for ϕ_r ranging from ± 1 to ± 5, and H ranging from 10^{-3} to 0.4. Electrophoresis component is now involved which has different impact upon positively or negatively charged particles. Three representative double layer thicknesses ($\kappa a = 0.1$, 1, and 10) are further picked to show the mobility profiles as a function of H for various ϕ_r ranging from ± 1 to ± 5, as shown in Fig. 16.15A–C. Combination of both the chemiphoresis and the electrophoresis components is underlying the particle behavior observed, which has been discussed earlier separately. These charts express the general behavior of particle diffusiophoretic motion in suspensions nicely. They are very revealing in terms of providing insightful observations on such a complicated system.

This concludes the discussion in this chapter.

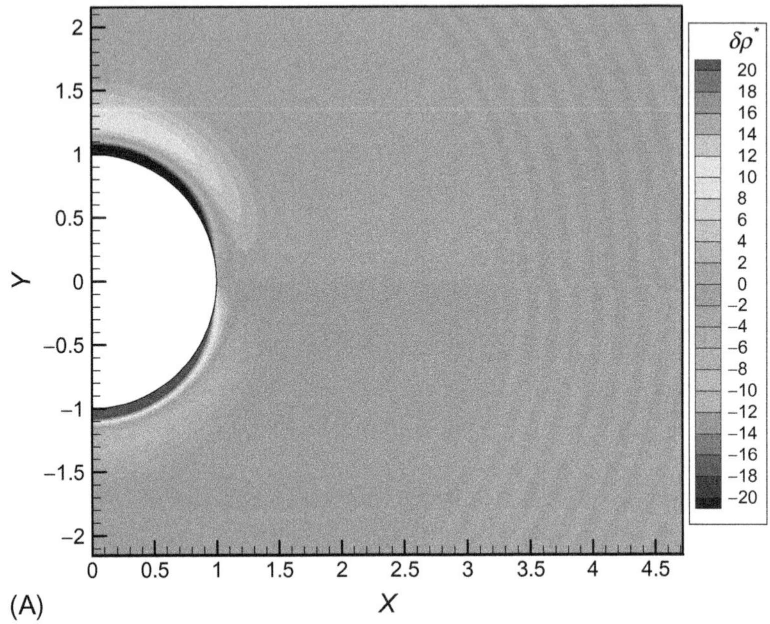

(A)

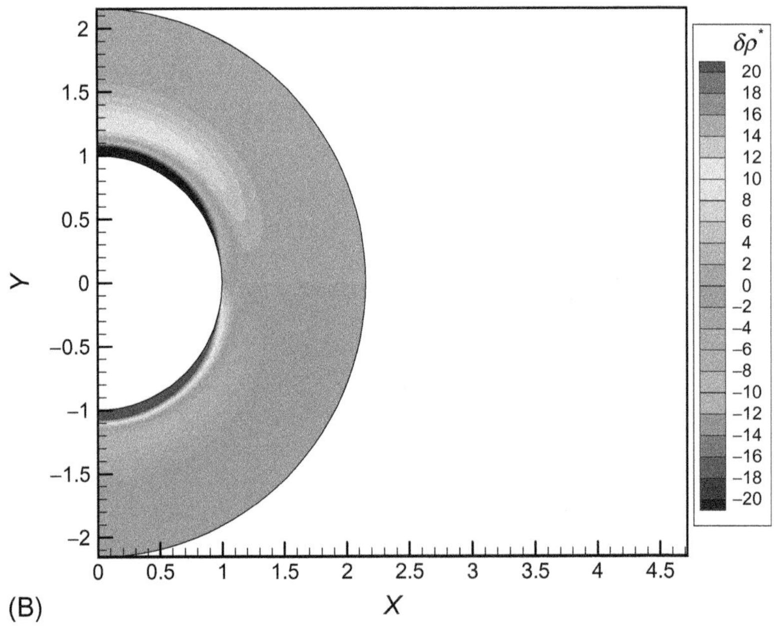

(B)

FIG. 16.12 Contour plot of the dimensionless disturbed space charge density $\delta\rho^*$ with $\phi_r = 5$, $\kappa a = 10$, and $\beta = 0$. (A) $H = 10^{-3}$; (B) $H = 0.1$; (C) $H = 0.4$.

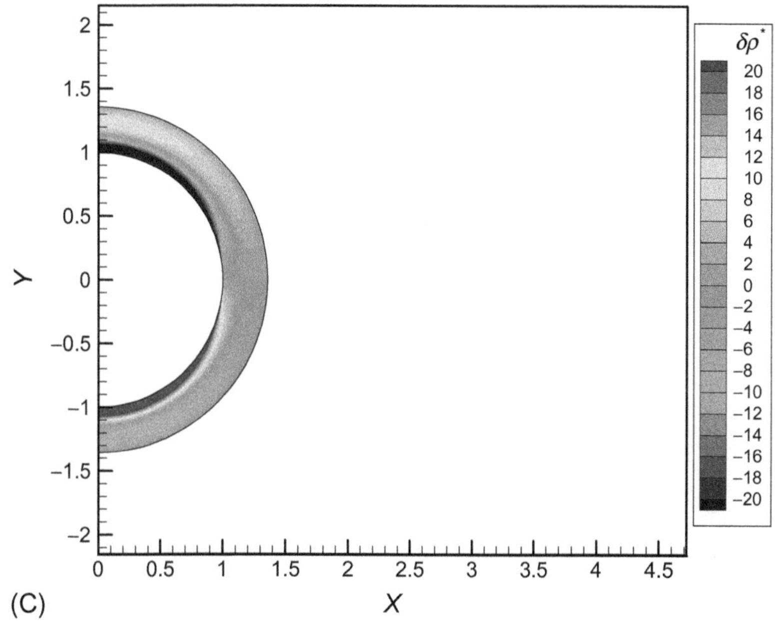

FIG. 16.12—CONT'D

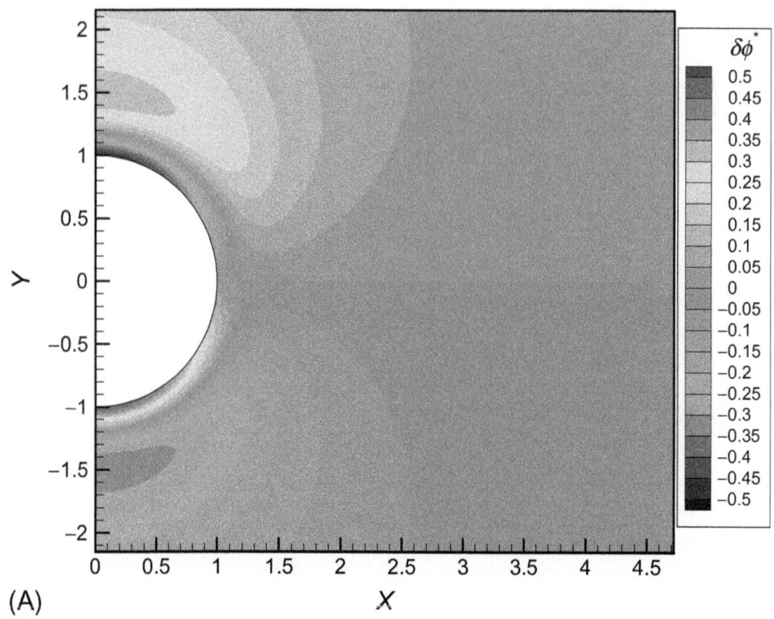

FIG. 16.13 Contour plot of the dimensionless disturbed potential $\delta\phi^*$ with $\phi_r = 5$, $\kappa a = 10$, and $\beta = 0$. (A) $H = 10^{-3}$; (B) $H = 0.1$; (C) $H = 0.4$.

Continued

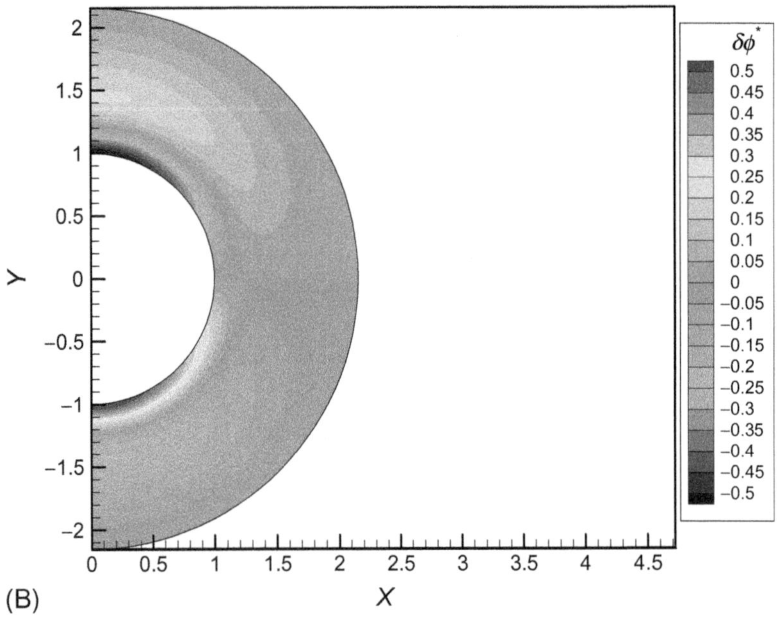

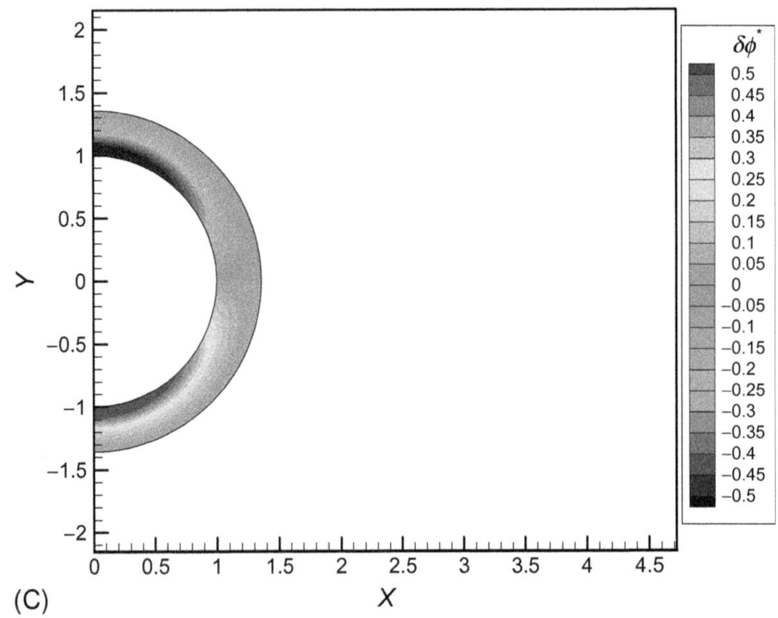

FIG. 16.13—CONT'D

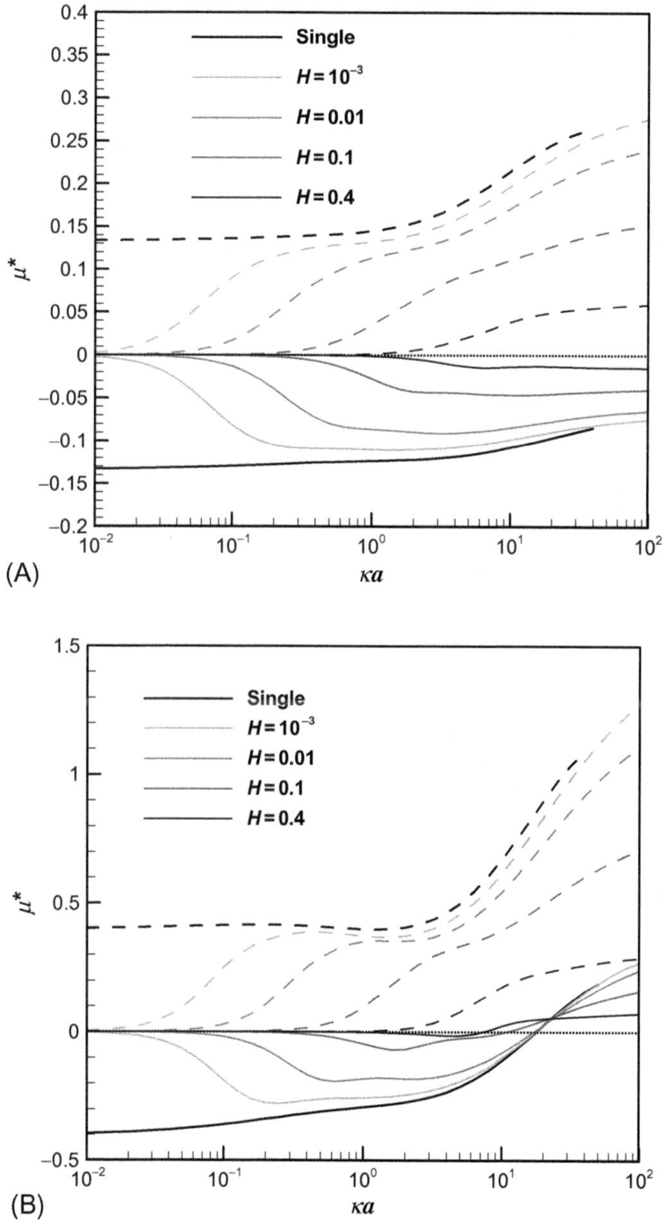

FIG. 16.14 Dimensionless mobility μ^* as a function of κa at various H with $\beta = -0.2$. *Solid lines*: positive ϕ_r. *Dashed lines*: negative ϕ_r. (A) $\phi_r = \pm 1$; (B) $\phi_r = \pm 3$; (C) $\phi_r = \pm 5$.

Continued

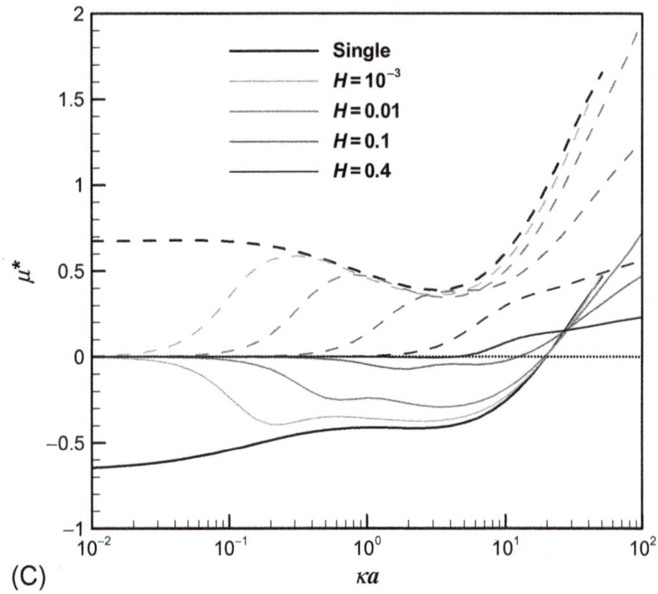

FIG. 16.14—CONT'D

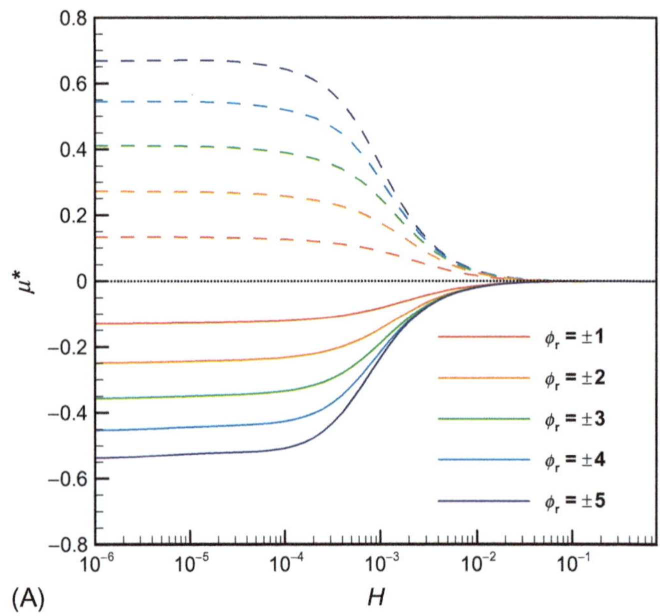

FIG. 16.15 Dimensionless mobility μ^* as a function of H at various ϕ_r with $\beta = -0.2$. *Solid lines*: positive ϕ_r. *Dashed lines*: negative ϕ_r. (A) $\kappa a = 0.1$; (B) $\kappa a = 1$; (C) $\kappa a = 10$.

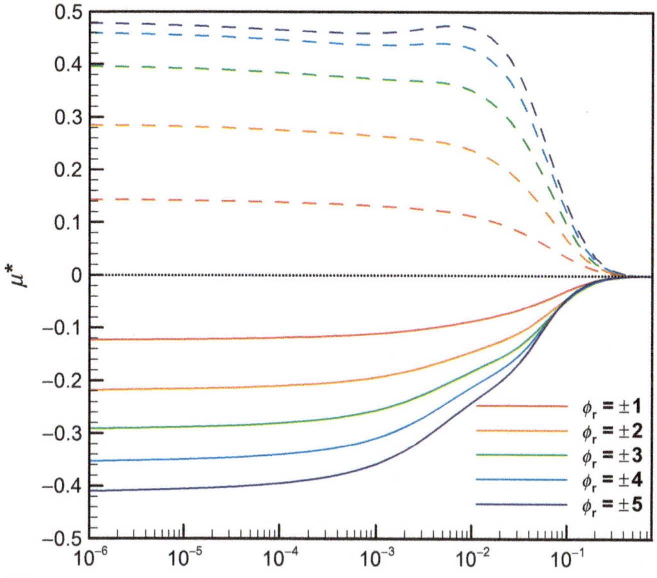

(B)

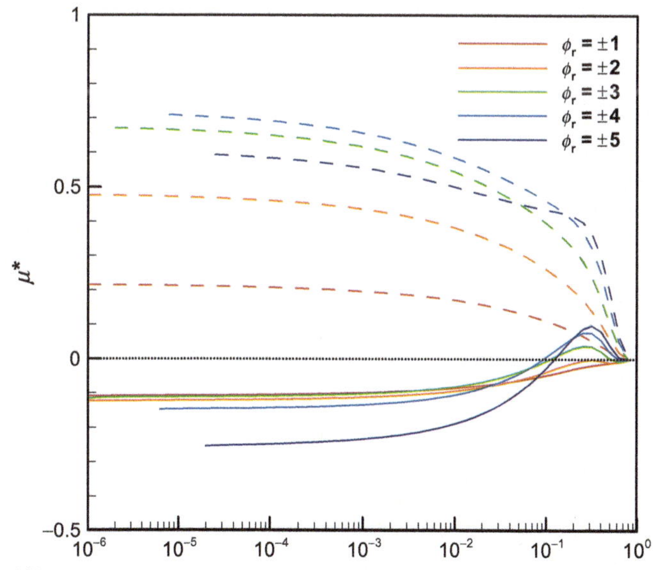

(C)

FIG. 16.15—CONT'D

REFERENCES

[1] Delgado AV, Arroyo F. Electrokinetic phenomena and their experimental determination: an overview. In: Delgado AV, editor. Interfacial electrokinetics and electrophoresis. New York: Marcel Dekker; 2002. p. 1–54.

[2] Derjaguin BV, et al. Kinetic phenomena in the boundary-layers of liquids. 1. The capillary osmosis. Prog Surf Sci 1993;43(1–4):138–52.

[3] Anderson JL, Lowell ME, Prieve DC. Motion of a particle generated by chemical gradients. 1. Non-electrolytes. J Fluid Mech 1982;117(APR):107–21.

[4] Prieve DC, et al. Motion of a particle generated by chemical gradients. 2. Electrolytes. J Fluid Mech 1984;148(NOV):247–69.

[5] Dukhin S, et al. Diffusiophoresis in electrolyte solutions and its application to the formation of surface coatings. Russ Chem Bull 1982;31(8):1535–44.

[6] Golestanian R. Anomalous diffusion of symmetric and asymmetric active colloids. Phys Rev Lett 2009;102(18):188305.

[7] Moran J, Wheat P, Posner J. Locomotion of electrocatalytic nanomotors due to reaction induced charge autoelectrophoresis. Phys Rev E 2010;81(6):065302.

[8] Velegol D, et al. Origins of concentration gradients for diffusiophoresis. Soft Matter 2016;12(21):4686–703.

[9] Zhou C, et al. Photochemically powered AgCl Janus micromotors as a model system to understand ionic self-diffusiophoresis. Langmuir 2018;34(10):3289–95.

[10] Singh DP, et al. Photogravitactic microswimmers. Adv Funct Mater 2018;28:1706660.

[11] Solovev AA, et al. Catalytic microtubular jet engines self-propelled by accumulated gas bubbles. Small 2009;5(14):1688–92.

[12] Cho SK. Mini and micro propulsion for medical swimmers. Micromachines 2014;5(1):97–113.

[13] Mou F, et al. Autonomous motion and temperature-controlled drug delivery of Mg/Pt-poly (N-isopropylacrylamide) Janus micromotors driven by simulated body fluid and blood plasma. ACS Appl Mater Interfaces 2014;6(12):9897–903.

[14] Agarwal A, Hess H. Biomolecular motors at the intersection of nanotechnology and polymer science. Prog Polym Sci 2010;35(1-2):252–77.

[15] Sánchez S, Soler L, Katuri J. Chemically powered micro-and nanomotors. Angew Chem Int Ed 2015;54(5):1414–44.

[16] Katuri J, et al. Designing micro-and nanoswimmers for specific applications. Acc Chem Res 2016;50(1):2–11.

[17] Golestanian R, Liverpool T, Ajdari A. Designing phoretic micro-and nano-swimmers. New J Phys 2007;9(5):126.

[18] Kolmakov GV, et al. Designing self-propelled microcapsules for pick-up and delivery of microscopic cargo. Soft Matter 2011;7(7):3168–76.

[19] Ebbens SJ, Howse JR. In pursuit of propulsion at the nanoscale. Soft Matter 2010;6(4):726–38.

[20] Fukuyama T, et al. Directing and boosting of cell migration by the entropic force gradient in polymer solution. Langmuir 2015;31(46):12567–72.

[21] Paustian JS, et al. Direct measurements of colloidal solvophoresis under imposed solvent and solute gradients. Langmuir 2015;31(15):4402–10.

[22] Shin S, et al. Size-dependent control of colloid transport via solute gradients in dead-end channels. Proc Natl Acad Sci 2016;113(2):257–61.

[23] Niu R, et al. Microfluidic pumping by micromolar salt concentrations. Soft Matter 2017;13(7):1505–18.

[24] Hartman SV, Božič B, Derganc J. Migration of blood cells and phospholipid vesicles induced by concentration gradients in microcavities. New Biotechnol 2018, in press.

[25] Staffeld PO, Quinn JA. Diffusion-induced banding of colloid particles via diffusiophoresis. 2. Non-electrolytes. J Colloid Interface Sci 1989;130(1):88–100.

[26] Anderson JL. Colloid transport by interfacial forces. Annu Rev Fluid Mech 1989;21:61–99.

[27] Pawar Y, Solomentsev YE, Anderson JL. Polarization effects on diffusiophoresis in electrolyte gradients. J Colloid Interface Sci 1993;155(2):488–98.

[28] Gao W, Uygun A, Wang J. Hydrogen-bubble-propelled zinc-based microrockets in strongly acidic media. J Am Chem Soc 2011;134(2):897–900.

[29] Dukhin SS. Nonequilibrium electric surface phenomena. Adv Colloid Interf Sci 1993;44:1–134.

[30] Sandler SI. Chemical, biochemical, and engineering thermodynamics. Hoboken, NJ: John Wiley & Sons; 2017.

[31] Hsu JP, Hsu WL, Chen ZS. Boundary effect on diffusiophoresis: spherical particle in a spherical cavity. Langmuir 2009;25(3):1772–84.

[32] Hsu JP, et al. Diffusiophoresis of a sphere along the axis of a cylindrical pore. J Colloid Interface Sci 2010;342(2):598–606.

[33] Hsu JP, et al. Diffusiophoresis of a charge-regulated spherical particle normal to two parallel disks. J Phys Chem B 2010;114(8):2766–78.

[34] Liu K-L, et al. Diffusiophoresis of a polyelectrolyte in a salt concentration gradient. Electrophoresis 2012;33(6):1068–78.

[35] Hsu J-P, et al. Separation of charge-regulated polyelectrolytes by pH-assisted diffusiophoresis. Phys Chem Chem Phys 2017;19(13):9059–63.

[36] Lou J, He YY, Lee E. Diffusiophoresis of concentrated suspensions of spherical particles with identical ionic diffusion velocities. J Colloid Interface Sci 2006;299(1):443–51.

[37] Hsu JP, et al. Diffusiophoresis of concentrated suspensions of spherical particles with distinct ionic diffusion velocities. J Phys Chem B 2007;111(10):2533–9.

[38] Lou J, Lee E. Diffusiophoresis of concentrated suspensions of liquid drops. J Phys Chem C 2008;112(32):12455–62.

[39] Lou J, Lee E. Diffusiophoresis of a spherical particle normal to a plane. J Phys Chem C 2008;112(7):2584–92.

[40] Lou J, Shih CY, Lee E. Diffusiophoresis of a spherical particle normal to an air-water interface. J Colloid Interface Sci 2009;331(1):227–35.

[41] Lou J, Shih CY, Lee E. Diffusiophoresis of concentrated suspensions of spherical particles with charge-regulated surface: polarization effect with nonlinear Poisson-Boltzmann equation. Langmuir 2010;26(1):47–55.

[42] Fang W, Lee E. Diffusiophoretic motion of an isolated charged porous sphere. J Colloid Interface Sci 2015;459:273–83.

[43] Haynes WM. CRC handbook of chemistry and physics. Boca Raton, FL: CRC Press; 2014.

[44] Glawdel T, et al. Microfluidic system with integrated electroosmotic pumps, concentration gradient generator and fish cell line (RTgill-W1)-towards water toxicity testing. Lab Chip 2009;9(22):3243–50.

[45] Jeon H, et al. Quantitative analysis of single bacterial chemotaxis using a linear concentration gradient microchannel. Biomed Microdevices 2009;11(5):1135–43.

[46] Nery-Azevedo R, Banerjee A, Squires TM. Diffusiophoresis in ionic surfactant gradients. Langmuir 2017;33(38):9694–702.

[47] Voldman J. Electrical forces for microscale cell manipulation. Annu Rev Biomed Eng 2006;8:425–54.

[48] Yadav V, et al. Bone-crack detection, targeting, and repair using ion gradients. Angew Chem Int Ed Eng 2013;52(42):10997–1001.

[49] Kar A, et al. Enhanced transport into and out of dead-end pores. ACS Nano 2015;9(1):746–53.

[50] Fang H. Electrokinetics of soft matter: diffusiophoresis and charge-regulation phenomena. PhD thesis, National Taiwan University; 2015.

[51] Prieve DC, Roman R. Diffusiophoresis of a rigid sphere through a viscous electrolyte solution. J Chem Soc Faraday Trans II 1987;83:1287–306.

[52] Anderson JL, Prieve DC. Diffusiophoresis: migration of colloidal particles in gradients of solute concentration. Sep Purif Meth 1984;13(1):67–103.

Chapter 17

Diffusiophoresis of Liquid Droplets and Gas Bubbles

17.1 INTRODUCTION

Compared with the rigid particles considered in Chapter 16, diffusiophoretic motion of a liquid droplet or a gas bubble is faster. In general, the role of the mobile droplet or bubble surface is similar to the corresponding electrophoresis of droplets or bubbles [1–4] in terms of the generation of a motion-enhancing electroosmotic flow, which can also be analogized as an effective excess surface conductivity. This increase is drastic when κa is large due to the double layer suppression effect. The only question here is by what extent this increase is.

As for its applications, essentially all the systems discussed in Chapter 12 for the electrophoresis of liquid droplet and gas bubble systems are also possible stages for diffusiophoresis. In practice, however, it is the biosystems such as the human stem cells or other mammal cells encountered in various situations in biochemical and biomedical fields that are particularly of interest [5–8]. Certainly, it would not be a good idea in general to apply a measurable electric field constantly and maybe over a long time upon a living biosystem, such as the human body. Diffusiophoresis, on the other hand, does not have this adverse impact upon biosystems. All one needs to do is to establish a concentration gradient of electrolytes to guide the motion of the droplet, or simply let nature take over. For instance, extraction of cells or droplet type of substances from a microcavity can be easily done with a constant flow of electrolyte solution passing over the entrance to the microcavity [9–11]. A concentration gradient is thus established across the longitudinal direction of the pore interior, which drives the droplet type of colloidal entities out from the dead end pore, as we introduced in Chapter 16. Moreover, the application in extracting surfactant-covered oil droplets from an oil-rich porous reservoir in the technology of the enhanced oil recovery [12] is very obvious. With the injection of electrolyte solution to the system, preferably the one which can result in large droplet velocity by establishing a large diffusion potential

Interface Science and Technology, Vol. 26. https://doi.org/10.1016/B978-0-08-100865-2.00017-5

spontaneously, i.e., large β, the production rate of oil can be drastically increased. The mechanism is essentially the same as those mentioned above in the dead-end-pore type of application for cells, where conventional pressure-driven type of flow is proven to fail in terms of providing the driving force for the desired droplet motion [10, 11, 13–16].

Moreover, nanoemulsion and microemulsion are very promising drug delivery systems in biomedical applications due to their various merits, such as better thermodynamic stability, optical clarity, and ease of preparation [17–26]. Optimum targeted drug delivery can be achieved in principle with these nano-reservoir systems by diffusiophoresis, if some specific chemicals are released from the injury wound or infected area [27–30]. A concentration gradient would be thus established in its neighborhood, which provides the guiding signal as well as the driving force for the diffusiophoretic motion of the drug-carrying nanoemulsions. It functions like the detection and remedy of the bone-crack case by rigid quantum dots introduced in Chapter 16 [31], but here the liquid droplets of nanoemulsions or microemulsions carry the workload instead.

Here, in this chapter, we focus on a sample liquid droplet system with viscosity ratio $\sigma_H=0.5$ as well as a gas bubble system with $\sigma_H=0.01$, as we did in Chapter 12 for the electrophoresis situation. The results are provided in Section 17.3.

17.2 THEORY

The system diagrams for a single droplet or a gas bubble and their suspensions are shown in Fig. 17.1A and B, respectively. Note that both the governing equations and associated boundary conditions on the droplet or bubble surface are the same as those in Chapter 12 where electrophoresis is considered. The only difference is the boundary conditions at infinity or the outer virtual surface in the cell model selected, which is again the same as those selected in Chapter 12. Hence, simply replace the boundary conditions there in Chapter 12 by those in Chapter 16 where a concentration gradient is applied and we are done.

17.3 RESULTS AND DISCUSSION

We examine the chemiphoresis component first, i.e., $\beta=0$. Fig. 17.2A and B show the corresponding general Henry's charts in diffusiophoresis for a sample liquid droplet with $\sigma_H=0.5$ in a KCl electrolyte solution, i.e., $\beta=0$, or only the chemiphoresis component is considered.

As shown in Fig. 17.2A for a close-up view of the mobility profiles when κa is relatively small, the droplet mobility increases with increasing κa in general. For a highly charged droplet ($\phi_r=5$), the behavior is kind of different

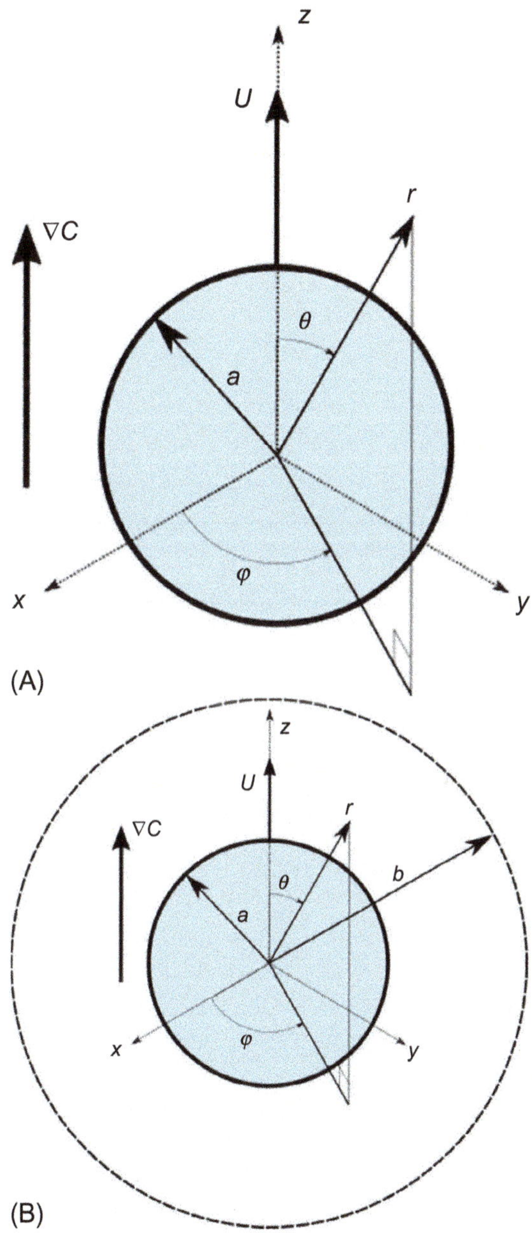

FIG 17.1 System diagram. (A) Single liquid droplet or gas bubble; (B) suspension of liquid droplets or gas bubbles.

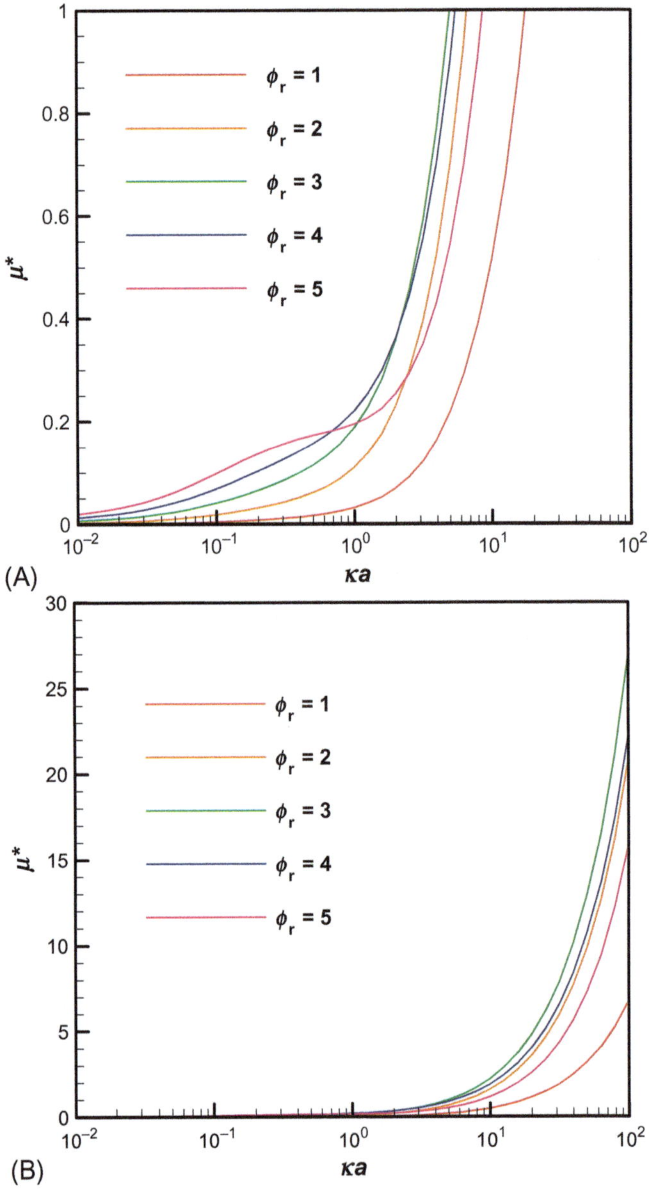

FIG. 17.2 Dimensionless mobility μ^* as a function of κa at various ϕ_r with $\sigma_H = 0.5$ ($\beta = 0$). (A) Small scale; (B) large scale.

from the rest, but the trend is still the same. Global view shown in Fig. 17.2B indicates the large droplet mobility when κa is around 100. This behavior of drastic mobility increase when κa is large is similar to the corresponding electrophoresis phenomenon observed in Chapter 12.

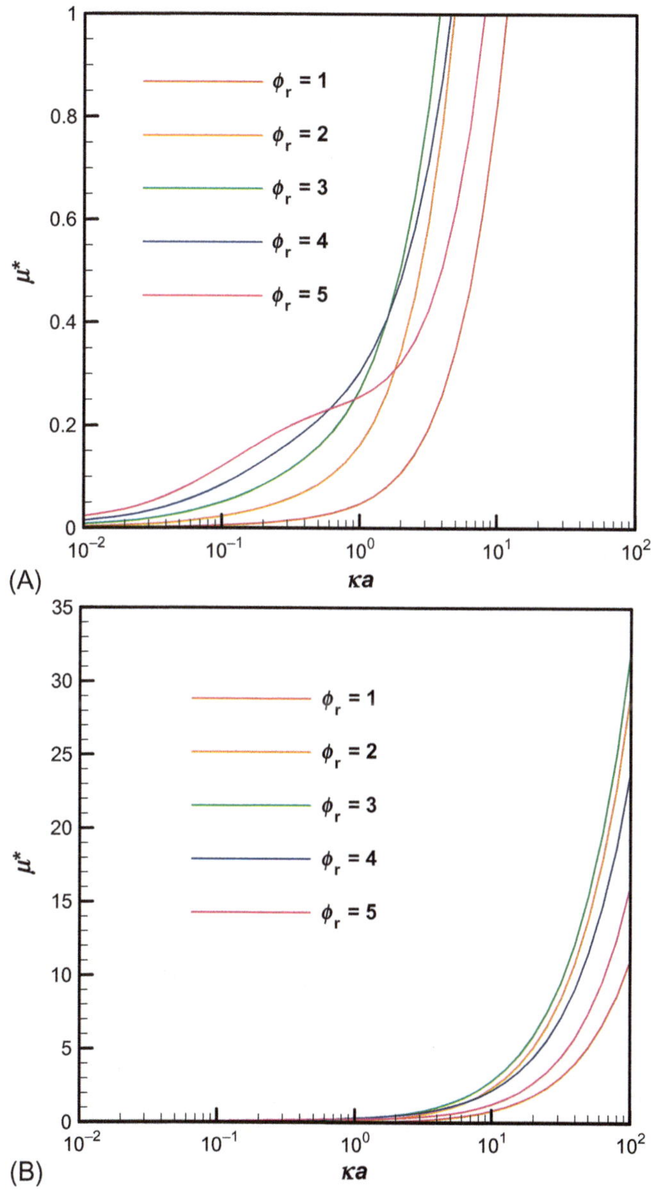

FIG. 17.3 Dimensionless mobility μ^* as a function of κa at various ϕ_r with $\sigma_H = 0.01$ ($\beta = 0$). (A) Small scale; (B) large scale.

Corresponding charts for the gas bubble ($\sigma_H = 0.01$) are presented in Fig. 17.3A and B. The behavior is essentially the same as the liquid droplet ($\sigma_H = 0.5$) considered above, except that the mobility is larger due to the smaller hydrodynamic drag force as well as the larger electric driving force, which is due to the faster velocity on the mobile surface as discussed in Chapter 12 for electrophoresis. Apparently, the basic scenario remains the same here.

The effect of droplet viscosity is shown in Fig. 17.4A–C, where the droplet mobility is expressed as a function of σ_H, the viscosity ratio of the droplet to the suspending medium, for various charge levels (ϕ_r ranging from 1 to 5) at three different double layer thickness ($\kappa a=0.1$, 1, and 10). In general, the larger σ_H is, the slower the particle moves, regardless of the double layer thickness. Moreover, mobility ratio to the corresponding rigid particle is presented in Fig. 17.5A–C to highlight the physical sense underlying it. Over 15 times faster motion is observed in Fig. 17.5C for a gas bubble ($\sigma_H=0.01$). Note that this is due to the joint effect of lower hydrodynamic drag force and larger electric driving force, as discussed in Chapter 12 for electrophoresis. Note that for $\sigma_H=100$, the droplet mobility reduces to the rigid particle situation.

Fig. 17.6A presents the particle mobility as a function of surface charge ϕ_r for the sample liquid droplet ($\sigma_H=0.5$) at various double layer thickness ($\kappa a=1$, 2, 5, and 10). Rigid particle situation is shown as well for comparison purpose. Other than larger mobility for a droplet, the behavior is essentially the same as the rigid particle system. The higher κa is, the larger the droplet mobility is. Moreover, a highly charged droplet may move slower than a lowly charged one with two local maxima in each profile. The profiles are symmetric with the vertical axis $\phi_r=0$. Corresponding profiles of a gas bubble are shown in Fig. 17.6B.

We now go on to examine the general case with the involvement of the electrophoresis component, i.e., $\beta\neq 0$. Figs. 17.7–17.11 show the overall results with the particles suspended in a NaCl solution, i.e., $\beta=-0.2$. Really, the general behavior is similar to the observation in Chapter 16 for the rigid particle situation, except with larger mobility for the sample liquid system ($\sigma_H=0.5$) and gas bubble ($\sigma_H=0.01$). No further comments are needed as these figures are self-explanatory. They are presented here as they provide various original information about the diffusiophoresis phenomenon, which still remains as an unfamiliar novelty to most people, especially the behavior of highly charged liquid droplets and gas bubbles. These figures shed light on the mysterious diffusiophoretic motion of liquid droplet and gas bubble systems. With their help, we now know what an elephant really looks like.

The impact of the β factor is shown in Fig. 17.12 for a representative liquid droplet ($\sigma_H=0.5$) and a gas bubble ($\sigma_H=0.01$), respectively. Again, these figures are self-explanatory with general behavior similar to the rigid particle systems treated in Chapter 16.

We now go on to examine the diffusiophoretic motion of liquid droplets and gas bubbles in suspensions. The results are shown in Figs. 17.13–17.16 for either sole chemiphoresis component ($\beta=0$) or with an extra involvement of the electrophoresis component as well ($\beta=-0.2$). Again, the general behavior is similar to the rigid particle situation in Chapter 16, except with larger mobility. The intriguing phenomenon is still observed that a liquid droplet or a gas bubble may move faster in a concentrated suspension than in a dilute one over

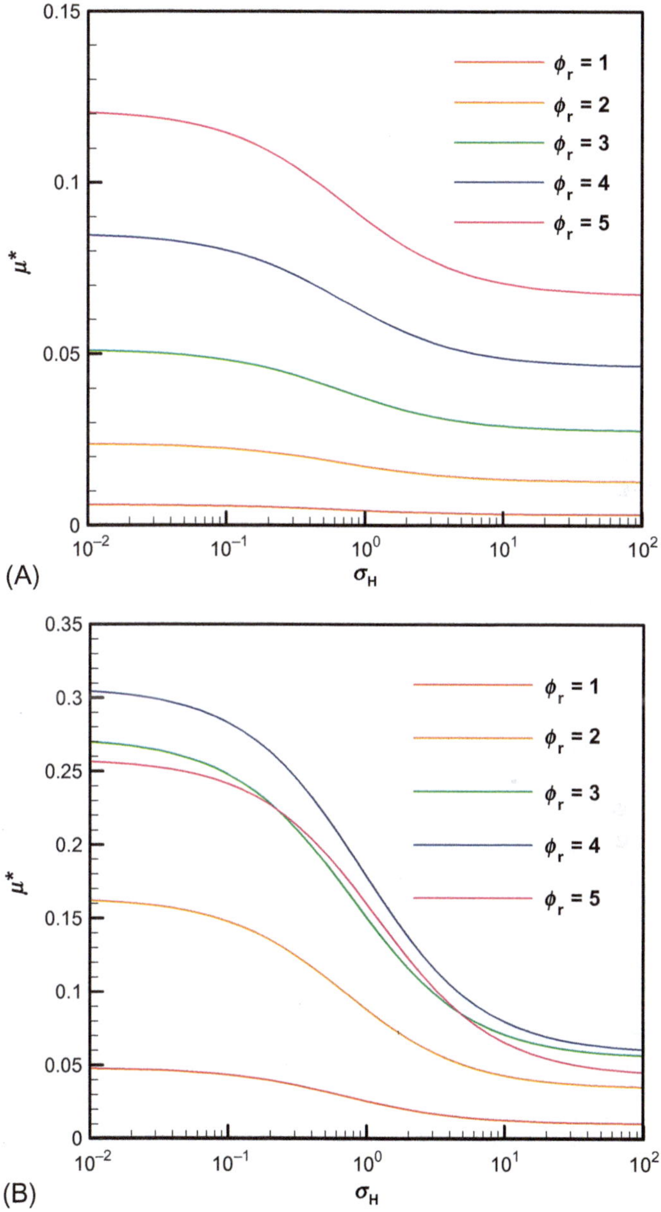

FIG. 17.4 Dimensionless mobility μ^* as a function of σ_H at various ϕ_r ($\beta=0$). (A) $\kappa a=0.1$; (B) $\kappa a=1$; (C) $\kappa a=10$.

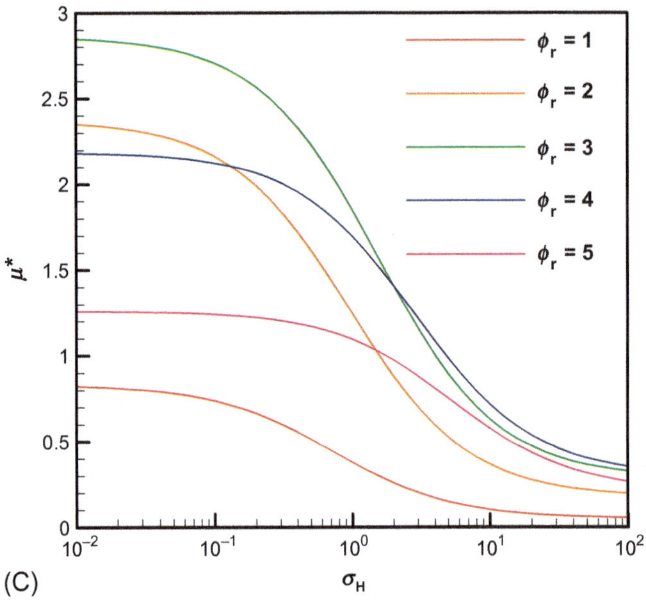

FIG. 17.4—CONT'D

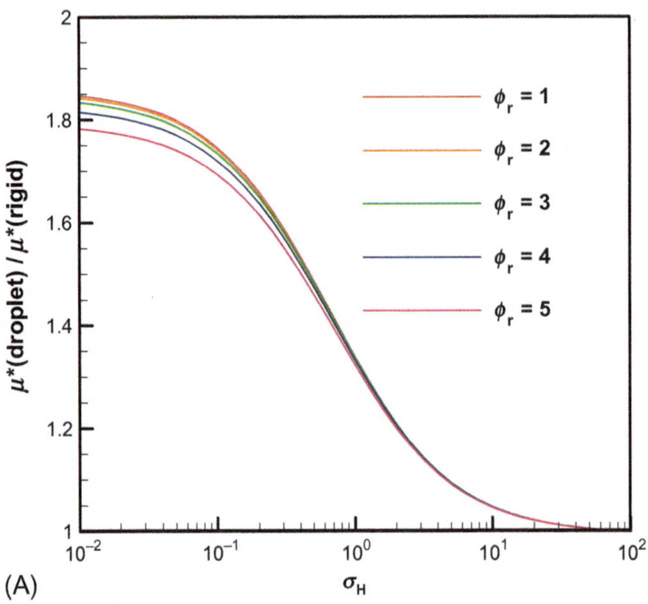

FIG. 17.5 Mobility ratio μ^*(droplet)/μ^*(rigid) as a function of σ_H at various ϕ_r ($\beta = 0$). (A) $\kappa a = 0.1$; (B) $\kappa a = 1$; (C) $\kappa a = 10$.

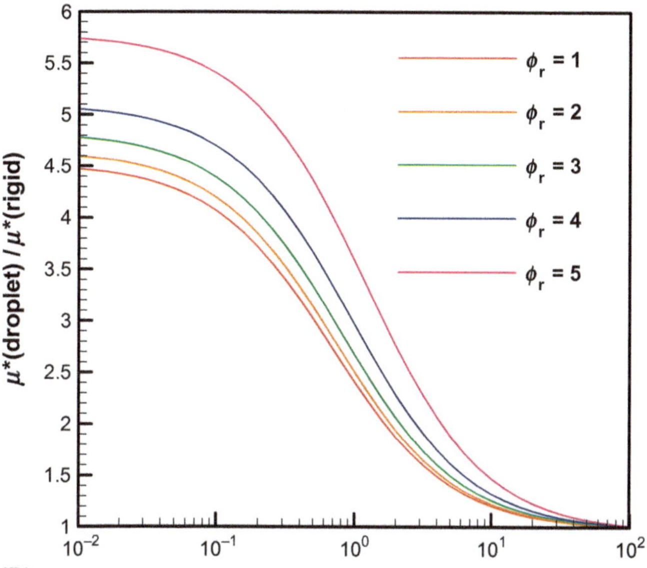

(B)

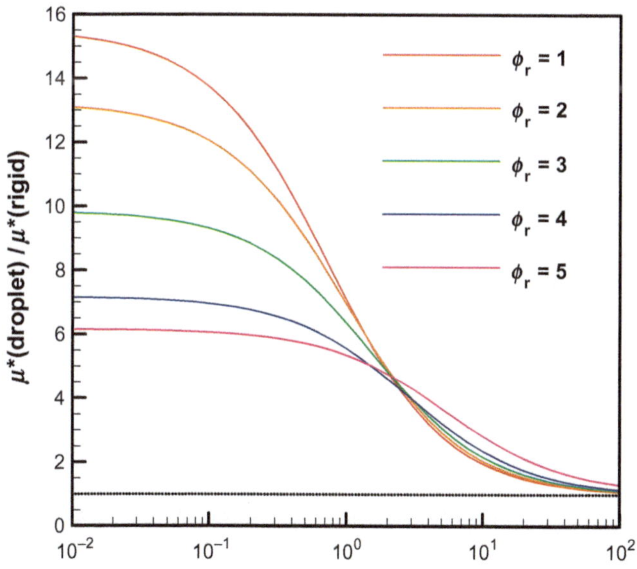

(C)

FIG. 17.5—CONT'D

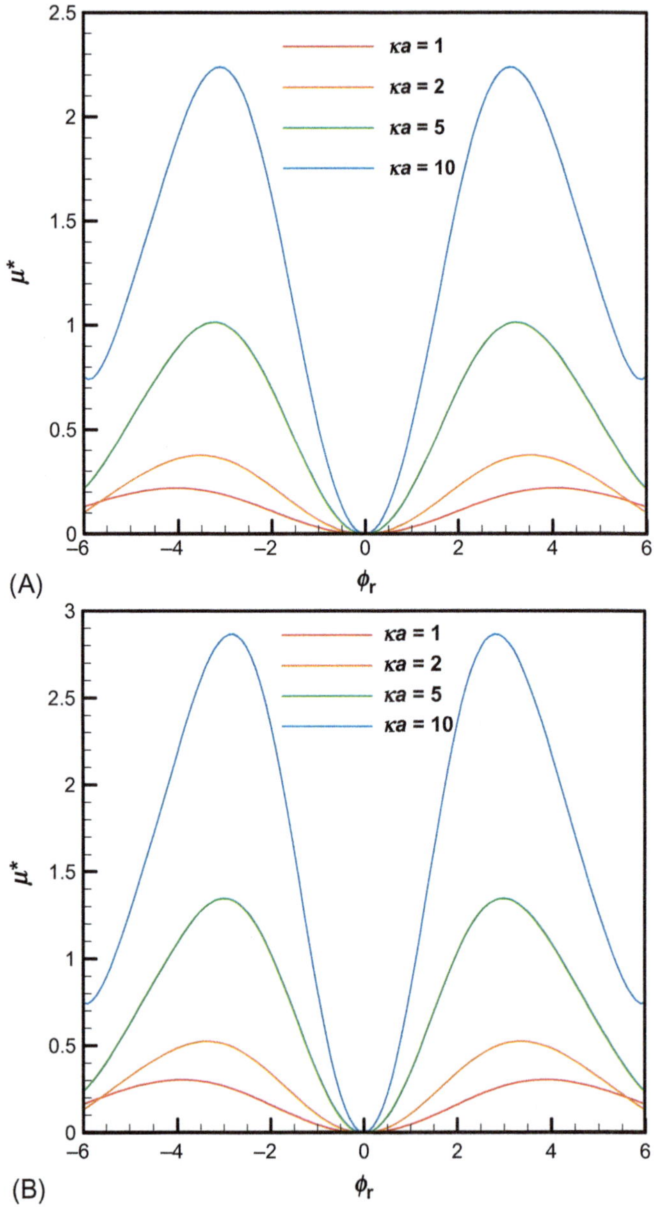

FIG. 17.6 Dimensionless mobility μ^* as a function of ϕ_r at various κa ($\beta = 0$). (A) $\sigma_H = 0.5$; (B) $\sigma_H = 0.01$.

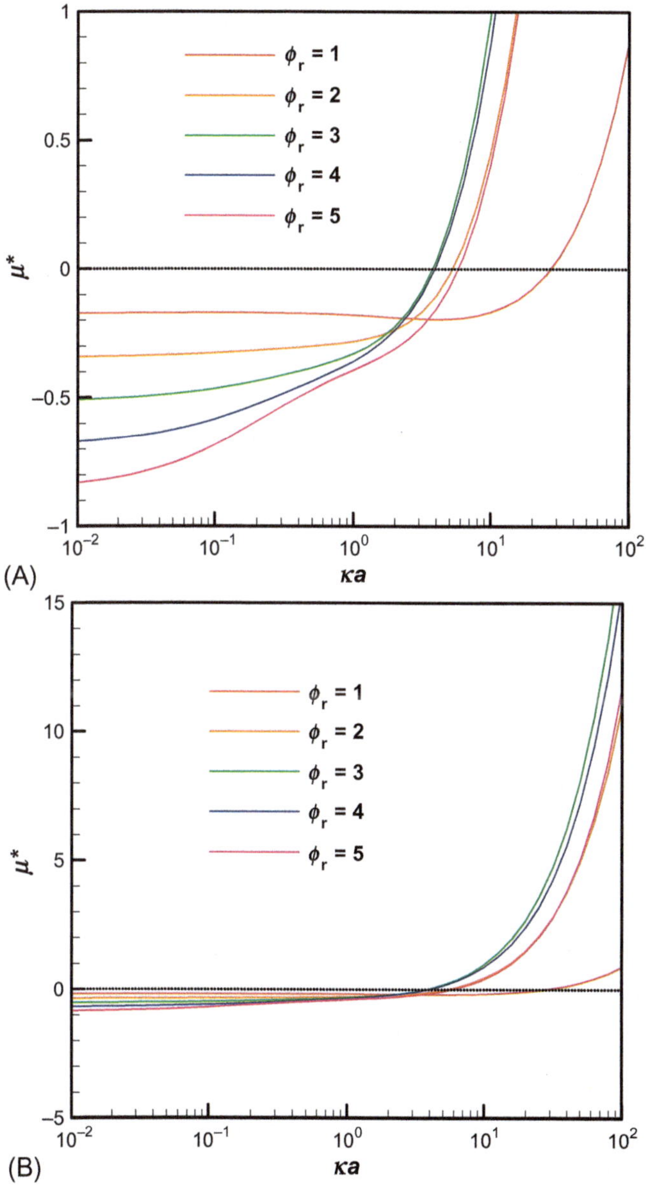

FIG. 17.7 Dimensionless mobility μ^* as a function of κa at various ϕ_r with $\sigma_H = 0.5$ ($\beta = -0.2$). (A) Small scale; (B) large scale.

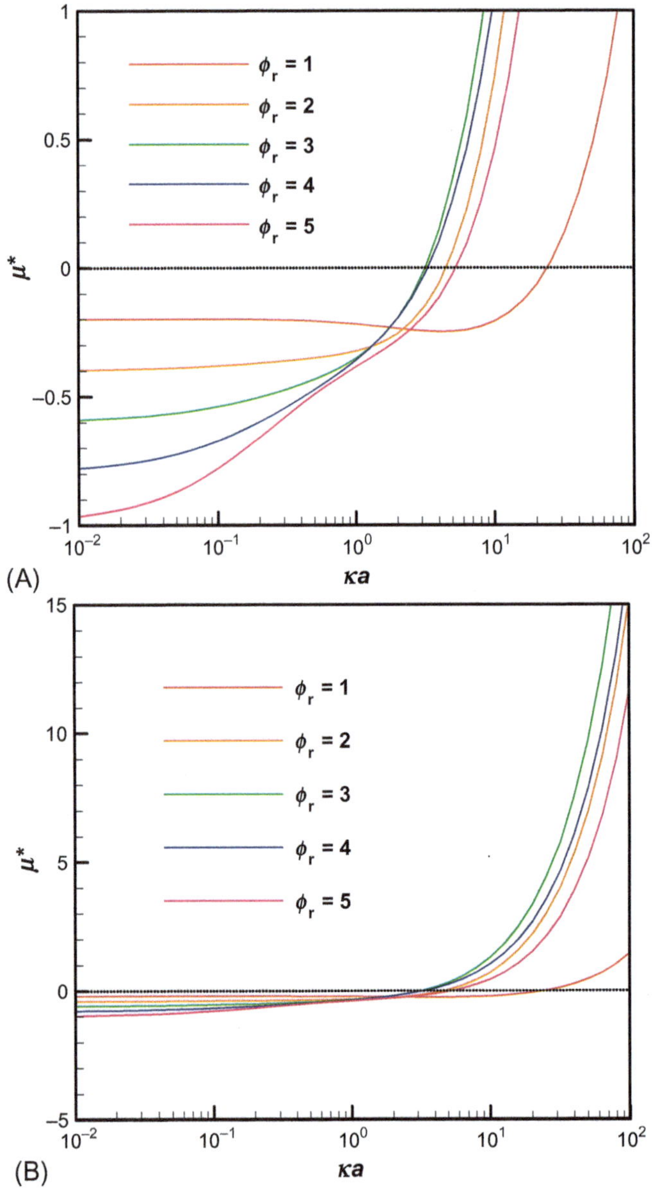

FIG. 17.8 Dimensionless mobility μ^* as a function of κa at various ϕ_r with $\sigma_H = 0.01$ ($\beta = -0.2$). (A) Small scale; (B) large scale.

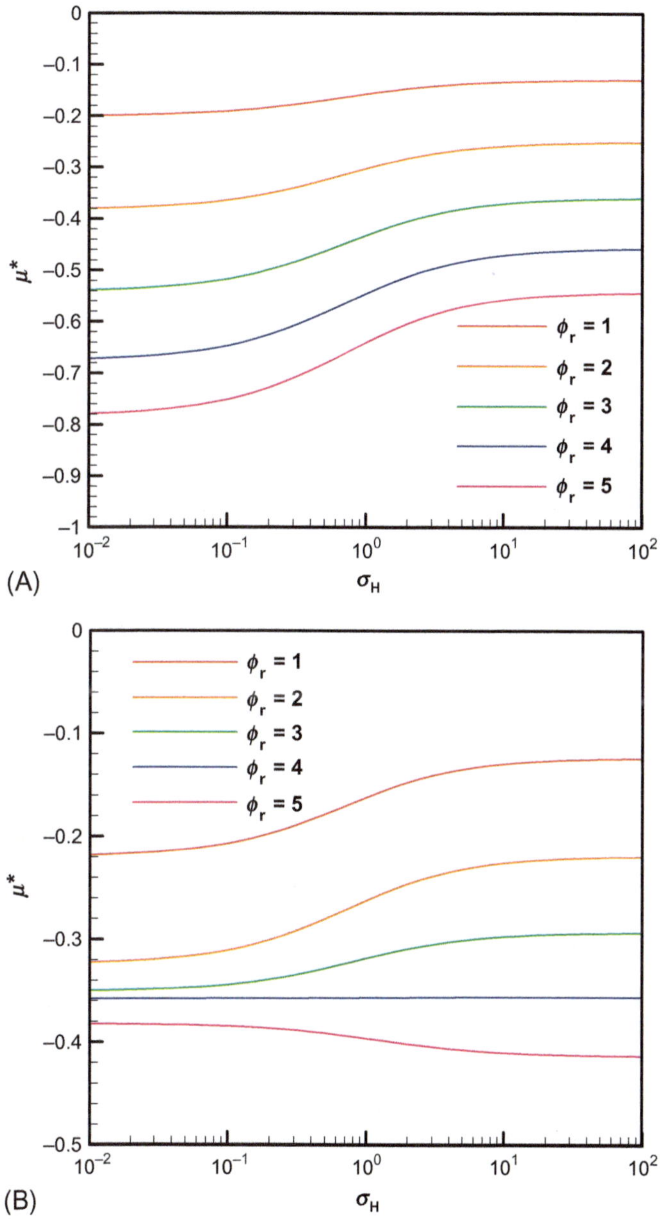

FIG. 17.9 Dimensionless mobility μ^* as a function of σ_H at various ϕ_r ($\beta = -0.2$). (A) $\kappa a = 0.1$; (B) $\kappa a = 1$; (C) $\kappa a = 10$.

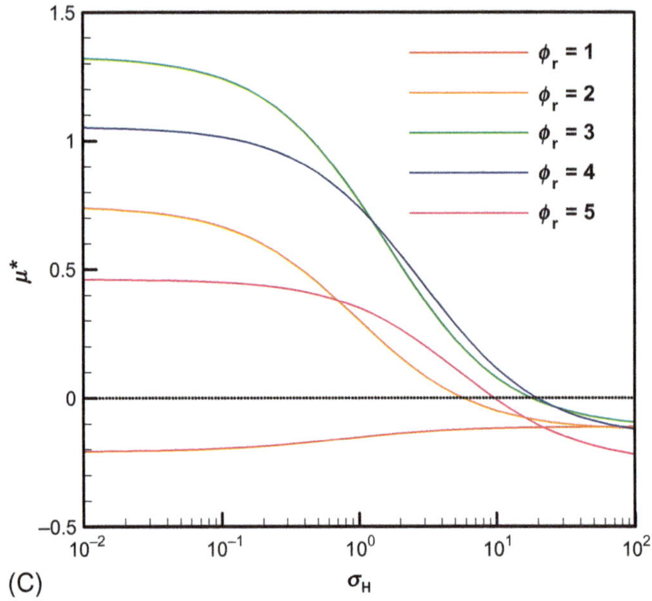

FIG. 17.9—CONT'D

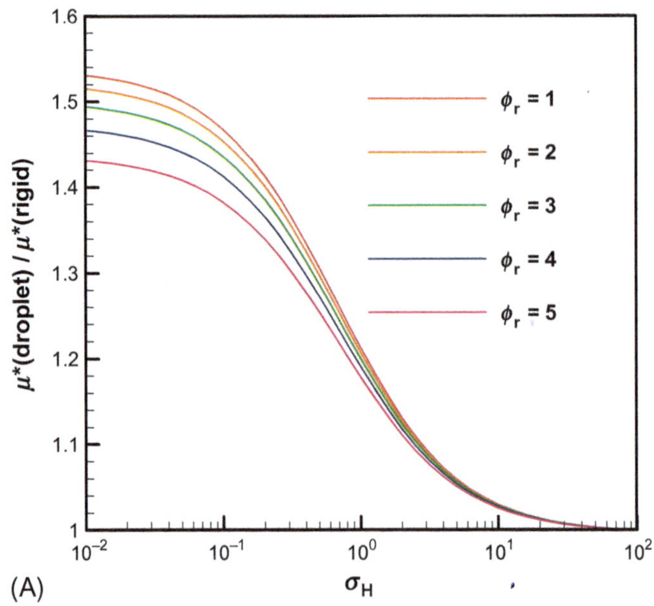

FIG. 17.10 Mobility ratio μ^*(droplet)/μ^*(rigid) as a function of σ_H at various ϕ_r ($\beta = -0.2$). (A) $\kappa a = 0.1$; (B) $\kappa a = 1$; (C) $\kappa a = 10$.

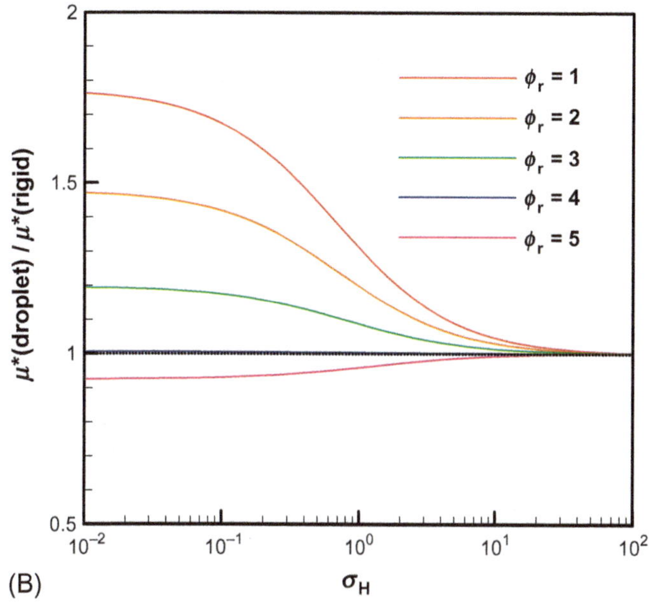

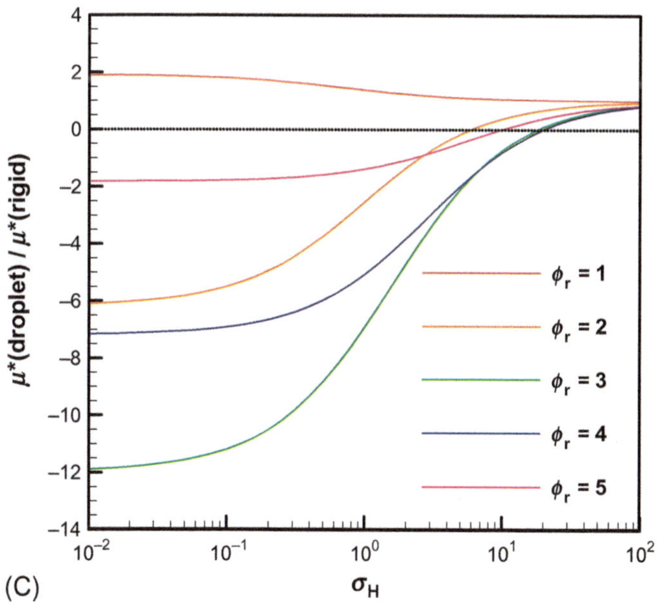

FIG. 17.10—CONT'D

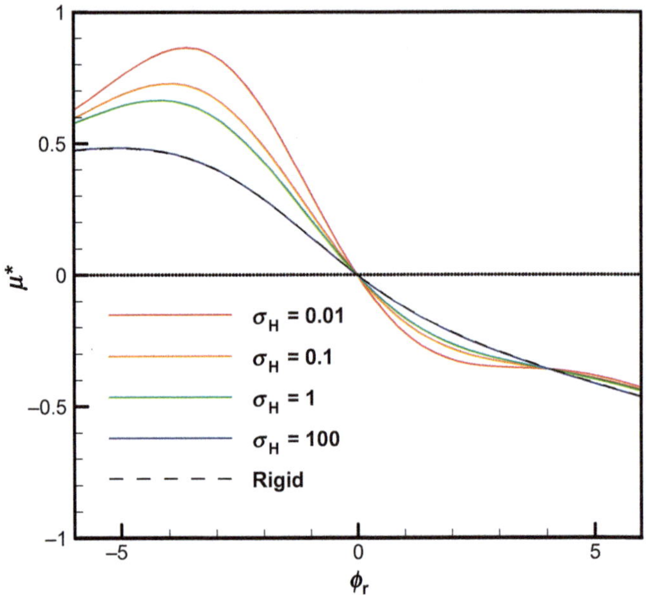

FIG. 17.11 Dimensionless mobility μ^* as a function of ϕ_r at various σ_H with $\kappa a = 1$ ($\beta = -0.2$).

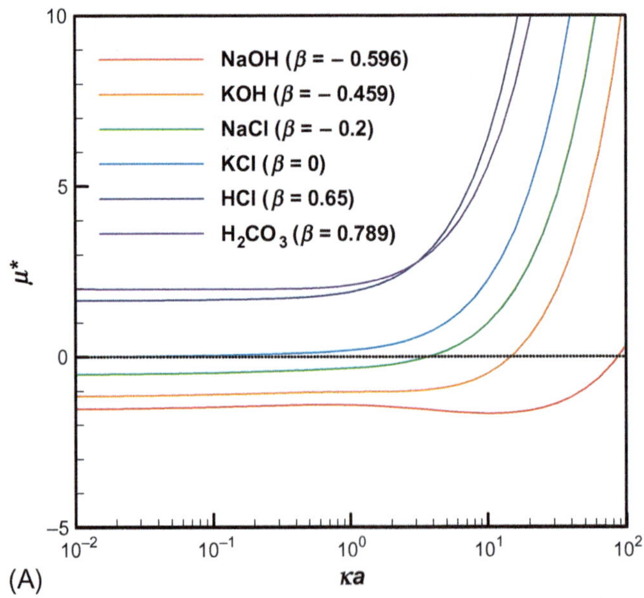

FIG. 17.12 Dimensionless mobility μ^* as a function of κa at various β with $\phi_r = 3$. (A) $\sigma_H = 0.5$; (B) $\sigma_H = 0.01$.

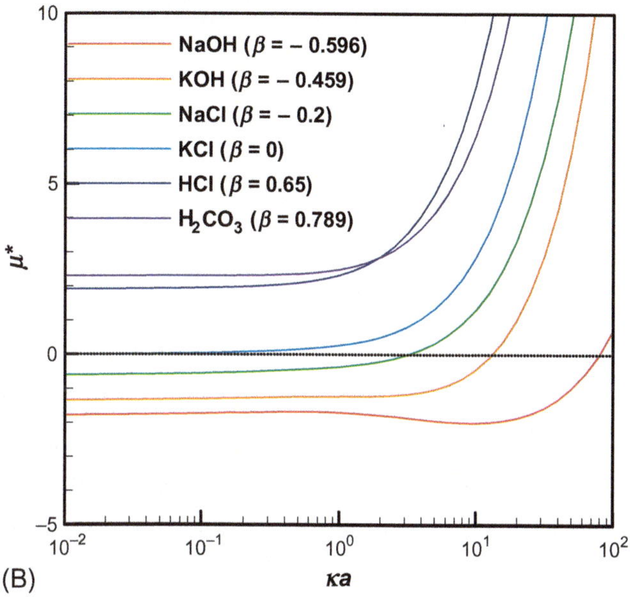

FIG. 17.12—CONT'D

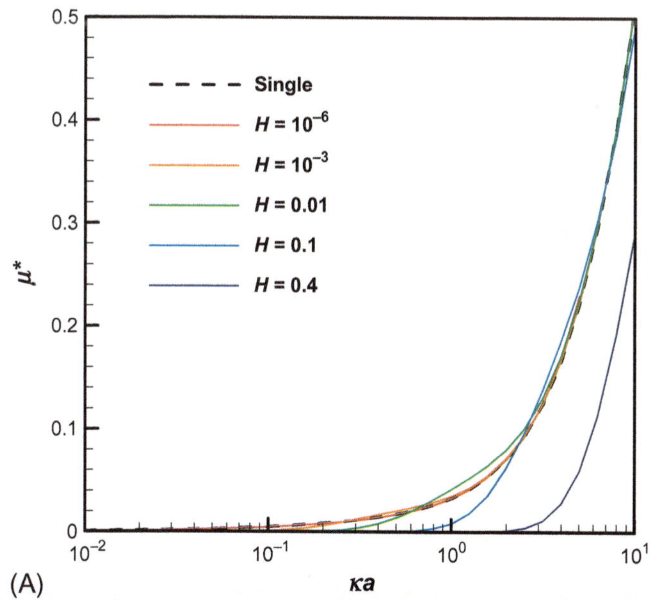

FIG. 17.13 Dimensionless mobility μ^* as a function of κa at various H with $\sigma_H=0.5$ ($\beta=0$). (A) $\phi_r=1$ (linear scale); (B) $\phi_r=1$ (log scale); (C) $\phi_r=3$ (linear scale); (D) $\phi_r=3$ (log scale); (E) $\phi_r=5$ (linear scale); (F) $\phi_r=5$ (log scale).

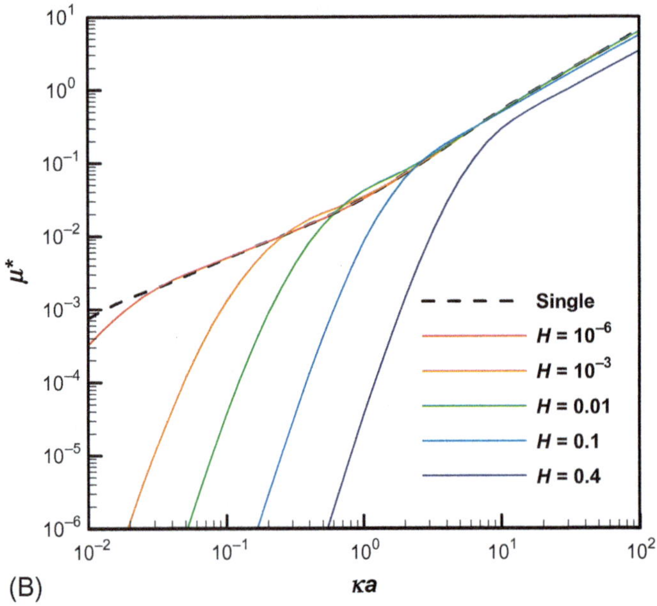

(B)

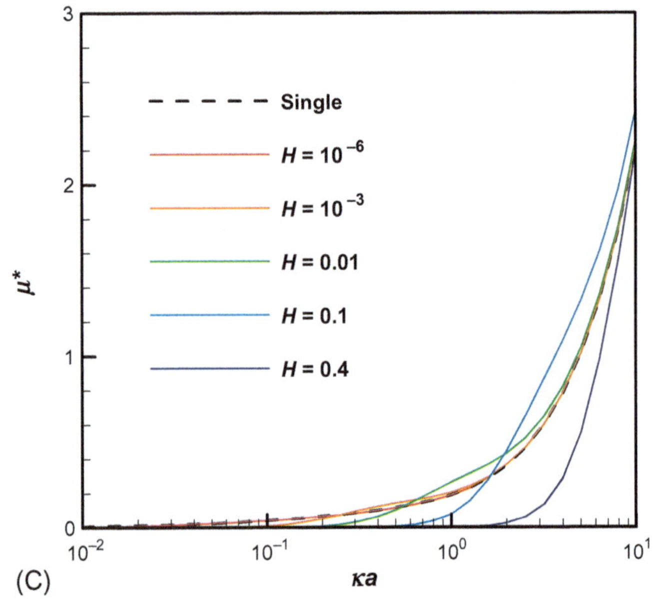

(C)

FIG. 17.13—CONT'D

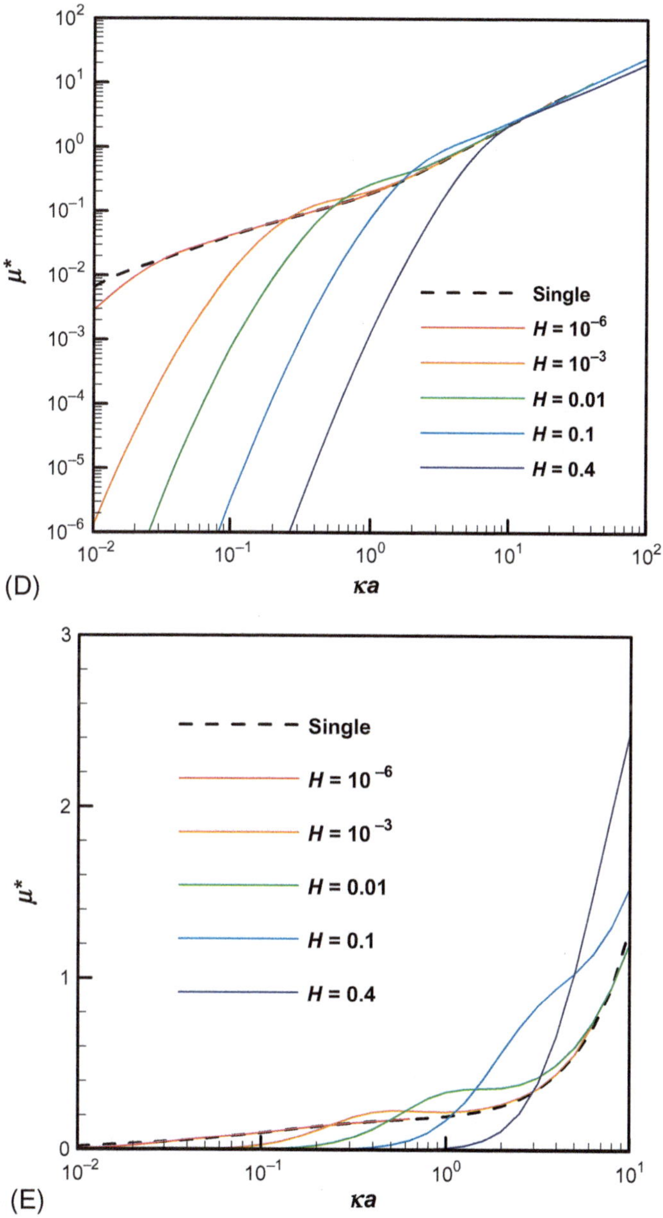

FIG. 17.13—CONT'D

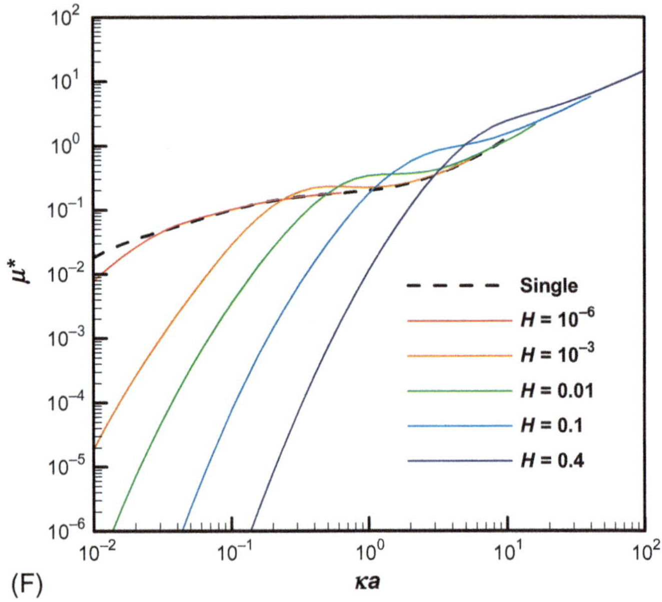

FIG. 17.13—CONT'D

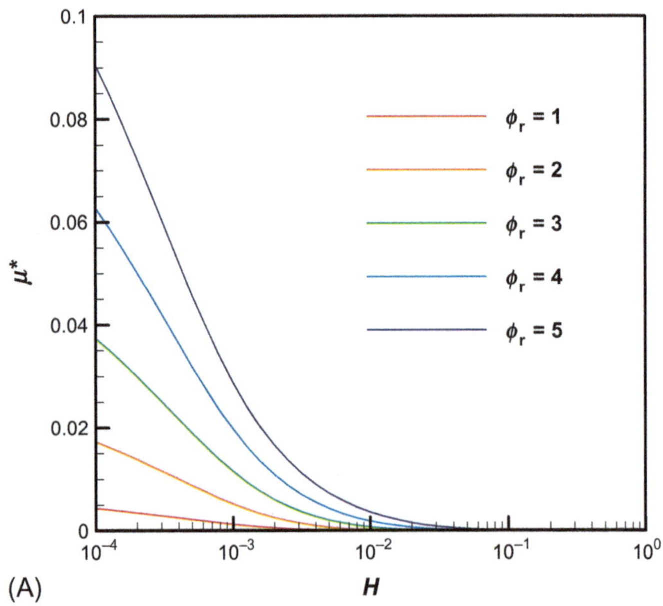

FIG. 17.14 Dimensionless mobility μ^* as a function of H at various ϕ_r with $\sigma_H = 0.5$ ($\beta = 0$). (A) $\kappa a = 0.1$; (B) $\kappa a = 1$; (C) $\kappa a = 10$.

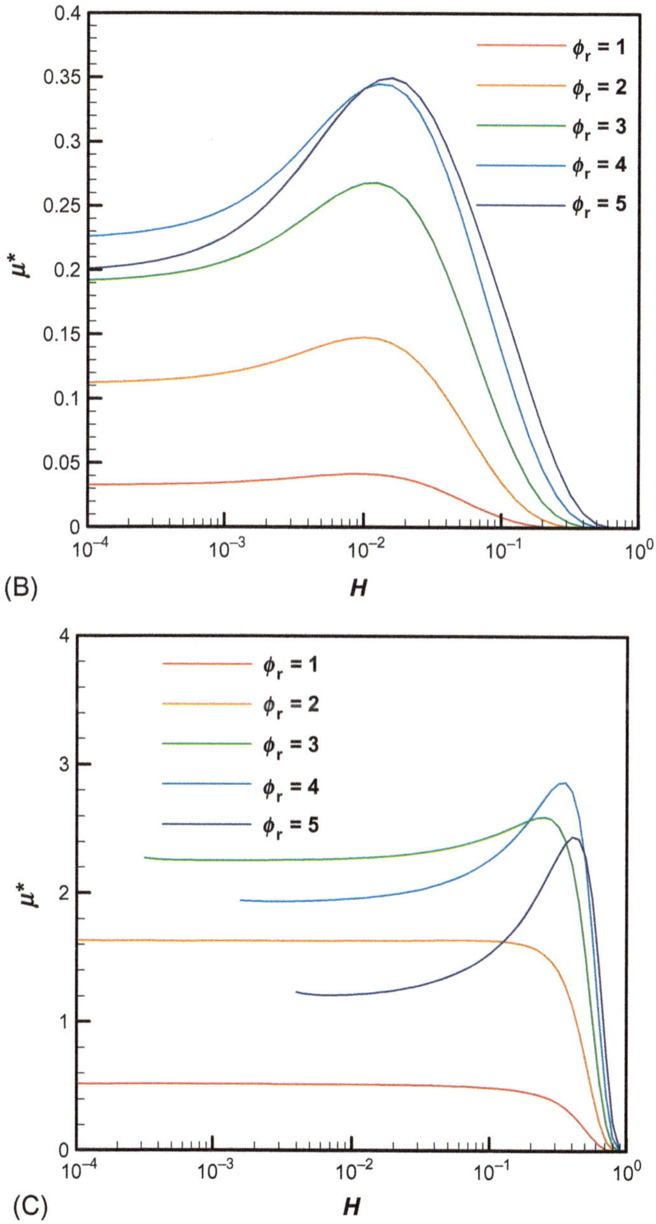

FIG. 17.14—CONT'D

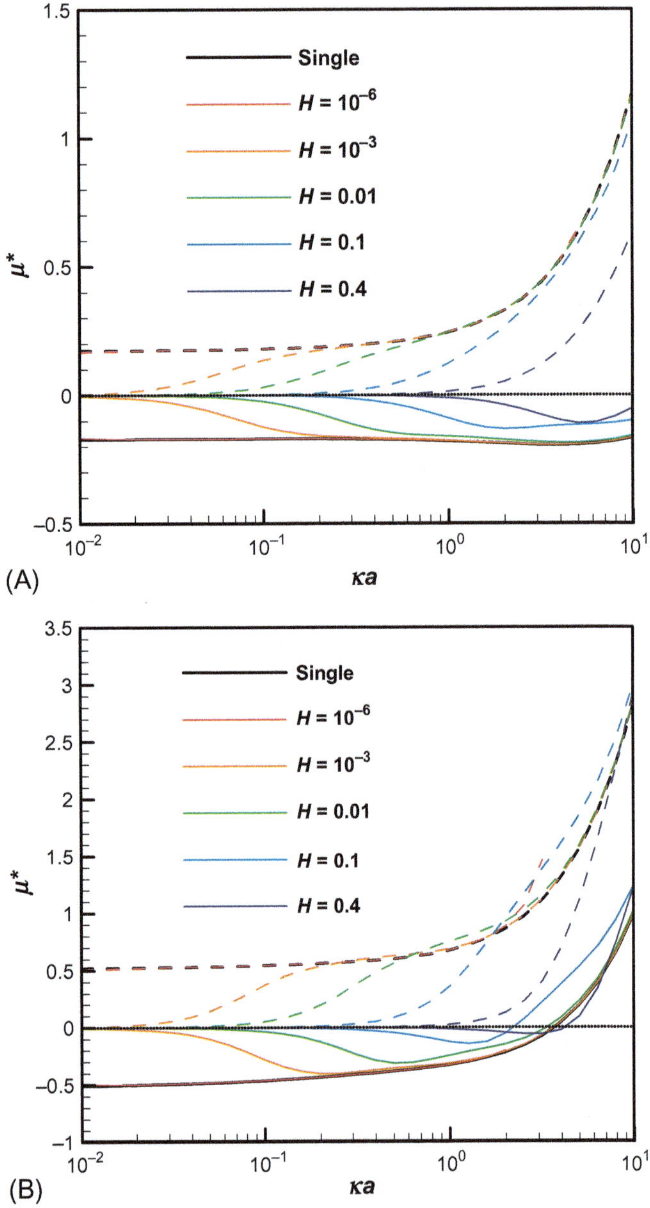

FIG. 17.15 Dimensionless mobility μ^* as a function of κa at various H with $\sigma_H = 0.5$ ($\beta = -0.2$). (A) $\phi_r = \pm 1$. *Solid lines*: positive ϕ_r. *Dashed lines*: negative ϕ_r. (B) $\phi_r = \pm 3$. *Solid lines*: positive ϕ_r. *Dashed lines*: negative ϕ_r. (C) $\phi_r = \pm 5$. *Solid lines*: positive ϕ_r. *Dashed lines*: negative ϕ_r.

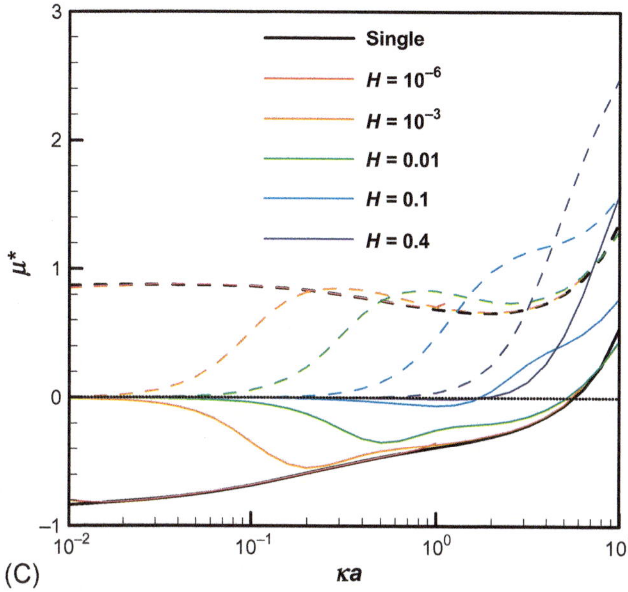

(C)

FIG. 17.15—CONT'D

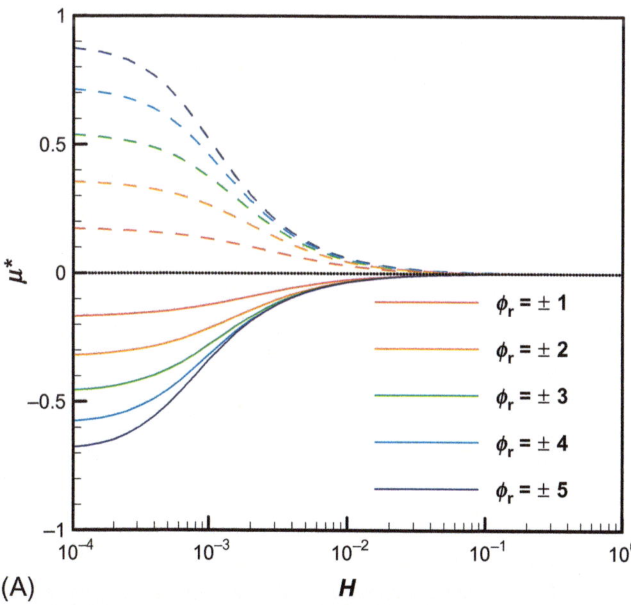

(A)

FIG. 17.16 Dimensionless mobility μ^* as a function of H at various ϕ_r with $\sigma_H = 0.5$ ($\beta = -0.2$). (A) $\kappa a = 0.1$. *Solid lines*: positive ϕ_r. *Dashed lines*: negative ϕ_r (B) $\kappa a = 1$. *Solid lines*: positive ϕ_r. *Dashed lines*: negative ϕ_r. (C) $\kappa a = 10$. *Solid lines*: positive ϕ_r. *Dashed lines*: negative ϕ_r.

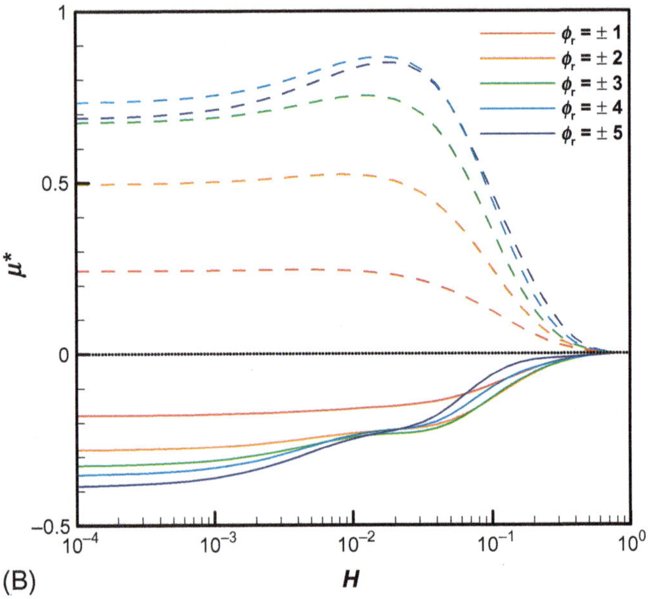

(B)

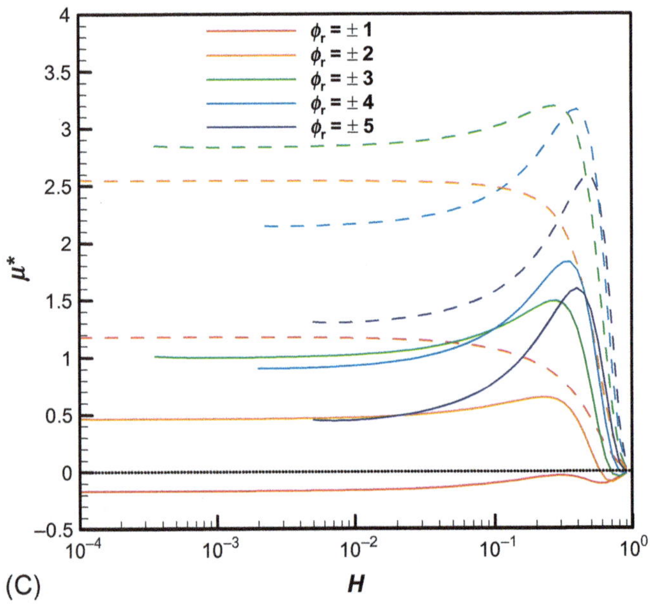

(C)

FIG. 17.16—CONT'D

κa range around unity. This is again attributed to the double layer suppression effect due to the boundary confinement of neighboring particles, which gets more profound in concentrated suspensions. This concludes the discussion in Chapter 17.

REFERENCES

[1] Lee E, Min WL, Hsu JP. Dynamic electrophoresis of a droplet in a spherical cavity. Langmuir 2006;22(8):3920–8.

[2] Lou J, Lee E. Dynamic electrophoresis of concentrated droplet dispersions at arbitrary surface potentials. Chem Eng Sci 2007;62(22):6234–45.

[3] Liao C, et al. Electrophoretic motion of a liquid droplet and a bubble normal to an air–water interface. Colloids Surf A Physicochem Eng Asp 2016;507:124–33.

[4] Huang CH, Lee E. Electrophoretic motion of a liquid droplet in a cylindrical pore. J Phys Chem C 2012;116(28):15058–67.

[5] Im DJ, et al. Influences of electric field on living cells in a charged water-in-oil droplet under electrophoretic actuation. Biomicrofluidics 2011;5(4):044112.

[6] Schnitzer O, Frankel I, Yariv E. Electrophoresis of bubbles. J Fluid Mech 2014;753:49–79.

[7] Vitale SA, Katz JL. Liquid droplet dispersions formed by homogeneous liquid–liquid nucleation: "The ouzo effect" Langmuir 2003;19(10):4105–10.

[8] Bouchemal K, et al. Nano-emulsion formulation using spontaneous emulsification: solvent, oil and surfactant optimisation. Int J Pharm 2004;280(1-2):241–51.

[9] Hartman SV, Božič B, Derganc J. Migration of blood cells and phospholipid vesicles induced by concentration gradients in microcavities. New Biotechnol 2018, in press.

[10] Shin S, et al. Size-dependent control of colloid transport via solute gradients in dead-end channels. Proc Natl Acad Sci 2016;113(2):257–61.

[11] Kar A, et al. Enhanced transport into and out of dead-end pores. ACS Nano 2015;9(1):746–53.

[12] Stubenrauch C. Microemulsions background, new concepts, applications, perspectives 2009, Wiley-Blackwell.

[13] Bear J. Dynamics of fluids in porous media. Dover Publication; New York; 2013.

[14] Fuerstman MJ, et al. Solving mazes using microfluidic networks. Langmuir 2003;19(11): 4714–22.

[15] Goodknight RC, Klikoff Jr WA, Fatt I. Non-steady-state fluid flow and diffusion in porous media containing dead-end pore volume 1. J Phys Chem 1960;64(9):1162–8.

[16] Dagdug L, et al. Transient diffusion in a tube with dead ends. J Chem Phys 2007;127(22):224712.

[17] Chhabra G, et al. Design and development of nanoemulsion drug delivery system of amlodipine besilate for improvement of oral bioavailability. Drug Dev Ind Pharm 2011;37(8): 907–16.

[18] Lawrence MJ, Rees GD. Microemulsion-based media as novel drug delivery systems. Adv Drug Deliv Rev 2012;64:175–93.

[19] Bhargava H, Narurkar A, Lieb L. Using microemulsions for drug delivery. Pharm Technol 1987;11(3):46–52.

[20] Kreuter J. Colloidal drug delivery systems. vol. 66. Marcel Dekker; New York; 2014.

[21] Lawrence MJ. Surfactant systems: microemulsions and vesicles as vehicles for drug delivery. Eur J Drug Metab Pharmacokinet 1994;19(3):257–69.

[22] Lawrence MJ. Microemulsions as drug delivery vehicles. Curr Opin Colloid Interface Sci 1996;1(6):826–32.
[23] Swarbrick J. Encyclopedia of pharmaceutical technology. CRC Press; Boca Raton; 2013.
[24] Tenjarla S. Microemulsions: an overview and pharmaceutical applications. Crit Rev Ther Drug Carrier Syst 1999;16(5).
[25] Constantinides PP. Lipid microemulsions for improving drug dissolution and oral absorption: physical and biopharmaceutical aspects. Pharm Res 1995;12(11):1561–72.
[26] Sarciaux J, Acar L, Sado P. Using microemulsion formulations for oral drug delivery of therapeutic peptides. Int J Pharm 1995;120(2):127–36.
[27] Mou F, et al. Autonomous motion and temperature-controlled drug delivery of Mg/Pt-poly (N-isopropylacrylamide) Janus micromotors driven by simulated body fluid and blood plasma. ACS Appl Mater Interfaces 2014;6(12):9897–903.
[28] Fanun M. Colloids in drug delivery. vol. 150. CRC Press; Boca Raton; 2016.
[29] Park K. Controlled drug delivery: challenges and strategies. American Chemical Society; Washington, DC; 1997.
[30] Sakeena M, et al. Effects of oil and drug concentrations on droplets size of palm oil esters (POEs) nanoemulsion. J Oleo Sci 2011;60(4):155–8.
[31] Yadav V, et al. Bone-crack detection, targeting, and repair using ion gradients. Angew Chem Int Ed 2013;52(42):10997–1001.

Chapter 18

Diffusiophoresis of Porous Particles

18.1 INTRODUCTION

Diffusiophoresis happens in systems of polyelectrolytes like proteins, polymers, and other macromolecules very often as well [1–4], in addition to the rigid particles and liquid droplets or gas bubbles considered in Chapters 16 and 17, respectively. As mentioned in Chapter 7 where the electrophoresis of charged porous particles is investigated, polyelectrolytes [5, 6] can be modeled very successfully with a charged porous sphere, indicated by Overbeek [7]. Hence, we will conduct a rigorous theoretical analysis on the diffusiophoresis of porous particles here as well.

Similar to the discussion in Chapter 7, the motion of a porous particle is very fast in general due to the large electric driving force from the fixed changes distributed all over the entire interior region of the particle, as well as the reduced hydrodynamic drag force due to the permeability of the particle. Orders-of-magnitude larger mobility compared with corresponding rigid particles could happen for porous particles both highly charged and highly permeable [8, 9]. This is the fundamental difference between the diffusiophoretic motion of porous particles and that of the rigid one. As a result of this fast motion, the convection-induced motion-deterring double layer polarization effect is profound [8, 10–12]. Any linear approximation neglecting this effect could lead to severe overestimation of particle mobility.

Moreover, there is a phenomenon called "counterion condensation" [13, 14] or "shielding effect" [14] referring to the fact that the fixed charges carried by the particle can be neutralized by the counterions driven into the interior of the porous particle when κa is large [14–16]. There are quite a few theoretical analyses on this topic in either electrophoresis or diffusiophoresis [8–12, 17–20], especially the ones from Lee's group [8–12, 17, 18], where complete solution of the general electrokinetic equations takes full account of the nonlinear double layer polarization effect and the electrostatic counterion condensation effect [10]. Among them, the one considering the

Interface Science and Technology, Vol. 26. https://doi.org/10.1016/B978-0-08-100865-2.00018-7

sole chemiphoresis phenomenon ($\beta = 0$) is so far the most extensive treatment on this topic, especially for highly charged porous particles.

In addition to the system of a single porous particle immersed in an infinite medium of electrolyte solution, suspensions of porous particles, either concentrated or dilute, will be treated in this chapter as well.

18.2 THEORY

18.2.1 Single Porous Particle

The system diagram for a single porous particle is shown in Fig. 18.1A. The governing equations, both within the porous particle and outside it, are exactly the same as the corresponding electrophoresis system considered in Chapter 7. The associated boundary conditions are the same as well, except that at infinity a concentration gradient instead of an electric field is applied. Thus, the boundary conditions at infinity are replaced with the ones adopted in Chapter 16, where the diffusiophoretic motion of a single rigid particle is considered.

18.2.2 Suspension of Porous Particles

The system diagram is shown in Fig. 18.1B, where Kuwabara's hydrodynamic cell model coupled with Shilov and Zharkikh (S-Z) electric boundary conditions is adopted. The governing equations as well as the associated boundary conditions are the same as above for a single porous particle, except that at the outer virtual surface, $r^* = 1/\sqrt[3]{H}$, corresponding boundary conditions for a suspension of rigid particles treated in Chapter 16 are used instead.

18.3 RESULTS AND DISCUSSION

18.3.1 Single Porous Particle

We examine the chemiphoresis phenomenon first with $\beta = 0$. Fig. 18.2A shows the general Henry's charts for a highly permeable particle with $\lambda a = 1$ in an electrolyte solution with $\beta = 0$, say, the KCl solution. Four different charge levels are examined with $Q_{\text{fix}} = 1$, 10, 30, and 50, respectively. The mobility is essentially null for $Q_{\text{fix}} = 1$ due to the poorly charged condition. Corresponding induced electric driving force by the diffusion of electrolyte ions across the interior of the porous particle is trivial as a result. Note that the electric driving force is explicitly proportional to Q_{fix}, as indicated by Eq. (7.23). When $Q_{\text{fix}} = 10$, on the other hand, significant particle mobility is observed with a peak value around $\kappa a = 1$. And the higher the charge level is, the larger the particle mobility is in general and the peak value in particular, as shown in Fig. 18.2A for $Q_{\text{fix}} = 30$ and $Q_{\text{fix}} = 50$. The peck value has a tendency to shift to right as κa increases. The reason for the appearance of the

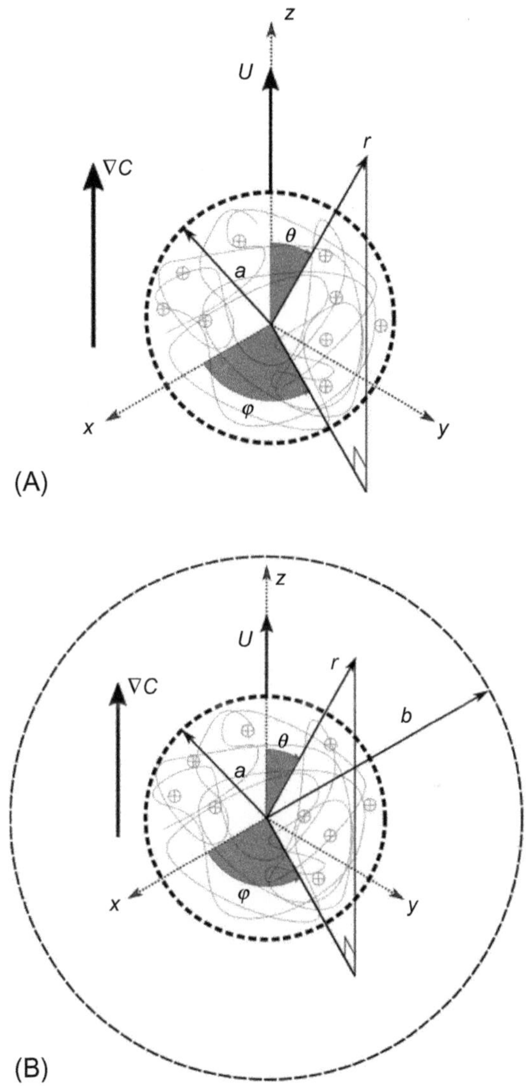

FIG. 18.1 System diagram. (A) Single porous particle; (B) suspension of porous particles.

peak is due to the double layer suppression effect which goes strong as κa increases. More counterions are driven into the interior of the porous particle to enhance the antisymmetric double layer polarization there as discussed in Chapter 7 for electrophoresis. The electric driving force and particle mobility are increased as a result. Once past the peak, however, these mobility profiles all tend to approach nullity as κa further increases to around 100. This behavior is contrary to the observation in Chapters 16 and 17, where rigid particle

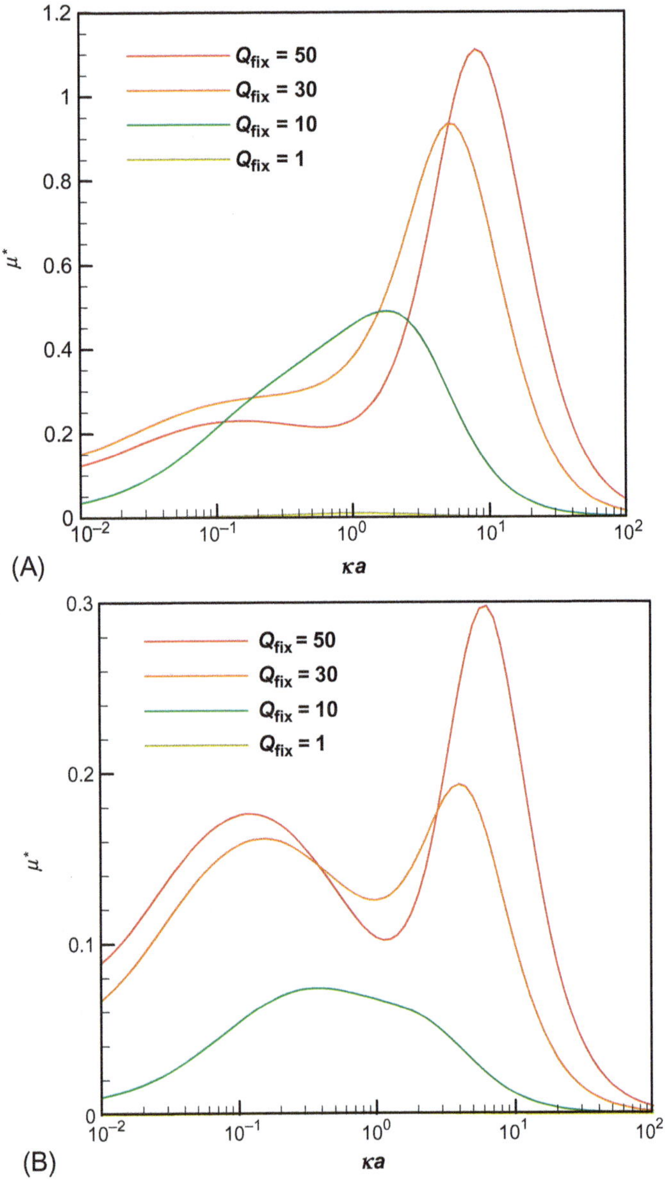

FIG. 18.2 Dimensionless mobility μ^* as a function of κa at various Q_{fix} ($\beta=0$). (A) $\lambda a=1$; (B) $\lambda a=3$; (C) $\lambda a=10$. *(Fig. 18.2(A) is cited from Ref. 10, with permission from JCIS.)*

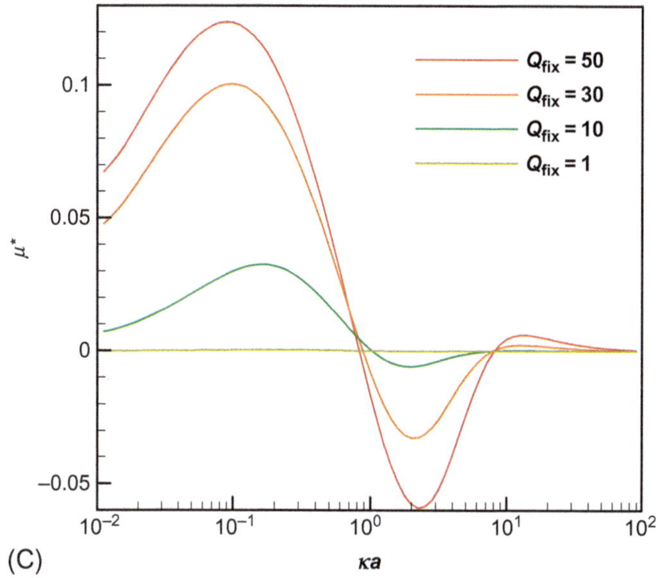

FIG. 18.2—CONT'D

and droplet systems are considered. The mobilities there increase with increasing κa in general and take the largest value when κa is very large, say $\kappa a = 100$. The underlying unique feature here is the particle permeability. At large κa, essentially all the counterions in the neighborhood of the particle will be driven into the interior of the porous particle by the strong double layer suppression effect, which leads to a perfect or complete shielding (or screening) of the fixed charges carried by the particle. A phenomenon is referred to as the "counterion condensation" effect in electrostatics. Hence, a porous particle at large κa is essentially chargeless. The ions migrate through the particle interior like through the bulk electrolyte solution. No double layer polarization is resulted; hence no electric driving force is generated. Unlike the corresponding electrophoresis phenomenon, the concentration gradient here apparently is not powerful enough to shake loose the strongly bounded ("condensed") counterions within the porous particle when κa is large (Fig. 18.3). The particle is motionless with a null mobility as a result. Corresponding contour plots of the dimensionless double layer polarization $\delta\rho^*$ and the dimensionless disturbed electric potential $\delta\phi^*$ are shown in Fig. 18.4A–C for a porous particle both highly charged ($Q_{\text{fix}} = 50$) and highly permeable ($\lambda a = 1$) at $\kappa a = 100$. Complete nullity of the double layer polarization is observed within the particle in Fig. 18.4A, hence complete shutdown of the local electric field $\delta\phi^*$ in Fig. 18.4B. The particle is, thus, motionless. Note that there is still some slight polarization as shown in Fig. 18.4A near

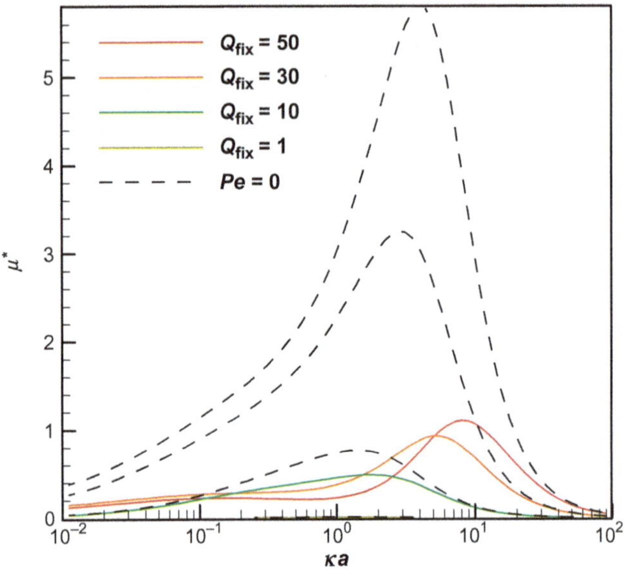

FIG. 18.3 Dimensionless mobility μ^* as a function of κa with $\lambda a = 1$ at various Q_{fix} ($\beta = 0$). *Dashed lines*: $Pe_1 = Pe_2 = 0$. *(Cited from Ref. 10, with permission from JCIS.)*

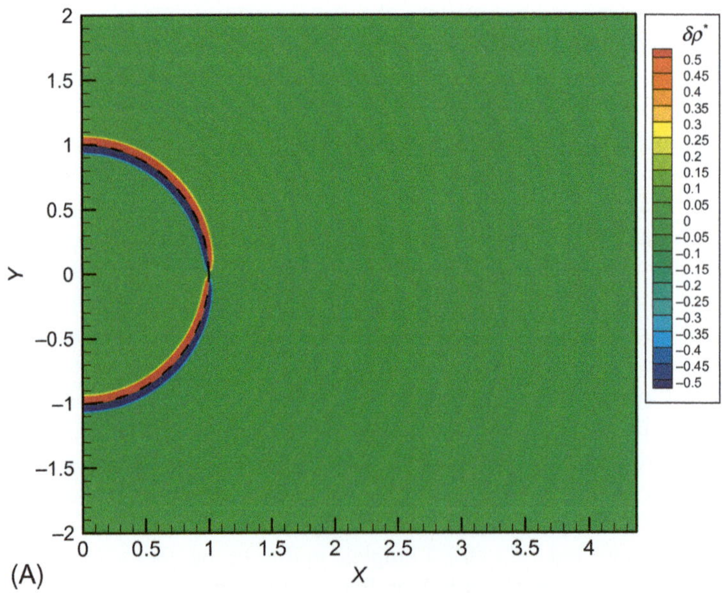

FIG. 18.4 Contour plot of the dimensionless disturbed quantities for a single porous particle with $\kappa a = 100$, $Q_{fix} = 50$, and $\lambda a = 1$ ($\beta = 0$). (A) Dimensionless disturbed space charge density $\delta\rho^*$; (B) dimensionless disturbed electric potential $\delta\phi^*$ (large scale); (C) dimensionless disturbed electric potential $\delta\phi^*$ (small scale).

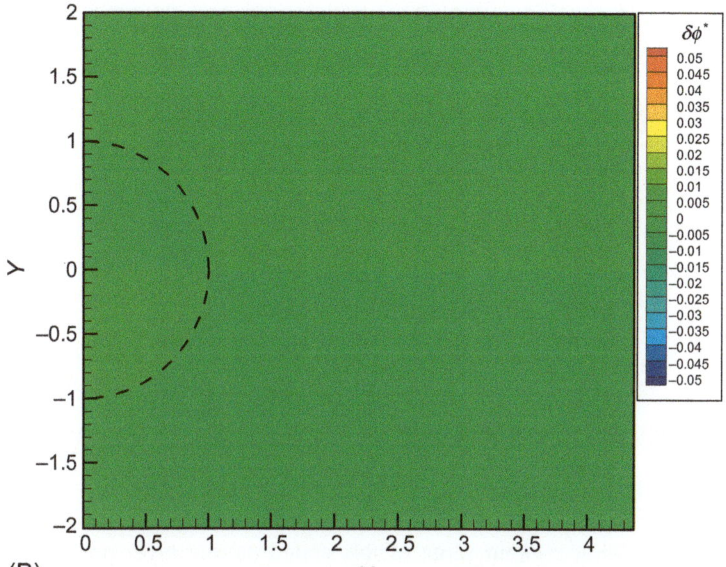

(B)

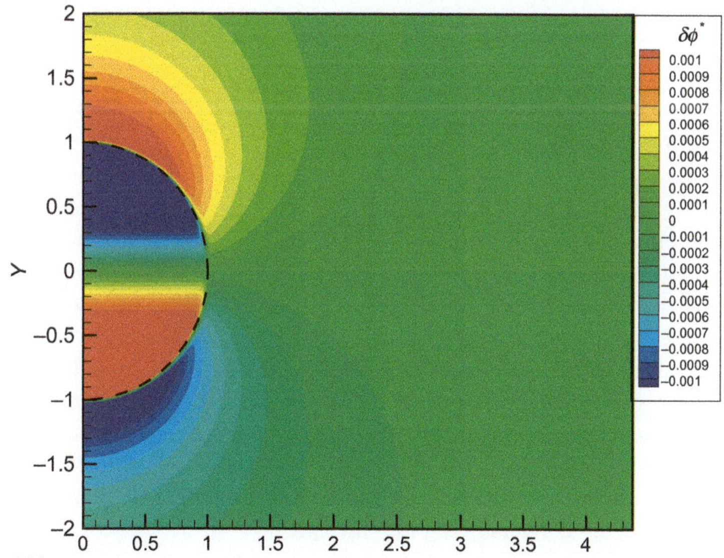

(C)

FIG. 18.4—CONT'D

the particle surface, as $\kappa a = 100$ is not infinity after all. Actually, the primary polarization of the counterions there induces a closely located secondary polarization layer of the coions outside. A close-up view in Fig. 18.4C depicts the corresponding $\delta\phi^*$ from this slight polarization near the particle surface.

The closely located distribution of counterions and coions in Fig. 18.4A induces electric fields of opposite orientation as shown, hence cancel each other.

Note that in the electrophoresis of porous particles, the entire space charges within the particle will respond to the superimposed external electric field no matter how large κa is. The shielding effect there does not keep the particle from moving. In diffusiophoresis, however, the perfect shielding effect when κa is very large completely shuts down the particle motion. They are horses of different colors. Here in diffusiophoresis, the impact of κa is twofold. The value of κa is preferred to be large enough to generate sufficient electric driving force, but not too large to shut down the motion mechanism. Indeed, it is a very subtle situation.

Note also that sometimes a highly charged particle may actually move slower than the less charged one, as shown in Fig. 18.2A at $\kappa a=1$, where the particle with $Q_{fix}=50$ moves slower than $Q_{fix}=30$, and even slower than $Q_{fix}=10$. This is similar to the electrophoresis behavior observed in Chapter 7.

Corresponding contour plots of the scaled double layer polarization $\delta\rho^*$ are shown in Fig. 18.5A–C for $Q_{fix}=10$, 30, and 50, respectively. The antisymmetric double layer polarization within the porous particle is most profound for $Q_{fix}=10$, where the particle moves fastest, and nearly completely driven out of the particle interior for $Q_{fix}=50$, where the particle moves slowest. The extent of polarization for $Q_{fix}=30$ is in the middle, and its speed is in the middle as well.

Moreover, corresponding contour plots of the scaled disturbed electric field $\delta\phi^*$ are shown in Fig. 18.6A–C. It is interesting to note that the extent of this antisymmetric distribution decreases from significant in Fig. 18.6A ($Q_{fix}=10$), to hardly noticeable in Fig. 18.6B ($Q_{fix}=30$), and finally to complete disappearance in Fig. 18.6C ($Q_{fix}=50$). This synchronizes with the trend of decreasing particle mobility.

Indeed, the diffusiophoresis phenomenon is very complicated as observed above due to the entanglement of the two underlying nonlinear factors: the counterion condensation and the double layer polarization effects, which are generally inseparable and only rigorous computation of complete governing electrokinetic equations will yield the ultimate particle mobility. Fig. 18.2B and C present the corresponding mobility profiles for lesser permeable porous particles, $\lambda a=3$ and $\lambda a=10$, respectively. Note that the extra hydrodynamic drag force is proportional to the square of λa. As a result, the overall particle speed decreases with increasing λa, hence the highest peak value of the particle mobility for $Q_{fix}=50$ is around 1.1 for $\lambda a=1$, 0.3 for $\lambda a=3$, and 0.12 for $\lambda a=10$. Moreover, deep valleys are observed for $Q_{fix}=30$ and 50 around $\kappa a=1$, which apparently is related to the motion-deterring double layer polarization effect as it usually manifests itself around $\kappa a=1$. Probably, the most intriguing phenomenon there is the negative value of the mobility in

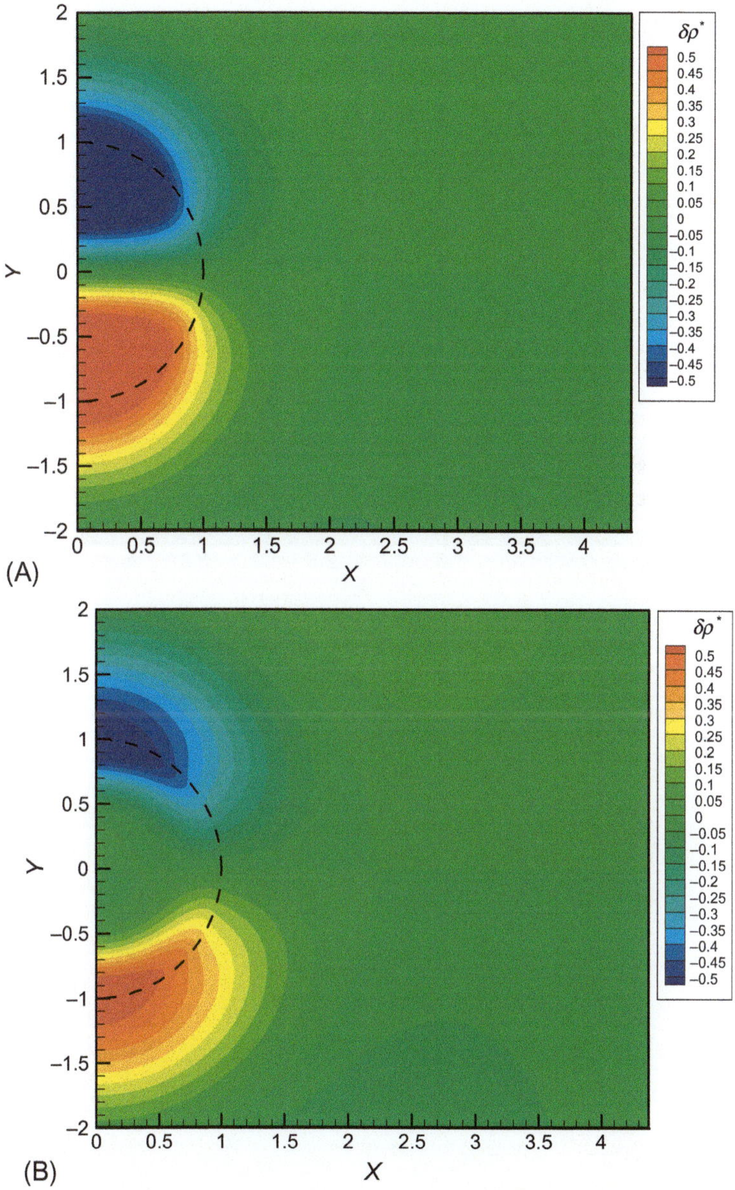

FIG. 18.5 Contour plot of the dimensionless disturbed space charge density $\delta\rho^*$ for a single porous particle with $\kappa a=1$ and $\lambda a=1$ ($\beta=0$). (A) $Q_{\mathrm{fix}}=10$; (B) $Q_{\mathrm{fix}}=30$; (C) $Q_{\mathrm{fix}}=50$.

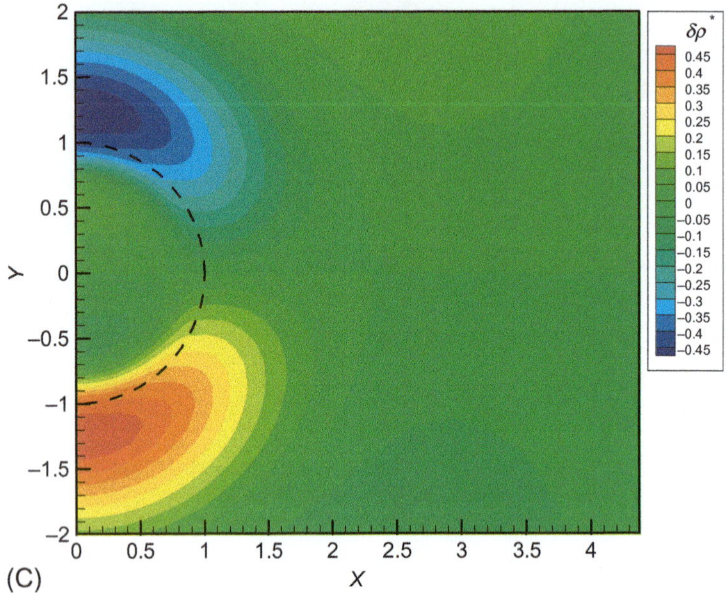

FIG. 18.5—CONT'D

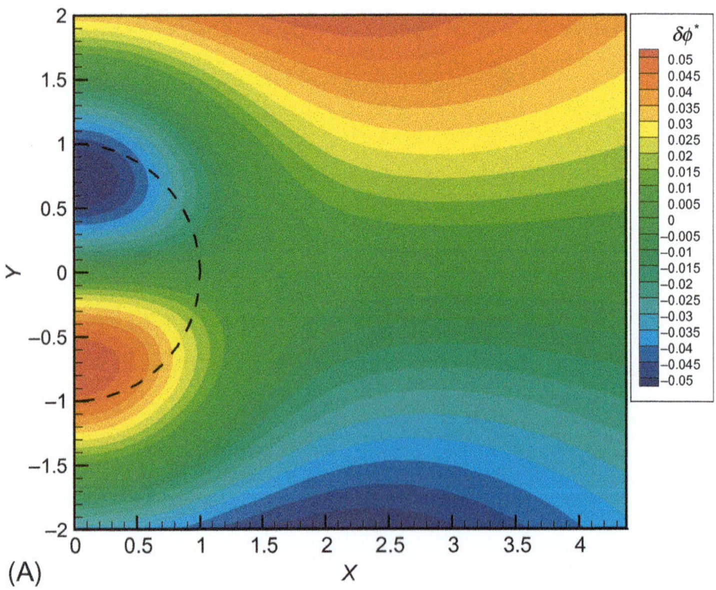

FIG. 18.6 Contour plot of the dimensionless disturbed electric potential $\delta\phi^*$ for a single porous particle with $\kappa a = 1$ and $\lambda a = 1$ ($\beta = 0$). (A) $Q_{\text{fix}} = 10$; (B) $Q_{\text{fix}} = 30$; (C) $Q_{\text{fix}} = 50$.

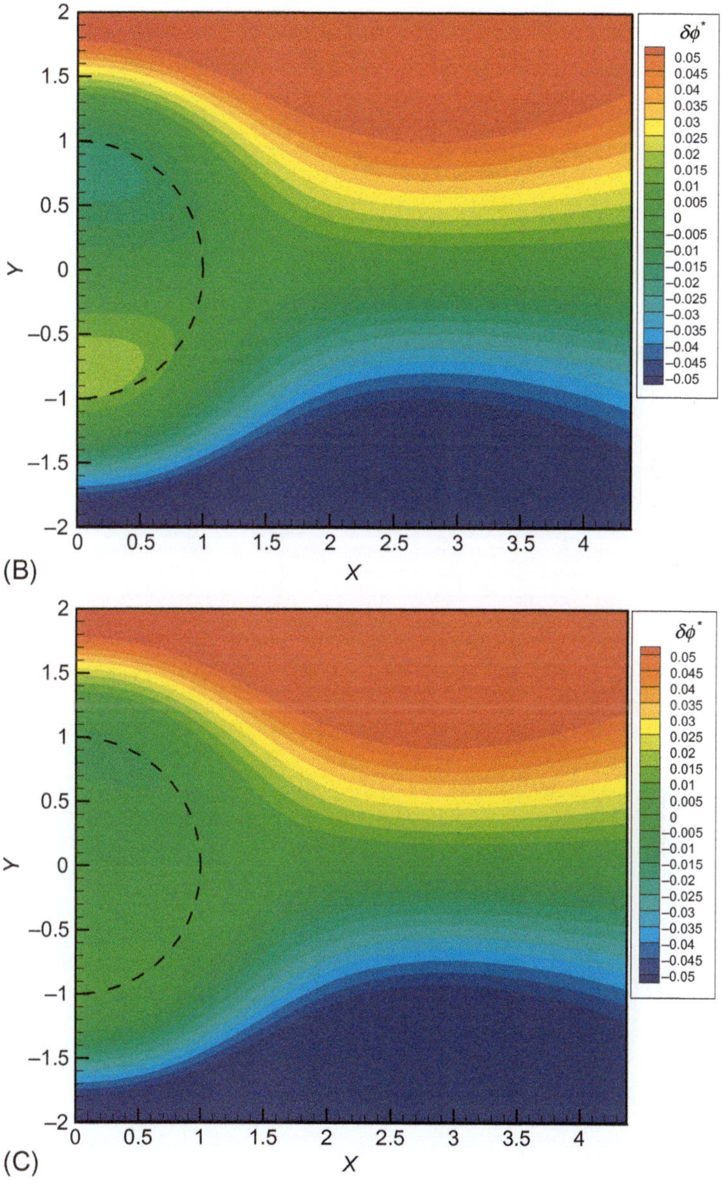

FIG. 18.6—CONT'D

Fig. 18.2C at κa between 1 and 10. This implies the porous particle actually moves toward the lower end of the electrolyte concentration, although the motion is very slow. This is against the general rule that, in chemiphoresis, particles move to the higher end of the electrolyte concentration mentioned

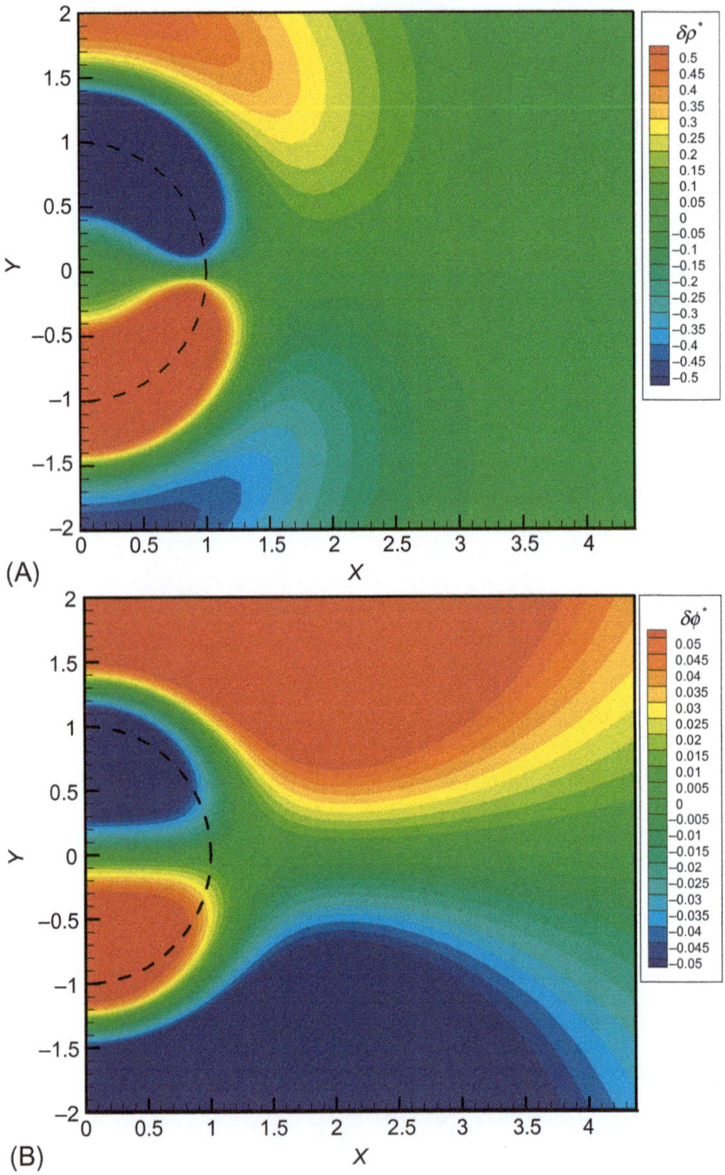

FIG 18.7 Contour plot of the dimensionless disturbed quantities for a single porous particle with $\kappa a=2$, $\lambda a=10$, and $Q_{\text{fix}}=50$ ($\beta=0$). (A) Dimensionless disturbed space charge density $\delta\rho^*$; (B) dimensionless disturbed electric potential $\delta\phi^*$.

earlier. This is due to the induced secondary polarization of ions outside the primary double layer polarization, as shown in Fig. 18.7A, where the negative particle mobility of $\kappa a=2$, $Q_{\text{fix}}=50$, and $\lambda a=10$ is pretty close to the local minimum in Fig. 18.2C. This secondary polarization generates an induced local electric field $\delta\phi^*$ against the particle motion outside the particle, as shown in Fig. 18.7B, thus results in a negative particle mobility.

Fig. 18.3 shows the impact of the convection-induced double layer polarization effect upon the particle motion for a highly permeable porous particle with $\lambda a = 1$. When Peclet numbers of the electrolyte solution are set to zero, the convection mechanism is eliminated from the Nernst-Planck ion flux equation. As shown by the dashed lines in Fig. 18.3, the particle mobility would be severely overestimated under this circumstance, about five times higher than the true mobility as shown in solid lines there. This is similar to the electrophoresis topic treated in Chapter 7, where more than 20 times overestimation is observed in some case.

Fig. 18.8 presents the permeability effect upon the particle motion for a representative highly charged particle with $Q_{fix} = 30$. Again, we focus on chemiphoresis by setting β to zero. Five orders of magnitude of the extra drag force in the Brinkman model are examined. In addition to the detailed observations in Fig. 18.2A–C, in general the higher the particle permeability is, the faster the particle moves.

Fig. 18.9 presents the mobility profiles as a function of Q_{fix} for a highly permeable porous particle with $\lambda a = 1$ at various κa. The profiles exhibit symmetric shape as observed in Chapters 16 and 17 for a rigid particle and a liquid droplet or gas bubble, respectively. When the double layer is very thick ($\kappa a = 0.01$), the space charges of counterions within the porous particle are

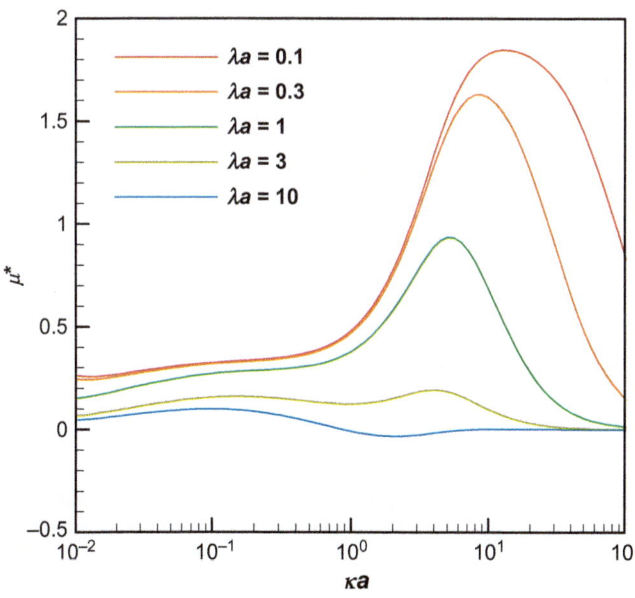

FIG. 18.8 Dimensionless mobility μ^* as a function of κa at various λa with $Q_{fix} = 30$ ($\beta = 0$).

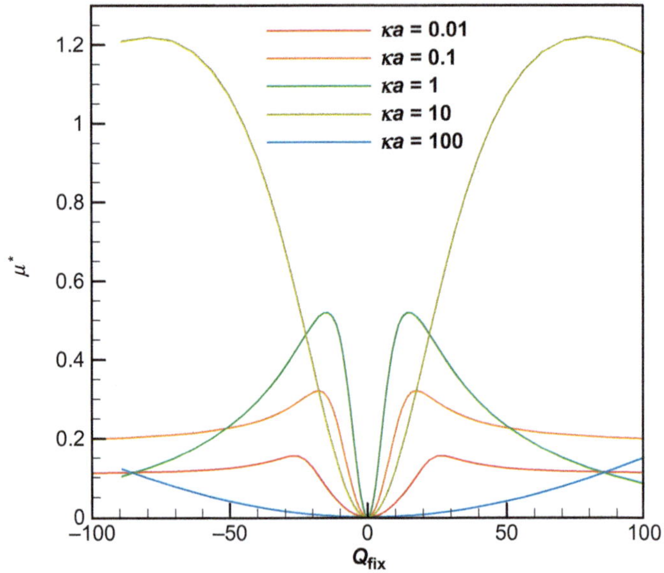

FIG. 18.9 Dimensionless mobility μ^* as a function of Q_{fix} at various κa with $\lambda a = 1$ ($\beta = 0$).

simply too sparse to yield significant particle mobility. The local maxima of the profiles observed indicate the fact that a highly charged particle may move slower than the less charged one, once the particle charges pass some threshold value. After that, the higher the particle charge level is, the slower it moves, according to Fig. 18.9. This threshold value is κa-dependent. This phenomenon is similar to that observed in Chapters 16 and 17. The highly charged particle simply does not allow the downstream counterions to leave the double layer due to its strong Coulomb attraction force. No exit, no entrance. Although some counterions do manage to leave eventually, the overall net migration process across the double layer is slowed down, which leads to a reduction of double layer polarization, and hence, the particle speed, as we discussed earlier in Chapters 16 and 17. Corresponding contour plots of $\delta\rho^*$ and $\delta\phi^*$ are shown in Figs. 18.10A–C and 18.11A–C, respectively, for three consecutive Q_{fix} levels ($Q_{\mathrm{fix}} = 30$, 50, and 100) with $\kappa a = 10$ and $\lambda a = 1$. Strong induced secondary polarization of $\delta\rho^*$ and $\delta\phi^*$ is clearly observed outside the particle due to the enrichment of counterions from the double layer suppression effect with $\kappa a = 10$. These figures shed light on the mysterious nonlinear diffusiophoretic behavior of porous particles in general.

We now go on to examine the general diffusiophoresis situation when both the chemiphoresis and the electrophoresis components are involved, i.e., $\beta \neq 0$. The electrolyte solution chosen is again the NaCl solution with $\beta = -0.2$. Fig. 18.12 depicts the general Henry's charts with the

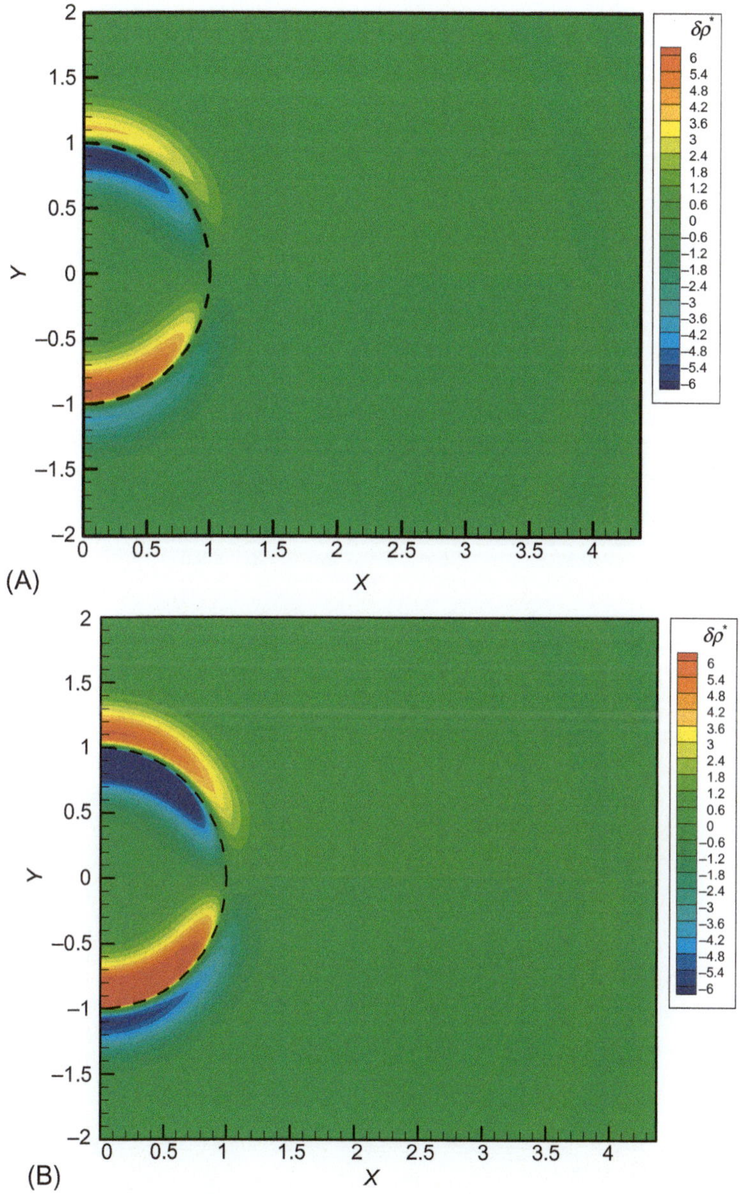

FIG. 18.10 Contour plot of the dimensionless disturbed space charge density $\delta\rho^*$ for a single porous particle with $\kappa a = 10$ and $\lambda a = 1$ ($\beta = 0$). (A) $Q_{fix} = 30$; (B) $Q_{fix} = 50$; (C) $Q_{fix} = 100$.

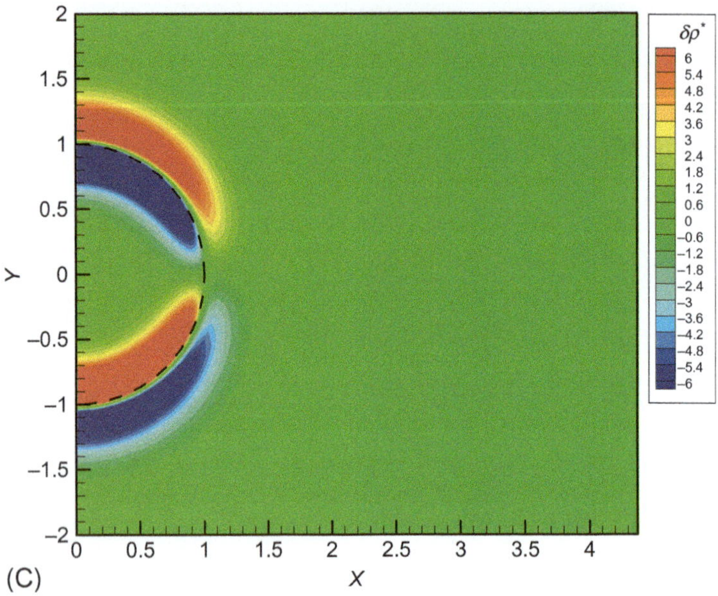

FIG. 18.10—CONT'D

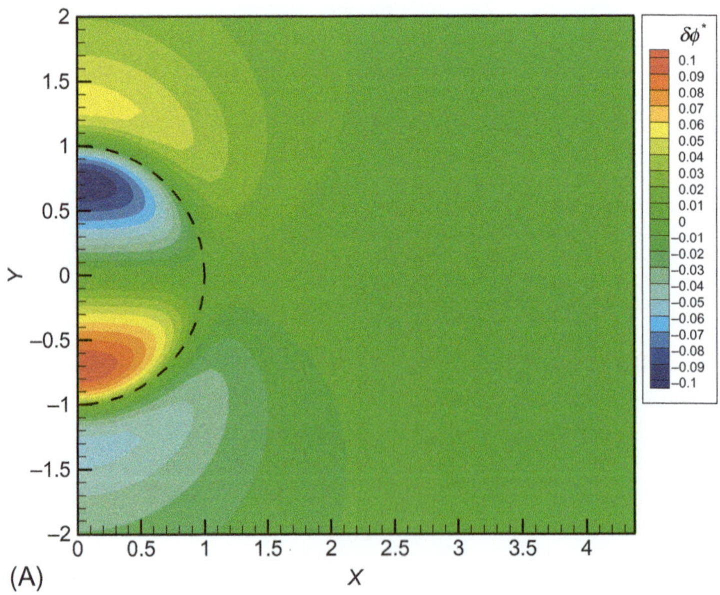

FIG. 18.11 Contour plot of the dimensionless disturbed electric potential $\delta\phi^*$ for a single porous particle with $\kappa a = 10$ and $\lambda a = 1$ ($\beta = 0$). (A) $Q_{\text{fix}} = 30$; (B) $Q_{\text{fix}} = 50$; (C) $Q_{\text{fix}} = 100$.

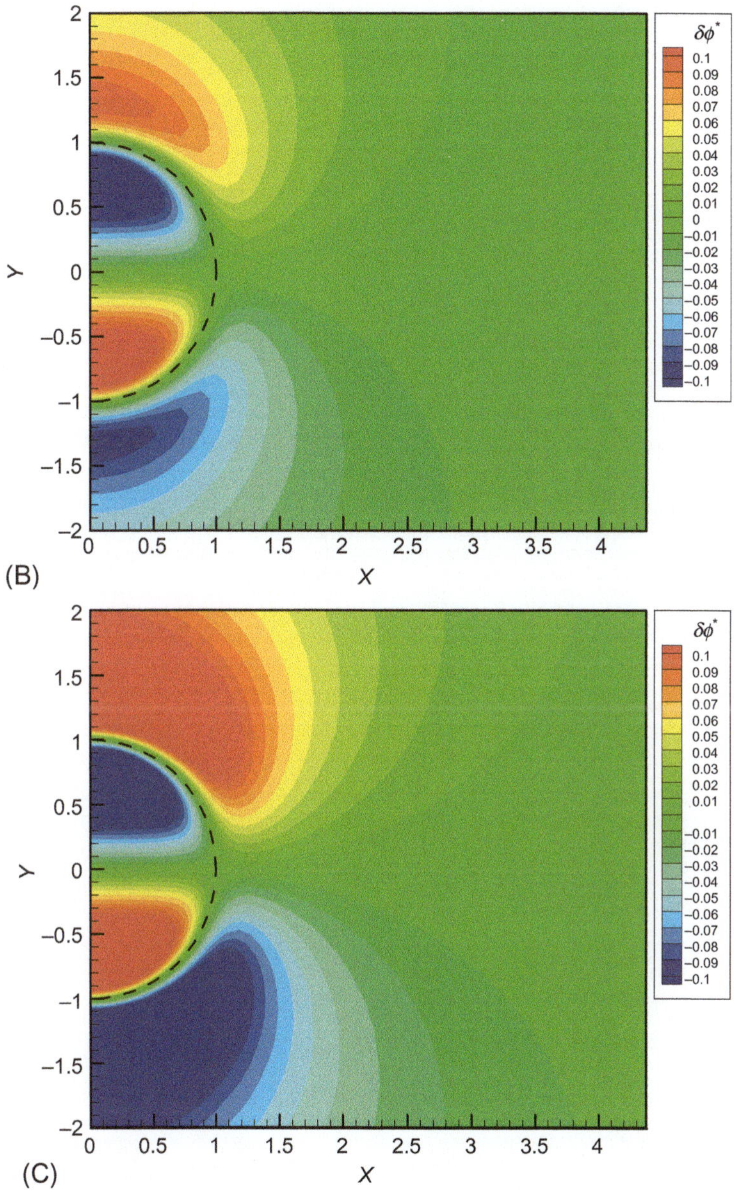

FIG. 18.11—CONT'D

dimensionless mobility μ^* further divided by a dimensionless quantity $(Q_{\text{fix}}/\lambda a^2)$, i.e., $\mu_s{}^* = \mu^*/(Q_{\text{fix}}/\lambda a^2)$. Positively charged particles are considered with $Q_{\text{fix}} = 1$, 10, 30, and 50, respectively. Similar to the discussions in Chapters 16 and 17, the particle now moves toward the lower concentration end of

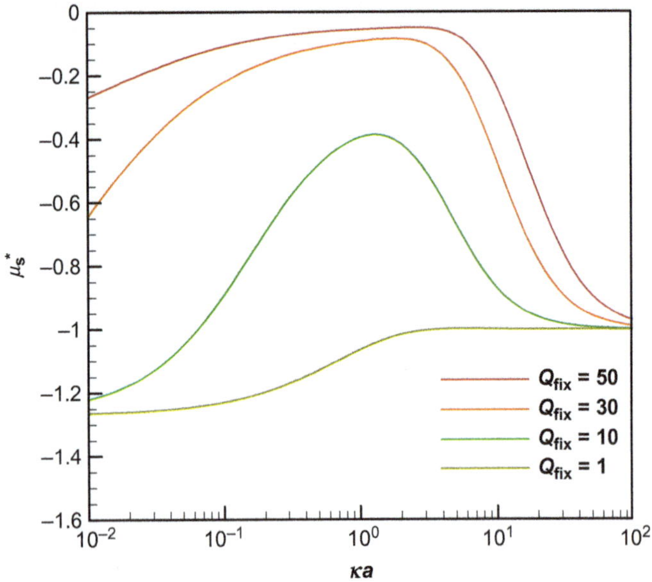

FIG. 18.12 Scaled mobility μ_s^* ($= \mu^*/(Q_{\text{fix}}/\lambda a^2)$) as a function of κa at various positive Q_{fix} with $\lambda a = 1$ ($\beta = -0.2$).

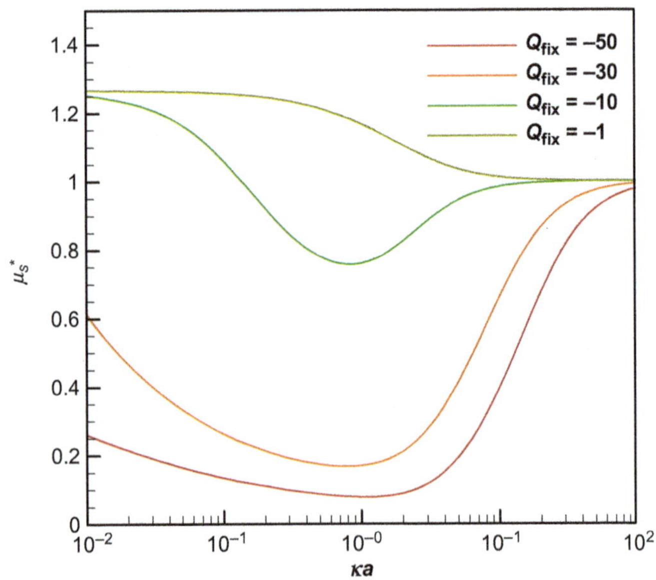

FIG. 18.13 Scaled mobility μ_s^* as a function of κa at various negative Q_{fix} with $\lambda a = 1$ ($\beta = -0.2$).

electrolytes, due to the dominance of the electrophoresis component and the direction of the electric field induced by the diffusion potential. Fig. 18.13 shows the corresponding situation when the particle is negatively charged. Now the directions of electrophoresis and chemiphoresis are the same, and the particle moves toward the higher concentration end of electrolytes with a faster speed compared with Fig. 18.12, where the directions of the two components are opposite.

Fig. 18.14 shows the mobility profiles as a function of Q_{fix} at three different κa levels for a highly permeable particle with $\lambda a = 1$. Asymmetric profiles are now observed compared with the situation of the sole chemiphoresis in Fig. 18.9. Note that the particle speed decreases once Q_{fix} passes some threshold value. As we discussed in Chapter 7 on electrophoresis, this is due to the strong Coulomb attraction force between the highly charged porous particle and the counterions within it and that sometimes the latter apparently are really condensed that the weak external electric field can no longer shake it loose. No local polarization, no local-induced electric field. Here, the strength of the concentration gradient is even weaker than an applied electric field.

Fig. 18.15 shows the impact of diffusion potential, i.e., the effect of the β factor. Five frequently encountered electrolytes are picked as sample solutions. When $\beta < 0$, the chemiphoresis component is against the electrophoresis component in the direction of resulted driving force; smaller magnitude of mobility is observed in general compared with the profiles with $\beta > 0$, where

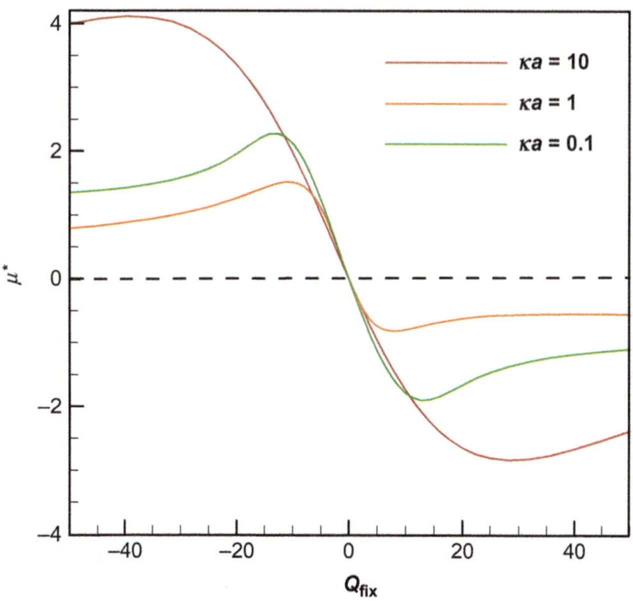

FIG. 18.14 Dimensionless mobility μ^* as a function of Q_{fix} at various κa with $\lambda a = 1$ ($\beta = -0.2$).

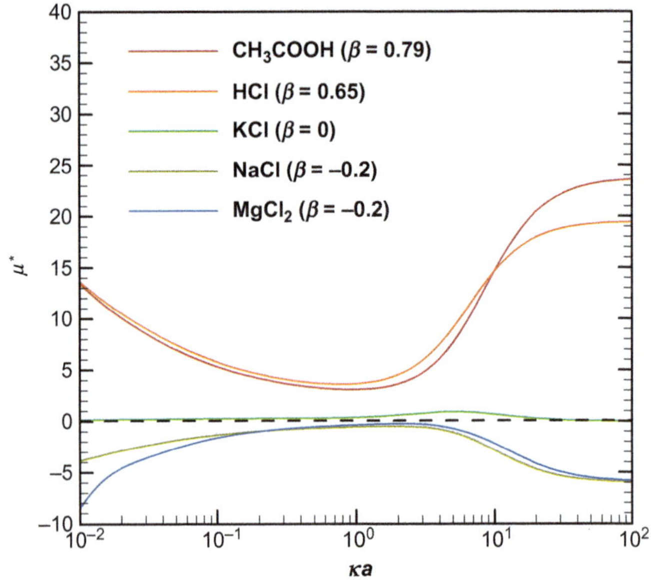

FIG. 18.15 Dimensionless mobility μ^* as a function of κa at various β with $Q_{\text{fix}}=30$ and $\lambda a=1$.

both components assume the same direction. Note also that the electrophoresis component by far is the dominant factor compared with chemiphoresis. The highest value of particle mobility is about 1.1 (Fig. 18.2A) in chemiphoresis ($\beta=0$), whereas the electrophoresis component alone can yield a particle mobility as high as about 24 in an electrolyte solution with $\beta=0.79$ and $\kappa a=100$ in Fig. 18.15. Note that at such a large κa, the chemiphoresis disappears due to the shielding effect, as shown in Fig. 18.2A. Hence, the mobility in Fig. 18.15 at $\kappa a=100$ can be regarded as purely from the electrophoresis contribution component.

Fig. 18.16A and B shows the β effect with particle mobility expressed as a function of Q_{fix} at $\kappa a=1$ and 10, respectively. The particle mobility is much larger with $\kappa a=10$ in general, as the electrophoretic driving force increases with increasing κa, as discussed in Chapter 7.

18.3.2 Suspensions of Porous Particles

Certainly, it is the chemiphoresis that is of major interest. Hence, we focus on the chemiphoresis here to begin with, i.e., $\beta=0$. As shown in Fig. 18.17A, the mobility profiles as a function of κa are shown for five different volume functions ($H=0.1$, 10^{-2}, 10^{-3}, 10^{-4}, and 10^{-6}, respectively) of a porous particle both highly charged ($Q_{\text{fix}}=30$) and highly permeable ($\lambda a=1$). A peculiar behavior is observed that, over some κa range at least, the more concentrated the suspension is, the faster the particle moves. Corresponding system in an

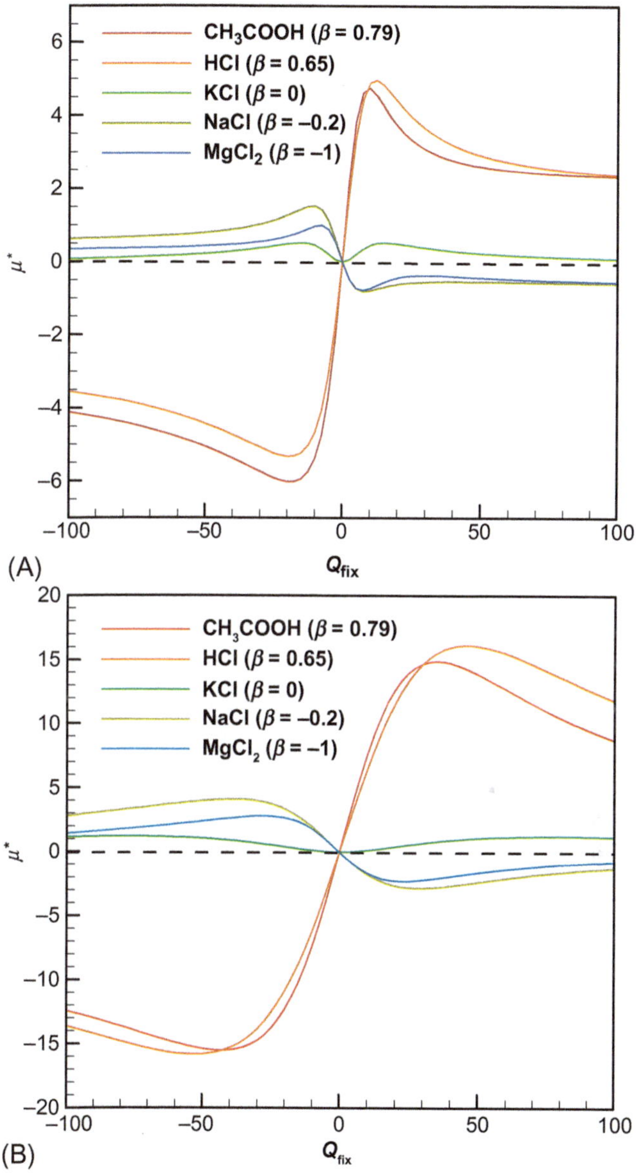

FIG. 18.16 Dimensionless mobility μ^* as a function of Q_{fix} at various β with $\lambda a = 1$. (A) $\kappa a = 1$; (B) $\kappa a = 10$.

electrolyte solution with $\beta = -0.2$ exhibits this behavior essentially over the entire range examined, as shown in Fig. 18.17B. The peculiar behavior is observed regardless of the particle permeability, as shown in Fig. 18.18A and B, and the double layer thickness, as shown in Fig. 18.19A and B.

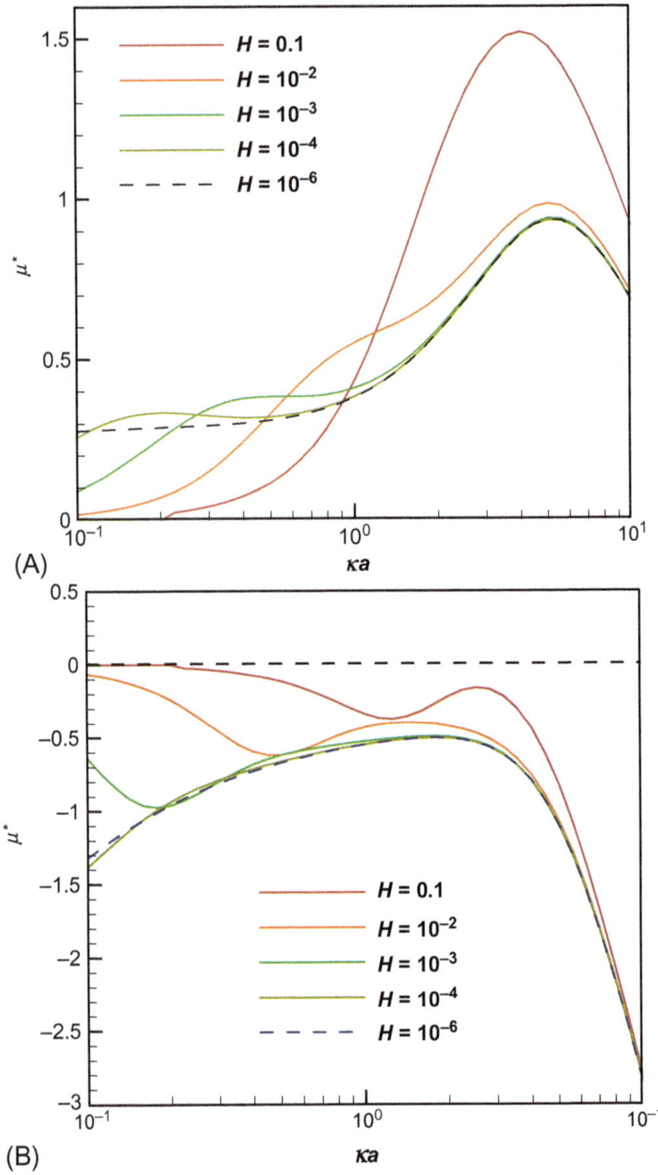

FIG. 18.17 Dimensionless mobility μ^* as a function of κa at various H with $Q_{\mathrm{fix}} = 30$ and $\lambda a = 1$. (A) $\beta = 0$; (B) $\beta = -0.2$.

The readers by now should be very familiar with this phenomenon, as it is observed in corresponding electrophoresis systems discussed in Chapter 7 as well. The increasing volume fraction implies closer presence of neighboring other porous particles, which attract and drive the counterions lingering

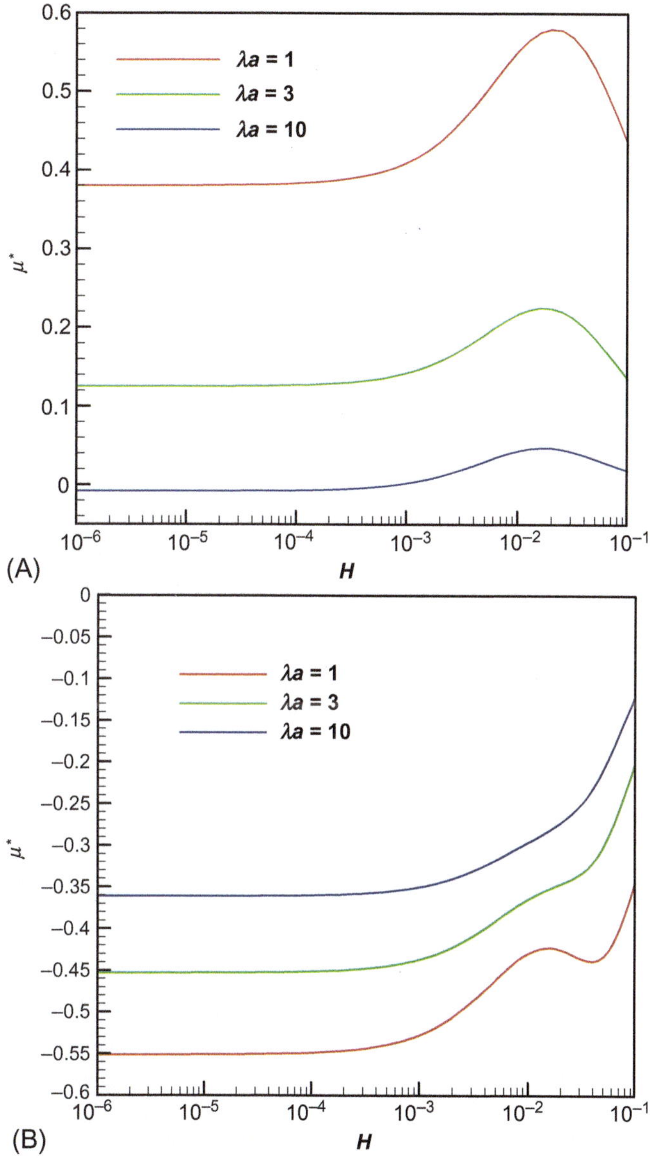

FIG. 18.18 Dimensionless mobility μ^* as a function of H at various λa with $Q_{\text{fix}}=30$ and $\kappa a=1$. (A) $\beta=0$; (B) $\beta=-0.2$.

around outside into their own interiors to enhance the double layer polarization there, and hence, the strength of the local electric field strength. The mobility of the particles in the suspension as a whole is thus increased. In short, the enhancement of the antisymmetric distribution of the double layer

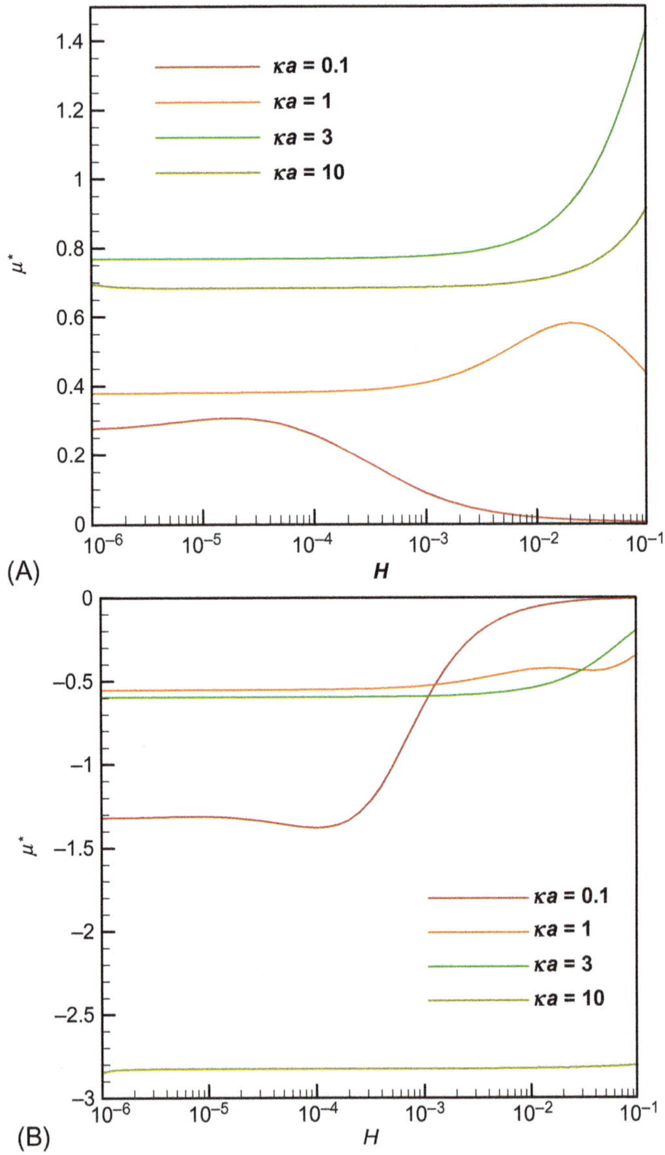

FIG. 18.19 Dimensionless mobility μ^* as a function of H at different κa with $Q_{\text{fix}}=30$ and $\lambda a=1$. (A) $\beta=0$; (B) $\beta=-0.2$.

polarization, which assumes the magnitude of perturbation level, determines the ultimate particle mobility. The increase of volume fraction provides the opportunity for this subtle physical mechanism to work out.

And that concludes the discussion in this chapter, and the entire book.

REFERENCES

[1] Annunziata O, Buzatu D, Albright JG. J Phys Chem B 2012;116:12694–705.
[2] McAfee MS, Annunziata O. Langmuir 2014;30:4916–23.
[3] McAfee MS, Zhang H, Annunziata O. Langmuir 2014;30:12210–9.
[4] McAfee MS, Annunziata O. Langmuir 2015;31:1353–61.
[5] Overbeek JTG, Stigter D. Recueil des Travaux Chimiques des Pays-Bas 1956;75:543–54.
[6] Fuoss RM. Science 1948;108:545–50.
[7] Overbeek JTG. Pure Appl Chem 1976;46:91–101.
[8] Hsu HP, Lee E. Phys Chem Chem Phys 2012;14:15729–38.
[9] Hsu H-P, Lee E. J Colloid Interface Sci 2013;390:85–95.
[10] Fang W, Lee E. J Colloid Interface Sci 2015;459:273–83.
[11] He YY, Lee E. Chem Eng Sci 2008;63:5719–27.
[12] He YY, Wu E, Lee E. Chem Eng Sci 2010;65:5507–16.
[13] Hsu H-P, Lee E. Electrochem Commun 2012;15:59–62.
[14] Manning GS. J Chem Phys 1969;51:924–33.
[15] Hsu HP. National Taiwan University Doctoral Dissertation 2012. p. 1–199.
[16] Yeh L-H, Hsu J-P. Soft Matter 2011;7:396–411.
[17] Tsai P, Huang CH, Lee E. Langmuir 2011;27:13481–8.
[18] Huang CH, Hsu HP, Lee E. Phys Chem Chem Phys: PCCP 2012;14:657–67.
[19] Wei YK, Keh HJ. J Colloid Interface Sci 2004;269:240–50.
[20] Liu K-L, Hsu J-P, Hsu W-L, Yeh L-H, Tseng S. Electrophoresis 2012;33:1068–78.

Index

Note: Page numbers followed by "*f*" indicate figures, and "*t*" indicate tables.

A

AC electrophoresis, 124–125
Air-water interface, 87, 183–184*f*, 184, 192–197, 192–196*f*, 275, 277*f*, 278, 284, 287–290*f*
  colloidal operations use, 88
  theory, 88–90, 89*f*
Aqueous electrolyte solution, 3–4
  charged colloidal particles in, 4–6

B

Bispherical coordinates, 67–70, 72, 276
Boltzmann distribution, 16–17
Boundary conditions
  at infinity, 18–20
  on particle surface, 18
  on the outer virtual surface, 48–49, 51–52
Boundary confinement effect, 115–117, 181–182, 184–187
  geometric, 74–75, 76–77*f*, 98, 99*f*
  geometric deformation of double layer, 66–67
  lateral, 204–206, 205–206*f*
Boundary effect, 101–104, 201
  as correction factor, 284–290, 285–286*t*, 290–291*t*
  correction factor of separation distance, 79
  expressed as correction factor, 80–82*t*, 191*t*, 197–198, 198*t*
Brinkman equation, 147–148, 234
Brinkman model, 202
Brownian motion of particles, 3
Bubble surface, 359

C

Capillary diameter, 108–109
Capillary electrophoresis (CE), 107–108, 293
Capillary gel-electrophoresis (CGE), 233
  polymer gel in, 234
Catalytic reaction, 323
CE. *See* Capillary electrophoresis (CE)
Cell model, 216
  electrostatic, 49–51
  hydrodynamic, 49, 125, 386
  unit, 48–49
CGE. *See* Capillary gel-electrophoresis (CGE)
Charged colloidal particles
  in aqueous electrolyte solution
    double layer structure, 4–6, 5*f*
    shielding/screening effect, 4
Charge-determining ions, 23–24
Charged particle, electrophoretic motion of, 88
Charged pore, electroosmosis flow of, 117–118, 118*f*
Charged porous particle, 181, 214
Charged porous sphere, 146
Charge-regulated particle, 83
Charge-regulation phenomenon, 40–42, 41*f*, 83, 101*f*
Chemiphoresis, 325
Coions, 4
Colloidal entities, 278–279
Colloidal industry, 247
Colloidal operations, use air-water interface, 88
Colloidal particle, 3, 87–88, 145, 325
  approach, 87
  in electrolyte solution, 4–5
  in electrolyte suspension, 4
  motion of, 6
Colloidal suspension, 3–4, 47
  model, 48–49
Colloidal vibrational potential (CVP), 123
Colloid, definition, 3
Colloid science
  charged colloidal particles, 4–6
  colloid, 3
  colloidal suspensions, 3–4
  double layer polarization, 6
  electrokinetic phenomena, 6–7
  particle phoretic motions, 6
Concentrated suspension, 124–125
Concentration gradient, 386–392, 403
Confinement effect, of cylindrical pore, 109

Contour plots
of electrokinetic interest, 118–120, 119–121*f*
for gas bubble, 298–299, 299–300*f*
for moving liquid droplet, 298–299, 299–300*f*
Correction factor
boundary effect expressed as, 191*t*, 197–198, 198*t*
at separation distances, 98–100, 102–104*t*
Coulomb force
attraction force, 5–6, 145, 213, 324–325
electrostatic, 181–182
electric force, 131
electrostatic, 4, 33
electrostatic force, 5–6
repulsive force, 324–325
Counterion condensation, 145, 147, 159, 163–165, 385–386
Counterion condensation effect, 151–154, 156–159, 213, 226, 386–392
Counterions, 4
Crystallization of proteins, 293–294
CVP. *See* Colloidal vibrational potential (CVP)
Cylindrical pore, 107–109, 115–117, 293–294
confinement effect of, 109
dielectric solid, 110
Henry's charts in narrow, 203
phenomena in, 241–242, 241–242*f*
rigid particle in, 113–115
rigid spherical particle in, 109

D

Darcy's law, 146–147
DBB. *See* Debye-Bueche-Brinkman (DBB) equation
DBB equation, 147
DC electrophoresis, 124–125
Henry's charts in, 131
Debye-Bueche-Brinkman (DBB) equation, 234–235
Debye-Bueche-Brinkman polymeric fluid, 109
rigid spherical particle in cylindrical pore filled with, 236*f*, 237
single rigid spherical particle in, 235, 236*f*
suspension of rigid spherical particles in, 235–236, 236*f*
Debye-Bueche screening length, 237
Debye-Hückel approximation, 9, 146, 156–159
Dielectrophoresis (DEP), 294, 303
Diffusion mechanism, 136
Diffusion potential, 324–326, 328–329, 337, 340, 398–404
Diffusiophoresis, 6, 323–324, 326, 359–360, 385
element in, 325
Henry's charts in, 360, 362*f*
Diffusiophoretic motion, 6
Double layer overlapping effect, 53–58, 57–58*f*
Double layer polarization, 7–8, 30–37, 31–33*f*, 131, 136
Double layer polarization effect, 36–39*f*, 38, 124–125, 136–137
Drag force, 65
Droplet. See also *specific types of droplet*
definition, 247–248
electrostatic feature of, 278–279
suspension system, 303
Droplet, liquid, 247, 257, 293–294. *See also* Single liquid droplet
contour plots for moving, 298–299, 299–300*f*
diffusiophoretic motion of, 364–383, 375–382*f*
electrophoretic motion of, 275–276
generation of
high energy methods *vs.* low energy methods, 248–249
surface tension consideration, 248
thermodynamic consideration, 248
Henry's charts for, 257, 258–259*f*, 295–298, 295–298*f*
motion of, 275, 294
suspensions of, 312–319, 313–317*f*
Droplet viscosity
effect of, 364, 365–367*f*
impact of, 262
Dynamic electrophoresis, 124, 213
evaluation of dynamic particle mobility, 128–129
governing equations for, 214–216
single particle, 126–128, 130–137, 132–135*f*
suspension system, 128, 137–141, 138–140*f*
system diagram, 126*f*
Dynamic particle mobility, 128–129

E

Einstein's formula, 146–147
Electric driving force, 233
Electroacoustics, 123

Electroacoustic spectrometry, 216
Electrokinetics, 87
equation, 47
interest, contour plots of, 118–120, 119–121*f*
parameters, contour plots of, 207–210, 208–211*f*
phenomena, 6–7
potential, 7–8
Electrokinetic sonic amplitude (ESA), 123
Electrolyte ions, shielding/screening effect by, 4
Electrolyte solutes, 324
Electrolyte suspension, colloidal particle in, 4
Electro-migration, 38, 131
Electroosmosis, 6–7
Electroosmosis flow (EOF), 109, 253, 294
of charged pore, 117–118, 118*f*
of charged wall, 300
effect of, 300, 300*f*
impact of, 206–207, 207*f*
Electrophoresis, 6–11, 107
classic analytical/semianalytical formulas, 9–11
component, 364, 369–375*f*
theory, 11–25, 11*f*
boundary conditions, 18
charge-regulation phenomenon, 23–25
electrical potential, 12
evaluation of particle mobility, 22–23
flow field, 13–18
ion concentration, 12–13
reduction to one-dimensional problem, 20–21
Electrophoretic behavior
of porous particles, 173
of single rigid particle, 28–30, 30*f*
Electrophoretic deposition, 65
Electrophoretic motion, 6, 47, 65, 147
of charged particle, 88
of liquid droplet, 275–276
Electrophoretic separation, of polyelectrolytes, 233
Electrostatic attraction effect, image-charge force analog, 66
Electrostatic cell models, 49–51
Electrostatic Coulomb force, 4–6, 33
attractive force, 181–182
Electrostatics
counterion condensation effect in, 386–392
interaction, 87
Electrowetting on a dielectric (EWOD), 294
Emulsion, oil-in-water (O/W), 276
Emulsion systems, 249–251
Enhanced oil recovery (EOR), 275
EOF. *See* Electroosmosis flow (EOF)
EOR. *See* Enhanced oil recovery (EOR)
ESA. *See* Electrokinetic sonic amplitude (ESA)
EWOD. *See* Electrowetting on a dielectric (EWOD)
Excess surface conductivity, 359

F

Fixed charge density, 148, 151–153, 152–155*f*, 165–166*f*, 174–175*f*, 177–178*f*
Free electrophoresis, 233

G

Gas bubbles, 251–254, 304–305, 304*f*, 309–311*f*
contour plots for, 298–299, 299–300*f*
diffusiophoretic motion of, 364–383, 375–382*f*
generation of
high energy methods *vs.* low energy methods, 248–249
surface tension consideration, 248
thermodynamic consideration, 248
suspensions of, 257, 312–319, 313–317*f*
system diagrams for, 360, 361*f*
theory, 254–257, 255*f*
Gel electrophoresis, 233
polymer gel in, 233–234
Geometric analogy, 146
Geometric boundary confinement effect, 74–75, 76–77*f*, 98, 99*f*
Geometric deformation of double layer, 66–67

H

Hadamard condition, 276. *See also* Rybczynski and Hadamard condition
Helmholtz linear equation, 156–159
Henry's chart, 28–30
in DC-electrophoresis, 131
in diffusiophoresis, 360, 362*f*
in gas bubble, 363–364, 363*f*
for liquid droplet, 257, 258–259*f*, 295–298, 295–298*f*
in narrow cylindrical pore, 203
for particle suspension, 239, 240*f*
for porous particle, 156–159, 157–158*f*
for rigid particle, 113–115, 114–117*f*

Henry's chart (*Continued*)
at separation distance, 72–74, 73–75*f*, 94–98, 96–97*f*
for single spherical particle, 237, 237–238*f*
for suspension concentrations, 53, 53–56*f*
Homogeneous polymeric solution, with linear polymers, 234
Hückel's formula, 257
Hydrodynamic cell model, Kuwabara, 49, 386
Hydrodynamic drag effect, impingement flow, 65
Hydrodynamic drag force, 128–129, 145–146, 214, 252
Hydrodynamic mixing, 107
Hydrodynamic motion-deterring drag force, 213
Hydrodynamic shear stress, 252
Hydrodynamic swarming effects, 47
Hydrodynamic unit cell model, Kuwabara, 151

I

Image-charge force analog, 66, 87–88, 184–186, 192–197
repulsive, 90–94, 91–95*f*
Impingement flow, hydrodynamic drag effect, 65
Induced electroosmosis flow, 253
Intriguing phenomenon, 186–187

J

Janus particles, 323–324
Joule heating effect, 107

K

Kuwabara's hydrodynamic cell model, 49, 125, 151, 386
Kuwabara's unit cell model, 125, 128

L

Lab-on-a-chip (LOC), 108, 293
Lateral boundary confinement effect, 204–206, 205–206*f*
Levine and Neale model (L-N model)
boundary conditions, 58–61, 59–61*f*
evaluation of particle mobility, 50–51
merits and shortcomings, 50
vs. Shilov and Zharkikh model (S-Z model), 49–51
Linear polymers
homogeneous polymeric solution with, 234
polymeric solution of, 234
Liquid droplet, 247, 257, 293–294. *See also* Single liquid droplet
contour plots for moving, 298–299, 299–300*f*
diffusiophoretic motion of, 364–383, 375–382*f*
electrophoretic motion of, 275–276
generation of
high energy methods *vs.* low energy methods, 248–249
surface tension consideration, 248
thermodynamic consideration, 248
Henry's charts for, 257, 258–259*f*, 295–298, 295–298*f*
motion of, 275, 294
suspensions of, 312–319, 313–317*f*
LOC. *See* Lab-on-a-chip (LOC)

M

Macroemulsion, 250
Microemulsion, 250, 360
Microfluidic device, 293–294
Microreactor, 247
Mobile droplet surface, 253, 259–260, 359
Motion-deterring effect, 35
Motion-deterring polarization effect, 67
Motion-enhancing electroosmotic flow, 359
Motion of liquid droplet, 275, 294

N

Nanoemulsion, 251, 360
Nanofluidic device, 293
Navier-Stokes equation, 13–14
Neighboring particles, in suspension, 47
Nernst-Planck equation, 12–13, 15, 33, 333

O

Oil-in-water (O/W) emulsions, 276
Optimum targeted drug delivery, 360
Ouzo effect, 247, 250

P

Particle migration, in porous medium, 234–235
Particle motion, 47, 323
effect of Peclet numbers upon, 38–40, 40*f*
Particle permeability, effect of, 203–204, 204*f*
Particle phoretic motions, 6

Particle surface
boundary conditions on, 18–20
charge density, 25–28, 26–29*f*
Particle suspension, Henry's chart for, 239, 240*f*
Passive-diffusiophoresis, 323–324
Patched pseudo-spectral method, 112
Permeability of porous particle, 213
Phase lag, 124, 129, 136, 141
Phoretic motion, 6–7
particle, 6
Poisson-Boltzmann distribution, 110–111
Poisson-Boltzmann equation, 19, 25–27, 156–159
Poisson-Boltzmann theory, 87–88
Polarization effect, 9–11, 25–27, 30–35, 146, 203
double layer, 36–39*f*, 38, 295–298
Polyelectrolyte, 145–146, 385
electrophoretic separation of, 233
molecules, 181
Polymer gel
in capillary gel-electrophoresis, 234
in gel-electrophoresis, 233–234
Polymeric Debye-Bueche-Brinkman (DBB) fluid, 109
rigid spherical particle in cylindrical pore filled with, 236*f*, 237
single rigid spherical particle in, 235, 236*f*
suspension of rigid spherical particles in, 235–236, 236*f*
Polymeric solution, 234–235
of linear polymers, 234
Pore radius, effect of, 207, 208*f*
Porous medium, particle migration in, 234–235
Porous particle, 201–204, 213
electrophoretic behavior of, 173
Henry's charts for, 156–159, 157–158*f*
motion of, 385
permeability of, 213
suspension of, 149*f*, 151, 386, 387*f*, 404–408, 406–408*f*
Proteins, crystallization of, 293–294
Pseudo-spectral method, 20–21, 111

R

Repulsive image-charge force analog, 90–94, 91–95*f*
Rigid particle, 47–48
in cylindrical pore, 113–115
Henry's charts for, 113–115, 114–117*f*
suspensions of, 329–332, 330*f*
Rigid sphere, 109
Rigid spherical particle, 11, 11*f*, 22, 67–68, 67–68*f*
in cylindrical pore, 109
in Debye-Bueche-Brinkman polymeric fluid, 235–236, 236*f*
Rybczynski and Hadamard condition, 254–255

S

Screening effect, 145
Self-diffusiophoresis, 323–324
Shielding effect, 151–154, 156–161, 163–165, 385–386
Shilov and Zharkikh model (S-Z model)
boundary conditions, 58–61, 59–61*f*, 71–72, 303
evaluation of particle mobility, 50–51
vs. Levine and Neale model (L-N model), 49–51
merits and shortcomings, 50
Single liquid droplet, 304–305, 304*f*, 306–308*f*
system diagrams for, 360, 361*f*
theory, 254–257, 255*f*
Single porous particle, 214–226, 217–225*f*, 386–404, 387–391*f*, 393–397*f*, 399–405*f*
Single porous sphere, 148–166, 149*f*, 152–155*f*, 157–158*f*, 161–168*f*
Single rigid particle, 327–329, 328*f*, 330*f*, 333–342, 334–342*f*
spherical particle in DBB polymeric fluid, 235, 236*f*
Smoluchowski formula, 325–326
Smoluchowski-type behavior, 136, 305
Solid plane, 182–191, 182–183*f*, 185–190*f*, 276–284, 277*f*, 280–284*f*
Solution electrophoresis, 233
Sonic wave, 123
SPB. *See* Spherical polyelectrolyte brush (SPB)
Spherical coordinates, 202
Spherical polyelectrolyte brush (SPB), 147, 214
Stern layer, 5–8
Stokes equation, 14, 146–147
Stokes law, 253, 275
Surface polarization, 56–58
Surface tension force, 293–294

Surfactants, 249–251
Suspension, 216, 226–229, 227–229*f*, 342–355, 343–355*f*
  dynamic electrophoresis, 128, 137–141, 138–140*f*
  gas bubble, 257
  of gas bubbles, 312–319, 313–317*f*
  of liquid droplets, 312–319, 313–317*f*
  neighboring particles in, 47
  of porous particles, 149*f*, 151, 166–177, 168–178*f*, 386, 387*f*, 404–408, 406–408*f*
  of rigid particles, 329–332, 330*f*

U

Unit cell model, 48–49
  Kuwabara, 125, 128

V

Viscosity ratio, effect of, 300
Vortex flow induced, 75–79, 78–79*f*, 98, 100–101*f*

Z

Zeta potential, 7–8

CPI Antony Rowe
Chippenham, UK
2019-01-17 19:19